The Yankee Road

Tracing the Journey of the New England Tribe that Created Modern America

VOL. 1: EXPANSION

James D. McNiven

The Yankee Road: Tracing the Journey of the New England Tribe that Created Modern America

Copyright © 2015 James D. McNiven. All rights reserved. No part of this book may be reproduced or retransmitted in any form or by any means without the written permission of the publisher.

Published by Wheatmark®
1760 East River Road, Suite 145
Tucson, Arizona 85718 USA
www.wheatmark.com

ISBN: 978-1-62787-141-9 (paperback)
ISBN: 978-1-62787-142-6 (ebook)
LCCN: 2014922457

To Jane, who could have shown Job what real patience is.

Contents

Acknowledgments	xiii
Some Notes to Readers	xv
Introduction: Values and Roads	xvii
1 Boston, Quincy Market: Larry Bird's Shoes Who the Yankees are and why they differ from other Americans.	1
2 Boston, Beacon Hill: The Lost Tribe The early experiences that formed the Yankee "tribe," and how the tribe disappeared.	15
3 Boston: The Yankee Road from Kenmore Square to Newport, Oregon Who put the 20 in US20, and where it leads.	30
4 Watertown, Massachusetts: Fred Taylor, "Mother" Perkins, and the "System" Why Frederick Taylor was the most important man of the twentieth century, and the identity of that lady with him.	46
5 The Blackstone Valley, Waltham, and Lowell, Massachusetts How the interaction of textile manufacturing and the frontier led to a new method of organizing people.	66
6 Route 128: The First Computer Revolution How research universities and leftover airbases created the computer industry.	84
7 Concord, Massachusetts: Walden Pond and the Idea of Wilderness Thoreau's dinner habits and how we got blank spots on the map.	99

8   Shrewsbury, Massachusetts: Pincus and the "Pill"   121
How two old women changed the world.

9   Springfield, Massachusetts: The Quest for Interchangeable Parts   141
How a 25-year government research project begun almost two
hundred years ago helped create modern America.

10 Albany, New York: Democratizing Capitalism   154
Corporations as a left-wing device.

11 Cooperstown, New York: Heroes or Winners?   169
The happy myth of American individualism.

12 Ilion, New York: Typewriters and the Second Computer Revolution   181
How Mr. Remington's rifle barrels ended up as word processors.

13 Cazenovia, New York: Organizing the Early Frontier   196
The lone family searching for new land in its Conestoga wagon
and the settlement of America.

14 Syracuse and Utica, New York: The American Religion   204
The Bible is all you need, as well as some good advertisements and
some serious organization.

15 Seneca Falls, New York: Women's Rights   227
How Mr. Blackstone's strictures failed to stand up to the quiet lady
from Nantucket and her friends.

16 Palmyra, New York: Saving the World, One Utopia at a Time   245
The persistence of Yankee religious separatist groups despite the
failure of the Pilgrims' early commune.

17 Rochester, New York: Telegraphy and the Five Internets   264
How Mr. Sibley's Western Union morphed into Facebook and
Twitter.

18 Batavia, New York: The World's Greatest Land Sale   279
How the West was won, one quarter-section at a time.

19 Niagara Falls, New York: Mr. Love's Model City   294
How Mr. Love dug a canal and a Yankee found a big, blue marble.

20 Buffalo, New York: Expanding the Yankee World through Air Conditioning   308
How Mr. Carrier completed General Sherman's work a hundred
years later.

21 Dunkirk, New York: The Yankees and the Railroad   321
Helen of Ramapo and the launch of a thousand railcars.

22 Chautauqua, New York: The Democratization of Education   340
How a Sunday School teachers' institute brought education and
culture to millions of American farmers and their small towns.

Epilogue   359

Notes   363

Bibliography   493

Index   519

Acknowledgments

Many others have said writing is a lonely activity, but that nobody writes alone. I have benefited from support and useful criticism from many people and institutions over the years it has taken to get this part of what is a large project through to completion.

Most integral to its production have been the support, suggestions, and criticism of my wife, Jane. She has traveled the road with me to many strange and interesting places. She read the whole thing as each piece was finished, and made a host of valuable comments and corrections. She is a real advocate for travelling with a purpose.

My daughters and son been involved directly in the production of the piece: Joan lining up some technical work, Steve with the cover design, and Christine, from a distance, with encouragement and help with our travels.

My old friend, Dr. Barry McGhan, read most of the pieces and offered detailed editorial criticism and a lot of conceptual refinement well into the night on many an occasion. Professor Michael Stamm of Michigan State University (MSU) also provided both criticism and ideas.

In Nova Scotia, I am thankful for advice from Emery Fanjoy, Joe and Roseanne Foy, Rob Peters, the late Charles Yeates, and JoAnn Napier-Chaisson.

I am especially grateful to Fulbright Canada for funding a Research Professor appointment at MSU in 2010–11. AnnMarie Schneider, Director of the Canadian Studies Center at MSU, offered both an enthusiastic invitation and continual support for my research, all the while getting me to line up Canadian speakers for her program. It has been a fruitful partnership, and has led to continuing relations between MSU and universities in Nova Scotia. AnnMarie also put me in touch with Frederick Bohm, former director of the MSU Press, who looked at the project from his unique vantage point and provided useful advice.

Dalhousie University provided two small travel grants at different times to support the research. As well, Professor Iraj Fooladi invited me to make two presentations in his long-running MacKay Lecture series on topics related to Yankee innovations in incorporation and banking. B. Anne Wood, a retired Dalhousie professor, now in Victoria BC, has been a useful source on how to approach historical subjects. Another BC resident, Prof. John Clements, provided useful science and technology comments.

I also want to thank Professor Stephen Blank of Pace University in New York City, now retired, for many discussions about American industrial history mixed in with our combined interest in the North American Free Trade Agreement and the North American communities. My interests also intersected those of my brother-in-law, David Spille, who lives part of the year in Maine and who has an interest in Maine and New England history. My brother, Professor Tim McNiven of Ohio State University, pointed out some Ohio resources and arranged for me to meet academics at that institution with interests similar to mine.

I want to recognize the help and access I received from a variety of libraries in Canada and the United States, from Nova Scotia to Arizona and places in between. My wife and I also visited many local history societies along or near US 20, which provided a living context to the topics and ideas featured in the book. They were also a source of local historians' work that has enriched some of the stories.

I cannot pass by the important help related to bibliographic and research organization provided by staff at Dalhousie's School of Public Administration — in particular, the late Cecilia MacDonald. Dolene LaPointe and Krista Cullimore were also of cheerful help over the time it took to get this project done.

Finally, Barry Norris edited a paper I wrote many years ago that made me seem a better writer than I was, and when it came time to have this book edited I sought him out. Thanks, Barry, for all your work.

Some Notes to Readers

- I have used the terms 'Indians' and 'Blacks' because I feel they carry more dignity than the shifting euphemisms developed over the past few decades. 'Yankees' has been a derogatory term since it was invented in the 1630s. I use 'Great Britain' or 'Britain' for events after the 1707 Act of Union and 'England' for the period before that. My apologies if any of these terms irritate some readers.

- I have not provided a lot of maps and diagrams. Google search and maps and Wikipedia should be considered as adjuncts to this book.

- A companion website for this book is in the production stage as *theyankeeroad.com*. I can be found at *j.mcniven@theyankeeroad.com*.

- Most of my sources should be easily available. I relied upon secondary material for the most part; otherwise this work would take a hundred years. I used online genealogy sites to confirm Yankee biological connections, something that would have been impossible a few decades ago.

Introduction: Values and Roads

*Not only for Cyrus' sake did Xenophon march up toward the Persians.
But in search of a road which led up to Zeus.*

— Diogenes Laërtius[1]

We are restive entirely for the sake of restiveness. Whatever we may think, we move for no other reason than for the plain, unvarnished hell of it.

— James Agee[2]

I like roads. I like to move.

— Harry S Truman[3]

Whither goest thou, America, in thy shiny car in the night?
— Jack Kerouac, On The Road[4]

What a long, strange trip it's been.
— The Grateful Dead, "Truckin'"[5]

This is a road book — a book not about a road but about encounters while traveling what I call "the Yankee Road," US Route 20, which snakes across the continent from Boston to Oregon. Along it, or near it, so-called Yankees developed ideas that underlay much of how the world works today. "Yankees" are not just a biological group but an American "tribe" that made America what it is.

A Greek Road Book

The road book is almost as old as the written word. Although the *Book of Exodus* and Homer's *Odyssey* came first, Xenophon's *Anabasis*, perhaps the classic road book, dates from 2,400 years ago, after the Peloponnesian War ended with the defeat of the Athenians by the Spartan navy. One of history's great struggles, the conflict, like the Cold War in our own time, had gone on for decades.

A couple of years after the end of the war, an Athenian officer named Xenophon found himself in one of the Greek cities on the Asian coast of the Aegean Sea. Cyrus, the brother of then Persian Emperor Artaxerxes, had raised a force, ostensibly to attack some mountain raiders, but he actually intended to move east toward the Persian capital in order to overthrow the emperor. Xenophon joined a 10,000-man Greek mercenary force that made up a part of Cyrus's army. They trekked "up country" from the western edge of the empire and, in early September 401 BC, after a march of 1,265 miles, met and defeated Artaxerxes's army just north of Babylon.[6] Cyrus, however, was killed in the battle, and his army dispersed, leaving the Greek mercenaries to the tender mercies of the emperor.

Greek discipline, tactics, and equipment had played a disproportionate part in the defeat of Artaxerxes's force, and the emperor was not inclined to let them return home with this intelligence, but neither did he have the strength to defeat them. Instead, he offered to guide them home with a Persian escort, which, through treachery, trapped and killed the Greeks' generals, hoping thereby to disorganize the "10,000." But Greek democratic practice threw up new leaders, including Xenophon, who led them north into the mountains of what is now eastern Turkey and away from the Persian forces. They crossed in midwinter, emerged at the Greek colony of Trebizond, on the Black Sea, in March 400 BC, and finished their march in Cotyora, on the western coast of the Black Sea in May. In fourteen months, Xenophon had walked 4,000 miles "up" to Babylon and "down" to the Black Sea. Twenty years and more later, he wrote *Anabasis*, or "march up country," about this exploit.

Unlike most Greek historical writing up to that time, *Anabasis* is not a mixture of gods and humans or of wonders and ordinary practice. Instead, it is a careful account of a military man who is concerned with the accurate portrayal of the people, cities, military, and geography of the western Persian Empire. The young general kept such meticulous track of the distances and the landforms of the areas he saw that British explorers as late as a century ago were able to retrace most of the route of the Greeks with relative ease. Xenophon, in fact, provided an intelligence manual for Alexander's campaigns forty years later. Persian army organization and tactics had not changed in the interim, and this time the Greek force was strong enough to go on and take over the Persian Empire.

Xenophon's implicit theme 2,400 years ago was that the Greeks' ability to meld personal freedom, organization, and the use of new technology gave them superiority over the Persians, despite the Persian superiority in people and money. This is not too far from the essence of the course the Cold War took. One cannot draw too many parallels, but Xenophon's journey does somehow relate to a trip "up country" along the Yankee Road, or US 20. My aim is not to provide a geographic guide; rather, it is to take stock of the three elements that Xenophon saw as the ingredients of future Greek successes, and that I see as the same ingredients — freedom, organization, and technology — that ultimately gave us victory in the long contest with the Soviet empire that took place in our own time.

American Road Books

Americans, have added, in great numbers, to the road book literature. American stories of travel and roads have been conditioned by the general habit of a people constantly on the move. The road book, both fiction and non-fiction, was standard fare in twentieth-century writing. Toward the end of the century, Kris Lackey attempted, in vain, to categorize this literature. Road books — those that use the road as a metaphor for life or as a means to connect with life — tend to defy pigeonholing, although Lackey manages to discern three major road themes: rediscovery of the self, observing the American condition, and the transcendental connection to the broader universe.

Whether it is Robert Pirsig, crossing the Great Plains to Montana to find his shattered former self; or Jack Kerouac as Sal Paradise, looking for God and girls; or Sallie Tisdale, *Stepping Westward* in search of her West; or William Kinsella as Ray, leaving his field of dreams to look for his dead father in Fenway Park or Hibbing, Minnesota, the goal is the same. Somewhere on the road an experience will allow them to realize who they really are. The road that takes us *west*, especially, takes us to our real selves.

The road also allows us to see and meet others. If the quest for self-realization is a solitary journey, one can observe others alone or in company, as Bill Bryson did in his reminiscences about family trips in all directions from Des Moines, Iowa. Steinbeck, with his dog, Charley, was technically solitary on the road, and William Least Heat-Moon was quite alone, having left a failed marriage and a career to visit small towns and rural areas on the "blue highways." Dayton Duncan, too, traveled solo to western counties that, by the standards of the 1890 census, have reverted to the population density of the frontier. Others look for adventures on the road — to ride with truckers, to visit sports halls of fame, or to comment on the various sociologies of suburbs, gated communities, ethnic neighborhoods, farmers, and latter-day pioneers.

There is also the desire, like Laertios in interpreting Xenophon's experience, to travel the road to connect to the universe, to Something Greater. Often, this means walking, as Xenophon did, a more personal experience than the windshield can provide. The "road to Zeus" now leads to the vortex at Sedona, to the edge of the Grand Canyon, along the Appalachian Trail and the paths of Cape Cod, to Mesa Verde and around Niagara Falls, and back to Walden Pond. America might not have miraculous shrines such as Fatima or Lourdes or, except for the Latter-Day Saints, great religious centers such as Mecca or Rome, but it is infused with the Yankee transcendental spirit, and its "shrines" are legion.

So the transformation of life into space leads naturally into the crossing of space — travel — as a means of signifying life's changes. Telling a tale about road books that have affected or symbolized a life's phases seemed a good route to personal rediscovery. Creating a road book about a country one loves offers a chance to observe its condition. Why is America the way it is? Following Lackey, we can then get on with these transcendental considerations. But, first things first, let's get "me" out of the way.

My Road Books

AT NINETEEN: KEROUAC

Once, while in Lowell, Massachusetts, I had breakfast at the Club Diner, in operation since 1933.[7] In the 1930s and '40s, Lowell was a blue-collar town and a good part of its population came from Quebec, Jack Kerouac's parents among them. The old mill town, where Jack Kerouac was born and brought up, has become, in large part, a historical center, and the Diner attracts a lot of tourists and kids off school buses.

Despite this new focus, the town doesn't seem to know what to do with Jack Kerouac. It recognizes his fame, but his image represents neither the old blue-collar mill town nor its new technology and history-oriented profile. Kerouac was one of those visionary, restless Yankees, not hardworking, rational Yankees like those Lowell thought itself to represent. Even though, as William Burroughs, Jack's poet-friend and the model for "Old Bull Lee" in *On The Road*,[8] claimed, "Woodstock rises from its pages" and "Kerouac opened a million coffee bars and sold a million Levis."[9]

There is a Kerouac Park in Lowell, but it reminds me of the memorial to the 1970 massacre at Kent State in Ohio. It's there, just off the main street, but it's not easy to find — the perfect way to memorialize someone who is a bit of an embarrassment, but who is too famous, or infamous, not to acknowledge.[10] Dedicated in 1988, and a little unkempt when I last visited, the site features a sculpture of pillars on which are Kerouac quotes about Lowell or, less often, about himself. The pillars sit on a base that incorporates both the Catholic cross and the Buddhist circle. On one pillar, Kerouac refers to himself as a crazy, Catholic mystic. He was, as he put it elsewhere, "waiting for God to show His face."[11]

Douglas Brinkley, taking some students across America to visit cultural symbols and people, observed about Kerouac and Lowell: "On this trip we also visited Thoreau's Walden Pond, but it was the Merrimack River in pre-Christmas twilight condensing toward darkness that caused us to shudder, to feel the shroud presence of Dr. Sax, from Kerouac's eponymous novel, whose mysterious spirit still seems to pervade Lowell, to understand at last that *On The Road* was not a novel, but a religious poem for America, a gift to us all."[12]

The Kerouac family was part of a close-knit ethnic group in Lowell, and Jack, born in 1922, was brought up on a diet of American comic books and French-Quebec Catholicism. Jack was a good football player — good enough to get a scholarship to play at Columbia, in New York City — but a knee injury soon put an end to his athletics, and he began a life of drifting and writing.

During and after the Second World War, Kerouac fell in with a group of poets, novelists, and assorted others who would become the core of the "Beat" Generation and its "weariness with all the conventions of the world."[13] He published a long novel about his upbringing in Lowell and New York in 1950. Sales were respectable enough to encourage him, and he produced a number of other autobiographical novels before turning away from his "Beat" friends and drinking himself to death at the age of forty-four.[14] Then there is "the rest of the story," as radio's Paul Harvey built a career telling us.

Kerouac's autobiographical *On The Road* came out in 1957, but the events that inspired it had taken place a decade or so earlier.[15] I first read it in college in northern Michigan around 1960. It had an amazing effect on me and millions of others around my age then. I was on my own for the first time, at a remote engineering school where social relations were crude at best, and along came this paean to a life of travel, freedom, and irresponsibility. Had I never heard of the book, I might still have gone on drinking bouts on the shores of Lake Superior, roared down to Wisconsin to chase the girls at Oshkosh State Teacher's College, or over to Hurley, Wisconsin, where Chicago gangsters used to unload their booze from Canada during Prohibition and where there were a hundred bars for the town's thousand residents. But Kerouac gave it all a romance that no one could ever take away.[16]

Kerouac, like Elvis Presley, apparently never saw himself as a social revolutionary. He was irritated by being blamed for the sex, drugs, and rock and roll of the 1960s.[17] Yet, Bob Dylan is said to have cherished his copy of *On The Road* like a bible,[18] and claimed "it blew my mind" as a teenager in northern Minnesota. John Lennon, Van Morrison, David Bowie, and Bruce Springsteen all were influenced by it.[19] Listen to Springsteen's sing about him and his buddy Wayne going down to Darlington County from New York City, and you can see Cassady and Kerouac in the car.

Kerouac documented his restlessness as his character, Sal Paradise, whose five near-penniless travels with his friend Dean Moriarity (in real life, Neal Cassady) from New York to Denver and San Francisco and back between 1947 and 1951[20] form the story. They were traveling in search, not of a new home or a new life, though both romanticized about these goals, but of freedom. Freedom meant leaving somewhere, especially to Cassady, whose character, Dean, as one critic put it, "spends the book running away from messes he makes of his life," while Sal tries to figure out how to live amid the mess handed him. Both of these impulses are amazingly powerful when one is eighteen.

For decades the road book stereotype had been the Depression-era migration of the Oakies to the West Coast. But Kerouac was not a refugee from poverty or dust storms, he was just "on the road." He might have had a lot in common with Depression hoboes,[21] but unlike them he wasn't out of work — he was just leaving his friends in one town to visit others elsewhere, where they spent their days talking about art and philosophy. All the while, he was taking notes for his writing.[22] Despite its omnipresence, the Cold War gains Kerouac's attention only once, when he describes passing through Washington, DC, on the eve of Harry S Truman's inauguration in 1949 and seeing the preparations for the military parade.

AT TWENTY-NINE: PIRSIG

A decade later came Robert Pirsig's *Zen and the Art of Motorcycle Maintenance*.[23] Pirsig, his eleven-year-old son, and a couple, friends of Pirsig's, motorcycle across the northern Great Plains toward the Montana Rockies and the university town where he once taught. Instead of Kerouac's hectic bouncing between two coasts, there is only this meditative trip from Minnesota to Montana. Where Kerouac's journey was almost frantic, Pirsig's is dignified, almost stately.[24] Along the way, Pirsig discusses the approach of a meticulous person who tries to understand his machine and do repairs and adjustments correctly. He ruminates on the ideas that once led him to a serious mental breakdown, especially the nature of "quality."[25] Pirsig refers to his former self as "Phaedrus," a personality that is trying to reconstitute itself in opposition to the "new" Pirsig. Leaving the couple in Montana, Pirsig and his son continue along back roads to California. Pirsig might have had his reasons for traveling, but so had his son, who was looking for the father who had disappeared behind the hospital doors years ago.

This late 1960s road trip was about Pirsig's healing an injured self and carefully picking up the philosophical threads of his former quest to understand value and quality. Even though he was doing it for at least the second time, there is a feeling in *Zen* that Pirsig was just then starting a new life and a career, something that was meaningful at a time when I was doing the same thing. In 1968, I took my wife, Jane, and three kids on a road trip northwest from Michigan across the border to Winnipeg, Manitoba. A college friend had written to see if I was interested in teaching at the University of Winnipeg. He knew that I had piled up debts as a graduate student with a family, and the tax treaty between the United States and Canada would enable me to save on paying taxes for the first two years of my stay in my father's native land. The temporary stay turned out to encompass a whole career.

Somehow, *Zen and the Art of Motorcycle Maintenance* also resonated with me in other ways. Pirsig began his university years in biochemistry; I began in engineering physics. He flunked out because some philosophical problems had begun to intrigue him; an interest in politics and economics did it for me. He could tie a meditation about quality to fixing a system called a motorcycle engine; I wanted to know how the system called society worked, and worked best. Pirsig wandered off into philosophy, and went to India to study religions: I got bored with the sedateness of being a university faculty member, and left for think tanks and government.

People still retrace the route Pirsig, his son, and his friends took.[26] Websites and discussions are devoted to him and his ideas. In his mid-eighties he was offered an honorary doctorate in philosophy by the university where Phaedrus taught and met his fate. Phaedrus had something in his grasp....

AT THIRTY-NINE: LEAST HEAT-MOON

Then, William Least Heat-Moon left a Missouri university to take a lengthy journey around America's small towns. He titled his book *Blue Highways*, because

> [o]n the old highway maps of America, the main routes were red and the back roads blue. Now, even the colors are changing. But in those brevities just before dawn and a little after dusk — times neither day nor night — the old roads return to the sky some of its color. Then, in truth, they carry a mysterious cast of blue and it's that time when the pull of the blue highway is strongest, when the open road is a beckoning, a strangeness, a place where a man can lose himself.[27]

Least Heat-Moon, who is part Indian and whose real surname, Trogdon, is that of an old North Carolina family, apparently lost his wife and his job on the same day. He then set out to find himself, visiting the people in the small towns on the blue highways. In his own way, he was looking for the nostalgic values that Ronald Reagan extolled on his path to the presidency: uniqueness and perhaps eccentricity, as well as the "rootedness" of the places he visited, something that Kerouac rebelled against and in which Pirsig had no interest.[28]

The blue highways are mainly a thing of the past now; most of them connect suburbs to cities and the Interstate highways that have reduced the frontier to a settled, more compacted country. One writer has called the blue highways the residue of an old America,[29] and Least Heat-Moon's book seems to give off that mood.[30] The pure open road has sunk into myth.

Like Least Heat-Moon, I had also left university life, in my case to work in a government economic development department in mainly rural Nova Scotia, so his journey resonated strongly with me.

At Forty-nine: Kinsella

Around 1990, I discovered another resonant road book, this time by a Canadian, William Kinsella. Kinsella's fiction focuses on two subjects: the lives of Alberta's aboriginal population and baseball. His novel, *Shoeless Joe*, was made into the movie, "Field of Dreams." *Shoeless Joe* concerns a not-very-successful Iowa farmer who hears a voice in his cornfield that leads him to construct a baseball field there that becomes populated by the spirits of the game's stars of fifty years before. (The disembodied voice is not unlike those in the Christian tradition, but that is a subject I will get back to in a later chapter.) The protagonist then goes off on a journey to collect a famous author he read in his youth and a failed major-leaguer who had only one lifetime at-bat and bring them back to his field. In doing so, he saves the family farm, reconciles with his deceased father, and creates a new tourist destination. Kinsella, to me, captures better than anyone else the complex of values, hopes, and dreams of Americans at that time, especially those who were on the forward slope of the demographic hill called the "baby boom," for whom roots and mortality were beginning to emerge as important.

As an aside, two of the shooting locations for "Field of Dreams" were Galena, Illinois, and a farm northeast of Dyersburg, Iowa, both but a short distance from US 20. The picturesque downtown of Galena was a movie stand-in for Hibbing, Minnesota, as it might have looked in the early twentieth century. The farm outside Dyersburg, when we saw it in midwinter, seemed a lot less imposing than on film, if only because there was nothing at its margins except plowed ground.

At Fifty-nine: Brinkley

My road book for the millennium was *The Majic Bus*, written by Douglas Brinkley before he became better known as a presidential historian. In its opening pages Brinkley relates how Kerouac affected him. A couple he had been waiting on in a Holiday Inn restaurant gave him their copy of *On The Road*

> as if it were the *Gideon Bible* and instructed me to read it...I had never heard of Jack Kerouac...but the paperback cover immediately caught my eye — a ribbon of America's highway disappearing into a blinding orange sun...From the first page, I was hooked. Here was the book I had intuited

and craved, just as an earlier generation needed *The Sun Also Rises* and *The Great Gatsby*...I finished *On The Road* in two sleepless nights, wishing I were Dean Moriarity (Neal Cassady) or Sal Paradise (Jack Kerouac), blazing across America.[31]

The Majic Bus is the story of Brinkley, then a young Long Island professor, and a group of college students touring around America in a bus to get a feel for its history and culture, especially its modern popular culture. Brinkley's misspelled bus was in homage to "Furthur," the bus that took Ken Kesey's Merry Pranksters across the country in 1964 to the World's Fair in New York, coincidentally driven by none other than Kerouac's old buddy, Neal Cassady. Brinkley advocates surrendering to the American predilection of "moving on" to find the real or ideal America, which is just over the next hill or around the next bend. "The road unshackles the American psyche like nothing else."[32] "Majic" took them to Lowell in search of Kerouac, to a Cherokee settlement in the South, to the Mississippi Delta to look for blues players, to Chicago to find writers and architects and St. Louis to find Chuck Berry, to Denver to appreciate football and boxing, to Taos, New Mexico, to look for writers and Los Alamos for nuclear scientists, and to NikeTown in Oregon.

Brinkley's journey was more educational than Kerouac's or Kesey's, and it caused me to reflect on how diverse are the ways he and, by implication, I were spending our careers trying to help students understand the complexities of life.

At Sixty-nine: The WPA Guides

By 2010, older and nostalgic, I became attracted to the series of travel guides created between 1935 and 1943 by the Federal Writers Project of the Works Progress Administration (WPA).[33] A make-work project for unemployed writers during the Great Depression, guides were published on the states, some cities and towns, some old "named" highways, ethnic and racial points of interest, and local historical sources. They constitute, for me, a panorama of life and travel that existed around the time I was born.

Although the WPA Guides were meant for automobile tourism, they were edited with an eye to doing more than just promoting places and attractions. They provided comprehensive coverage of a state and some critical commentary on its history and places. They were not meant for the "good-time" tourist looking merely for amusement and a place to eat. Instead, they leave the modern reader with a "feel" for early mid-century America, before the Interstates were built. Hardly literary masterpieces, they were written by young aspiring writers and "washed-up" editors, with varying approaches to history, fact gathering, and style. Their success prompted successors such as John Gunther's *Inside U.S.A.* (and the rest of his extensive *Inside* series) and Jamie Jensen's more contemporary *Road*

Trip USA,[34] but, as one commentator says, "they were something special." When John Steinbeck set out on the road with his dog Charley, he noted:

> If there had been room in Rocinante [Steinbeck's camper] I would have packed the W.P.A. Guides to the States, all forty-eight volumes of them. I have all of them, and some are very rare....The complete set comprises the most comprehensive account of the United States ever got together, and nothing since has ever approached it. It was compiled during the depression by the best writers in America, who were, if that is possible, more depressed than any other group while maintaining an inalienable instinct for eating.[35]

Following the Yankee Road in the WPA Guides reminds me of the world I knew when growing up. The romance of the road was in my blood early on. When I was four and a half and the Second World War had been over for a month, my parents loaded up the family's Oldsmobile and left Michigan on a three-month odyssey through the Midwest, Southwest, and California, looking for new opportunities. My father was a printer, and in those early postwar days newspapers all across the country needed his skills. We worked our way west, my mother entertaining two boys and making a household in dozens of motels and cottages. Eventually, just before Christmas 1945, we headed back to Michigan at top speed, barring the inevitable "blowouts." For more than sixty years, my parents never moved farther than five miles from where they lived after they were married.

Yankee Values and the Cold War

If travel is critical to a road book, then the selection of route gives structure and purpose to travel. In my case, the genesis of the selection was somewhat complicated. It began in 1987, when I was in charge of economic development for the government of Nova Scotia, in Canada. At the time, the so-called Toshiba Scandal filled the international media. Officials of that company apparently had agreed to the sale of metal-machining equipment to Scandinavians who, in turn, rerouted the shipment to Leningrad (now St. Petersburg), where it was to be used to improve the quietness of Soviet nuclear submarine propellers, an act that was prohibited by a Cold War international treaty intended to deny communist states access to sophisticated technology.

What struck me about the scandal, however, was something different. The supposedly sophisticated technology that ended up in Leningrad was "5-axis" CNC (computer numerically controlled) milling machines, but I knew of an aircraft company in Nova Scotia that was just then building an experimental machining facility employing 7-axis machines. Why were the Soviets trying through undercover means to buy less-sophisticated equipment than was already coming onto the market, when either industrial espionage or their own sophisticated computer scientists could allow them to make their own equivalent? The answer I formed had more to do with how physical equipment is priced and sold than with how it is made.

The Soviets' centrally planned economy had an extremely limited demand for any kind of technological goods outside a few military uses. They then had to budget carefully for the kinds of equipment the economy might produce, since the heavy development costs were borne up front. In effect, every technological breakthrough in this economic context led to an almost one-off production of equipment. A mass-market economy such as America's, however, spreads these considerable costs over a large number of units and buyers, allowing for more development of different technologies. Where one-off government spending exists in market economies, the technology has often found its way into the private market — Tang orange juice crystals and the Internet are the famous examples.[36] Quite simply, their economic structure did not allow for spreading costs, so the Soviets could not afford to build their own versions of all the equipment available in the West, even if their scientists and engineers were capable of designing them. From my point of view, this meant the Cold War would end; it was only a question of when.[37]

I had occasion to revisit this observation derived from machine tools and the mass market when I was back at university and teaching courses about the relationship between business and government. I was curious about *who* had devised the broader notion of applying capitalist principles to fulfilling the needs of the large mass of society. For centuries, if not millennia, capitalism has focused on the elite — Mercedes-Benz over Chevrolet — a trite and unsatisfying answer, but one that seemed to me a part of the American character. Did this character proceed from Divine Revelation, or did it simply exist in the air of the frontier forest? Neither was likely.

Looking for an answer led me into culture and attitudes. In Canada, where I had spent virtually all of my career, the multicultural notion of a "mosaic," with a lot of identities living alongside each other, is a now-familiar pattern. In America, however — and reflected particularly in much of American historical literature — the "melting pot" notion assumes that everyone in the country *at any time in its history* was an "unhyphenated" American. "Ethnic" geography, as it might be applied to different groups of Americans of British descent, generally has faded as a subject of interest. In like manner, the existence of different geographic types of Americans, such as the "Yankee" tribe, has gone unnoticed.[38] Only those topics that relate to the short period before the Civil War manage to escape being treated as "American," since the distinctions between North and South were well known. One was the North's economic and technological dynamism, fueled by the immigrants who were pouring in, unlike the sleepy rural South.[39] Yet history textbooks usually treat immigration as a process affecting an undifferentiated "America." Why?

Invoking the disparities underlying the Civil War then reminded me of the familiar terms, "Yankees" and "Rebels." In the 1860s Yankees, generally, were all those people north of the Mason-Dixon line — the southern border of Pennsylvania — and north of the Ohio and Missouri rivers. Significantly, the term seems to have been used more in the South, which claimed to have a "Southern" culture. But why did they not refer, in turn, to a "Northern" culture? Why the epithet "Yankees" to stand for all Northerners?[40]

Suppose one were to accept the notion of a Yankee-dominated North and begin to push it back in time from 1861 to 1831, 1801, to a time when America was, in a much more literal way, the United States.[41] Then the geographical focus of what some have called "the universal Yankee nation"[42] recedes first to upstate New York and then to New England. I satisfied myself that the demise of the Soviet Union in 1991 could be traced to things the Yankees did in the early 1800s and somehow imposed on or transmitted to the rest of the North.

The Rebels' use of "Yankee" implied that they, at least, believed this was so. New England was putting its stamp on the myth of American origins. Garry Wills quotes Edmund Morgan: "Long before 1860 New Englanders laid claim to the national consciousness and gave their own past as a legacy to the nation, whether it wanted it or not."[43] The characteristic bedrock of the Northern culture that imposed itself on the South went on to dominate all of what we call American culture. To Europeans in two World Wars, all Americans were "Yanks," and the influence of Yankee culture is now to be found in every corner of the globe.

It seems to me there were three ways to investigate the making of the "universal Yankee nation": through *geography*, through *historical timelines*, or through the *selection of topics*. Tying these avenues together, I found, was a highway designated as US 20, which follows the general route Yankees and their descendants took as they expanded westward. My using US 20 also points to a second, more literal interpretation of *The Yankee Road*: the geographical path that ties into a cultural path, as Senator Jim Webb has done for the Scots-Irish South in *Born Fighting*.[44] The book thus consists of a series of essays about the large-scale and diverse ideas that came out of Yankee conditions, thoughts, and practices. The topics range from conservation to telegraphy (and incidentally, the Internet), from mass production to management, from roads and railroads to interchangeable parts, religion, organized land sales, and women's rights, each contributing its own piece to the Yankee pattern. As we proceed, we meet hundreds of Yankees, including such characters as Thoreau, Lowell, Remington, Hall and Lee, Pierson and Corning, Carrier, Olmstead, Pinchot, Cleaveland, Beecher, Dow, Finney and Moody, Perkins, McCormick, Mott, Stanton and Anthony, to name some of the most prominent.

Introduction: Values and Roads

I have nicknamed US 20 the "Yankee Road." Naming roads is a hit-and-miss process. Tourism agencies still refer to the National Road, created in the late eighteenth century to connect the Potomac River with the lands north of the Ohio. The Lincoln Highway is still lamented and promoted, even in television documentaries. US 66, popularized as the "Mother Road," had been designated as an afterthought near the end of the highway numbering process, and led not to new settlement but to fame as the route settlers and descendants of settlers of the Great Plains took to seek a new life in California in the 1930s and 1940s.

In this first of three volumes of exploration along the Yankee Road, we start in downtown Boston, at Quincy Market, and finish where the New York-Pennsylvania border runs into Lake Erie. On the way, we make twenty-two short side trips. What we find are the foundations of modern America.

Boston, Quincy Market: Larry Bird's Shoes

*Yankee Doodle went to town, a-riding on his pony.
Stuck a feather in his cap, and called it macaroni.*

— Traditional[1]

He [Cornelius Vanderbilt] made an art out of hiding his hand, having risen with the original generation of New York and New England's smart men, the wily pioneers of free-for-all commerce who knew how to speak and say nothing. He would make this connection himself in dodging an inquiry from a New York assembly committee. "Let me answer your question by asking another," he would say, "as the Yankee does."

— T.J. Stiles[2]

In the American League [in 1908], John Taylor, owner of the Boston team, decides to redesign the team's uniform, switching from light blue stockings to red ones. Taylor jokes, "You newspapermen will have to pick a new nickname for my team, then known as the

> Pilgrims, and previously as the Collinsites, Puritans, Somersets and even Yankees (!). He modestly proposes one possibility: Red Sox.
> — Cait Murphy[3]

> [W]hoever has been down to the end of Long Wharf, and walked through Quincy Market, has seen Boston... the more barrels, the more Boston. The museums and scientific societies and lyceums are accidental.
> — Henry David Thoreau[4]

In the Quincy Market/Faneuil Hall shopping area of Boston's downtown a bench sits near a bronzed replica of Larry Bird's shoes. Alongside is a memento to Red Auerbach, longtime coach of Bird's team, the Boston Celtics. It's fun to sit there, as I have done many times, and watch people and daydream.

Boston's professional basketball team is naturally called the Celtics. The great wave of Irish emigrants forced out of their homes by the famine of 1846–50 hit all the East Coast ports, but nowhere harder than Halifax and Saint John, and especially Boston, which became almost half Irish in a very short time, with long-lasting effect. Boston's professional football team, which plays in the suburb of Foxboro, is called the Patriots. No surprise here, either, given the close connection of the American Revolution to Boston and New England. But — and here is the puzzle — why is the professional baseball team named the Red Sox, while the team called the Yankees has resided in New York City for a century and more? Surely the Boston Yankees and the New York Red Sox is a better allocation of team nicknames. Everyone knows that the word "Yankee" was once a synonym for New Englander.

Well, yes and no. "Yankee" is a corruption of a derogatory term used by the Dutch of their seventeenth-century colony of New Netherlands to stereotype the English Puritan settlers who had moved onto Long Island. They apparently provided the Dutch with agricultural products, including cheese, and so were called "Jon Kees," or "John Cheeses."[5] The townspeople of New Amsterdam, now New York City, considered them real bumpkins, and applied the term to them with particular scorn for intruding into fertile land claimed by the Dutch. In short, the Dutch needed the Puritans' produce but resented their presence.[6]

Finding some group that deserved to be looked down upon is a constant human tendency. The English bumpkins were labeled Yankees, and Yankees they are.[7] Yet the story does not stop there.

Over the centuries since the 1650s, the definition of who is a Yankee has broadened. In the late eighteenth and early nineteenth centuries, Yankees were people who lived in New England. Then, descendants of Yankees spread westward across upstate New York into Michigan, northern Pennsylvania, Ohio, Indiana, and Illinois. Even Uncle Sam looked like a Yankee patriarch in artists' renditions — accurately so, as the man whose nickname began this tradition was a Yankee, an Albany-area meatpacker from Massachusetts who supplied the army in the War of 1812. During the Civil War, as we have seen, the Confederates applied the term to anyone who lived in the non-slave states. In the First World War, "The Yanks are coming" meant American troops, no matter where they might live or what their ethnic or racial backgrounds were. Still later, Yankee and American — as in "Yankee Imperialism" or "the ugly American" — became fixed as synonymous. Yankee, as it had been in seventeenth-century New Amsterdam, is still an epithet meaning common, uncouth, and rather pushy. After 9/11, the French president, in sympathy, said that "we are all Americans," though maybe humanity is not all Yankees — yet.

My point is that what began as an identifier of a people who originated in a particular part of America eventually came to represent a set of qualities and ideas originally propagated by that people. These attributes subsequently were absorbed by larger groups of Americans — immigrants who came in the millions in the nineteenth and twentieth centuries,[8] and then by groups of people all over the world, a point that this book explores, perhaps not to prove, but to make it seem plausible.

Yankee Origins

England in the 1500s and 1600s was wracked by almost continuous intrigue and warfare based on religious differences. Most agreed that the traditional Roman Catholic Church needed reforming: the question was how far this reform was to be pressed. The reformers varied in their ideas from simply severing the connection to Rome to a more radical "purification" of the Church by establishing new practices based on those that were thought to have existed in the first century or two after Christ.[9] The established Church of England persecuted both the radicals, the austere Puritans to the "left" as we might say today, and English Catholics to the "right," who sought to restore the primacy of the pope.[10]

In the early 1600s, two groups, commonly called the Pilgrims and the Puritans, left Europe and settled in New England. There were few significant differences between them, the Puritans having come directly from England and the Pilgrims having first emigrated to more liberal Holland before moving on to America. After some decades of struggling to maintain their identity there, many were ready to leave — the tolerance of the Dutch proving, in its own way, to be as great a threat to their faith as the intolerance of the English had been. Accordingly, influential Puritans in Holland and their sympathizers in England obtained a charter of incorporation for the Plymouth Company,[11] and in 1620 a ship full of colonists made its way to Massachusetts Bay. After an initial probe at Cape Cod, the Pilgrims landed at what we now know as "Plymouth Rock,"[12] south of present-day Boston.[13]

Conveniently, the Pilgrims found fields and clearings devoid of people, as the Indians had been largely killed by a plague acquired from other Europeans a couple of years before. Still, almost half the Pilgrim party died in the first winter.[14] The Pilgrims had pledged themselves to live as a communal society, but soon finding the arrangement to be a threat to their survival they turned to individual ownership of land and an internal market.[15] Even so, the ideal of an American separatist, communal utopia has come down to us through the centuries. Over the next few years, the Pilgrims struggled to develop their colony through farming, fishing,[16] and the beaver fur trade.

The Puritans who stayed on in England found their situation deteriorating as the seventeenth century moved on. A civil war eventually would lead to the triumph of Puritanism, but in the 1620s they could not know that. Then, in 1628, a second colonial charter of incorporation was issued for the Massachusetts Bay Company, which was to establish itself to the north of the Plymouth Company's holdings.[17] Over the next decade, the prominent Puritans who acquired the charter sent ships full of emigrants, Puritans, indentured servants,[18] and "adventurers" better prepared than the first Pilgrims to settle in the region. And, unlike the Plymouth Colony, they did not consider themselves separatists but Englishmen and women, in the political sense.[19] Massachusetts and its capital, Boston, were founded.

The next three colonies in New England grew out of Massachusetts: Rhode Island to the south and two in present-day Connecticut. Rhode Island was begun by a Puritan named Roger Williams, who, in the four years after his landing in Massachusetts in 1631, had been so contentious that he had to flee from the authorities in a blizzard and find a home with some Indians. The following summer of 1636, he bought land at the head of Narragansett Bay and founded a settlement he named Providence, which became a kind of multi-religious home for dissidents, including other Puritans, Baptists, and Quakers, from other colonies. In 1652, he tried and failed to get a charter for his colony, and only in 1663 did one of his sympathizers succeed in acquiring a charter for Rhode Island and the Providence Plantations.

Two other colonies were created out of the desire of orthodox Puritan families to find better land. As early as 1635, settlers were making their way west and south toward the Connecticut River, and soon two new colonial governments were established, one at New Haven, on Long Island Sound, and another upriver, called Connecticut. At that time, the coast of today's Maine and New Hampshire was lightly settled and Vermont, though claimed by the Dutch in New Netherlands, virtually uninhabited. The Dutch also claimed Long Island, but its good farmland was attracting settlers from Connecticut — the "Yankees."[20]

During the latter half of the Seventeenth century Plymouth and Massachusetts merged, as did the Connecticut and New Haven colonies. Massachusetts gained control of the area known as New Hampshire in 1641, lost it in 1679, regained it in 1686, and lost it again in 1691. Massachusetts also managed to gain control of what would later become Maine in 1658, but control went back and forth as different claims were advanced in London. The situation was complicated until the Revolution by the presence of hostile French and Indian forces in the wilderness behind the coast.[21] Unlike the rest of New England, Maine and New Hampshire attracted some non-Puritan settlers, such as Anglican English and Presbyterian Scots.

During the 1630s, nearly 25,000 people emigrated from England to Massachusetts and Connecticut.[22] With the Puritan revolution and civil war in England in the 1640s, emigration to New England dwindled as the victorious Puritans in England saw no reason to leave, while other potential immigrants were discouraged from going to New England because their faith and practices were not seen as "pure" enough to allow them to be made welcome.[23] Indeed, for the next two centuries, most of New England's population growth came through natural increase alone.

The Rise and Fall of the "City on a Hill"

Originally, the Puritan idea of settlement had to do with creating a garden in the wilds, a "city on a hill,"[24] where the faithful could practice their religion in peace, separated from both temptation and oppression by the surrounding forest. This ideal of living in an English village or town did not survive contact with the wilderness.

The Americanization of English Puritans created Yankees,[25] and Americanization meant first and foremost living with the shock of the frontier. The pressures of colonization and survival, the real and imagined threats coming from the wilderness, and the presence of aboriginal people already adapted to the environment all contributed to this mutation.[26] The reduction of the wilderness to a garden became part of the Puritans' sacred mission. That wilderness which made Massachusetts a sanctuary also became its enemy.[27]

Living on the edge of the wilderness might have been daunting, but then there was the economic attraction of new land. Almost from the time the Puritan colonists landed, some moved west into more promising country.[28] The colonial leaders, however, had begun settlement as a compact colony of believers, located in English-type villages, each around a Puritan (Congregational) church, the whole expanding gradually, town by town, as the population grew.[29] This attempt to promote a gradual, collective expansion failed on a number of counts. First, the Massachusetts colonies were soon plagued by religious dissenters. They refused to return to Europe, so they were simply expelled, making their way west or south into new lands, such as Rhode Island. As well, land fever began to take hold, along with the pressures of new immigrants, and new townships were forming faster than expected. Some people simply moved on to the Connecticut River, where good land was to be had. The tight politico/religious organization of the Puritan "city on a hill" thus dissipated into an expanding number of "towns," where lands were owned by individuals or families, rather than church or government allocations. Individual ownership also led to land speculation, where a person might purchase land from a new town, even though he had no intent of living there, but only to hold it for an eventual profit.[30]

The conflict over how to settle the wilderness persisted over the next two centuries, with some seeing it the wilderness as a force for the moral degeneration of a heretofore civilized people, while others saw it as a redemptive force, following the Bible's prescription to exercise dominion and to turn the dark tangle into a new Eden. Along with these conflicting perceptions, there emerged a desire for a symbolic hero who could take civilized values into the wilderness and bring order into it. Literary attempts to provide the archetype of this "frontiersman" came as early as the latter seventeenth century and continue to this day.[31]

A second major cause of the downfall of the idea of the "city on a hill" was wealth, which turned Puritans into "Yankees." At the beginning of settlement in Massachusetts Bay in 1629–30, the middle-class Puritan settlers expected they could live in the wilderness as comfortably as they had in England, but they misunderstood colonial economics.[32] Almost at once, they found it necessary to learn to cultivate strange, local plants that were already adapted to New England conditions and to work with imported seeds and animals so they could thrive. They also found that paying for English goods drained the money they had brought with them, and that they had to produce things that were in demand in Europe.

They were lucky, as Providence favored them, and "every thing in the country proved a staple commodity."[33] Furs were a useful export, as everyone else settling in North America found. In competition with the French to the north, the Puritans began to trade with the Indians for beaver pelts, which could be sold on the English market.[34] Then they realized that the forest could also produce useful products to sell abroad, such as "pot ash," lumber, and timber, the production of which also led to shipbuilding and shipping. Fish were saleable as well, and the Puritans, to supplement their diet and their incomes, took to the sea for food.[35] They named Cape Cod after one species, and Boston soon became the provisioning port for the large English North Atlantic fishery. As well, almost from the beginning, attempts were made to develop manufactures — a foundry to provide iron for nails and tools, for example — that could substitute for imports from England and allow some hard cash to remain in the colonists' hands. It did not take long for English merchants, who already dominated the European end of the Atlantic cod fishery, to realize there was a market in New England for almost everything they could provide and another market in England for American raw materials. New England found it could more than pay its way in the world.

The wealth gap between rich and poor settlers widened, however, in the second and third generations, rows and splits weakened church authority, the social atmosphere became more secular and mercantile, and the Puritan gradually became the Yankee, "a race whose typical member is eternally torn between a passion for righteousness and a desire to get on in the world."[36] Prosperity created a group of merchants and producers of consumer and export products. Traders bought from local distillers and tanners and encouraged farmers to plant wheat and produce lumber, all of which might enter the trade streams going to the West Indies and to England.[37]

A third major cause of the evolution of Puritan society had to do with the assertion of English control over the colonies.[38] During and after the English Civil War of the 1640s, there was no real attempt to exercise serious political control over the American colonies, which developed political institutions that generally resembled English ones, though without the aristocracy that characterized the home country. Then, England (and after the Acts of Union in 1707 that joined the separate crowns of England and Scotland, Great Britain) became involved in a series of wars extending into the 1700s that led to a cautious assertion of control of the colonies.[39] The first step came in 1664, during the Restoration, when the English captured the Dutch colony of New Netherlands and, in contradistinction to New England practice, the colony's governor was appointed by London.

In 1686, a Dominion of New England was created, with an authoritarian and Anglican governor, again appointed by London in the face of Puritan resistance. In 1688, hearing of the Glorious Revolution that brought William and Mary to power in England, the colonists had their own revolution, driving the Royal governor out. London temporarily acquiesced in the resumption of the *status quo ante*. Even so, as the British and French engaged in a series of wars until 1763, relations between Britain and America once again grew tense. The British government reasserted its right to appoint senior officials in the colonies and to pay them from London instead of leaving them dependent upon the local legislatures. Further, the British government insisted on its right to legislate taxation in the colonies and for the proceeds to go to the Treasury in London. These controversies tended to erode the communal town-to-capital political ties that had characterized the Puritan colony.[40]

A legal problem also made the colonists uneasy. The charters Charles I provided in the 1620s and 1630s had time limits, and by the 1660s, with the restoration of the monarchy under Charles II, their expiry dates were coming up. Not only would the ending of these charters affect the legitimacy of the colonial governments; it would also call into question whether the land given or sold to settlers would have to revert to the Crown.[41] If it did, the king's favorites then could be granted title to these lands and demand rent, similar to common practice in England. In one stroke, such a policy could disestablish Puritan churches and lead church members to penury, land being their prime capital asset. The Glorious Revolution of 1688 and the cementing of a constitutional monarchy put an end to such fears, but concerns continued that Royal prerogatives might be resurrected.

The fourth impetus for the transformation of Puritan into Yankee came with the passage in 1689 of the *Act of Toleration*, which gave limited rights to English dissident religious groups, except for Catholics. Had the Act been applied strictly in the colonies, it would have denied Puritans the right to attend or teach at their own Harvard and Yale universities or to vote in elections, but its application was modified, though limited rights were extended to non-Puritans. These were becoming so numerous, however, that the change risked swamping Puritan and Congregational control of the political process and putting it on a more secular basis — over time eroding the Puritan colonial tradition and replacing it with a more diverse and practical Yankee attitude.

Enter the Yankee

Thomas Carlyle called Benjamin Franklin "the father of all Yankees"[42] and, though he was more famous as a resident of Philadelphia, Franklin did exemplify the Yankee story. His father, Josiah, migrated to Boston in 1682 during a renewed persecution of Puritan believers in England. Benjamin was born in 1706, the youngest of seventeen children by Josiah's two wives, the second a descendant of one of the earliest Puritan settlers. He grew up in a literate, if not bookish, family and tried a number of trades, including working for one of his brothers as a printer's apprentice. At seventeen he ran away from his indentureship in Boston and went to Philadelphia.[43] His career as a writer, scientist, diplomat, businessman, and tinkerer exemplifies in many ways the character of the Yankee, as the Puritan descendants shook off most of the priorities of their forefathers. Frederick Jackson Turner, the great historian of the frontier, noted this mutation:

> After the War of 1812, New England, its supremacy in the carrying trade of the world having vanished, became a hive from which swarms of settlers went out to western New York and the remoter regions. These settlers spread New England ideals of education and character and political institutions, and acted as a leaven of great significance in the Northwest. But it would be a mistake to believe that an unmixed New England influence took possession of the Northwest. These pioneers did not come from the class that conserved the type of New England civilization pure and undefiled. They represented a less contented, less conservative influence.[44]

Henry Adams observed that Thomas Jefferson saw some differences that emerged between northern Americans and southern Americans. As Jefferson put it:

IN THE NORTH, THEY ARE	IN THE SOUTH, THEY ARE
• cool	• fiery
• sober	• voluptuary
• laborious	• indolent
• persevering	• unsteady
• independent	• independent
• jealous of their own liberties, and just to those of others	• jealous of their own liberties, but trampling on those of others
• interested	• generous
• chicaning [sic]	• candid
• superstitious and hypocritical in their religion	• without attachment or pretensions in any religion but that of the heart[45]

This process of partial cultural mutation continued as the Yankees spread west. Novelist James Fenimore Cooper, no friend of the Yankee farmers pouring across the Hudson River into and past the Cooperstown his father had founded, has one of his characters, Vermonter Billy Kirby, say:

> Why, I don't know, Judge...It seems to me, if there's plenty of anything in this mountainous country, it's the trees...I know you kalkilate greatly on the trees, setting as much store by them as some men would by their children, yet to my eyes, they are a sore sight at any time...I have heern the settlers from the old countries say that their rich men keep oaks and elms, that would make a barrel of pots [potash] to the tree, standing round their doors and humsteads, and scattered over their farms, just to look at. Now I call no country much improved, that is pretty well covered by with trees. Stumps are a different thing, for they don't shade the land.[46]

The Yankees' temperament was practical and unromantic, for the most part, seeing their task as one of subduing the wilderness and making a profit from doing so. This eye toward profit did not go down well with those on its receiving end. In the 1830s, a half-Yankee Nova Scotian[47] took a look at his home province through a Yankee character that came to contribute to American self-perception. Judge Thomas Chandler Haliburton published a series of articles in a local newspaper, collected in 1837 into a book of sketches entitled *The Clockmaker, or, the Sayings and Doings of Samuel Slick of Slickville*. Haliburton went on to publish six volumes of Sam Slick stories. A sympathizer of the Loyalists, or Tories, who emigrated to Canada after the Revolution, he had aristocratic sympathies, favored the South in the Civil War and, leaving Nova Scotia, spent much of his later life in England, where his writings were a great hit and he himself was elected to Parliament.

Sam Slick was not a flattering portrait of the Yankee businessman but part of a tradition that went back to pre-Revolutionary days of presenting the Yankee as a clever and unscrupulous judge of human character.[48] After the War of 1812, a number of actors had successful runs in British playhouses giving shows based on Yankee peddlers and their adventures. The Yankee was a salesman, jocular, ingenious, appealing, and prone to tall tales. Davy Crockett, a Tennesseean and a hero of the Alamo and no shrinker from the tall tale himself, described Yankees as "[a]n itinerant class of gentry, now identified with every new country, whose adventures are as amusing as they are annoying to its inhabitants. I allude to the tribe yclept *Clock Pedlers*, which term implies shrewdness, intelligence, and cunning."[49] Crockett continues elsewhere: "Reader, did you ever know a full-blooded yankee [sic] clock peddler? If not, imagine a tall, lank fellow with a thin visage, and small dark grey eyes, looking through you at every glance, and having the word *trade* written in his every action and then you will have an idea of Mr. Slim."[50] In 1816, British comic actor John Bernard made an American tour. In a conversation, he supposedly

> discovered that the Southern trader regarded the Northern peddler "in the light of a visitation," and looked upon a "Connecticut chap as a commercial Scythian, a Tartar of the North whose sole business in life is to make inroads on his peace and profit. He ranks him in the list of plagues next to the yellow fever, and before locusts, taxation, and a wet spring; indeed some go so far as to suppose that the shower of Yankees was the crowning pestilence which made Pharaoh give up the Israelites....There is no getting rid of them."[51]

There were traditional aspects of individual and community life that remained rather constant and gave Yankees an identifiable character. An example is the development of a form of town planning on the frontier. As a visitor to America would note: "Making governments and building towns are the natural employment of the migratory Yankee. He takes to them as instinctively as a young duck to water. Congregate a hundred Americans anywhere beyond the settlements and they immediately lay out a city, frame a state constitution and apply for admission into the Union, while twenty-five of them become candidates for the United States Senate."[52] As well, the scarcity of labor led to an interest in labor-saving devices, encouraging Yankee cleverness. When associated with significant investment and rational calculation, this produced the "American System" of manufacturing. Mark Twain put the Yankee mechanical aptitude this way:

> I am a Yankee of the Yankees — and practical, yes, and nearly barren of

sentiment, I suppose — or poetry, in other words. My father was a blacksmith, my uncle was a horse-doctor, and I was both, along at first. Then I went over to the great arms factory and learned my real trade; learned all there was to it, learned to make everything: guns, revolvers, cannon, boilers, all sorts of labor-saving machinery. Why I could make anything a body wanted — anything in the world. It didn't make any difference what; and if there weren't any quick, new-fangled way to make a thing, I could invent one.[53]

Interpreting Yankee Culture

I want to put my own spin on the epithet "Yankee." The expansion of Yankeedom and the location and identity of the original people whom this term described began to part company after the Civil War, but not before the Yankee ideal had spread further nationally and eventually internationally. De Tocqueville, in his travels through America in 1831–32, stopped at Baltimore and noted a part of a conversation he had with a local businessman:

> Yes, from certain points of view. This [Southern aristocratic] class was, in general, a nursery for distinguished men for the legislature and the army. They formed our best statesmen, our finest characters. All the great men of the Revolution, in the South, came from this class. And yet I am brought to believe that, taken altogether, the new order of things is better. The upper classes with us are less remarkable now, but the people is more enlightened; there are fewer distinguished men, but a more general happiness. In a word, every day we are growing more like New England. Now New England, in spite of what I have just told you about it, is far superior to us in all that constitutes the economy of society. I believe that the whole American continent is destined one day to model itself on New England, and what is hastening the movement is the continual importation of Northern men which is taking place in the South. Their desire for wealth and their enterprising spirit thrust them constantly among us. Little by little all commerce and all the control in society are falling into their hands.[54]

How can one analyze this Yankee phenomenon?[55] The core is one of values that include the virtual democratization of everything, the underpinning of which is the mass market, tied to mass production and interchangeable parts and an attachment to systematic organization. Capitalism tied to the mass market was noticed by the British paper *The Observer* as early as 1851, following the Great Exhibition at the Crystal Palace in London:

> Our cousins across the Atlantic cut many degrees closer to the ground than

we do in seeking for markets. Their industrial system, unfettered by ancient usage and by the pomp and magnificence which our social institutions countenance, is essentially democratic in its tendencies. They produce for the masses, and for wholesale consumption...With an immense command of raw produce they do not, like many other countries, skip over the wants of the many, and rush to supply the luxuries of the few.[56]

Further, Yankees organized everything. They, and now Americans as a whole, are the most organized people on the planet, and it is a key Yankee value. Millions of immigrants were organized into a growing economy; much of the West was organized, first by government surveyors and then by groups of Yankees moving into these ordered lands. This was not the Southern experience, until air conditioning and federal law brought the North south. After that, the whole Yankee process went global. That this group had a particular set of values is one interpretation of my title, *The Yankee Road*.

Some other Yankee characteristics began to emerge: the use of incorporation as an economic vehicle, the idea of individual ownership of land as opposed to company or Royal plantations, a basic capitalist structure, and a slow, evolution toward the equality of all people, not just of white men. By the Columbian Exposition in Chicago in 1893, which was designed by Yankees and featured America's commercial and industrial prowess, the Yankee ethos was dominant in the country. But by then the Yankees themselves were being swallowed up in the great migration from all of Europe.

It is hard to see these Yankee ideas as something original today: we live in them, as did our parents and grandparents. They are as "natural" as the sun coming up in the morning. As well, most of these ideas began as inventions by different people and cultures over centuries and millennia. For instance, Fernand Braudel defines capitalism as the injection of money and organization between producers and consumers — the next time you are at a farmers' market, think about the difference between it and a supermarket. Early capitalism provided Indonesian pepper to Roman markets. Clearly, the pepper producers and the Imperial consumer never met. Somebody risked fortunes to buy, preserve, and transport that pepper two thousand years ago. A supermarket is the same thing, only more complex. Yet, behind the creation of the complex supermarket and department store chains were four Yankees named Woolworth, Sears, and the Hartford brothers. Wal-Mart is an example of the Yankee ethos moving into the South, into Arkansas.

Production systems for products other than agriculture remained rather basic until about three hundred years ago. Anyone who has tried to encourage artisans to produce in volume for tourists or as new exporters knows that the artisanal temperament is susceptible to novelty. Making something new is often preferable to the monotony of making another thousand similar items just because they are popular. Guilds and apprenticeships were used in the past to get around this problem, to a degree. Today, automation and factories do it better, and we call it "mass production." Yankees named Tracy and Lee, Whitney and Colt, were central to this process. The Yankee genius was to marry these concepts to the needs and wants of the average person, or the "mass market." Capitalism up to the late eighteenth century primarily served the rich, providing them with spices, furs, arms, and other articles; most people in any society were too poor to be included even in the money economy.

The British flirted with the notion of capitalism, but the Yankees perfected it. The British caught a glimpse of the potential of a wider economy with the productivity gains in textile manufacturing that came about in the factory system of the late 1700s. After the Revolution, Yankee capitalists basically stole the plans for textile factory construction and machinery design from the British, and then made significant improvements to their processes. Their initial impulse was to stop the export of specie, or money, through import substitution, while the idea of then selling their products, not so much to the elite, but to every American farm and town family appealed both to their democratic sensibilities and to their wallets. They concentrated on textiles, shoes, clocks, and other mass market items, which were then distributed and sold throughout the new States by the "Yankee peddlers" Davy Crockett mentioned. Yankees were as much the improvers of products as inventors of them.

Back to Larry Bird

Even though what characterized Yankees in the eighteenth century is different from what is associated with the term today, there is a distinct relationship. The spread of the term as an identifier coincided with the spread of a set of ideas and ways of doing things that grew out of the original practices of the people who became "Yankees." Sally Marston describes the process partially when discussing the Yankees and Irish immigrants in the Lowell, Massachusetts, mills in the 1850s:

> [I]t is important to recognize that, in general, the gulf between the Yankees and the Irish was vast. One way of characterizing the relationship between them is to see it as one of competing meaning systems where one, the Yankee, is clearly dominant, while the other, the Irish, is subordinate. Within the social context of Lowell both of these systems were in competition for expression, though the native American system held considerable and, for many years, irresistible sway. In Lowell, because the Yankee value system was dominant, its social and political definitions became part of the

major institutional order. Conversely, because the Irish system was subordinate, it was accommodating, emphasizing various ways of negotiating improvements or at least maintaining the status quo.[57]

Though the original Yankees might be surprised to see how their ideas turned out today, the words used to describe them, in the sense of the original mocking "Yankee Doodle Dandy" of the Revolutionary War ditty, can still be found on a hundred placards in a hundred demonstrations around the world each year. The notion of bringing a vision of a better life to the bottom half of traditional social structures across the world is fundamentally subversive of the traditional privileges and status of the top half.[58] Regardless of the vagaries of American foreign policy, this subversion has been embraced, frustrated, or fiercely opposed, but the process still goes on. It might be a misuse of Marshall McLuhan's famous dictum, but "the medium (McDonald's or Nike) is the message."

All this brings us back to Larry Bird's shoes. Bird wasn't a great military hero or a king or an emperor, but he could throw a ball through a hoop better than almost anyone. He did it inside arenas full of average people and in front of TV cameras that carried his feats, and sponsors' commercial messages, to millions more. He is one of today's Yankees.

Boston, Beacon Hill: The Lost Tribe

A people's point de depart is an immense thing. Its consequences for good or ill are perpetually surprising in their scope.
— Alexander Everett, American diplomat,
to Alexis de Tocqueville, 1831[1]

If we look into history, we shall find some nations rising from contemptible beginnings and spreading their influence until the whole globe is subjected to their ways... Soon after the Reformation a few people came over into the New World for conscience sake. Perhaps this (apparently) trivial incident may transfer the great seat of empire into America. It looks likely to me.
— John Adams to Nathan Webb, October 1755[2]

The live Yankee is but the eccentricity of a truly wonderful people; the moral and physical impress of New England is stamped upon the universe; we owe her our nationality, and the world owes her admiration and respect.
— Anonymous churchman, 1862[3]

A mile or so west of Quincy Market is the Massachusetts State House. Originally, US 20 began here (or ended, depending on your travel direction) and went west along Beacon Street past Kenmore Square, where it begins today. The State House sits on the edge of Beacon Hill. For a long time this hill was known as the redoubt of the Massachusetts financial and social elite, the Brahmins. They were descendants of the Pilgrims and Puritans, of true Anglo stock. Now almost nobody cares and a lot of Beacon Hill has been given over to upscale Yuppies whose names do not connect to the passenger list of either the *Mayflower* or the *Arabella*. The same is true of most of Yankee New England. As in Britain, where those of Celtic descent are most visible on the fringes of the Atlantic, Yankee names are today most obvious only in rural, northern New England.[4] So, by inference, what happened to the Yankees?

The novelist Gore Vidal had an answer when he had one of his characters say:

> In his old age, [Aaron] Burr[5] was walking down Fifth Avenue with a group of young lawyers, and one of them asked him how he thought some aspect of the Constitution ought to be interpreted. Burr stopped in front of a building site, and pointed to some newly arrived Irish laborers, and he said, "In due course, *they* will decide what the Constitution is—and is not." He understood, wicked creature, that the immigrants would eventually crowd us out and re-create the Republic in their own image.[6]

Yes, but not quite. They learned to imitate their Yankee predecessors.

In this vein, consider what happened to those people of Roman culture after the Western Roman Empire faded away.[7] Surely, they were not all killed by Goths and Vandals and the like. Yet history books seem to imply that the barbarians successfully invaded the Empire and the Romans just disappeared. But the Romans more or less stayed on — they just got a lot of new neighbors and some rulers who were more interesting to later historians. Eventually, these Romans' sons and daughters married into their new neighbors' families and an amalgam of Roman-barbarian civilization took over. Even today, pieces of Roman civilization are in the bedrock of our culture. Take religion for instance: the Roman Catholic Church is the most obvious remnant, but many Protestant churches still employ rituals and music that emerged in Roman times. The calendar is another example: September, October, and November are derivatives of the Roman words for "seven", "eight," and "nine," while July and August are named for Roman leaders dead two millennia and more.

Basically, the biological Yankees in America encountered a similar fate as the Romans in Europe 1,500 years earlier, though in a more compressed time frame. In the cities and much of the rural areas of southern New England, Italians, Irish, French Canadians, and mixed refugees from New York City seem to dominate.

Farther west, millions of central, northern, and eastern European immigrants who came after 1860 were transported by rail to the cities and farms of the northern states, swamping the original Yankee settler population. Like me, they grew up in Yankee-named towns, which generally had meaning only to people who are long gone.

For instance, my father was Canadian and my mother Pennsylvanian-Slovak. I was born and raised in Michigan lands that were named after places in western New York, many of which were named in turn after people and places from New England. My hometown of Flint, Michigan, is named after Flint, New York, a small village just east of Geneva, New York, on US 20. In the late 1830s, Yankee settlers from this area and surrounding communities in New York were attracted to Michigan by the promise of lumber acreage covering good farmland. These original settlers gave much of southeastern Michigan its geographic names.[8] Flint is in Genesee County, along with Mt. Morris. Nearby is Livingston County and near Detroit are Rochester and Bloomfield Townships. All are western New York names, although the flat Michigan geography hardly resembles the steep lakeshores and river valleys the settlers left.[9] The growth of a wagon manufacturing industry and then the spread of the automobile industry attracted other migrants to the area, including my grandparents' families in the 1920s. In 1831, when deTocqueville passed through southern Michigan looking for authentic Indian culture, he visited some of the Yankee families clearing ground around what are now Pontiac, Flint, and Saginaw. A lot of their descendants are still there, lost among the later mixture of immigrants, foreign and domestic.

The point is simple. The Yankee tribe disappeared into the millions of immigrant families that came mostly to the northern part of America. Whether they arrived in New York City, Chicago, or California, they were employed by Yankee businessmen, bankers, and manufacturers or by other immigrants who had learned from them. After all, these millions mostly came from farms and villages, but not that many ended up on American farms or in American villages. Immigrants were not pioneers; they had to learn to live in cities for the most part. And there, they learned Yankee skills and attitudes. The tribe had disappeared, but the culture, though always changing, persevered.

But this is getting ahead of the story. First, the Yankees had to prepare the way for the later European immigration.

Five Streams of Internal Migration

Once the Atlantic coast was settled, pioneers flooded into the interior of America in five broad movements.[10] New England and tidewater Virginia were settled early, but the former was bottled up until almost 1800, while the latter was not particularly expansionist. Some areas encouraged immigrants to become frontierspeople, while others did not. Although all of these streams tended to consist of people who came from the British Isles, differences in local origins as well as American geography and climate resulted in different cultural variations.[11] As well, diverse historical experiences and different cross-Atlantic migration times affected these American migration streams and caused them to behave differently. At the start, there was no obvious reason one stream, the Yankees, eventually would come to characterize the rest.[12]

Figure 1 is a representation of the streams of ideas and people as they moved west. The map shows two centers of migration in New England, but to most people both are taken together to represent Yankees. Another center is in eastern Pennsylvania, a third in tidewater Virginia, a fourth in the Carolinas, and a fifth in Louisiana. The equal size of the streams is deceptive. Philadelphia probably attracted the lion's share of immigrants before the opening of the Erie Canal in 1825, while New Orleans became a center only as the Mississippi River territory was finally absorbed into America in the 1820s. The Yankee stream operated in somewhat the same manner as did those crossing other parts of the country, even though it came later than some. Yankee population growth was generated internally after the exodus of English Puritans to the Plymouth Colony and to the Massachusetts Bay Colony in the first half of the seventeenth century. For a number of reasons, few emigrants to America after that chose to come to New England before 1825.[13]

For decades, as the population grew and outgrew decent land, the Yankees were cooped up east of the Hudson River and Lake Ticonderoga. While Scots-Irish settlers moved southwest from Philadelphia, armed force in northern Pennsylvania stopped a Connecticut settlement, the large *patroons* of the Hudson valley would only rent, not sell, land, and beyond them lay the powerful Iroquois in what is now upstate New York. Fitful attempts were made to transplant New Englanders to the coast of the Gulf of Mexico and to the banks of the Ohio River. By the time land was opened for sale in New York, the natural Yankee outlet, Daniel Boone had already arrived in St. Louis via the Appalachians, a thousand miles to the southwest.

But then, New Englanders began to pour into the west and across the north of the United States. The whole of upper New York State was populated after 1800 largely by Yankees who left their family farms and townships in New England to find land farther west. People who immigrated to New York City did not move north from there in any numbers because of the persistence of the larger estates held on the Hudson River by former Dutch families, who kept their "feudal" privileges until the mid-1840s. As well, fertile land in New Jersey beckoned. The Yankees had only to leap past the Albany area to reach land that was for sale.

This Yankee stream then populated the northern parts of Ohio, Indiana, and Illinois as well as southern Michigan and parts of Wisconsin and Minnesota. De Tocqueville, who traveled through much of this area in 1831, was struck by the

Figure 1. Migration streams in the Early America

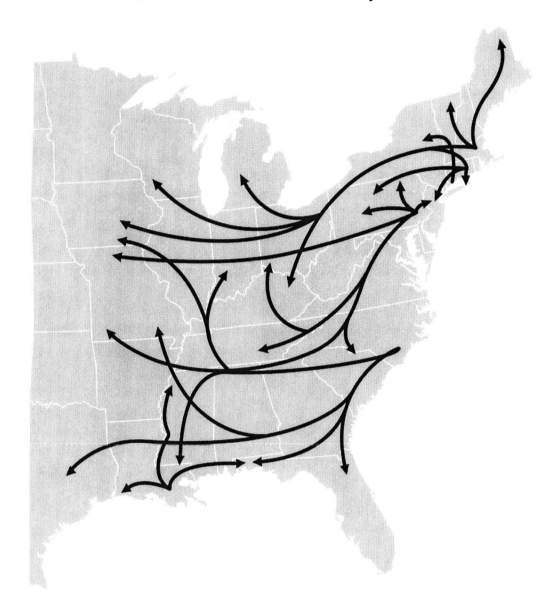

Source: Adapted from Wilbur Zelinsky, *The Cultural Geography of the United States*, Englewood Cliffs NJ: Prentice-Hall, 1973. P.81. Zelinsky adapted an earlier map published by Henry Glassie.

realization that the earliest settlers in these areas were the sons and daughters of farmers in upstate New York, not new immigrants as he had originally supposed. Upon reflection, it is not surprising that the real work of pioneering was done by descendants of the families that had started the process generations before along the coast. The same pattern held true in the movements farther south, especially among the Scots-Irish of the Appalachians. Figure 2, based on linguistics, but concerned with house construction practices, illustrates the direction and spread of Yankee language and associated values. The source of the map, William Labov, even titles the chapter in which this and others like it appear, "Yankee Cultural Imperialism and the Northern Cities Shift."[14]

Figure 2: Linguistic Migration Streams in the American North

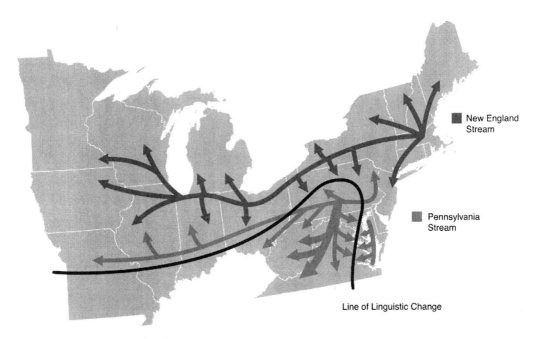

Adapted from William Labov, *Principles of Linguistic Change*, vol. 3 (New York: Wiley-Blackwell, 1994; reprinted 2010) at p.212 http://www.ling.upenn.edu/phonoatlas/PLC3/Ch10.pdf

An interesting value that was transplanted to the Yankee frontier was that of community organization. The New England "town" experience was replicated across the Old Northwest and was at odds with the more individualistic experience in the mountains and valleys of the Kentucky-Tennessee frontier.[15] This community experience has, with difficulty, melded with Southern individualism; a combination that drives American thinking and political conflicts to this day.

A second stream flowed from Baltimore and Philadelphia west along the Potomac River and over the seemingly endless ridges of the Appalachians to the headwaters of the Ohio River. Although road traffic was greatly more expensive than water traffic, as it still is today, the creation of a good road across these ridges helped to open up western Pennsylvania and the good farmlands of the Ohio Valley. The nation-builders of Virginia, among them George Washington and Thomas Jefferson, were especially keen on the Potomac River canal, plus a road connection to the Ohio.[16]

A third stream left Pennsylvania for the valleys of southwestern Virginia and North Carolina and then on to Kentucky and Tennessee. It was made up primarily of Scots-Irish settlers from Ulster.[17] They were Presbyterian, but the Scots-Irish did not find New Englanders partial to their practices,[18] although some did settle in New Hampshire. The bulk, having left Ireland in the early 1700s because of English laws that discriminated against Catholics and "Dissenters" alike, focused on the more welcoming surroundings of Quaker Philadelphia. The desire of Virginian planters to sell off landholdings in the Shenandoah and other mountain valleys attracted these immigrants toward the southwest, their descendants eventually spilling over into relatively unpopulated Kentucky. They found the land south of the Ohio River relatively unpopulated by Indians.

A fourth stream was the movement of the plantation economy slowly across the Deep South. This movement was not encouraging of significant population growth, since the demand for labor was taken up by slaves and the nature of the slaveholding economy argued for large farms on the best land. It was also inhibited by the presence of foreign territories in Florida, the Gulf Coast, and the lower Mississippi Valley until almost 1820, and by the strength of the Cherokees and Creeks.

A fifth stream was made up of immigrants, largely Germans, who came through the port of New Orleans. Once this city became American territory in 1804, it served as a major entry point for immigrants settling in the river towns and farmland at the center of the continent, especially after the introduction of steam power on the Mississippi in the 1820s. Water transportation downriver by barge had always been cheap, but upstream travel, against the flow, was difficult. Steam power made it possible for immigrants to transfer from a trans-Atlantic ship to a riverboat and go relatively comfortably and quickly to St. Louis, Galena, and Cincinnati.

Yankee Colonization Patterns

The Yankee expansion began as population pressure increased, in part because of the combination of seasons in New England, which limited the effect of many illnesses prevalent in the South, especially malaria and other mosquito-borne fevers. Large families thus had a greater chance of surviving *en masse*, leading to rapid growth. Boston-born Ben Franklin, for instance, was prompted to note that all sixteen of his brothers and sisters lived to start families and that his parents died in their mid- to late eighties. New England's population in 1650 was about 30,000; a century later it grown close to a half-million. The growth continued, even though immigration slowed and more people left the region between 1650 and 1700 than came in.[19] By the middle 1700s, even though birthrates were beginning to decline,[20] numbers were still increasing.

Population pressure was coupled with a desire to find better farmland. Little of the New England soil was good for intensive farming and it grew worse as settlers moved north from southern New England:

> An eighteenth-century traveler summed up all that statisticians have told us when he described the migrations from his native New England: "Those, who are first inclined to emigrate, are usually such, as have met with difficulties at home. These are commonly joined by personas, who, having large families and small farms, are induced, for the sake of settling their children comfortable, to seek for new and cheaper lands. To both are always added the discontented, the enterprising, the ambitious, and the covetous... others, still, are allured by the prospect of gain, presented in every new country to the sagacious, from the purchase and sale of new lands; while not a small number are influenced by the brilliant stories, which everywhere are told concerning most tracts during the early progress of their settlement."[21]

So, by 1775, the typical New England farm of fifty acres could hardly support a family, as soil exhaustion, overgrazing, deforestation, and overcrowding onto poor land took their toll.[22] Yankees had to leave; as one Connecticut family told the naturalist-painter John James Audubon in Louisiana, "the people are too numerous now to thrive in New England."[23]

Those Yankees who went south and encroached on Dutch lands on Long Island were perhaps the most fortunate in terms of the quality of their holdings, since New Hampshire, Vermont, and the part of Massachusetts that later became Maine had little more to develop than river bottoms among stony and mountainous land. Some eight thousand were attracted to the new colony of Nova Scotia to replace the French Acadian settlers who were deported in 1755; others moved north from Vermont into what are now the Eastern Townships of Quebec. New England businesspeople and land speculators created other alternatives with such vehicles such as a colony in West Florida in 1773[24] and the Ohio Company of Associates, which in 1786 created a model New England town at Marietta, the first town in the future state of Ohio.[25]

Movement directly west began in earnest once the ownership of New York territory was settled in the early 1790s. The Iroquois Confederacy was broken during the Revolution, and the lands owned by departing Loyalists passed to speculators. Farther west, Indian resistance in the Ohio Valley dissipated at the end of the War of 1812.[26] Once the region was opened to settlement, Yankees moved westward in great numbers relative to their Dutch and English predecessors around the Albany area. The movement was noticeable not only to census officials, but also to the diarists of the time. Rural New England began to depopulate. When author Bill Bryson traveled the Appalachian Trail in the 1990s, he came upon a hilltop field in Vermont where Daniel Webster had addressed ten thousand people in the 1840s, now so densely forested that Bryson could see little more than a few yards in any direction.[27] Among the Yankees who moved west were the Woolworths and the Rockefellers (at least the Mrs.), the families of Joseph Smith and Brigham Young, Susan B. Anthony, and John Seward.

A new wave of Yankee emigration was spurred by the opening of the Erie Canal in 1825 and the drop in transportation costs of goods and travelers. Yankee settlers, especially from western New York, pushed into Ohio, Michigan, Indiana, Illinois, and Wisconsin. As diplomat George Kennan notes in his *Sketches From a Life*, "as far as I knew, the family — the paternal line at least — had never lived any longer than that in any other place, at least since they left Scotland, which was presumably in the time of Cromwell. They had lived successively in Ireland, in Massachusetts, in Vermont, in upper New York state, and finally in Wisconsin, but never anywhere longer than two generations."[28]

Around 1850, there were final distinct movements of people of Yankee descent into Kansas as an ideological response to the slavery provisions of the *Kansas-Nebraska Act*, into Oregon as part of the migration to that territory, and into Utah as part of the Mormon migration. All of these, save for the Mormon lands, were then overwhelmed by immigrants from Europe, who began arriving in huge numbers in the mid-1840s, around the time of the famine in Ireland. Many did not stop in New York, Boston, or Philadelphia, but made their way to the towns and cities of the old Northwest, especially as the cost of transport was reduced by canals and railroads.

Pioneering for Yankees meant transforming forested land into farmland. The New Englanders who took the Oregon Trail in the 1840s and 1850s crossed over what they called "The Great American Desert" to reach and clear the forests of the West Coast. Only the Yankee colonies planted in "Bloody Kansas" to stop the spread of slavery and the Yankee Mormon exiles who chose the open land around the Great Salt Lake deviated from traditional Yankee-style pioneering.

De Tocqueville and the Yankees

Probably the best-known observers of early America were two Frenchmen, Alexis de Tocqueville and Gustave de Beaumont, who spent the better part of two years noting American manners and morés, but from Boston to Buffalo and in New York City and Cincinnati, the Americans they were observing were Yankees.

Let us start west of Syracuse, New York. The town of Auburn is one of many that grew up between the northern shores of the Finger Lakes and the Erie Canal. What attracted us to Auburn is situated a few blocks north of downtown off US 20: a state prison. It is an odd place for a prison — these days such institutions are generally placed in rural areas well away from population centers. It is also an odd facility, with a couple of old crenellated towers and antique military statues on the walls. The whole site is relatively compact — about the size of a good warehouse complex in an industrial park.

The prison also attracted de Tocqueville. Built in 1819, the prison and its rules were generally regarded as models at the time.[29] French and British prisons then were places of simple detention: people who were convicted of crimes were simply locked up, often in large collective halls that bred disease and more crime. There, the prisoners were flogged or waiting to be hanged or transported to penal colonies overseas. Whatever their fate, few French or British prisoners expected to be held for long. In America, however, where people were needed to roll back the frontier and labor was valuable, a different ethos began to develop. The prison was replaced by the penitentiary, a place where the convict would be given "time" to think about the nature of his crime and do penance through labor. At the same time, he would learn skills or a trade and, it was hoped, rejoin society ready to make a more valuable contribution.[30]

De Tocqueville was a young government official who had got his job through aristocratic connections. Then, the success of the Revolution of 1830 placed his career in jeopardy as the new regime came in with different people in charge.[31] He and his friend de Beaumont managed to get the revolutionary government to agree to accredit them as a voluntary commission to go to America to investigate the new penal philosophy, of which the warden at the Auburn prison was one of the most notable proponents.[32] For the Frenchmen, the notion of going to America to see some new idea in action was fairly unusual, as Europeans were convinced, with some reason, that their sophisticated societies had little to learn from a nation of farmers and small-town folk. For its part, the French government might have felt that this was an easy way to get rid of a couple of junior legal bureaucrats for a while, at no cost.[33]

De Tocqueville was not entirely forthright in his plans, either.[34] In the guise of prison reformer,[35] his real goal was to investigate American society to see if there were elements, democratic or otherwise, that could be transplanted into France or other European societies.[36] His aristocratic background made him skeptical of its desirability and he remained so, even though convinced of its inevitability: "We are traveling towards unlimited democracy, I don't say that this is a good thing, what I see in this country convinces me on the contrary that it won't suit France; but we are driven by an irresistible force. No effort made to stop this movement will do more than bring about brief halts."[37]

In the eyes of his eventual American readers, however, de Tocqueville gave the world a view of a brash and proud people who preferred a government that kept out of the way, were intensely religious, and concerned with industry and education in the North and with aristocratic touchiness in the South. He and de Beaumont, in fact, came to America during a period when the country, and parts of Europe, were in political upheaval. It was the time of the Jacksonian Revolution (1828) in the United States, the July Revolution (1830) in France, and the passage of the *Reform Act* (1832) in Britain. What really seemed to interest them was not so much America's democratic mechanisms as the notion of equality. They arrived at the floodtide of democratic change in America; had they come a decade earlier or later, their perceptions might have been different.

Twenty-six-year-old de Tocqueville and his twenty-nine-year-old friend landed at Newport, Rhode Island, on May 9, 1831, and made their way by steamship to New York, where they were welcomed by the local society leaders.[38] During their stay in New York, which lasted almost two months, they also went upriver to see the new prison being built by its future inmates at Ossining, and spent nine days studying its organization — Sing Sing was designed and managed along the Auburn model — and its results.[39] They then went on to Albany by steamer, where they participated in the July Fourth celebrations, marching in the parade alongside the governor. Unfortunately for their work, they were visiting during a holiday period when the legislature was not in session.[40] They also attended a Shaker service, among other activities.[41]

As well as missing some of New York's political structures, de Tocqueville and de Beaumont largely overlooked, or at least failed to appreciate, the important new technology around them and that was the real symbol of the state government.[42] First was the steamship they had taken up the Hudson, a technology that was opening up large parts of the continent to settlement and exploitation. Second was the Erie Canal, then the world's longest canal, constructed by a New York State corporation through mostly unpopulated country and leading to Buffalo, a port connected to even more unpopulated country. Six years old at the time, the canal would reconfigure the country over the next three decades.[43] Third was a new railroad that was under construction between Albany and Schenectady and that was the talk of the state.[44] Although the line had been designed only to cut time off the canal passage through the many locks between Albany and Schenectady, railroads would lead to a second reconfiguration of the country just before and after the Civil War. Later in his trip, however, de Tocqueville did note that he met the last surviving signer of the Declaration of Independence, John Carroll, who claimed that turning the sod for the start of the Baltimore and Ohio Railroad ranked in his mind with that signature.

Rather than taking the canal westward, the two Frenchmen took a stagecoach over rough roads up the bank of the Mohawk River. Traveling along the Erie Canal's route, they passed through Utica and Rome, eventually reaching Syracuse, where they met an ex-warden of the Auburn prison. In Auburn itself, they visited the prison and its variety of manufacturing shops, and encountered the governor again at their boardinghouse and went to his farm.[45] They then traveled to Canandaigua, where they met the legal expert John C. Spencer and were both smitten by his daughter.[46] Lingering a few days, they then went on to Buffalo.

De Tocqueville and de Beaumont visited four of the five Great Lakes looking for the true, natural savage that figured so much in French literary and philosophical thought in those years. They went inland from Detroit to the village of Pontiac and north to the ford of the Flint River, where three houses stood and just a few miles from where I was born 110 years later. They then pushed on through the swamps to the trading post at Saginaw, the site of just five houses.[47] Returning to Detroit, they took a tour boat that steamed up Lake Huron to Fort Michilimackinac, on Mackinac Island, where Lakes Huron and Michigan join.[48] They then went on to Sault Ste. Marie, where Lake Superior empties into the lower lakes. They topped off their northern tourist experiences with a trip down Lake Michigan to Green Bay, where they got as close to the real frontier as they could. They never found the "noble savage" of Rousseau, Fenimore Cooper, Chateaubriand, and the Romantics.[49]

The two French travelers went back east in the fall via Montreal and Quebec City to Boston,[50] where they were regaled by society and visited more prisons. After a month in New England, they moved on to Philadelphia for a month to study another prison intensively. They then visited Baltimore briefly and called on the president in Washington before heading west again. In Cincinnati and Louisville, they found themselves facing the slavery question. They mainly felt that slavery dulled the work and intellectual ethics of Southerners and that eventually they would be dominated by the North. Salmon Chase, who had moved from New Hampshire to Cincinnati, and eventually would be a member of Lincoln's cabinet, was a key informant of theirs, as were a couple of other Yankees.[51]

In December, the two French visitors traveled cross-country to Memphis, the Ohio River being impassable with ice floes. After waiting for a week, they were able to take a steamer down the Mississippi River to New Orleans, then one of the three or four largest cities in the country. While onboard, they met Sam Houston, former governor of Tennessee, who was on his way to Texas and glory, and accompanied by some of his Indian "relatives," who both attracted and horrified the two Frenchmen. From New Orleans, where they attended New Year's Day celebrations, they traveled overland northeast, eventually reaching New York, from where they departed for home.

Their tour of the South was made in haste, and their research could hardly be said to have extended south of the Ohio River, but what de Tocqueville had to say about the land and people along the Yankee Road is of enduring interest. He was one of the first to note that the frontier was settled in the first instance by people who had grown up on it, or near it, rather than by immigrants.[52] He also noted the proclivity of the Yankees to organize — indeed, some of his observations suggest that the interaction among individual liberty, organization, and technique go far back in American history, especially in the Yankee north. He and some of his prominent American conservative acquaintances were appalled by the way Americans formed private conventions or associations to discuss political matters. Most interpreters of his remarks about how quickly settlers set up local governments and replicated political structures stop there, but in his journals de Tocqueville recognized the tendency of Yankees to want to organize other things as well. Towns and businesses sprang up practically overnight, and the frontier was alive with improvisation.

De Tocqueville also noted the near universal literacy of Americans, commenting that, in Michigan, even the rudest shack in a forest clearing had some books in it:

> [A]n Englishman might fancy himself in the vale of Stroud. But, mark the difference: at the next crack of the whip…he is again in the depths of the wood at the other extremity of civilized society, with the world just beginning to bud, in the shape of a smoky log-hut, ten feet by twelve, filled with dirty-faced children, squatted round a hardy-looking female, cooking victuals for a tired woodsman seated at his door, reading with suitable glee in the *Democratic Journal of New York*, an account of Mr. Canning's campaign against the Ultra Tories of the old country.[53]

The American myth of young Abraham Lincoln studying his lessons by firelight may or may not have some truth to it, but it was taken as a truism on the northwestern frontier. De Tocqueville was more critical of literacy below the Ohio River, where he found a tendency to mythologize the man of action on horseback over the literate, organized pioneer. Despite his own aristocratic background, he did not like what he saw there.

Equality, a subject of great interest to de Tocqueville,[54] clearly did not extend to everyone in America. Relations among whites, blacks, and Indians were a particular concern of de Beaumont, who later wrote a treatise in a novelistic format about the racial problems the two Frenchmen had encountered. Essentially, American responses to their questions about race were consistent no matter where they went. Indians were born free and could only live free; they could not be organized or settled, and were doomed to be pushed back as American civilization pushed forward. Blacks could be organized but were not free; they could function in American society but in a dependent way, as their slave heritage disqualified them for individual liberty. In essence, where others could come to America and be assimilated into the dominant culture, Indians and blacks were unwelcome.[55] The rights of women were hardly mentioned.

Finally, de Tocqueville was alternately fascinated and repelled by American commercialism.[56] He noted that, in New York, concern over material prosperity dominated the after-dinner talk at the functions he attended. In Cincinnati, people remarked about the fast pace of growth on the northern side of the Ohio River compared with that of Louisville on the southern side.[57] Part of the innovative Auburn penitentiary system included establishing workhouses where the prisoners could earn a small amount of money learning a trade, while the prison arranged to cover many of its expenses by selling the products they made.[58] De Tocqueville also noted how fairly uniform American society was in its tastes, products, and information sources.[59] In short, de Tocqueville's observations are both perceptive and of lasting relevance; sometimes it takes an outsider to point out what is so obvious to insiders that it is invisible to them.

The Disappearance of the "Tribe"

Back in New England, immigrants began to arrive in large numbers in the 1840s, just as the great migration of Yankees west reached a peak. They flooded into Boston, Lowell, Providence, Hartford, and other "Yankee" cities, with each European calamity seeming to bring a fresh wave: Irish fleeing famine; Germans and Italians fleeing political repression; later, Russians and Poles, especially Jews, fleeing ethnic repression. In North America, French Canadians fled their farms and looked for work in New England; many stayed, replacing Yankee mill girls, shoemakers, and railroad navvies.[60] The complexion of New England changed, and those Yankees who stayed gradually became a minority in their own homeland, though a minority often found at the top of the economic and political worlds for a long time to come.[61]

The same process went on in the west. Yankees cleared the good land, then the chain of later purchasers began to include Scots and Germans. In cities such as Buffalo, Irish came in as day laborers and household servants; others brought specific skills, or acquired them, in services such as barbering and locksmithing. All in all more than 30 million immigrants came to America in the nineteenth century after 1840, the vast majority following the rough geographic dimensions of the Yankee stream west. Some went to Pennsylvania and on into the Ohio Valley, but most stayed north. The sheer volume of immigrants overwhelmed the Yankees and their descendants. What the new arrivals entered, however, was a society that was fundamentally Yankee.[62] They called it "American." The new Americans were presented with Yankee rules about childhood education, methods of work in the shops, attitudes about success, and approaches to religion, business activity, banking, and life in general to which they were expected to adapt.

Sherwood Anderson, in his novel *Winesburg, Ohio*, about a town in the northern, or Yankee, part of the that state, has his character mulling over the Yankee/Puritan dilemma:

> That is what Jesse hungered for and then also he hungered for something else. He had grown into maturity in America in the years after the Civil War and he, like all men of his time, had been touched by the deep influences that were at work in the country during those years when modern industrialism was being born. He began to buy machines that would permit him to do the work of the farms while employing fewer men and he sometimes thought that if he were a younger man he would give up farming altogether and start a factory in Winesburg for the making of machinery. Jesse formed the habit of reading newspapers and magazines. He invented a machine for the making of fence out of wire…Faintly he realized that the atmosphere of old times and places that he had always cultivated in his own mind was strange and foreign to the thing that was growing up in the minds of others. The beginning of the most materialistic age in the history of the world, when wars would be fought without patriotism, when men would forget God and only pay attention to moral standards, when the will to power would replace the will to serve and beauty would be well-nigh forgotten in the terrible headlong rush of mankind toward the acquiring of possessions, was telling its story to Jesse the man of God as it was to the men about him. The greedy thing in him wanted to make money faster than it could be made by tilling the land.[63]

So, for the most part, the Yankees disappeared into the mix of immigrant nationalities that came to America after about 1845, but not before turning most of them into cultural Yankees themselves. Wilbur Zelinsky sums it up: "[T]he region was the most clearly dominant in the century of national expansion following the American Revolution. New England may have been strong, if not preeminent in the fields of manufacturing, commerce, finance and maritime activity, but it was in the spheres of higher social and cultural life that the area exercised genuine leadership — in education, politics, theology, literature, science architecture, and…technology."[64] The now lost tribe of Yankees grew from a kind of pseudo-ethnic/cultural group identification into a broad idea about how the world, not just America, should work.

Boston: The Yankee Road from Kenmore Square to Newport, Oregon

Wisdom sits in places.
— Apache proverb[1]

The architectural historian Vincent Scully.... cites two examples of American peoples who voluntarily gave up their sedentary life in the face of a new mode of transportation: the Plains Indians, when confronted by the horse, and twentieth-century Americans, when confronted by the automobile.
— Witold Rybczynski[2]

[A] horse was the poorest and most expensive motor ever built: he eats ten pounds of feed for every hour he works. Motor trucks, on the other hand, need feed (fuel) supply only when they are working.
— Thomas Edison[3]

US 20 stretches across the United States nearly 3,400 miles from Boston to Newport, Oregon. It is a designation rather than a singular highway; the numbered US highway system was created in 1926 as a result of the *Federal Aid Road Act of 1916*, which called for the classification of highways by their importance and provided federal funds to the states to improve the more important ones.[4] Unlike most of the east-west numbered highways, US 20 did not lose its identity in the Interstate system — it shares only about seventy miles (2 percent of the distance) *in toto* with a couple of Interstate highways on its way from coast to coast.

I call US 20 "the Yankee Road." In the late eighteenth century, the National Road was created to connect the Potomac River and, by implication, Virginia and Pennsylvania with the lands north of the Ohio River. Other roads were "named" before the numbering system was imposed, but they mostly died out, although, in the 1930s, US 66 became the "Mother Road," the route that settlers and descendants of settlers used to leave the Great Plains Dust Bowl for a new life in California. More recently, the old transcontinental "Lincoln Highway" has been resurrected as a tourist and nostalgic route. So, I guess I can invent one as well.

For me, US 20 is the Yankee Road because, at least into eastern Iowa, it blazes through the center of lands that Yankees and their descendants first settled. US 20 has no particular identity of its own, since its historical antecedents are a series of turnpikes and roads connecting one city with another. Today, it is paralleled by an Interstate from Boston to Chicago, but from there to the West Coast at Newport, Oregon, the route is nearly all its own. Fifty miles into Nebraska, a stretch of the highway seems more like a simple country road. The value of traveling US 20 lies not so much in its geography as in the significance of the people and events along its route for the creation of America.

The Creation of US 20

The story of American roads is a complex one. In the early nineteenth century, the federal government developed the National Road from Maryland to Illinois and contributed to the building of other roads.[5] Later, it subsidized the states by allotting them a small proportion of the receipts from the sale of federal lands for road building.[6] Then it left the task of road building to private entrepreneurs and local governments, and the various states gave the task of providing roads to turnpike companies. By 1810, there were at least 180 such companies in New England and in 1821 there were 278 in New York State.[7] For the most part, the roads they built were rough and pockmarked with low stumps and passable only in dry or freezing weather.[8]

Between 1800 and 1850, mail-coach roads grew from 21,000 to 170,000 miles in length,[9] but canals and then railroads retarded future growth of the road system, or even caused parts of it to fall into decay, especially where the road competed with rail for long-distance traffic.[10] Only those local roads that fed farm produce to canals or railheads were maintained in decent condition through much of the latter half of the nineteenth century.[11] Nearly all road construction and maintenance was a local responsibility, and road workers often were people who needed temporary jobs or were willing to work off their tax bill.[12]

These early roads were devised for wagons and carriages and were too narrow, too steep,[13] or improperly surfaced for any kind of mechanical travel, which began in earnest with the rising popularity of the bicycle. In 1878, a Connecticut Yankee, Colonel Albert Augustus Pope, began importing a high-front-wheel design from England often referred to as a "boneshaker" from its uncomfortable ride,[14] and in the 1880s, Pope, by then the largest bicycle manufacturer in America, began producing his own version of the new, modern-looking "safety" bicycle, which reduced injuries from falls and was designed to be accessible to both men and women. Meanwhile, in 1880, to promote his business, Pope formed the League of American Wheelmen to press for better roads.[15] He also financed courses on road engineering at the Massachusetts Institute of Technology and had a short demonstration stretch of macadamized road surface built in Boston.

The 1890s witnessed the growth of a bicycling fad based on the safety bicycle. Women were especially avid consumers. Women's rights activist Susan B. Anthony called the bicycle "the freedom machine": "Let me tell you what I think about bicycling. It has done more to emancipate women than anything else in the world."[16] By the turn of the century, more than three hundred companies were manufacturing bicycles, and ten million bicyclists were making good roads a hot political issue in the cities.[17]

By then, some states had begun to finance local roads through property taxes and later, as automobiles gained in popularity, through licensing receipts.[18] In 1893, the federal Department of Agriculture created an Office of Road Inquiry (later the Office of Public Roads) to collect data, study road construction methods, and encourage road improvement across the country.[19] As the states and the federal government began to become involved in road design and technology, considerable testing of materials and techniques was done under public rather than private auspices.[20]

A second force was the farm community, which began to see roads and trucks as a counterweight to the power the railroads held over them.[21] This revival of interest was prompted as well by improvements in technology, especially the gradual introduction of the telephone and electric power. Service could not be provided or maintained without road access. The innovation of Rural Free Delivery (RFD), which enabled the growth of catalogue shopping, also required better roads.[22] City people were already benefiting from cheap and frequent postal service, and the times were right for politicians to push for expanding such services equally to rural people. The reform began as an idea by President Benjamin Harrison's postmaster general, Philadelphia department store owner John Wanamaker, whose motives were not altogether altruistic.[23]

Begun in 1896, RFD had the effect of making the Post Office into a national development tool. First, standard postage within the United States was fixed (at 2 cents per ounce), so that rural people were not penalized by their remoteness. Then came a fixed low rate for magazines and newspapers, then comprehensive free delivery to homes in towns and villages of fewer than 10,000 people. These innovations, of course, revolutionized catalogue sales and goods distribution;[24] eventually, even house and automobile kits were sold by mail. But this form of postal service required roads that allowed the mail to be delivered. In 1900, four years after the first experimental RFD routes were started, the postmaster general of the day could say:

> "Rural free delivery brings the farm within the daily range of the intellectual and commercial activities of the world, and the isolation and monotony which have been the bane of agricultural life are sensibly mitigated. It proves to be one of the most effective and powerful of educational agencies. Wherever it is extended, the schools improve and the civic spirit of the community feels a new pulsation."[25]

In the early twentieth century, the introduction of the automobile and, by extension, the truck, tractor, motorcycle, and bus, created even more pressure for good roads. Like bicycles, automobiles generated a national organization to pursue users' interests. The Automobile Club of America was created in 1899 and became the American Automobile Association in 1902. A year later, in 1903, a Vermont physician, Horatio Nelson Jackson, and his mechanic, Sewell K. Crocker, accompanied by a dog named "Bud," drove a gasoline-powered Winton automobile from San Francisco to New York on a journey that had to go north to Oregon to avoid the Sierras and the Nevada desert and took sixty-three days through often roadless territory.[26] By 1905, more than twenty states were funding the construction of highways that could accommodate automobiles and trucks.[27] Each state tended to develop its own road system, most often concentrating on a network connecting the state capital with county seats; little thought was given to connections between states.[28]

Trails across America

A variety of innovations accompanied the development of the automobile and the truck.[29] The first state maps of improved roads were devised in 1895. Lacking route numbers and local signage, these had to be accompanied by fairly detailed instructions on how to access the roads through and between towns. One logical extension of these maps was the designation of "trails" by clubs of automobile enthusiasts.[30] By 1924, there were 250 marked trails in the country, sponsored by 100 road and tourist associations,[31] including the [Theodore] Roosevelt from Washington, DC, to Los Angeles; the Jefferson from Winnipeg to New Orleans; the Old Spanish Trail from St. Augustine, Florida, to San Diego (a remnant in Tucson, Arizona, still goes by this name); the Pacific Highway from British Columbia to Mexico; the Atlantic Highway from Maine to Florida; the Yellowstone Highway from Plymouth, Massachusetts, to Seattle; and the Dixie Highway from Duluth, Minnesota, to Miami Beach, Florida.[32]

Another Dixie Highway, the one that still passes near my birthplace, was a series of roads marked by red and white stripes from the Strait of Mackinac in northern Michigan, or perhaps only from Bay City, to Miami, Florida, in 1915.[33] The driving force behind this Dixie Highway was J. Dallas Dort, an early Flint, Michigan, automaker, whose business was absorbed by General Motors after the First World War. In Flint, where I grew up, it was the main thoroughfare, known as Saginaw Street, but north and south of town it was still the Dixie Highway to everyone even into the 1960s.

The problem with these trails in the early years of the century was their chaotic spontaneity: they often overlapped, and there was a lot of competition between their respective boosters, so that signage, which tended to be colored rings painted on telephone poles, was erratic and sometimes sabotaged. On one 1,500-mile stretch of road in the West, the markers of fifteen routes were painted on poles, barns, and the like.[34]

The most famous of these routes was the Lincoln Highway, which was promoted by a group of businessmen around 1913 as a means of linking New York with San Francisco.[35] The group was led by auto and auto parts manufacturers and was seen as a private sector venture. Carl Fisher, an auto parts manufacturer who had just completed a new speedway in Indianapolis, stated that "the highways of America are built chiefly of politics, whereas the proper material is crushed rock or concrete."[36] The Lincoln Road eventually fell to the government-mandated numerical grid system, as most of its length was designated as US 30, and by 1927 Henry Joy, one of the original sponsors of the idea, would write to a colleague: "We tried to put on the map of the United States a wonderful main arterial route from New York to San Francisco as a memorial to Abraham Lincoln…The effort has resulted in total failure. The government, so far as has been within its power, has obliterated the Lincoln Highway from the memory of man."[37] In fairness, a lot of the trails were not maintained properly, and the Lincoln Highway became the exception that proved the rule. As Wisconsin state highway engineer, Arthur Hurst, commented in 1918: "The ordinary trail promoter has seemingly considered that plenty of wind and a few barrels of paint are all that is required to build and maintain a 2000-mile trail."[38]

War and Chaos

The first real military use for cars and trucks came in 1916, when General John "Black Jack" Pershing was sent to avenge a cross-border raid on Columbus, New Mexico, by Mexican revolutionary Pancho Villa. A contract to build fifty-four special trucks to haul munitions and supplies was given to the Packard plant in Detroit on a Monday morning; they left by rail the next afternoon for El Paso, Texas, where they were put to use along with horses and mules as the army moved south into Mexico.[39] This example led entrepreneurs to consider the possibility of using commercial trucking as an alternative to railroads. The first commercial trucking service was apparently set up the next year as a promotional idea by Goodyear Tires to ship its products. In the spring of 1917, it took twenty-one days for the first truck to go from Akron, Ohio, to Boston, with weather the cause of many delays. Another Goodyear transcontinental trip in 1918 took twelve days to go from Boston to San Francisco.[40]

The need for more transport than the railroads could provide in the First World War demonstrated the potential of trucks, but the uncoordinated and underbuilt system of highways made trucking movements difficult and somewhat undependable. In desperation about the congestion on the railroads, the Council on National Defense asked that military trucks made in the Midwest not be shipped by rail, but filled with war materiel and driven to East Coast ports. The first convoy left Toledo, Ohio, for Baltimore in December 1917 and took three weeks to get there.[41]

In 1919, a military experiment in moving a motorized convoy across the country had a galvanizing effect on federal interest in roads.[42] After a number of send-off speeches by politicians, a three-mile convoy of military vehicles, accompanied by a young observer named Dwight Eisenhower, set off from Washington, DC, to cross the country to San Francisco. What with frequent stops to hear more speeches, attend dinners and dances, have breakdowns, and overcome the lack of roads as they moved west, the convoy took sixty-two days to make the trip. Considering the problems of relying on rail for military mobilization that had arisen in the previous two years and the need to defend two coastlines, the five-mile-an-hour average convoy speed was unacceptable. Eisenhower described the trip as a "journey through darkest America with truck and tank" and the roads as "average to non-existent."[43] One result of the convoy, however, was the donation by the federal government of war surplus construction equipment to build roads.[44] In 1920, a second convoy was dispatched to Los Angeles across the South. It took sixteen days just to reach Atlanta and three months to arrive at the West Coast.[45]

Federal Money

Direct federal involvement in roads had ceased in the early nineteenth century. It resumed with a $500,000 congressional appropriation in the Post Office budget of 1912,[46] followed by the *Federal Aid Road Act of 1916*, which was passed ostensibly to improve rural mail delivery.[47] The money, $75 million over five years, was provided under the principle that the funds were to be administered by the state governments, a testimonial to the success of the new (1914) American Association of State Highway Officials (AASHO).[48] The roads were to be classified in terms of their importance, a step toward the national system a decade later, but in practice the expenditures followed political pressures rather than a rational plan.

State activity can be seen in the case of Michigan. In 1905, there were about a thousand cars in the state and horse-drawn vehicles still outnumbered them a decade later.[49] An alliance developed between the bicycle associations, which had teamed up with farmers in the 1890s in the "Good Roads Association,"[50] and automobilists, this time to oppose farmers, who were in favor of radial roads to and from market centers. Instead, the new bicycle-auto alliance favored intercity travel based on a grid system. Michigan bicycle activist Horatio "Good Roads" Earle successfully promoted an investigation into the state's road conditions and then led a move to amend the state constitution to allow it to support trunk road construction. Earle was then appointed the state's first highway commissioner. In 1913, Michigan committed itself to fund a three-thousand-mile trunk road system, just as federal funds started to become available.[51]

This new government action reflected the growing importance of motorized vehicles in the country. For instance, the average automobile in 1916 cost about half a year's wages for a clerical worker, about the same proportion as in 2006. The number of cars and trucks in 1916 was about 3.5 million, but grew to 4.8 million just a year later. Forty thousand garages and repair shops had sprung up to service these vehicles. Yet the state of the roads was the weak link in a rapidly growing important new industrial sector.[52]

In the *Federal Highway Act of 1921*, the federal government began to consider the uses of a road system for the country. To the officials of the time, roads should serve four functions: provide better communication between farm and city, provide recreation opportunities for citizens, improve commercial flows,[53] and make all parts of the country accessible for defense. These considerations led to the development of a rough outline of a national system of improved highways. Part of the accompanying policy mandated that 7 percent of the highways it helped finance must connect across state lines.[54] The next year, 1922, saw a federal-state conference set up a grid of national highways in the Northeast that would receive support for construction and maintenance from Washington.[55] Agreement by all forty-eight states to its expansion came in late 1923. This was followed by the creation of a federal-state commission to designate and number a national grid. In 1925, the AASHO decided to move from designating routes by names to a numerical system, and routes and their numbers were approved by the states in 1926,[56] effectively ending the informal "named" system that had grown up in the previous decade. That year, the head of the federal Bureau of Public Roads declared: "There have been just three great programs of highway building within recorded history: that of the Roman empire, beginning with Julius Caesar and ending with Constantine; that of France under the Emperor Napoleon; that of the United States during the last decade."[57]

The grid was numbered, with even numbers for roads running east-west (the highest number was 96) and odd numbers for those running north-south (to 101). Further, the numbering system provided for smaller numbers starting in the northeast corner of the country and ending in larger numbers in the southwest.[58] Supplemental highways were given related but even higher numbers, such as 301, which roughly parallels US 1 along the East Coast. The most important from a strategic and use perspective were numbered as multiples of 10 (10, 20, 30, ...) and 1 (1, 11, 21, ...).[59] The system was quite rational and was a great assistance to long-distance travelers. Later, the Interstate system was numbered in similar fashion, though the numbers themselves were reversed to avoid much confusion. Low Interstate numbers begin in the Southwest, for instance, and for half its distance US 20 is paralleled by Interstate 90. The old US highways developed some romance even though the numbering system was criticized as "soulless":[60] Route 66 became famous in song and literature, Highway 61 became part of the legend of jazz and folk music, and US 95 became the loneliest road in the country.

The Interstate System

In a foreshadowing of the Interstate highways, both Philadelphia (1903) and New York City (1906) authorized the construction of "parkways," though they were not completed until after the First World War I. In New York City, parkways were conceived as early as 1868 by Frederick Law Olmstead and his partner Calvert Vaux, the designers of Central Park,[61] but the city's first, the Taconic Parkway, was begun only in 1940 and finished in 1950.[62] A limited access toll thruway across the state was decided upon in 1942, but work on it did not start until 1946.[63] Later, E.M. Bassett would contrast the parkway, a road for recreation where trucks were banned, with the freeway, a road for movement and commerce. California began to promote freeways with a 1939 law, and these soon became the model for the Interstate system.[64] The landscape designers were marginalized to building tourist parkways, such as in the Blue Ridge and Great Smokey Mountains. Other political opponents to these new limited-access roads were communities that feared being cut off from passing commerce and city neighborhoods that feared they might be split or isolated. In the late 1930s and '40s, however, state and federal legislation was passed to standardize the concept of limited access.[65]

A nationwide system of limited-access roads was not seriously considered until 1938.[66] Then, in 1940, Pennsylvania completed a 160-mile turnpike through the Allegheny Mountains, offering a vision of the ties the country might aspire to in the future.[67] A 7,500-mile system of defense highways was legislated in 1941, but though initial funding was provided in 1944, it was not until 1956 that work on the system began in earnest.[68] As was the *Federal Highway Act of 1921*, the *Defense Highway Act of 1956* was prompted by the need to get war materiel and troops quickly across the country in quantities and timeframes that existing rail systems could not meet. There was also the undertone of nuclear evacuation and rescue behind the 1956 act.[69] As before, the system was to be developed and maintained by the States under a cost-sharing agreement (90/10) paid for by a Highway Trust Fund built by taxes on transportation material and fuel.[70] The system was not intended at first to affect the structure of cities, but urban lobbying resulted in about a seventh of the mileage and a considerably higher proportion of the costs being allocated to urban transportation.[71]

The new Interstate highway system was to be 42,500 miles long, about a quarter the length of the initial 167,000-mile federal system.[72] Its gradual realization over the next twenty years overshadowed the older federal highways, reducing them to a system of Main Streets, regional feeder lines, and scenic "blue highways" for tourists and automobile romantics.[73] In the West, the Interstates even obliterated parts of federal highways. In the 1970s and '80s, the Interstates, along with air conditioning, helped to create the Sunbelt.[74] They hurt the rail-based industries of the North and transferred activity to the more lightly populated cities of the South, such as Charlotte, Atlanta, Dallas, and Phoenix, where it was cheaper and politically feasible to create new highway grids than in the crowded and developed cities of the North. The South became Yankee of sorts.

A Fast Trip across the Continent on US 20

In Boston, US 20 used to start near the harbor, but over time the beginning point has moved west, from Beacon St. to Commonwealth Ave. and now out to Kenmore Square.[75] It is still near tidewater, however, as the basin of the Charles River is just to the north. From Boston, the road goes through a string of towns and suburbs, generally taking the route of the northern, or Upper, branch of the old Boston Post Road, one of the earliest of the intercolonial roads.[76]

The origins of this route date back to the 1635, when some sixty Puritan settlers elected to push westward from Massachusetts Bay toward more fertile lands along the Connecticut River. A thousand or so others elected to go by sea. The land travelers went along an Indian path known as the "Connecticut Trail,"[77] part of which became the Upper Branch of the Boston Post Road.[78] The Post Road was created as a result of King Charles II's 1672 order for a mail service to connect New York through Hartford to Boston.[79] For a while the Post Road was made up of a variety of bad and not-so-bad pieces, but in 1703 the New York Assembly ordered its portion to be widened from a horse path into a road capable of carrying coaches, and the other colonies followed suit.[80]

The mail service, in fact, predated the road. The first mail delivered at a distance in America accordingly was carried from New York to Boston in January-February 1673. It took the post rider about two weeks to make the trip via New Haven, Hartford, Springfield, and Worcester. Unfortunately, the experiment was soon halted due to the temporary reoccupation of their former New Amsterdam colony by the Dutch and then to the outbreak of the Indian conflict known as King Phillip's War. Only in 1691 were attempts made to revive this overland route. Mail service along the Atlantic coast was not put on a sound footing, however, until the appointment of Benjamin Franklin as one of two postmasters in North America in 1751. Franklin managed to cut the post time between Philadelphia and Boston from weeks to days.[81]

Going westward through the Boston suburbs, US 20 crosses the Blackstone River. It was on the Blackstone, at its mouth in Pawtucket, Rhode Island, that a British immigrant named Samuel Slater started a water-powered textile mill in 1790. Industrial growth subsequently took place upriver as well, and in 1986 Congress designated the Blackstone River Valley National Heritage Corridor, stretching from Worcester, Massachusetts to Narragansett Bay, which claims to be the birthplace of the American Industrial Revolution.[82]

The highway then continues to the south of Worcester,[83] past the farm where Robert Goddard launched his first liquid-fueled rocket, and on through built-up areas without a real break until it is past Springfield and its historic armory. US 20 then leaves the route of the Old Post Road and begins to climb into the Berkshire Hills of western Massachusetts. Even there the population density along the road is considerable. Here and elsewhere, US 20 is Main Street for every town it passes through, with the exception of large cities on its route, such as Buffalo, Toledo, and Chicago, where it has been rerouted through suburbs.

The main route west from Hartford, just south of Springfield, once climbed through a low spot in the Berkshire Hills, but the current US 20 grew out of one of a number of roads built in the 1700s to give settlers access to the northwestern part of Massachusetts. The portion that comes down into the Hudson valley west of Albany connects that city with New Lebanon, New York, on the border, which grew into the main Shaker center after the American Revolution. Roads had to go from there into New England by then for it to function in this role.

Through the Berkshire Hills US 20 becomes the main artery for the tourism industry of the area, with the Berkshire cultural industries of Tanglewood drama and music sites, the Rockwell Museum, and a Shaker village/farm located along it. This popular section gives way to the environs of Albany, which do not fade on the western side of the city for about twenty miles. The highway west from Albany follows the route of one of the original pioneer roads. In 1794, a rough road was constructed from Albany to Cooperstown, then extended west in 1796. In 1802, a toll road called the Great Western Pike was begun, which improved on this roadway.

There were apparently as many as three different Great Western Pike toll companies. William Cooper, the founder of Cooperstown, led the group that incorporated the second Great Western Pike in 1801, only to have it fail by 1806. The last company, owned by the Holland Land Company promoters, was incorporated in 1803 and maintained until 1858, when the railroad running along the Mohawk River and the Erie Canal finally destroyed the business. Although less used than the Seneca Pike down by the canal, it was touted as being healthier than the swampy lowlands with their yellow fever and malaria.[84] By 1830, there was daily stagecoach service along its eastern length, which also eventually fell to the competition from the canal-side railroad.

US 20 misses all the Mohawk valley population centers, but as the land companies that built it hoped, it did open up the large fertile acreages above and behind the river.[85] Except for occasional daunting ravine slopes, the road is mostly straight and flat all the way to Buffalo. It is paralleled by strip development, and strings together numerous small towns, some of which, such as Seneca Falls, the birthplace of the women's rights movement, lay at the outlets of the Finger Lakes. West of the Genesee River, the road runs through what was once a large tract belonging to the Holland Land Company that extended to Buffalo. Just before its designation as a federal highway in the mid-1920s, the eastern portion, encompassing Cooper's old route and the Great Western Pike, was called the Cherry Valley Turnpike, probably for tourism reasons, and local promoters began to press for it to be hard surfaced.[86] The growth of traffic and the need for public works in the Depression led to its being paving across the State. It was then touted as the most direct route from Albany to Buffalo.[87]

Today US 20 swings south and west through the suburbs of Buffalo. Near Lake Erie, the highway follows an old Indian trail that went into present-day Ohio, which was improved in 1806 by the Connecticut Land Co. from Buffalo to the Ohio border to allow settlers easier access to its Western Reserve lands in that state — eleven northern Ohio counties that once were "reserved" for the state of Connecticut in compensation for revolutionary war damages.[88] One of these settlers was the father of Willis Carrier, the man who developed air conditioning. Here, US 20 became the main lakeshore route connecting Buffalo and Cleveland, but in later years Interstate 90, built a few miles inland, assumed much of the traffic load. From US 20, roads go inland to towns such as Chautauqua, New York, and Titusville, Pennsylvania, famous, respectively, for education and petroleum.

From Buffalo to Cleveland, the road is flat and once more runs through nearly continuous strip development and small towns. This feature continues as the road passes through the Cleveland area and then goes somewhat inland through the farm, small town, and industrial country between there and Toledo. This area is also the southernmost served by US 20 along its whole route. In the nineteenth century, this segment was known as the Western and Maumee Pike and was infamous for its poor quality. Later, the road became the major connector between Cleveland and Chicago before the development of the Interstate system.

The inland plateau disappears with the end of the Appalachians in eastern Ohio, and the land near Toledo is very flat. It reminds me of Winnipeg, where we lived for four years in the late 1960s and early '70s — both areas were lake bottoms at some prehistoric time. After Toledo the population and strip development along US 20 start to drop off a bit. In-between the small towns, cropland comes right to the roadside instead of being hidden by houses on acre plots fronting along the right-of-way. Along most of its route, though, the highway's features remind me of fractals: mathematically based diagrams and figures that have continuously recurring features so that it is possible to reconstruct the whole fractal diagram from any part of it. US 20's continuous flow of fast-food franchises, gas stations, shopping malls, doctors' offices, and suburban homes seem like an endless repetition of a single theme, whose entirety could be reconstructed from any given part. Only the Berkshire Hills and the downtown cores of Albany and Cleveland provide any diversion.

From Toledo, the road heads across northern Ohio and into Indiana The land remains quite flat but the poverty of the more widely spread farms and villages tends to increase as one approaches the Indiana border. Into Indiana, the scene remains one of poorly maintained farms until the Mennonite farms east of Elkhart suddenly appear, lacking the clutter and junk that seem to characterize the previous seventy-five miles. From Elkhart, past Chicago to west of Elgin, Illinois, the continuous belt of population and industry resumes for almost 150 miles. South Bend is a university town, Michigan City is lakeshore, Gary and Hammond are huge industrial complexes, some of them abandoned, and in the westernmost portion of this segment US 20 goes through some of the worst housing imaginable in the United States. The land is still quite flat, but its composition changes from farmland to sand as one approaches the south shore of Lake Michigan.

The road cuts across the south side of Chicago through a number of relatively poor districts, then acts as the main street for a number of better-off suburbs before turning north and repeating this task in the westerly suburbs. Finally, almost due west of downtown Chicago, it turns northwest toward Elgin, goes through some very affluent areas, and then on into farmland once again. From here to the West Coast, US 20 is dominated by increasingly open spaces.

East of the Mississippi, the land starts to become increasingly hilly, and the road follows the crest of a series of high ridges, the most impressive landforms since the highway came down out of the Berkshire Hills in western Massachusetts. Then, passing through Galena, which claims Ulysses S. Grant as its favorite son, the highway reaches the Mississippi River at East Dubuque, Illinois. On the Iowa side of the river, the ridges continue for ten miles or so, then the land levels out into gently rolling farmland. The farms here are much larger than those in western

Ohio and Indiana, and the houses much farther apart. US20 becomes a major four-lane divided highway in a recent realignment here as well, so there is no strip development along it as there is farther east. But there was not a lot of strip development on the old route either.

Near the Missouri River, the land becomes hilly again. Some of this is due to the erosion of the plain by tributaries of the Missouri, but there are also eroded loess hills near the river. The Missouri has created a narrow valley about ten miles wide, and for much of its length south of Sioux City, Iowa, where US 20 meets the river, the flow hugs the bluff on the Nebraska side. In the afternoon, the eroded loess on the Iowa side is a striking scene after so much landscape that was gently rolling to flat. The Missouri's narrow stream is disappointing, but Sioux City marks the approximate halfway point in US 20's route across America.

Leaving the Missouri river and entering Nebraska, the highway increasingly loses its function as a trunk highway, and for a hundred miles is more of a local farm road. The land use pattern gradually shifts from farming to hay production and cattle grazing. Toward north-central Nebraska, the road enters the Sand Hills, an enormous area of dunes stabilized by grass and used for pasture. The dunes persist for forty miles before the road encounters the weathered rock and small canyons of the Pine Ridge. In the far west of the State, past the fabled 100th meridian that marks the limit of practical dryland cultivation, rolling grassland begins again and the land becomes increasingly drier. About here sagebrush appears, and remains a constant feature of the scenery through to west-central Oregon. The road then crosses into Wyoming and follows the North Platte River upstream for fifty miles, half of which is also Interstate highway, until it reaches Casper.

Casper, an oil service and shopping center of 50,000, sits on a plain created by the North Platte River and was once where the Oregon Trail crossed the river. The road then parallels a railroad track through low buttes and mesas populated almost entirely by cattle and antelope. Mountains begin to appear in the distance. Turning northwest, US 20 enters the deep and narrow Wind River Canyon. The Wind River rises on the southern fringes of Yellowstone Park and runs southeast until turning north into the Bighorn Basin. Because of confusion between two groups of early explorers, the upper part of the river was named the Wind and the lower part was named the Bighorn. The name change occurs as the river emerges from the canyon. Highway 20 then turns into an access road for the central and east parts of the Bighorn Basin, until it reaches Cody, named after William "Buffalo Bill" Cody.

From Cody, the road rises up a canyon made by a branch of the Shoshone River for fifty miles until it reaches a high pass (8,500 feet) at the eastern end of Yellowstone Park. Theodore Roosevelt considered this the most scenic fifty miles of road in America. Yellowstone itself is a high alpine plateau surrounding a huge volcanic caldera or crater. At seven to eight thousand feet, the plateau towers above the potato country a hundred miles farther on to the southwest, where the land drops to four thousand feet above sea level. Technically, US 20 is interrupted when the road reaches the eastern end of Yellowstone Park, where it was originally designated to end because there was a toll on cars entering the park. The existing western portion of the highway was designated from Yellowstone to Albany, Oregon, in 1940, once that state committed to improve a road across its central desert. Then, in the 1950s, US 20 was extended to the coast as the final western portion was upgraded.[89]

The highway leaves the western entrance to Yellowstone Park — where the grades are the easiest in the Yellowstone country, making the area favored by the early explorers and tourists[90] — and turns southwest, crossing and descending from the Continental Divide, which is also the border between Idaho and Montana. The road is 392 miles long through Idaho, but the Montana portion is only 14 miles long. After passing through potato country north and east of Idaho Falls, the road enters a desert that contains a huge nuclear laboratory. In doing so, it bypasses much of the irrigated and populated eastern Snake River Basin. Farther west, the road skirts a huge, uninhabitable area of basalt rocks, lava flows, and ash, much of which is known as Craters of the Moon National Monument. A range of high mountains to the north waters a long stretch of lightly populated ranch country that has been described as "the road to nowhere."[91] Here US 20 follows the general route of the Oregon Trail's Goodale Cutoff, developed to avoid Indian attacks on wagons taking the more commonly used trail along the Snake River.

At Mountain Home, Idaho, US 20 merges with an Interstate highway and proceeds northwest toward Boise, thirty-five miles away. Boise, the largest city on the route west of the Mississippi River, sits on the Boise River, a tributary of the Snake, which in turn is a large tributary of the Columbia; it became the state capital when the population grew in southwestern Idaho as a result of mineral discoveries.

From Boise to the border with Oregon, about sixty miles away, the highway connects a series of farm towns, occasionally joining the Interstate system for a mile or so each time. Once in Oregon, US 20 begins to ascend the valley of the Malheur River until, after a hundred miles, it climbs into a series of buttes, mesas, and low mountains. Eventually, it descends until it reaches the Harney Lakes region. The lakes exist in this dry region because there is no outlet for the water that comes off the mountains to the east and west and collects there. The lakes thus expand and contract according to the seasons, but there is enough water to support a good-sized cattle range, some mixed farming, and a couple of small towns. Here, the highway crosses US 395, one of the loneliest highways in the country, running north from San Bernardino, California, to the east of the Sierra Nevada range, through Reno, Nevada, and north into the deserts and mountains of Oregon. Only its sister road, US 95, follows such an unpopulated route.

Leaving the Harney Lakes, US 20 enters a wide, flat valley rising between two mountain ranges that extend until about a hundred miles east of Bend, Oregon. The road then rises and falls in long, sweeping grades, and the Three Sisters peaks in the Cascades become visible seventy-five miles ahead. The land is deserted except for a rare ranch house, a few cattle, and an occasional "settlement" such as Brothers, about fifty miles from Bend, consisting of little more than a gas pump and a store.

The only two towns on the road on the eastern side of the Cascades are Bend and Sisters, named for the Three Sisters peaks. Both are shown on maps from the beginning of the century and probably started as cow towns. Today, they are tourist towns, trying the Western formula of appealing to skiers with their access to mountains in winter and to summer vacationers with their sunny days. West of Sisters, the road ascends the Santiam Pass to nearly five thousand feet, with vegetation becoming increasingly plentiful as the high desert is left behind. The pass itself was discovered by Oregon settlers looking for a way to bring their cattle up from the Willamette Valley to these dry ranges on the Cascades' eastern slopes.

Once over the pass, the road follows the Santiam River down into the Willamette Valley. The clear, dry desert weather disappears as Pacific clouds, held up by the mountains, accumulate overhead. The Cascades are wet and heavily wooded. The Willamette Valley was the destination of 350,000 emigrants who took the Oregon Trail from Independence, Missouri, two thousand miles westwards from the 1840s until the arrival of the railroads in the 1880s. The valley seems more like the Midwest than somewhere near the Pacific coast. US 20 crosses the valley about midway between the river source and its mouth at the Columbia River, and then passes through a series of farm towns. Then, rising into the coastal range from the Willamette Valley, the road runs through a low pass and then down through a series of river and creek valleys to sea level, and its western terminus at Newport, which is also the northernmost point on the highway. The road ends at its junction with US 101, about half a mile from the Pacific Ocean. William Least Heat-Moon describes the last few miles of US 20:

> U. S. 20: a scribble of a road, a line drawn by a palsied engineer. The route was small farms — one with a covered bridge — and small pastures and mountainsides of maple and fir and alder and wet green moss. Oblique sunlight turned blossoms of Scotch broom into yellow incandescence that illumined the highway; settlers brought the plant from Scotland to use in broom-making, but it had escaped cultivation and is now a nuisance to coastal farmers as well as a fine ornament of spring.
>
> The road squeezed through a narrow pass, then dropped to Yaquina Bay with its long arches of bridging. In the distance, the blue Pacific shot

silver all the way to the horizon. I had come to the other end of the continent.[92]

For us, the drama at the western end of the road was a few miles north of Newport, at Cape Foulweather. This is where Captain James Cook first sighted the Pacific Coast of North America in 1779. He planned to return the next year to explore it further, but his destiny was to be murdered in Hawaii. Cook got his big break as an explorer far away on the northeast coast of North America, where, working out of the Royal Navy's base in Halifax, Nova Scotia, he mapped the Newfoundland coast and in 1759 led British troopships up the St. Lawrence River to attack and take the French fortress of Quebec during the Seven Years' War. Yankee Hawaii began to take shape a century after Cook died there. US 20, the Yankee Road, may end at Newport, but Yankee culture has continued on across the world.

Watertown, Massachusetts: Fred Taylor, "Mother" Perkins, and the "System"

This is the Arsenal. From floor to ceiling,
Like a huge organ, rise the burnished arms;
But from their silent pipes no anthem pealing
Startles the villagers with strange alarms.
                     — Henry Wadsworth Longfellow[1]

In the past the man has been first; in the future the system must be first.
                     — Frederick Winslow Taylor

The most widely discussed topic in Europe [Taylorism] is the latest methods [sic] of exploiting the workers... It is sweating in accords with all the precepts of science. [1913]

> [T]he decree [on labor discipline] should definitely provide for the introduction of the Taylor system.... without it, we shall not be able to introduce socialism. [1918]
> — V.I. Lenin, before and after the Russian Revolution[2]

> I'd rather get a law than organize a union, and I think it's more important.
> — Frances Perkins[3]

A few miles northwest of Boston, on the north bank of the Charles, lies the suburb of Watertown. It was once known for its armory, where, among other things, cannon were manufactured for coastal defense.

One day, I went out on US 20 to find the Watertown Arsenal. I should have found it right away, as the highway borders its grounds for a few blocks, but I didn't. It was only while on another street, also bordering the site, that I realized that the old government facility, with only a small exception, had been converted into a shopping mall (appropriately named the Arsenal Mall) and condos. Today, the Arsenal property is an example of a late-twentieth-century version of the transformation of swords into ploughshares. Only a portion of the original buildings still house military activities.

In 1776, it was clear that the British had no intention of giving up their American colonies to an independent government. General Henry Knox, on George Washington's staff, recommended that ordnance dumps be established at points, generally far inland, that were likely to be inaccessible to British forces, and that these establishments should be centers for the production of arms and ammunition. Carlisle, Pennsylvania, and Springfield, Massachusetts, were the first sites to be selected. At the end of the war, the Springfield[4] and West Point, New York, sites were kept in operation.[5] Later, at Washington's insistence, another, now called an arsenal,[6] was created at Harpers Ferry, Virginia, on the upper Potomac River.[7] Others were gradually established in places such as Watervliet, New York, during the War of 1812 and at Watertown, Massachusetts, in 1816, which focused originally on gunpowder manufacture and the repair of military wagons.[8]

I had gone looking for the Watertown Arsenal because, in the summer of 1911, it had gained a brief measure of celebrity. The working people in the Arsenal, all government employees, had walked off the job for the best part of a week. There was no violence and the strike was only notable because US government employees were not known for their propensity to strike.[9] Many military leaders had been impressed by the work of Frederick W. Taylor and his associates in increasing productivity and production in foundries, factories, and other businesses and institutions. Shortly after management brought Taylor's "time-study" people into the

Watertown Arsenal, the workers walked out. The strike prompted a congressional investigation into Taylor's scientific management movement and brought him to the attention of people far beyond those who knew of him as a consultant on shop productivity problems.

Taylor, who died in 1916, was in a way the most influential person of the twentieth century. We all live in a Taylorite milieu, though we don't know it. In the 1970s, the playwright and political activist Barbara Garson spent months interviewing workers in tuna-processing plants and automobile factories, and working as a keypuncher herself. "Today," she wrote darkly in *All the Livelong Day*, "we type, assemble and even talk under routines like those worked out for Schmidt [an immigrant trained in a new method of moving pig iron, described in a famous anecdote told by Taylor]. Since Frederick Taylor's day, the division of labor has continued in its soul-destroying direction. Some large categories of work seem difficult at first to reduce to small job modules, but gradually stenographers are replaced by tapes, secretaries are relegated to the typing pool, and look what McDonald's has done to the highly individualized jobs of waiters and short order cooks." As Garson's language suggests, dark clouds have all along threatened Taylorism's sunlit vision, but they little interfered with its spread. It was as if, early in the twentieth century, a concentrated gas of immense potency, until then trapped within a corked vial, had suddenly been released and had ever since been wafting across the world.[10]

Me and Fred

I first met Fred Taylor in 1965, even though he had been dead almost fifty years. I was a married graduate student with kids, and that summer was fortunate to get hired on at the General Motors Corvair plant in Ypsilanti, Michigan. In the morning, I would attend classes and at 3 p.m. I'd show up for the second shift at the plant. It was a rough schedule, but the money was good.

My job was to help unload boxcars. The new guys, like me, got the afternoon shift because, even though the sun was going down and it became cooler in the plant, the boxcars had sat all day in the sun and inside they remained very hot. The foreman assigned us to different cars: there were "two-hour cars," "four-hour cars," and the occasional "eight-hour car," depending on the difficulty of emptying them. It did not take too long to get used to the rhythm of unloading. You got used to the heat that rolled out as the doors were opened. Then, you had to carefully unblock the hot metal parts and carry out the heavy wood "breaker bars." I was especially careful with the breaker bars because one had dropped on my brother's foot and broken it when he worked on a similar job a few years earlier.

One day the team the foreman assigned me to had to unload what was called a "one-hour–and-forty-eight-minute car." Not two hours. The experienced guys on the shift stayed away from there — the new guys got these particular cars, whenever they arrived. We started in on the car and found it no more difficult than others, except that the foreman came along at certain points and told us to rest then and there for no more than ten minutes. Then it was up again and back to work. In an hour and forty-eight minutes, we were done — and we were not happy. The work was not overly tiring, in comparison to the other cars, but we had lost all control of our time. We had to work at a constant rate to finish on time, and we had to stop and rest even when we didn't want to. On the other cars, we would often work hard and fast just to have a longer break at the end, but after we had emptied this car we were put to work for twelve minutes sweeping up the area. We found out afterwards that cars like this had been "time-and-motion studied" by Taylor's intellectual descendants.

Taylor, Machine Tools, and Interchangeable Parts

Taylor has to be seen in the context of two factors. First there is his Yankee heritage, through his mother and his wife. Educated in New England as well, he seemed to have been well equipped with Yankee inquisitiveness and the tendency to apply logic to problems. He wasn't born there, didn't die there, and for most of his life didn't live there, but New England occupies a central place in Fred Taylor's story. Taylor was said to have, in the words of his first biographer, "one whale of a New England conscience."[11] At least in broad brushstrokes, he did possess those Puritan sensibilities of self-restraint and inner mission that we associate with New England, and that Max Weber had in mind when he wrote *The Protestant Ethic and the Spirit of Capitalism* in 1904.[12]

The second factor in Taylor's context was the ongoing quest for interchangeable parts. The idea that one might make tools that contained many parts goes back at least to the metal or stone-tipped weapons of prehistory. Making a tool out of one part was a clear enough task, but once craftspeople began to fashion multipart tools, the problem of "fit" between the parts emerged. How could you make parts that would be interchangeable — that would fit any other companion part or parts to make a whole tool? No problem. Just make each part identical to its twins. For the craftsperson before the nineteenth century, however, this was a virtual impossibility to perform.[13] One could, the industrial archeologists have shown, make a set of nearly identical parts by painstaking handiwork, but it was impossible to make these parts in any volume.[14] In 1760, British steam-engine inventor James Watt struggled to get tolerances of 1/100th of an inch, and another British inventor, Henry Maudsley, got to 1/1000th of an inch.[15] Maudsley was also involved in the development of lathes — tools that eventually enabled workers to make more tools. (I tell the story of the development of interchangeable parts in Chapter 9.)

By the beginning of the nineteenth century, inventors began to attach cutting tools to water-powered belts. Using harder metals such as iron or steel, they found they could cut out parts to relatively tight tolerances as long as the parts were made from softer materials such as wood or brass — hence the growth of New England industries such as shoes (leather cutting) and clocks (wooden parts) as a spin-off of the water-driven textile mills. Armories, like that in Springfield, Massachusetts, contracted for musket and pistol parts that were interchangeable, and men like Eli Whitney and Samuel Colt in Connecticut and Eliphalet Remington in New York attempted to provide them, with mixed success. The existence of standardized weapons led to standardized cartridges. One standardization led to another, and another. During the Civil War, for example, Union forces (known as Yankees regardless of origin) used prefabricated wooden bridge components to replace damaged railroad bridges so fast that a Confederate joke circulated that Sherman's troops on their way to Atlanta even carried along some spare tunnels to replace ones damaged in the Georgia mountains.

By Taylor's time, after the Civil War, experimentation with various alloys had led to the discovery of harder steel that could be used to cut metal parts harder than brass or bronze. Taylor and his associates adapted cutting tools, first to steam power and later to electricity, to provide the power to cut iron and steel parts to tolerances that, for most products, fit the concept of true interchangeability. By the 1880s it became possible to assemble tools and products from piles of parts, rather than laboriously match parts or sand or grind them to fit a unique combination. Assembly lines, like those popularized by Henry Ford, their "inventor," became a logical next step.

Scientific Management

Fred Taylor was employed in a small foundry in Philadelphia just as the country slid into depression after 1873. Like others, he had to turn to cost-cutting actions to keep the company alive. And like others, the sixteen-year-old looked at possible technology solutions, Unlike a lot of others, part of his focus turned to the organization as well.[16]

Until the 1880s, workplaces tended to be relatively small in both size and workforce. The modern corporation existed primarily in the form of railroads. Factory buildings tended to be poorly lit, dirty, cluttered, and disorganized. Hiring and training were in the hands of foremen, while skilled and semi-skilled workers carried production knowledge around in their heads. One of Taylor's first moves was to organize his facility in a rational manner, to clean it up and properly light it, and pick up debris from the floor. Workstations were rationally placed and productivity increased.

Taylor then began to study how tools were made. Through experiments, he developed ways of making hardened steel from which high-speed cutting lathes made tools and dies that were less susceptible to wear and kept their tolerances longer. He also worked out mathematical tables to indicate what kinds of metals, gauge settings, and processes were best for different tools. Thus began the process of "deskilling" (as it is called today) the shop floor, moving production knowledge to management and away from the workers.[17]

By 1883, Taylor had made another leap. Once he had satisfied himself of the possibility of truly interchangeable physical parts, he then wondered if the idea might not be extended to manufacturing tasks and organization. Could people be made "interchangeable"? The labor force in large industries increasingly consisted of immigrants and domestic migrants, both looking for a better life. The key problem relating to such workers was one of productivity: how to absorb them quickly and how to get good production from them.

The first task was to simplify jobs, thereby taking them out of the control of skilled workers, who required lengthy apprenticeships, and making it possible to employ many more people to do them.[18] This had been done in the early 1800s when textile production moved from home spinning and weaving to huge factories,[19] but the existence of close-fitting, interchangeable parts meant that this revolutionary process could now be extended into all kinds of production. Simplification of skills was followed by measurement and ergonomics — the design of equipment and its physical placement so as to improve performance and safety. Measurement, in turn, led to ideas of productivity, standardization, and the timing of actions. Measurement, beyond the conduct of a decennial census, had gained momentum in the Civil War, when recruits were measured for body sizes to be provided with uniforms and shoes that fit, later leading to standardized clothes sizes for the consumer market.

Gradually, Taylor and his associates, working through their connections with the American Society of Mechanical Engineers, began to measure the interaction between the worker and his tools.[20] These engineers noted that a tired worker was unproductive and that irrational management decisions about production also hampered productivity. In 1884, Taylor experimented with different methods of loading pig iron onto carts to see which would be the "one best way." He measured how many ninety-two-pound pig iron bars a worker could handle safely in a day, and looked at how to place the appropriate tools efficiently nearby, the timing of rest periods, and the statistical character of workers in terms of hiring needs for the job.[21] He also measured shovel sizes against the stamina of coal handlers, and determined the exact size of shovel that would maximize production over the long run. He transferred his techniques to other industries as well. Everything was to be rationalized.

Simplifying jobs meant that they could be learned quickly by immigrants and farm boys.[22] Productivity then depended upon properly relating the simple tasks to each other to make the final product. What was proper came from the systematic study of this relationship, just as what was simple came from the logical breakdown of production activities into their component parts. Proper engineering could develop systems that eliminated the need for skilled labor, as shown in the notes Willis Carrier made as he developed a system for drying macaroni at about the same time as the Watertown incident:

> Some Manufacturers have an idea that only the Italian understand[s] the drying of macaroni. That may have been true at one time, but today it is absolute rot. When I say it does not require macaroni skill, I mean that it does not require a man familiar with macaroni to dry it with our dryer. All a man needs is a regular amount of human intelligence. You may wonder what that may mean to the manufacturer. It means that instead of a small field of men to hire from, he can chose [sic] from a large field. Then too, if a man quits, there is no great loss, for a new one can be taught in less than a week to do with our dryer what in the past men have taken years to learn.[23]

Plants were reorganized physically so that production was not impeded by poor location of tools and work stations. Simplification, measurement, and ergonomics gave management a scientific aura. And by the turn of the century, Taylor had become an early management consultant, though that was not what these were called at the time:

> Back in the 1870s, Chordal [the pen name of James Waring See, an Ohio writer who did a column of "Letters from Chordal" in the 1870s for a periodical, *The American Machinist*] had limned an admiring portrait of a breed he called "Yankee contractors," men from out of town who would contract with a machine shop to take over its production. Introducing their own methods, which emphasized system and organization, they promised lower costs and higher profits, of which they took a cut. "He had on nice boots, and nice clothes, and a white shirt, which would do credit to a lobbyist," Chordal describes one "Doolittle," probably a fictionalized composite.
>
> Taylor was not one of Chordal's Yankee contractors. But as an outside hired expert, unhampered by tradition or personal history, and largely free to kick out old ways and introduce new, he was their lineal descendant. His letterhead read, *Frederick Taylor, Consulting Engineer in Management*. At first, he charged thirty-five dollars a day. Later he got forty, occasionally fifty. In today's money, he was getting something like a thousand dollars a day.[24]

Taylor thus transferred his experience in developing machine tools to managing foundry operations and organization: simplify the tasks, routinize them, measure and reduce the time necessary to accomplish them, and provide a work environment that makes it easy to do the task quickly, and the resulting efficiencies could vastly improve productivity.[25]

The Strike at Watertown

In 1903, with the publication of his "Shop Management,"[26] Taylor began to be taken seriously by the labor movement, whose problem seemed to be centered around opposition to any form of compensation that based wages on how many pieces of product a person might produce. Not without reason, unions felt that such schemes were always subject to reinterpretation by employers, such that if someone was able to produce more, then the "normal" rate would simply be raised and the worker would end up working harder for the same pay as before.[27] A key part of Taylor's system was to reward workers with productivity improvements; his first paper delivered to the American Society of Mechanical Engineers in 1895 was titled "A Piece-Rate System, Being a Step Toward Partial Solution of the Labor Problem."[28]

In the early 1900s, the federal government's Ordnance Department was managed by General William Crozier, who oversaw five manufacturing arsenals, including Watertown, which turned out a number of different armaments, but was primarily concerned with naval cannon and related equipment. The government did not give the arsenals a monopoly: they had to compete for many contracts with the private sector, and by 1906 Crozier was growing concerned that the arsenals' productivity was falling behind the competition. To some degree, this was due to insufficient funds appropriated by Congress for modernization, but Crozier was also concerned with the basic organization, management, and efficiency of the arsenals.[29]

With these latter concerns in mind, Crozier went looking for private management consultants. He approached Taylor about his system in 1906, and by 1909 had hired Taylor's team to replicate the improvements they had made in the Midvale Steel and Bethlehem Steel factories. His choice of Watertown as a first facility to be changed was obvious. As he stated to a congressional committee, "I have considered that the management of the arsenals has been good as compared with the management of the ordinary industrial establishments of the country"…but…. [f]or some time I have found that the cost at Watertown Arsenal compared less favorably with the cost of the same material procured from private manufacturers than at any other arsenal. It has often made the poorest showing."[30]

Taylor and his associate, Carl Barth,[31] traveled to Watertown to study the organization of the arsenal in preparation for performing time-study measurements of its employees' work. In this he had the support of the new commanding officer, Colonel Charles B. Wheeler, who was already at work reorganizing the plant's shop management along Taylorite lines. The reorganization began without problems, and by mid-1911 another of Taylor's associates, Dwight Merrick, began to do time-studies there to measure the actions workers took in the performance of their tasks. These studies presumed that the workflow and environment had been optimized and that what was left to be learned was the best way that laborers' time could be used. Merrick worked without incident in the machine shop, but when he began using his stop-watch in the foundry on August 10, things went wrong. While he was measuring one activity of a worker at twenty-four minutes, another workman surreptitiously measured it at fifty minutes. The workman was a good foundry molder but not a good time-study person. Merrick was the opposite. That evening the workers met and signed a petition asking Colonel Wheeler to stop the time-study.[32] When Merrick showed up in the foundry the next day and began to measure another task, the workman refused to proceed unless he left. The workman then refused his foreman's orders and was let go. Meanwhile, another workman delivered the petition to Wheeler, but he did not see it until all the employees had walked off the job in support of their comrade. They were gone for a week, and the sight of government workers on a wildcat strike became a *cause célèbre*.

The workers' union was appalled, since there were specified procedures for going on strike and they were not followed. Still, the union had to support its members. Taylor and others accused organized labor of fomenting the strike by their opposition to Taylor's methods; in fact, not only were the foundry workers unfamiliar with union materials, but a number had read Taylor's "Shop Management." They had understood his ideas and mistrusted the piecework portion, not believing management would keep its word, as it had not elsewhere. The rest of the reorganization did not seem so bad to them.

The Aftermath

In the end, the workers got their jobs back and time-study was halted, but the rest of the reforms went ahead. In 1912, Colonel Wheeler could boast that the reorganization spearheaded by Barth had raised output 150 percent. However, Congress was pressured to look into the Taylor system, and eventually it decreed that time-study was not to be used on federal employees, a prohibition that lasted into the 1940s. What was fascinating was that the controversial piece of Taylor's system proved to be the time-study, because of Taylor's insistence on coupling it to a piece-rate pay system. Once his successors had abandoned that particular connection, Taylor's system was adopted in whole or in part by almost all social organizations, even today where the "shop" has largely disappeared.

In 1912, with the publication of *The Principles of Scientific Management*, which is still in print, Taylor expanded on his ideas. The congressional inquiry into his system served only to widen the awareness of scientific management. It also led to his book's being a hit with people trying to manage public and private organizations, though it was condemned by the very mass labor unions it had indirectly helped to create by eliminating the old crafts. In his appearance before the congressional committee, Taylor summed up his approach. It was about changing the organizational culture — changing the world — not just about measurement. He was a revolutionary:

> Taylor's performance [at the hearing] included some of his most oft-quoted utterances on scientific management. "I want to clear the deck," he said at one point, and "sweep away a good deal of rubbish first by pointing out what scientific management is not." Then commenced the astonishing litany:
>
> Scientific management is not any efficiency device, not a device of any kind for securing efficiency; nor is it any group of efficiency devices. It is not a new system of figuring costs; it is not a new scheme of paying men; it is not a piecework system; it is not a bonus system; it is not a premium system; it is no scheme for paying men; it is not holding a stop watch on a man and writing things down about him; it is not time study; it is not motion study[33] nor an analysis of the movements of men; it is not the printing and ruling and unloading of a ton or two of blanks on a set of men and saying, "Here's your system; go use it." It is not divided foremanship or functional foremanship; it is not any of the devices which the average man calls to mind when scientific management is spoken of.
>
> These were merely tools, "adjuncts" he called them. Scientific management was more, demanding "a complete mental revolution" on the part of workers and management. "And without this complete mental revolution on both sides scientific management does not exist."[34]

All through his testimony, Taylor repeated or alluded to this need for a new understanding. For instance: "The great revolution that takes place in the mental attitude of the two parties under scientific management is that both sides take their eyes off of the division of the surplus as the all-important matter, and together turn their attention toward increasing the size of the surplus until this surplus becomes so large that it is unnecessary to quarrel over how it shall be divided." His dream implied that personal greed and desire for control and/or autonomy would be subordinated to a neutral arbiter called "science." As his biographer put it, "[t]his mental revolution required a radical change in viewpoint, one 'absolutely essential to the existence of scientific management' — the sovereignty of science. 'Both sides must recognize as essential the substitution of exact scientific investigation and knowledge for the old individual judgment or opinion,' in all matters bearing on work. 'This applies both as to the methods to be employed in doing the work and the time in which each job should be done.'"[35]

Taylor zeroed in on the idea that the theory of the great leader or manager somehow was worth more to a company than a well-organized workforce:

In the past the prevailing idea has been well expressed in the saying that "Captains of industry are born, not made"; and the theory has been that if one could get the right man, methods could be safely left to him. In the future it will be appreciated that our leaders must be trained right as well as born right, and that no great man can (with the old system of personal management) hope to compete with a number of ordinary men who have been properly organized so as efficiently to cooperate.

In the past the man has been first; in the future the system must be first [italics added]. This in no sense, however, implies that great men are not needed. On the contrary, the first object of any good system must be that of developing first-class men; and under systematic management the best man rises to the top more certainly and more rapidly than ever before.

This paper [*The Principles of Scientific Management*] has been written:

First. To point out, through a series of simple illustrations, the great loss which affects the whole country through inefficiency in almost all of our daily acts.

Second. To try to convince the reader that the remedy for this inefficiency lies in systematic management, rather than in searching for some unusual or extraordinary man.

Third. To prove that the best management is a true science, resting upon clearly defined laws, rules, and principles, as a foundation. And further to show that the fundamental principles of scientific management are applicable to all kinds of human activities, from our simplest individual acts to the work of our great corporations, which call for the most elaborate cooperation. And, briefly, through a series of illustrations, to convince the reader that whenever these principles are correctly applied, results must follow which are truly astounding."[36]

In the end, the congressional committee report did not recommend any legislative changes, simply settling for saying that Taylor's principles were useful to modern industry, but that any time studies, the action that had caused the uproar in the first place, be done with the consent of those being studied. The effect of the committee's deliberations was to publicize Taylors ideas and to some extent legitimize them.[37]

Taylor died in 1916, at age fifty-nine. He was found in his hospital room, said the novelist John Dos Passos, with his watch in his hand.[38] He knew the revolution was coming. He saw a world that was focused on the consumer. The mission of industry was to give the world all it needs. A history of scarcity and want was to be replaced by a future of prosperity and general well-being.[39]

Taylor's Reach

Fred Taylor's importance to the twentieth century is in the organization of what could be called the "shop floor," but to him every social organization was a shop floor: "The same principles of scientific management can be applied with equal force to all social activities: to the management of our homes; the management of our farms; the management of the business of our tradesmen, large and small; of our churches, our philanthropic institutions and our government departments."[40] Until the growth of computerized automation, whether in offices or factories or hamburger stands, Taylor's approach to work has dominated the lives of most Americans and others in industrialized and industrializing countries for the past century. Charlie Chaplin, Lucille Ball, and Dilbert all parodied scientific management and its offshoot, the assembly line.

As historians probed the early years of the twentieth century and critics and thinkers asked why we live as we do, their gaze fell again and again on Frederick Taylor, finding that his ideas had seeped into some of the narrowest, most seemingly remote niches of modern life. At Taylor's door they laid the emergence of modern time consciousness, leisure's transformation from genuinely free time to organized recreation, the Reagan administration's approach to managing the federal bureaucracy, and even what one critic, Christopher Lasch, has called "a new interpretation of the American Dream."[41]

In May 1908, the first dean of Harvard's graduate school of business, Edwin F. Gay, visited Taylor

> and was soon introducing "industrial organization" to the curriculum. Taylor and [his associate, Carl] Barth, among others, regularly lectured there. Penn State's was among the first engineering programs to absorb Taylorist ideas, through Hugo Diemer, a Taylor devotee formerly at the University of Kansas; by 1909 he had begun the country's first true industrial engineering department...From the 1910s to at least the 1940s a large percentage of business students and a smaller but not inconsiderable number of engineering students were exposed to the tenets of scientific management, whether they realized it or not.[42]

In 1911 Taylor recommended devotee Morris Cooke to the new reform mayor of Philadelphia.[43] Cooke applied scientific management to Philadelphia's department of public works, saving the city millions. His example led to another academic development: the teaching of scientific management was soon being applied to subjects in the public sphere, giving birth to a new discipline, public administration, with its own terminology, university courses, and journals.

The belligerents in the First World War, when their very fate as nations depended on replenishing the artillery shells and machine-gun ammunition they hurled at the enemy, turned to Taylorism, if only in bastardized form. In Britain, one scholar has written, management's prior coolness to scientific management was "transformed into a veritable heat wave of enthusiasm for it after the outbreak of the...[w]ar." In France, Le Chatelier wrote, "Taylor's books are being sold as though they were appearing for the first time." Late in the war, French premier Georges Clemenceau ordered that the Taylor system be introduced into all government departments. By war's end, French industry was primed for modernization.[44] In 1919, the French poet Paul Valery, looking out over a postwar Europe haunted by the ghosts of the dead and memories of mindless production in death's service, concluded a "Letter from France" by writing "[t]he world, which calls by the name of 'progress' its tendency towards a fatal precision, marches on from Taylorization to Taylorization."[45]

Even the Soviets came to appreciate scientific management. In 1913, Lenin wrote that it was one of the most effective means of capitalist extraction of productivity from workers. Once he came to power, he began to realize the enormous challenge of taking an almost medieval peasantry into an industrial economy. Marrying the "scientific socialism" of Marx to the scientific management of Taylor was irresistible, and a variety of western experts were invited to Moscow during the 1920s to expound its principles. Stalin then applied them with ruthlessness.

During the late teens and twenties, scientific management split off into pieces and parts, factions, specialties, new disciplines, nascent social movements. Harvey Gantt, the inventor of the ubiquitous Gantt chart, rounded up fifty engineers in a group he called the New Machine, which sought to establish an "aristocracy of the capable"; engineers, efficiency experts, and their friends would take the helm of society from know-nothing financiers and politicians. The New Machine proved short lived, but the young Columbia University engineering professor who was its secretary, Walter Rautensrauch, later helped fashion the much broader technocracy movement, to which we owe our word "technocrat."[46]

Efficiency reached into the doctor's office and operating suite, too. "Where could the rationality of business better be joined to the rationality of science," wrote one historian of medicine, "than in increasing the efficiency of the hospital?" — or, for that matter, that of medical care generally? Medical journals carried articles like "The Efficiency of Out-Patient Work" and "'Efficiency Engineering' in Pelvic Surgery: One and Two-Suture Operations." Frank Gilbreth, the leading exponent of measuring motions in work settings, went into the operating room himself to observe dozens of procedures. "If you were laying brick for me," he told one surgeon, "you wouldn't hold your job ten minutes."[47]

In his 1994 book *Post-Capitalist Society*, management guru Peter Drucker reasserted his view that Taylor, not Karl Marx, warrants a place in the trinity of makers of the modern world, along with Darwin and Freud. Even America's victory in the Second World War, he said, could properly be credited to Taylor's influence. After Pearl Harbor, Hitler declared war on the United States because he reasoned that, lacking a merchant marine, modern destroyers, and a good optics industry, America could scarcely wage war at all, much less wage it in Europe. Hitler was right, said Drucker, America did lack those things: "But by applying Taylor's Scientific Management, U.S. industry trained totally unskilled workers, many of them former sharecroppers raised in a pre-industrial environment, and converted them in sixty to ninety days into first-rate welders and shipbuilders. Equally, the United States trained the same kind of people within a few months to turn out precision optics of better quality than the Germans ever did." One of the hurriedly built Liberty ships, identical to hundreds of others churned out during the war, was the 7,176-ton SS *Frederick W. Taylor*, launched in South Portland, Maine, in 1942.[48]

Taylorite ideas are behind the whole ergonomics industry. They permeate the Watertown Armory more now than in 1911, as the whole layout of the Arsenal Mall's department store that occupies the original foundry building is designed along variations of time-and-motion and ergonomic study principles applied to shopping. We live in a world that is organized on the basis that the system is more important than the person, just as Taylor claimed. Another facet of Taylor's work that is still with us is the application of measurement to human behavior. Taylor's search for optimal work methods led him into the measurement of human motions, developed to its extreme by his followers, especially Frank Gilbreth and his reversed-eponymous "therbligs." Today, Taylor is seen as the "father" of ergonomics, and has even been associated with the Google focus on measuring search choices.[49]

The application of research and science to human work behavior is so much a part of our lives that we hardly notice it today. It has led to a system that produces and distributes more goods and services to everyone than has ever been seen before. Without Taylor's application of experimental research to the problems of production, Henry Ford could not have created his version of the assembly line, Ray Kroc could not have developed McDonald's systems, Ray Walton could not have developed Wal-Mart's logistics — and so on.

Finally, there is no critical review of Taylorism here. His ideas have been extended so far and into so many configurations that to discuss a vision from a century ago is, dare I say it, academic. We live in the revolution that Taylor symbolizes, we are used to its stresses and really wouldn't know how to live any other way. That said, factory managers were very slow to adopt Taylor's whole system. They largely resisted sharing productivity rewards with the workers, and when the economy turned down they had little compunction in relying on traditional methods in response, discarding workers who had to rely on charity or the help of friends and family to survive. In 1911, the same year as the Watertown strike, disaster struck a Manhattan that had begun the effective creation of a counterbalance to corporate power: regulatory power. Yankee organizational ideals were central to corporate productivity and to the use of public power to enable labor to achieve its share of the economic pie.

Frances Perkins and the Triangle Fire

Nineteen-eleven was not only an important year for Frederick Taylor, but it also changed Frances Perkins' life. If Taylor was the architect of productivity, Perkins was the architect of the democratization of its benefits.

In March 1911, Frances Perkins, the executive secretary of the Consumers' League in New York, was attending an afternoon tea in a Washington Square home along with some of the city's prominent society women. They became aware of a nearby commotion, and Frances went out to see what was happening. It proved to be a disastrous fire in a nearby sweatshop.[50] The fire at the Triangle Waist Co. killed 146 of its female workers[51] and, indirectly, led to the New Deal, some twenty-two years later. The road from Washington Square to the White House led through Albany.[52]

Frances was a real Yankee. She was born Fannie Perkins in 1880 in Boston, on Beacon Hill, across the Charles River from Watertown. She was a descendant of a Revolutionary War leader and of a Union general in the Civil War who ran afoul of too many postwar politicians because of his advocacy of the improvement of the conditions of freed slaves.[53] Raised in Worcester, Fannie was educated at Mount Holyoke College, taking an interest in the sciences. Inspired by one of her educators, she developed an interest in social causes, not unlike a lot of students at similar institutions over the decades since then. She turned to teaching science when she was rejected by a New York anti-poverty organization.[54] Eventually, she got a job teaching science in Chicago, where she fell in with people who were active with the Addams House settlement program. Her interests drifted back to the problems of working conditions and poverty. Leaving teaching, she then worked for settlement houses in Philadelphia and New York.[55]

Perkins always managed to maintain good relations not only with the kind of society that paralleled that of Brahmin Boston, but also with the kind that were clients of the settlement houses. She also enjoyed the company of the intellectuals who wrote about their situation. She began to alter her persona, becoming "Frances," instead of the more frivolous "Fannie," and altering her churchgoing from Congregational to "high" Episcopal. The combination of society connections and social activism were to stand her in good stead throughout her career. One biographer calls her "a natural actress."[56]

Her interest in working conditions, as opposed to other poverty problems, grew as a result of her research[57] in Philadelphia into the prostitution trade, which was fed by the immigrant women who were induced into it through force, drugs, or promises of luxury. She also met the men involved in the trade as well as the police and politicians. Gradually she began to understand how she might get them to make changes. Frances was also beginning to learn how she might achieve her social objectives in a man's world, and found that she was a "natural" at managing an organization and at fundraising.

Moving to New York, she became involved with social and artistic causes, and counted writers Theodore Dreiser and Sinclair Lewis, who was in love with her, and future infrastructure czar Robert Moses among her friends, as well as many women in the upper reaches of society. She gained a master's degree in political science from Columbia University in 1910, and started work with the Consumers' League.[58]

The Triangle fire occurred about a year after a successful strike by the female workers in New York's garment industry led by a combination of immigrant women and society women who were active in the progressive movement of the day.[59] For more than a century, New York had been a city of manufacturing as well as trade. As early as 1790, Samuel Slater had offered textile producers there his services in mechanizing the industry, but to no avail. In the 1850s, most workers were employed in relatively small firms of fewer than a hundred people — for example, 539 ironworking companies employed only 10,000 people, fewer than 20 per company. Powering factories, the key to scale production, was expensive, while immigrant labor was cheap. By 1860, New York City was producing 40 percent of the country's clothes.[60] It was still doing so fifty years later.

The strike had come about as a result of the formation of the Women's Trade Union League (WTUL) in 1903. Women were not regarded well by traditional labor leaders, hence the need for a gender-based union. The WTUL was founded by a man, but he soon passed its leadership on to the reformers in the settlement house movement, who, like Frances Perkins, came from a more affluent background. Eleanor Roosevelt, for instance, was a member of the WTUL.[61] The most active leaders in opposing the strike were two corporate entrepreneurs, Max Blanck and Isaac Harris, immigrants themselves and owners of the Triangle Waist Company, one of some five hundred blouse factories employing some 40,000 workers in the city.[62]

After the Triangle fire, the leaders of the most powerful Democratic faction in New York City, Tammany Hall, began to realize that their hold over immigrant voters was threatened by their inattention to conditions that led to such tragedies.[63] A week after the fire, a funeral march for the Triangle dead attracted 350,000 people as marchers and onlookers. Opponents of Tammany saw an opening here. In particular, they had to react to a Yankee parson's son and a Republican Progressive, Charles Whitman, who had managed to get himself elected district attorney and who needed a "big case" to make his reputation.[64] Prosecuting the owners of the Triangle Waist Co. was to be his big case.

The Triangle fire was a turning point for many concerned with the conditions most workers faced. It showed that even successful union organizing and tactics were not enough to make the necessary changes to most aspects of corporate behavior. Organized workers could improve their wages and hours, but either could not, or did not want to, involve themselves in improving workplace conditions that were seen as "management's" domain. Further, unions' democratic structures meant that they tended to reflect the (male) morés of their membership, something that was most obvious when it came to representing women's issues. For progressive women especially,[65] their only recourse was to get legislation passed requiring employers to improve specific conditions or face the might of the state. Unions were useful, but not as useful as legislation.

As executive secretary of the Consumers League,[66] Frances was expected to spend a considerable amount of time in Albany lobbying for such legislation. Prowling the lobbies of the state legislature after the Democratic victory in the 1910 election, she began to be accepted by two of Tammany's most powerful politicians, Al Smith and Robert Wagner, the leaders of the Assembly and the Senate, respectively, and both about her age.[67] She also met the new state senator from upriver, Franklin D. Roosevelt. Smith and Wagner realized that, to stay in power, the Tammany faction had to shift from providing individual and ethnic favors for constituents to embracing legislative reform that favored the whole workforce.[68] Reform Republicans of the Teddy Roosevelt stripe also saw this approach as the route to power. Gradually, Frances Perkins learned that she could be more effective if she looked more matronly and, at thirty-four and recently married, she changed her look to appear older than she really was. Politicians and reporters began to refer to her as "Mother" Perkins (she had kept her maiden name), but found they could now relate to her more easily, especially as she cultivated their wives and mothers.[69] She became a very effective lobbyist and a good public speaker.

Frances managed to develop a working relationship with Smith and Wagner as members of the Factory Investigating Commission (FIC), which the New York state government created in 1911 in response to the Triangle fire.[70] The two men were the driving force behind the FIC, which recommended fifteen new laws on fire safety, factory inspections, and health regulations, of which eight were enacted. Frances was "loaned to the FIC to help with its research program."[71] Next on her agenda was passage of a law regulating the hours that women in most industries could be required to work in a week,[72] which had languished for years in the state legislature. Frances was disconcerted by the terms of the bill, which exempted women in the cannery business, but even though her friends in the movement opposed passage, she decided to support it and to take what she could get; the cannery workers were added the following year.[73] New York State began to lead the nation in workplace reform, and Tammany realized that progressive politics could be a vote getter.[74] It began to advocate for a minimum wage and for giving the vote to women, which New York accomplished in 1917.[75]

In 1912, Frances was asked, at the suggestion of Theodore Roosevelt, to be the executive director of the Committee on Safety, a non-partisan, though Republican-supported, reform group. Its biggest success had been to get the New York state legislature to create the FIC. Soon, the offices of the FIC were located in the Committee's premises and Frances was effectively the executive director of both.[76] By the time the FIC wound up its work, a plethora of legislation regulating all kinds of labor practices from hours of work to sanitary conditions had been passed. Frances also became the first woman on the State's Industrial Commission after Al Smith was elected governor in 1918, defeating Charles Whitman.[77] For Smith, this was the first of four terms in office, and his progressive policies made him an eventual contender for the presidency. Gradually, the other commissioners began to let Frances do the Commission's administrative work, and she gained the respect of some upstate businesspeople and the unions. When Smith was defeated in 1920, she went back to non-profit work, returning to state government in 1922 when Smith was re-elected. One of her major accomplishments was to expand the new system of workers' compensation to more industries, especially hazardous ones that had complained of the cost of premiums.[78]

When Smith left the governor's mansion to run for president in 1928, his successor, Franklin Roosevelt, kept Perkins on, in part due to her growing friendship with his wife Eleanor. Frances's policy advice to him, especially on structuring a program of unemployment insurance when the economy fell apart after the Crash of 1929, made him the leading governor in the nation who was trying to blunt the effects of the Depression.[79] She also tried to modernize the state employment service to match jobs to qualified applicants more efficiently. As a US senator, Robert Wagner used her ideas on unemployment insurance and labor market efficiency in his 1931 federal stabilization bill.[80] Her advice on how to respond to the needs of the working class voter made the patrician Roosevelt a credible presidential contender and Wagner his potential future ally in the Senate.

The New Deal

Upon his election to the White House in 1932, Roosevelt asked Frances to work with him in Washington as secretary of labor. Frances agreed, provided he in turn would agree to her policy agenda. She thus became the first female US cabinet member, a post she held until Roosevelt's death in 1945. Her friend from the state legislature, Robert Wagner, became Roosevelt's connection to the US Senate and sponsored the New Deal's legislation in that body. She managed the creation of the first New Deal job-creation program, the Civilian Conservation Corps,[81] helped re-establish the strength of the labor movement, oversaw the development of unemployment insurance and social security, and encouraged States to pass workplace legislation that the courts felt was not within federal jurisdiction. By Roosevelt's death, much of the agenda she had developed while working in Albany had been enacted nationally, and the factory environment had changed for the better in a significant way.

Perkins was not the only person involved in the workplace legislation that characterized the New Deal, but for over three decades she was at the center of it all.[82] The Triangle fire had thrust a problem to the fore that all the politicians knew had to be addressed someday. Perkins, with her understanding of the male politician, managed to insert her ideas quietly into their calculations in such a way that credit went to them rather than to her. "Mother" Perkins chose to get the job done through legislation, and became one of the most effective politicians who never ran for office.

Conclusion

In a broad sense, Frances Perkins's career represented something that de Tocqueville saw with his prophetic eye. He noted in his look at American democracy that there was a tendency toward "aristocracy" in the capitalist part of society, and that this could be a "hard" aristocracy in its treatment of wage workers. He understood, however, that there was a corrective measure in that wage workers constituted the large majority of the voting public and, when aroused, as they were in the 1830s and again in the 1930s, they could elect politicians who would use the power of the state to counteract, to some degree, the power of owners and managers. It was not the organizing of the workers that attracted Frances Perkins, but the use of politics to counter the power of corporations.

Further, it has been argued that the New Deal was largely responsible for the growth of American capitalism after the Second World War:

> In historic measures regulating banking, the stock market, labor, public power, communications, airline travel, and in society-strengthening reforms from old-age pensions to rural electrification, FDR did more for the free enterprise system than Henry Ford, giving it a security against malfeasance and economic mischance that it had never enjoyed before and the lack of which had led to its shipwreck. Roosevelt haters could not be expected to see this at the time, but the postwar boom would rise on the foundation FDR established. Securing the right of collective bargaining for labor, the worst of the New Deal to big business, was in the larger view insurance against the overproduction haunting many industries in the 1920s. What corporations lost in the union wage (which also pushed up the wages of nonunion workers), they gained in customer purchasing power. "I'm pro-business," President Bill Clinton said on Labor Day 1999, "but I'm pro-labor too. I don't think you can help the economy if you hurt the working people." Still, in the 1930s, playing by the zero-sum rules of the age of

incorporation, the great corporation would be taught to see things Clinton's way in the postwar boom.[83]

Max Weber noted that capitalism is identical to the pursuit of profit and to *forever renewed profit* by means of a rational enterprise that organizes (formally) free labor. Neither greed nor acquisition is relevant to capitalism. Clearly, the short-term focus on greed and acquisition is antithetical to the long-term view Weber espoused. Making the system one that benefits owners, managers, and workers helps to keep profits rolling in over the long term.[84] At the same time, the use of law and regulation as a means of control can, over time and with overuse, have deleterious effects. By the twenty-first century, so much had been loaded onto the legal system as protections and restrictions that, as columnist David Brooks notes,

> [o]ver the past several decades, [Philip K. Howard] argues, a thicket of spending obligations, rules and regulations has arisen, which limits individual discretion, narrows room for maneuver and makes it harder to assign responsibility.
>
> Presidents find that more and more of their budgets are precommitted to entitlement spending. Cabinet secretaries find that their agenda can't really be enacted because 100 million words of existing federal rules and statutes prevent innovation this way and that. Even when a new law is passed, it's very hard to tell who is responsible for executing it because there is a profusion of agencies and bureaucratic levels all with some share of the pie.[85]

Frances Perkins probably would be upset at where the road to social reform has led in the century since the Triangle fire. The thicket of laws and regulations has become not only an impediment to social change, but the issue of government involvement itself is dividing the country and its politics. Fred Taylor's productivity thrust continues to change the economics of America and the world, but the distribution of the benefits it bestows has become mired in the political fight over the proper levels of regulation.

The Blackstone Valley, Waltham, and Lowell, Massachusetts

The economic history of New England is as dramatic as the transformation of any region on earth. Starting from an unpromising situation of losing population in the eighteenth century, New England came to lead the United States in its progress from an agricultural to an industrial nation.

— Peter Temin[1]

Cotton thread holds the Union together....Patriotism [is] for holidays and summer evenings, with music and rockets, but cotton thread is the Union.

— Ralph Waldo Emerson, 1846[2]

You know what comes between me and my Calvins? Nothing.
— Actress Brooke Shields in a 1980 advertisement for Calvin Klein jeans

This is a story about thread, yarn, and cloth. It is about the Yankee creation of America's Industrial Revolution,[3] featuring technology that was stolen and improved upon, using machinery made of wood and iron, powered by waterwheels, run by companies that were incorporated, had interlocking directorates, indulged in insider trading and dealing, sold stock, employed women and children, developed new management techniques, planned cities, watched over public morality, encouraged railroads, and showed the way to tap a large domestic market for low-cost consumer products. Nothing came between them and their customers.

The British Industrial Revolution

The keys to industrial growth were at least three.[4] One was the widening of the effective market, which encouraged productive innovations such as the division of labor. With its control of the seas from the 1700s[5] onward and its Empire, Britain's exporters could count on a good deal of security in their trading relations. The second key was the development of new tools, such as banks and insurance companies, for the accumulation of capital. These were begun in order to finance large-scale exports and imports, but they also could be redirected into supporting the upstream supply chain of British production, including factories. The third key was mass production in a factory setting, based on the substitution of machines for human dexterity and of waterpower, and later steam, for human and animal energy, and the expansion of raw materials in variety and quantity.[6]

The original notion of a factory has been the subject of considerable debate, but the creation of a place where tasks are specialized, machinery and tools are used, and the whole overseen by a hierarchical organization goes back at least to Roman times. The application of waterpower to produce a consistent movement of machinery dates back as long as mills have been used to grind grain. The first British textile mills were based on a water-powered machine patented in 1768 by Richard Arkwright, and were developed to spin yarn. Not until a generation later was a practical power loom devised to turn yarn and thread into cloth and move weaving into the factory environment.[7] Cotton was especially favored as it lent itself better to these mechanical processes than wool or flax.[8] Cotton fibers could be formed and cut using machinery that could be made from softer materials, such as wood and leather, along with some ironwork. The machinery also did not have to meet exacting tolerances, a quality that was impossible to achieve with consistency using such materials.[9] This comparative simplicity lent itself to the abilities of the water-powered processes of the time. An assembly of wooden wheels, posts, gears, and leather belts could draw power from waterwheels (not yet turbines!) and turn it into the spinning motion of the jenny.

All the cotton fiber imported into Britain in 1765 had been spun by hand into thread and yarn, but by 1785 thirty times as much thread was produced by machine.[10] England and Scotland were dotted with spinning mills (or factories) wherever suitable waterpower could be found and channeled to a site,[11] and local carpenters and blacksmiths could maintain the mill equipment. The biggest obstacle was the variability of the water supply, solved to a degree in the case of small mills by building millponds.

Textile mill yarn productivity was dramatic and disruptive,[12] being many times greater than that of the spinning wheels traditionally used in village homes. Yet, under the system of "putting out" weaving production, households could adapt fairly easily, and replace this loss with another source of income, muting social unrest. It was not until the end of the eighteenth century that the power loom became widespread, drastically reducing the use of home production.[13] This was met with considerable resistance from people in rural villages and towns who had already lost their farming privileges to enclosure movements by landowners who desired empty land for sheep, responding to the price of wool and agricultural products.

As cotton cloth dropped in price, demand grew in middle-class households in Britain and abroad, including America, where incomes were relatively high. The British focus became increasingly export oriented, ignoring the large, but poor, domestic market.[14] British colonial laws allowed its merchants, for instance, to control both the supply of (mostly) Indian cotton and the supply of product to consumers in America and elsewhere in the Empire. It was a convenient arrangement for British merchants, but it did help launch the American Revolution as well as Gandhi's nationalist "homespun" clothing campaign in twentieth-century India.[15]

This export focus became increasingly oriented toward Britain's American colonies, such that, by the 1750s, half of British shipping was devoted to this trade. Most of this activity was part of a triangular system that saw British imports of sugar from the Caribbean helping to finance Caribbean imports of foodstuffs and raw materials from North America, which, in turn, largely provided the American colonies the money to pay for the consumer manufactures they bought from Britain.[16] A commodity boom in the mid-1700s encouraged British merchants to extend more credit to American consumers, while the colonial wars with the French that lasted until the 1760s only injected more funds into the American economy. When the boom and the wars ended, British credit began to dry up, and when the British government added insult to injury by asking the American colonies to pay new taxes to cover the debts accumulated by the wars, American resentment exploded.

The American Industrial Revolution

By the origin of the Republic, it appears that the average American lived as well as the average European and maybe a bit better, since there were no exaggerated disparities of wealth and poverty. Americans' demand for products led to mass production and to new labor-saving devices. High wages, because of a widespread labor shortage, meant that consumer demand was high as well and that saving labor through mechanization was desirable. Adam Smith, writing in England at the outbreak of the Revolution, noted: "It is not, accordingly, in the richest countries, but in the most thriving, or in those which are growing rich the fastest, that the wages of labor are highest. England is certainly, in the present times, a much richer country than any part of North America. The wages of labor, however, are much higher in North America than in any part of England."[17]

The forces behind American industrialization were complex, but the most obvious seem to lie in the financial problems of colonial economies and the standard of living that pre-Revolutionary Americans were used to. The needs of Americans for capital to develop land, buildings, and transportation routes as well as imports of consumer goods meant that the early Republic was always short of specie, or cash. Banks could resort to printing paper money for internal use, but international trade required "hard" currency. There were only two ways to adapt: earn money from exports (wood, wheat, and other agricultural products) or produce some products locally (import substitution) and reduce the amount of "hard" money that had to be sent out of the country. The logical place to focus American efforts at import substitution was in the area of consumer goods, as the American household was a relatively heavy purchaser of these and most had to come from, or through, British merchants. Industrialization was meant, at first, to accomplish this objective.

Any arguments about the future destiny of the country revolved around this reality of relative equality. By 1776, the British rural reality was fast becoming one of large landowners and dispossessed or landless village people. The Highland clearances and the confiscation of village common land were beginning to become features of the changed relationship between rich and poor. Part of the driving force behind the Revolution in America had to do with the possibility that this creation of a landed aristocracy coupled with landless poor might be transplanted to North America, no matter how impractical that might be in reality.[18] Even after the Revolution, some of those who had now risen to the top ruminated publicly about re-creating an upper class in imitation of the "Tories" they had expropriated or thrown out of the country. By 1800, these interests were in political decline, however, and the debate shifted elsewhere.

Thomas Jefferson approached the question in paradoxical fashion. While a slave owner and plantation owner himself, in the 1790s he saw the future of the country as one of independent "yeoman" farmers, all landowners, existing by exporting the surplus from their farms. His vision was based on his reading of the history of the Roman Republic, which reached its zenith as a country that ran its affairs on a balance between plantation owners and free, small-landholding farmers.[19] In part, Jefferson's vision was fulfilled, in the next century, by the measuring and selling of frontier land for farm development. A generation later, Jefferson was still the proponent of a country primarily filled with small-holding farmers[20] — and to a great extent some variant of this vision maintained itself as a part of American ideals up to the end of the twentieth century. Jefferson felt that exports of agricultural products and raw materials could pay for most of the imported consumer goods.

The adversaries of Jefferson's vision were no longer those who looked to the creation of an American landed aristocracy. Instead, they saw a nation of cities and industry.[21] The British Industrial Revolution had sparked a degree of envy and desire for emulation in America. The British tried to prevent knowledge of industrial techniques from getting beyond their shores, though plans for water-powered "spinning jennies" were brought to Philadelphia around the beginning of the Revolution. War, inflation, and other factors, however, combined to abort early industrialization. After the war, the British increased their attempts to stop industrial espionage by Europe and America. These efforts did not work in the long run, though they did slow down the diffusion of technology to a degree and helped to prolong British superiority.

The most prominent spokesman for an urban industrial vision was Alexander Hamilton, a contemporary of Jefferson's and a fellow cabinet member in George Washington's first administration. Though his life was cut short in a duel in 1804, Hamilton exercised enough influence though the 1790s, the first decade of a united American state, to give inspiration to a generation and more of American business people. An effective state was, to Hamilton, the first step in promoting an American industrial policy. He produced two reports on public support for manufacturing and for the creation of a National Bank. In 1781, Congress had chartered the Bank of North America, which helped to finance the Revolutionary Army, but Hamilton wanted a new Bank of the United States that would act more like what we know as a central bank today.[22] In a famous deal in the first Washington administration, Hamilton gave Jefferson his prized national capital on the Potomac in return for Jefferson's reluctant help in settling Revolutionary War debts and putting the national finances on an even keel. However, Jefferson opposed Hamilton's call for what was, in effect, a national industrialization plan, though he saw some merit in temporary assistance for infant industries. Washington himself recognized the military importance of a manufacturing base, and supported Hamilton.[23]

Hamilton later tried to sponsor a manifestation of his dream at the falls of the Passaic River at the village of Paterson, New Jersey. The concept was to create a new town around factories that would be powered by the head of water above the falls. This was copied from the basic approach of British industrialization at the time. The site was then a day's journey west of New York City and close to the Philadelphia market as well. Nothing practical came of these plans;[24] instead, the real beginnings of American industry took place in a rough north-south line from Lowell, Massachusetts, through Waltham to Providence, Rhode Island.

New England and Industry

Less than two decades after the founding of the Massachusetts colony, some of its leaders incorporated the Saugus Ironworks near Lynn. This "Company of Undertakers" received considerable public assistance, in terms of land, tax abatements, and monopoly privileges. The works produced "bog" iron blooms and had a forge that turned the iron into pots, nails, and other products. The company could not just export its iron, but had to sell a part locally and take farm products in payment. This 1640s project failed, however, because of start-up costs, skilled labor shortages, and poor management.[25] Yet it led to the later establishment of ironworks throughout New England, and they and other works in America were producing more iron by the 1770s than England and Wales combined.[26] The Saugus Ironworks moreover established the principle that public support and involvement in economic development, especially in manufacturing, was an acceptable practice.[27]

In the 1700s, New England firms began to provide marine transportation services to all the colonies. They were active in the importation of sugar and molasses to be distilled into rum, which was then distributed throughout the American colonies and back into the Caribbean islands where the raw material had been produced.[28] Some vertical integration tied to this trade did occur, as when the Hutchinsons of Boston purchased nineteen sawmills in New Hampshire to provide them with a steady stream of lumber for their West Indian import-export business. Some of these enterprises also grew relatively large for their day, such as a sawmill in Maine that employed thirty people, including twenty-two women, in sawing and potash production There were even attempts in New England in the early 1700s to develop plantation-like farms employing African and Indian slave labor.[29]

Shipping and privateering also played a role in New England industry as, from Revolutionary times until about 1830, they were the means by which local entrepreneurs acquired their capital.[30] In the 1790s, as many as five hundred ships could be seen in Boston harbor at any given time. By the time of the US embargo on British and French shipping in 1807 during the Napoleonic Wars, Massachusetts (then including Maine) alone employed more shipping tonnage than any other State, and virtually all the American fishing fleet was based in its waters, sending whalers out as far as the Pacific Ocean. The money gained from this trade was not entirely reinvested in trade and the fishery; some was kept at home for other uses. This local investment only intensified as the embargo and then the War of 1812 disrupted American maritime commerce.

So-called cottage industries developed around other products that could be made from local resources. For instance, farmers clearing land could produce potash from the fallen hardwood trees and stumps. Unneeded and unsalable wood was burned and the ash further reduced in pots (hence the name) to produce lye and potash. These could then be sold to collectors who either provided the products to soapmakers or glassmakers, or exported it. The farmer not only got cleared land, but gained the equivalent of his wages for doing from the sale of "pot-ash." The availability of leather encouraged the production of shoes for local and regional use.[31] The town of Lynn, Massachusetts, became a center for shoe production, and by the late 1760s 80,000 pairs of shoes were being turned out annually in households and "manufactories" that were simply workshops where people could engage in specialized tasks, such as cutting and sewing.[32] The complicated process of producing shoes did not lend itself to mechanization until the 1800s.

Besides providing direct assistance, New England governments also established stable frameworks for business and innovation to proceed, including contract and patent laws. They also assisted in road building and education and training.[33] As a result of state tariff measures to protect homespun products following the Revolution, the first American cotton spinning factory was set up as the Beverley Cotton Manufactory in Beverley, Massachusetts, in 1789, halfway between Boston and Newburyport, by the Cabot family of merchants.[34] The first woolen factory was incorporated in Hartford, Connecticut, about the same time, with £1,280 in capital raised from selling £10 shares.[35] At the start of the American trade embargo against Britain and France in 1808, Massachusetts had fifteen spinning mills,[36] but by 1810 there were more than fifty.[37] The embargo served to depress the price of Southern raw cotton and raise the price of domestic cloth. New England businessmen were quick to take advantage of the opportunity. By 1800, in addition to cloth and shoes, there were a number of other small manufacturing businesses in New England, including a rolling mill for copper plate, and makers of glassware, hats, paper, shovels, axes, and mousetraps.[38]

In Connecticut, manufacturing tended, by 1820, toward wooden and brass clocks, whose parts were cheap to produce and could be produced in standardized pieces. This allowed assembly to be brought from the woodlots to the towns, where small streams could produce the power needed in assembly. Early clockmakers used traveling peddlers to sell and distribute their wares, although, after 1840, more formal distribution and sales methods came into use.[39]

By 1814, however, it was Pennsylvania, rather than a New England State, that ranked first in manufacturing output, supplying the growing number of settlers moving west across the state and southwest up the Shenandoah Valley to the Cumberland Gap.[40] But most of this manufacturing was done in artisan workshops, not factories, and by 1850 New England had nearly double the capital and workers in manufacturing employed by Pennsylvania and produced almost half again as much as the whole Mid-Atlantic region of the country, which had three times

New England's population.⁴¹ New England cotton textile production rose from 46,000 yards of cloth in 1805 to about 142 million yards in 1830.⁴²

The Blackstone Valley

The title of "first true factory" in the United States is generally accorded to Samuel Slater's mill in Pawtucket, Rhode Island, just north of Providence, at the mouth of the Blackstone River.⁴³ Slater was the son of a prosperous English farmer and timber merchant who had become involved with Jedediah Strutt and Richard Arkwright, the developers of British spinning mills. Strutt offered fourteen-year-old Samuel an indentured apprenticeship at one of his mills, and the boy spent seven years there, working with Strutt as what one might now call a management trainee. He did bookkeeping and mathematical calculations, and became familiar with the mill's manufacturing technology, which consisted of a series of continuously operating machines that created, rolled, and twisted strands of cotton fiber into yarn that could then be put out to households for weaving. When, in 1785, Arkwright's patents were overturned in the courts, production opened up to anyone who could afford to develop a mill. Slater, then seventeen and with ambitions to follow his late father's entrepreneurial footsteps, saw his opportunity slipping away.

Sometime between then and 1789, when his indenture was up, Slater determined to take his knowledge with him to America.⁴⁴ To avoid the British government's policy of preventing those who knew the business from emigrating with their knowledge, the then twenty-two-year-old Slater pretended he was a farmer and managed to convince the authorities at dockside that he should be allowed to emigrate.⁴⁵ He landed in New York in November 1789, and promptly began to advertise his expertise.

Earlier that year, some prominent New Yorkers had set up what was known as the New York Manufacturing Company to promote local factories, as opposed to the artisanal workshops common in the city.⁴⁶ Both President Washington and New York Governor George Clinton had words of encouragement for manufactures and textiles, and two hundred investors bought up the company's shares. By the end of the year, the Company had opened a textile factory, eventually employing 144 people to spin and weave using "jennies" powered by people and animals. It was said to have cost $57,500, almost as much as the renovations to New York City's Federal Hall, the first national capitol. Slater got work at this facility and offered to put his expertise to use and construct modern machinery for the plant, but was refused. While working there, Slater heard of the efforts of Moses Brown, in Providence, Rhode Island. He began to correspond with Moses, and soon decided to leave New York.⁴⁷ The New York Manufacturing Company, which had spurned him, was fated to fail within a couple of years due to poor management, the lack of adequate power, and competition from imports.⁴⁸

It was not for lack of trying that America had failed to build a spinning mill with up-to-date technology. With the release of Arkwright's patents in Britain, some knowledge crossed the Atlantic. The Massachusetts Assembly agreed to subsidize two Scottish brothers to develop models of Arkwright-type machines. Completed in 1786, the devices were not operated, however, but placed on display so that others might learn from them and improve on them. A number of entrepreneurs instead resorted to the older spinning jenny technology or tried to make some parts of the Arkwright process work for them.

In 1789, a prominent Providence merchant, Moses Brown, decided to buy up the various prototypes that had been produced in Rhode Island to try to use them to produce cotton yarn.[49] He put them in a facility he owned at the mouth of the Blackstone River, in Pawtucket, just north of Providence, and turned the place over to his son-in-law, William Almy, and nephew, Smith Brown; the resulting partnership became known as Almy and Brown. Almy and Brown were primarily interested in marketing cloth. They had other things to worry about besides inefficient yarn production, and were interested in finding someone who could come in and make their spinning machines work. They needed a manager who knew the technology, and along came Slater, with good timing, in early 1790.[50] He and Smith Brown disagreed about the usefulness of the local machines that Moses Brown the merchant had bought, but Slater worked with local machine builders, modified the machinery so it would work properly, then signed a contract to develop the spinning processes for Almy and Brown.

The management of the conversion of raw cotton into yarn and thread was the most important thing Slater brought with him. He pointed out that the locally built machines did not include among their number others that were vital to the continuous process that was key to British productivity. The key was physically to arrange a series of single-purpose machines into a production process.[51] One after another, the technical and mechanical hurdles were overcome by the area's shipwrights and mechanics until, ten days before the end of 1790, the small mill[52] was producing yarn. It was not until 1793 that a larger mill would be completed on the site, but the process and machinery had been successfully transferred to America. A reconstruction of the mill is on the original site today.[53]

Also transferred to America in Slater's mill was the employment of women and children. Alexander Hamilton's 1792 *Report on the Subject of Manufactures* suggested their use in manufacturing processes. He noted, "the husbandman himself experiences a new source of profit and support from the increased industry of his wife and daughters...women and children are rendered more useful, and the latter more early useful by manufacturing establishments than they would otherwise be." And two months earlier, when Slater established his first mill at Pawtucket, among his first workers were nine children ages seven to twelve.[54] Women and children had been commonly "employed" out of necessity on farms since time immemorial, of course, and modern notions of childhood and long schooling were unknown. Moreover, some of Slater's first child employees were not poor, but were the children of his partners and investors.

Over the years after 1792, Slater established the basic principles of mill management, such as maintenance of machines, controls on waste, supply management, and division of labor. He also narrowed the kind of production from Brown's early dreams of extensive competition with British imports to large-scale production of yarns and simple woven goods. In 1797, Slater left his partnership with Almy and Brown and went into business with his father-in-law, constructing a second mill in Pawtucket the following year. After that, the rush was on. In all, Slater and his competitors built eighty-seven other mills by 1810,[55] gradually turning the Blackstone Valley into the center of early American textile manufacturing.[56] The early mills were small, their equipment crude, and they depended on a local workforce of women and children — in some cases, entire families would contract to work for the mill.[57] Pay was often a mixture of cash, food, and yarn, which households could weave into cloth to use or sell. The mills sold their output locally to weavers or to wholesalers in larger cities from Boston to Baltimore.[58]

Ironically, Jefferson's, and then James Madison's, anti-British policy, including embargo and war, had the unintended consequence of promoting the very industrialization they philosophically opposed. In 1814 alone, 105 new textile (spinning) mills opened, and by 1815 New England cloth production was estimated to have grown fifty-fold.

Waltham

A few blocks south of US 20, the main street of Waltham, Massachusetts, is a group of large, old factory buildings on the banks of the flowing Charles River, not yet the tidal estuary that divides Cambridge from Boston. When we visited them, the buildings were tall wood and brick structures containing a mix of organization offices, a seniors' center, and a museum. They lie just downstream from a low weir on the Charles whose original had created the millpond for the factory. The original buildings housed the first integrated textile mill in the world. It was a key component in the Yankee Industrial Revolution.

The textile mill was the brainchild of a Newburyport/Boston shipowner and merchant, Francis Cabot Lowell. He was an importer of British textiles, and felt that producing competing products might be a good way to diversify his family's investments. He derived some inspiration from Rowland Hazard, who developed the first integrated woolen mill (spinning plus power loom) in Massachusetts, near Plymouth, in 1800.[59]

Lowell and his family went to England for his health in 1810 and returned home as the War of 1812 broke out. Lowell used his time in Britain to study spinning and weaving technology and methods in the textile industry around Manchester as well as social relationships in planned villages for workers in Scotland.[60] Nathan Appleton, a Boston merchant and future investor in Lowell's plans,[61] recalled that he had met Lowell in Edinburgh: "[Lowell] informed me that he had determined, before his return to America, to visit Manchester, for the purpose of obtaining all possible information on the subject, with a view to the introduction of the improved manufacture in the United States."[62] The idea of integrating spinning and weaving had also entered Lowell's thoughts, As he told Appleton, "a large establishment with all the branches connected would be sure to do well."[63]

British suspicions, as he left for home, resulted in Lowell's luggage being searched twice for illegal models and drawings related to textile production, Lowell memorized a lot of detail, however, and, once back in Boston, he committed what he knew to paper. Assisted by a very capable local mechanic, Paul Moody, in 1813 Lowell devised in a Boston attic the elements of a workable power loom[64] that was intended to function alongside the modern spinning equipment already in America.[65]

In late 1813, Lowell, his brother-in-law and other relatives, and some merchants received a charter of incorporation from Massachusetts for the Boston Manufacturing Company.[66] This was an unusual instrument at the time for a manufacturing activity, but probably reflected Lowell's concerns about his health, the expectation of longevity for the enterprise, and the amount of capital that would be tied up in it.[67] As well, this approach was dictated in part because the banks of the time would not lend to manufacturing concerns, so raising capital became a task of finding investors rather than bondholders.[68] The search for capital was assisted, in part, by the War of 1812, when the official prohibition on exports and imports to and from Britain meant a manufacturing boom in the Northeast as shipowners, faced with ruin, invested instead in domestic manufacturers and used the trade situation to their advantage.[69]

Because their eventual complex of corporations, starting with the Boston Manufacturing Company, had interlocking directorates and shareholders, the shareholders were traditionally called the "Boston Associates." The Associates set out to build a large new mill at Waltham at the falls of the Charles River that would use their technology to profitable effect. One of the Associates, Nathan Appleton, the Boston merchant in whom Lowell had confided when both met in Edinburgh, arranged for the mill's product to be sold through B.C. Ward & Co., a trading firm in which he was a major investor.[70] Most of the output of Waltham, and later Lowell, was distributed through B.C. Ward to New York wholesalers.

The first power loom became operational in the fall of 1814.[71] The importance of the Waltham mill lay in its groundbreaking integration of spinning and weaving, producing mechanically at one site what other British and American companies did in a combination of mechanized spinning and handcrafted weaving.[72] The true innovation that Lowell and his associates brought to the production of textiles was the development of a system that incorporated more production steps in the mechanized process than had been attempted up to then in either Britain or America. The British were hampered by the structure of their industry and by labor opposition. There, goods were produced in a discrete set of steps by different factories and methods. Lowell saw that, by housing these steps in one complex, he could increase productivity and lower costs, thus maintaining a competitive advantage over British goods far beyond the end of the war. The Associates' production system was characterized as "bale [of cotton] to bolt [of cloth],"[73] a continuous flow process devised by Moody. The product was a coarse, durable cotton cloth that appealed to the rough users on farm and frontier.[74]

A second feature of the Waltham mill was the scale of capital devoted to the enterprise ($400,000, or 400 shares at $1,000 each), which outstripped that of other mill owners.[75] It took five years and three share offerings to complete the financing.[76] Lowell estimated that only $100,000 would be needed for the plant and equipment, while the rest would be required for cash-flow purposes,[77] not unlike the same relative costs in a technology start-up today. Beyond this financial scope, in 1814 the Associates began to press for increased tariff protection from British goods by hiring a talented lawyer, Daniel Webster, to represent their interests. New England and Western interests managed to get a revised *Tariff Act* passed in 1816[78] that provided large producers like the Waltham mill enough room to price their goods profitably. Another effort to increase tariffs in 1820 failed, but this was not enough to discourage further growth and expansion.[79]

Because higher wage differentials between New England and Britain could not be offset by savings from the proximity of both the cotton producers and the cloth consumers to New England and the technological integration of its mills,[80] the 1816 US tariff was focused on hampering imports of cheap cotton goods. These were produced in British India and therefore were not made from Southern cotton exported to Britain. This meant that Southern planters did not have to worry that Northern textile producers were hurting their British sales, so they could then support, or at least not oppose, the tariff.[81]

Organizationally, the Waltham mill also deviated from the British norm in that the owners did not live at or near the site, but delegated management to an agent. Until his death in 1817, Lowell acted as what we would call today the chief executive officer, overseeing the construction and initial operations of the facility. He was succeeded by Patrick Tracy Jackson, another of the Associates, as "superintendent" of the mill. Given Jackson's other interests in Boston, the practical management of the mill was gradually left to others. This practice was later formalized at the Lowell mills.

As for the mill's workforce, the Associates eschewed the hiring of children, focusing instead on adults, especially educated and trainable young women, of whom there were eventually three hundred.[82] The site was too remote for this number of employees to commute from home, so the company began to build or encourage the building of boardinghouses in the town.[83] The company assumed responsibility for the workers' social and moral welfare, regulating dress, decorum, and even religious development — publicizing this last to reassure farm families that their daughters would not be led astray when living on their own.

The Waltham factory marked a transition point in early American manufacturing from individual mills to integrated factories with large numbers of workers and professional managers.[84] The mill also developed its own technical solutions to production problems, as Paul Moody was asked to create a machine shop at the site. No one in America had ever run a production facility of this size and complexity. It was a "proving ground" for what grew at the Lowell site, including building the equipment for the first Lowell mills,[85] which eventually sparked the New England machine-tool industry. The use of female labor housed nearby indirectly helped finance frontier development in a minor way through the payment of cash wages,[86] which were either sent home or saved as dowries. Their marriages, in turn, financed a large amount of emigration to the West.

By 1820, the Waltham facility had so expanded that it employed twelve times as many workers as the average plant in the State, including the long-established spinning mills to its south; it also had eight times the capitalization, with only five other companies reporting as much as $50,000.[87] Its scale, producing half a million yards per year, meant that it could compete on price with British imports. Its relatively low prices for durable coarse cotton cloth opened up a larger demand in the growing and relatively prosperous domestic market than could be filled by upscale British fine cloth.[88] Entering the coarse cloth market, "Lowell sheeting," named after the company's founder, also reduced British merchants' practice of bringing in flimsier coarse cloth from India.[89]

Lowell

The shareholders soon began to realize that the Waltham site on the Charles River, having grown to two mills by 1821, was on too small a stream for their aspirations.[90] They looked north to the larger Merrimack River, where Lowell's brother-in-law, Patrick Tracy Jackson, had inherited shares from his father in a company that operated a little-used canal.[91] The one-and-three-quarter-mile, four-lock canal had been built in 1796[92] to allow New Hampshire products to bypass the thirty-two-foot drop of the Merrimack's Pawtucket Falls.[93] The falls area was not unknown to earlier entrepreneurs, as a woolen mill had been built nearby in 1801, which failed in the aftermath of the War of 1812 and was resuscitated under new ownership in 1818.[94] An additional advantage was that the area was served by the then-longest canal in America, the Middlesex Canal,[95] finished in 1804 to connect the Merrimack to Boston, bypassing Newburyport. It could provide cheap bulk transport between Boston harbor and the proposed mill site.[96]

The Boston Associates, including Jackson, bought up the rest of the canal company shares and three hundred acres of farmland around the falls,[97] and arranged a stock swap among themselves that merged the canal works with their landholdings in the original canal company, now called the Proprietors of the Locks and Canals on the Merrimack River (PLC).[98] This became the Associates' real estate arm. It built and expanded canals, mills, and boarding houses, sold land for new mills, and leased waterpower rights. Paul Moody was asked to create a larger machine shop to fabricate and repair mill equipment. The Associates also chartered a second company, with $600,000 in capital (600 shares at $1,000 each), called the Merrimack Manufacturing Company, which was to operate a large, new mill on the site. In effect, the Associates had three interlocking companies in Waltham and the two Lowell organizations.[99]

In early 1822, the Associates began to build power canals and the Merrimack mill, and even laid out a town, named in 1826 after Lowell, who had died in 1817. The size of the site[100] and the large capital investment would give the Associates the capacity to produce what were then massive amounts of cloth — this time, printed calicoes rather than the rougher sheeting made in Waltham.[101] Company agent Kirk Boott, a retired British army officer,[102] planned and constructed the power canals, the mills,[103] a print works for the calicoes, housing, and even a church.[104] Power wheels first started turning in the mill in September 1823, and cloth was first shipped to Boston in early 1824.

In 1825, the company's capital was increased to $1.2 million, which was used to build three more mills.[105] The Associates incorporated the new mills separately, selling the sites to new companies, leasing water rights to them, and then building the mills with the assistance of their own machine-shop and construction divisions.[106] By 1836, there were twenty-six textile mills, two print mills, and the machine shops, comprising in all nine complexes running off the Merrimack's waterpower.[107] These were among the largest enterprises in America at the time, and their success led to the development of similar, but competing, complexes at Lawrence, Massachusetts, downstream from Lowell,[108] as well as at Manchester, New Hampshire, upstream. Others were built by competitors on the Connecticut and other rivers that had significant water and a good falls, including, eventually, at Hamilton's old site, Paterson, on the Passaic River in New Jersey.

In contrast to the different ownership structures of the mills, all of them depended on one central machine shop, which was tied to the ownership of the power source for all the mills. The advantage of this arrangement proved to be critical to the future technological development of the country. Unlike the smaller workplaces spread throughout New England and other parts of the country, the size of the Lowell complex meant that significant investment could be put into research and development. Early hydrological studies, manuals on waterpower transmission processes, the development of turbines, and the later development of railroad technology came from this large-scale machine shop.[109] The shop gradually expanded its sales outside the Lowell area, and so was eventually incorporated separately from the PLC in order to branch out[110] into metal tools and equipment and, eventually, locomotives.[111]

The problem of interstate incorporation, which would lead to the creation of the infamous trusts of the 1890s and beyond, was first encountered by the owners of the textile mills at Lowell and Lawrence, both dependent on Merrimack River water. Control of the sources of the Merrimack, necessary so that the supply of water to the mills could be regulated to provide a constant feed regardless of season, had to be acquired from the New Hampshire government. The mill owners got around the issue by forming a dummy "manufacturing" company in New Hampshire, which in 1846 asked that State's legislature for control over the water flows that created the Merrimack. Owned equally by shareholders in both mill complexes, the company was not intended to manufacture anything, but to regulate the water flowing downstream, acting on behalf of the Massachusetts corporation.[112] Henry Thoreau, after traveling the river in 1839, commented that "it falls over a succession of natural dams, where it has been offering its privileges in vain for ages, until at last the Yankee race came to improve them."[113]

As the complex of Lowell organizations grew, a model emerged that separated ownership from management, characteristic of the modern corporation.[114] Each mill company had a separate president, who was largely a figurehead and either one of, or a representative of, the owners; a treasurer who set policy and managed the finances; and an agent who was responsible for the actual plant operation. Both presidents and treasurers worked in Boston, while the agents were on site.[115] These Yankees also "invented" insider lending and trading, as well as the corporate trust structure, long before these came to public attention and condemnation in the railroad and oil manipulations later in the century.

New England production capacity, measured in installed spindles, increased sevenfold between 1820 and 1831, accounting for 59 percent of the national total in 1820 and 66 percent in 1831.[116] In 1836, the Lowell mills produced just under fifty million yards of cloth, and by 1850 Lowell alone was producing 20 percent of American cotton goods, making Massachusetts the most industrialized jurisdiction outside Britain.

The profits from textile manufacturing[117] allowed a tight-knit group of Boston families — the Appletons, the Cabots, the Lowells, the Lawrences, and the Jacksons — to build an economic, social, and political empire.[118] In 1832, Secretary of the Treasury Louis McLane commissioned a survey of American manufacturing. Of 106 firms with assets over $100,000, 88 manufactured textiles, and 85 percent of these made cotton rather than wool textiles. Of the 36 firms with 250 or more employees, 31 were textile firms,[119] and the Boston Associates probably controlled a dozen or more of them. Invested capital in the Lowell works grew to $8 million by 1840 and to $12 million a decade later.[120] The contrast between Lowell and the Southern cotton-producing economy is stark. In the South in 1850, there were fewer than a dozen plantations with a workforce of three hundred or, excluding slaves as property, assets worth at least $300,000, a figure comparable to just over a quarter of the investment in Lowell.[121]

Lowell, the "Manchester of America," was the country's first large-scale planned industrial community.[122] In a way, it was the creation of the town that Hamilton had dreamt for New Jersey almost thirty years earlier. In another way, it resembled New Lanark in Scotland, a place Francis Cabot Lowell must have visited around 1810. Visiting the Massachusetts mill town, English novelist Anthony Trollope called it" the realization of a commercial Utopia."[123] Over the years, Lowell grew. In 1826, it had a population of 2,500, almost all engaged in mill operations and construction. By 1836, the town had grown to 17,600 inhabitants, half of whom worked in the mill complexes.[124] By the census of 1840, Lowell had 20,000 inhabitants, making it the second-largest city in Massachusetts. By 1850, the town had grown to 33,000, many of whom were agents who supplied the mills and local merchants or other businesses who sought to tap the mill's skilled and disciplined workforce.[125] By 1870, the population reached 41,000, and by 1881, 60,000,[126] of which around a quarter were employed at the mills directly. After 1900, employment began a slow slide as textile production moved elsewhere.[127]

The Lowell model became known as the "American System," which was conditioned by a large, relatively free domestic market coupled with a scarcity of labor and its attendant costs relative to those of materials,[128] both of which encouraged mechanization and the standardization of products. Production was organized along what came to be seen as standard factory principles, but in Lowell and Waltham included a kind of paternalism for its workforce that disappeared as time went on.[129] The use of incorporation allowed for the pooling of capital without the complication of partners with differing objectives, as well as avoiding complications brought on by the death or withdrawal of investors. Soon, this model was applied to the manufacture of all sorts of other consumer and industrial goods, including watches, shoes, and handguns. While Frederick Taylor, a century later in Watertown, could extol the systematizing of organized activity, the American affair with applying rational concepts to physical and organizational life began almost a century before at Waltham and Lowell.

The Boston Associates' money was reinvested wherever a good return was promised, especially in railroad ventures. Patrick Tracy Jackson developed the Boston and Lowell Railroad, importing two British locomotives and having the Lowell machine shops build others on their design.[130] Boston investors financed other railroad lines in New England, as well in the northern Midwest. They owned controlling stock in a host of Boston financial institutions, allowing them to finance and insure ventures through their own companies.[131] In 1845, seventeen of the

Associates served as directors of the seven Boston banks that controlled over 40 percent of the city's financial capital, twenty were directors of insurance companies with 40 percent of the city's business, and eleven would serve as directors of five New England railroads.[132]

Decline

But change was coming to the industrial landscape in America. Waterpower eventually would be replaced by steam and then electricity, though for much of the first half of the nineteenth century, factories were tied to the geography that surrounded falls of river water. By 1830, when the American cloth industry was in its expansionary mode, Britain's water resources were almost fully utilized, and the British began to shift to steam made from burning coal.[133] Steam technology developed there and then crossed to America. In both countries, steam freed the location of plants to coastal cities, where shipping costs were less and to other market areas inland, where canals[134] and railroads could easily move raw materials and finished goods.[135] Even Lowell used steampower to supplement its water resources after the Civil War, and by 1880 steam provided more power there than water.[136]

Other changes affected labor relations. First, mill technology began to be developed elsewhere in the country. As well, a generation of "Brahmins" emerged who lived off their family capital, so retained earnings from Lowell mills necessary to upgrade facilities were forgone to ensure large dividends in the short term.[137] As product prices fell, labor relations focused on trying to get more production out of the same mills for less pay. This discouraged single, local women from going into the mills, and they were replaced by Irish and then French-Canadian immigrant families, who would work for less pay as part of two-wage-earner households. This squeezing of the labor force meant that successive waves of immigrants had to be tapped to replace those whose children had moved on West.

The whole market structure began to change after 1840, as more ready-to-wear clothes began to be sold, especially in cities. Up to this time, such clothes were made only for people such as slaves, sailors, and frontiersmen, who needed them for rough and distant work. New York City became the leader in ready-made clothes, including "knock-offs" of European ladies' wear. The 1849 Gold Rush also touched off a demand for ready-to-wear clothing, leading to the production of Levi Strauss's jeans in California. By the Civil War, the standardization of sizes had proceeded far enough that Union Army recruits were measured and issued uniforms based on a size chart.[138]

Then, as a result of determined attempts by Southern States to industrialize after the Civil War, textile production itself began to move south.[139] By 1900, 97,000 Southerners worked in more than four hundred 400 mills, mostly in the Piedmont country of North Carolina.[140] Between 1912 and 1918, the Lowell and

Middlesex companies, now owned by Bigelow, moved production to the South. More mill facilities closed in the 1920s and '30s and the machine shop was razed. Lowell's population peaked around 120,000, while manufacturing employment fell by half between 1924 and 1932.[141] Decline continued through the next decades, with the area unable to diversify, until the 1970s, when much of the remaining facilities were given over to museum space and to technology companies looking to expand beyond the Route 128 corridor.[142]

Conclusion

The workers who populated Lowell and other mill towns in New England explain some of the story of the "disappearance" of the Yankees. The first thirty years of millwork in New England was dominated by a 3:1 ratio of female to male workers.[143] In large part, these were the daughters of farmers in the region, who would come to the mill towns and live in dormitories for an average of four years, under strict supervision, sending much of their earnings back to their parents or saving it for a dowry.[144]

As the supply of farmers' daughters dwindled, they were replaced in the mid-1840s and after by a wave of immigrants,[145] first from Ireland as a result, in part, of the potato famine and, in the 1860s, from Quebec to the north.[146] The impact of the Irish on Boston and its environs is a well-known story, but that of the French-Canadians is not. In 1860, there were about 37,000 French-Canadians in New England, but through recruiting agents and the pull of family ties some 600,000 more migrated, primarily to the mill towns, by 1900,[147] when the mills themselves began to decline and close because of the cost pressures from new mills closer to the cotton-producing areas of the South. By 1900, people of French-Canadian ancestry made up 10 percent of New England's population and 20 percent of Lowell's,[148] while Boston became America's most Irish city. By the early years of the twentieth century, three-quarters of the city's population was of non-Yankee origin.[149] This "disappearance" of the Yankees happened in all the industrial cities and towns found in the Yankee migration path from Boston to the Mississippi River.

Like the Watertown Arsenal, the textile mills in New England are gone now, turned into shopping complexes, trendy office facilities, and other uses. The mills at Lowell were transformed in part into the first Urban National Historic Park in 1978, and the downtown has been turned into a tourist destination. The Blackstone River Valley is designated as a National Historic Corridor, and the facilities at Waltham include the Charles River Museum. The Industrial Revolution moved on, not just to new places, but across industrial sectors, spawning machine tools, advertising, new modes of financing, and new corporate structures, management techniques, and labor relations. Yankee modes of thought and attitudes are, for better and worse, still at the core of American productive behavior.

Route 128: The First Computer Revolution

It is a fact not generally realized that the American Philosophical Society at Philadelphia, the Royal Society of Great Britain and the Royal Institution of London, are all of them in a measure indebted for their birth and first foundation to natives or inhabitants of New England.

— Jacob Bigelow, 1816[1]

I wish to God these calculations had been executed by steam.
— Charles Babbage, designer of the first computer, about 1830[2]

It is bad engineering to assume that a thing is perfected.
— Willis R. Whitney, founder of the General Electric lab, 1935[3]

Life was simple before World War II. After that, we had systems.
— Rear Admiral "Amazing" Grace Hopper,
author of COBOL, the pioneering computer language[4]

The Road to Nowhere

Early roads tended to radiate outwards from the centers of major cities and capital cities. When traffic was light, as in the days of horse-drawn vehicles and early automobile traffic, going through a city to get to the other side was not a problem. As cities grew larger and traffic increased, however, traffic planners started to consider the advantages of building circumferential highways around city centers. In Boston, this idea first gained interest in 1912. Nothing was done about it, other than to suggest that the road should go around Boston about twenty miles from its center.[5]

By the middle 1920s, road construction began to take off with the agreement on the national highway grid, connecting the States from east to west and north to south. The popularity of the automobile meant that better roads had become a political necessity. Massachusetts brushed the dust off the idea of a circumferential highway around Boston and designated a number of local streets and roads about fifteen miles out as a state highway[6] from the sea, around Boston, and back to the sea. To steer clear of the new national grid numbering system[7] and because this highway would go in three directions, not one, over its length, it was given the number 128 in 1925. It began to appear on maps as a crooked collection of streets and roads by 1929.

From the beginning, Route 128 was controversial. Benton MacKaye, a Connecticut native and Harvard-trained forester, put forward the idea that the road should be developed as a kind of wilderness "dam" against the spread of "metropolitanism." It would serve as a "townless" highway, be divided, and have turnoffs for rest areas. Roads like this would control the automobile, billboards, commercialism, and urban sprawl. MacKaye was, as well, starting to formulate his concept of what later became the Appalachian Trail. He went on to become one of the founders of the Wilderness Society in the 1930s, along with Aldo Leopold and Bob Marshall.[8] Route 128 didn't turn out as he had proposed, but instead became popularized by its other critics as "the road to nowhere."[9]

In 1934, Massachusetts' Commissioner of Public Works, William F. Callahan, proposed the construction of a limited access "circumferential highway" that would connect the roads radiating out from Boston, including US 20. In that, he was ahead of most highway officials in the country in recognizing the link between highway development, increased traffic, and economic development.[10] Taking advantage of Depression-era Works Progress Administration funding from Washington, Route 128 became the first such highway in the country to move from plan to construction.[11] Callahan saved money for the project and muted public protest by choosing farmland and swamps for the routing where he could. In 1939, the state government changed hands and Callahan was out, which slowed construction on 128. By 1941, when the war put an end to local road construction, only two stretches, north and south of Boston, had been constructed. In the gap, local roads had to be used.[12]

After the war, new federal money was made available on a 50/50 basis and, in 1948, Callahan was reappointed. Construction began anew on Route 128, now named the "Yankee Division Highway," and the highway was basically finished by 1953, although extensions and local renovations continued until 1959.[13] The main East Coast expressway in the Interstate system, I-95, was supposed to run into and through the center of Boston, but in 1973 the governor, pressured by public opinion,[14] cancelled these plans, and I-95 was connected to 128 and routed around the city. Even though 128 was subsumed into I-95, this segment is still known as Route 128 today, and the joint designation remains on maps and signage.[15] US 20, one of the original radial roads, was relegated to regional status by the construction of the Massachusetts Turnpike (I-90), which runs parallel to it a few miles to the south. Both cross Route 128 at about its midpoint, just west of Waltham.

The "road to nowhere" exploded with activity precisely because it went nowhere through nowhere, avoiding town centers and built-up areas. Just as the highway was being restarted in 1948, a planner whose father was a Massachusetts Institute of Technology (MIT) professor joined a commercial real estate and development firm, Cabot, Cabot, and Forbes, armed with a new idea. Edward Blakely proposed that the firm buy land along Route 128 near its interchanges and develop it for companies that were growing out of their facilities near MIT. Blakely noted that the first question outside industry executives asked him was "What's the driving time to MIT?" He could show that 128 connected to the radial access routes that went near the institute.

A modern industrial park had been built in 1947 along the old 128 roadway, but no one had considered putting one along a divided expressway.[16] Cabot, Cabot, and Forbes provided a package that would include sites and permits, design and construction, and short-term financing. The concept was especially appealing around Waltham, whose largest industry, the Waltham Watch Company, had recently closed, putting a large number of skilled workers on the street. Canada Dry, Polaroid, Sylvania, and others made the move to the edge of Waltham, and the development rush was on its way.[17] The federal government added to the development when Hanscom Field, located near 128, was closed and turned over to the Air Force to be the location for its new electronics lab, operated in conjunction with MIT.[18] Through the 1950s, "Callahan's folly" ran past a half a billion dollars' worth of new industrial and residential investment.

Blakely's office park concept was an industrial park in a new guise. Industrial parks themselves were not new — they had been originally designed around the turn of the century in the Chicago area as places where companies could manufacture products for widespread rail distribution.[19] During the Second World War, contractors found that a low-slung steel building could be built more efficiently and at less cost than traditional a brick multi-storey factory building. Further, the land along Route 128 was cheap compared with land near MIT and Harvard, so there was adequate room for parking and landscaping.[20]

Route 128's parks thus were highway based, and their "products" were both intellectual and physical.[21] They became the premier intellectual and research base for America's Cold War effort, offering pleasant, campus-like amenities while providing easy connections to the brainpower nearby. The pressure of all the new companies along 128 acted as a magnet for others, and the region continued to grow of its own accord in the manner of what are called "agglomeration economies."[22]

By 1961, eight years after the completion of the highway, 169 businesses with 24,000 employees had located along it, with many more nearby.[23] Four years later there were 574 technology companies located along 128.[24] By 1967 this number had grown to 729 enterprises employing 66,000 people,[25] and in 1973 there were 1,212.[26] Shopping malls were built near busy interchanges between 128 and the radial expressways. The road was full of traffic, and sections were continually being widened. By the 1980s, a second circumferential highway, I-495, was started farther out from Boston to relieve congestion on I-95/128. Other companies moved farther out, to restored old mill buildings in Lowell or to sites in Worcester and southern New Hampshire.[27]

Years later, California followed a similar process when Route 101 was made into an expressway bordering on the communities south from San Francisco to San Jose. Stanford University acted the role of MIT and Harvard, and nearby small farming and suburban communities were gradually taken over by technology-based office parks and their professional employees. As the twentieth century came to a close, Route 128 and Silicon Valley became the role models for similar developments in Raleigh-Durham, North Carolina, Austin, Texas, Boulder, Colorado, Maryland northeast of Washington, DC, and a variety of other "Valleys" across the country and abroad.[28] UNESCO now estimates there are over 400 "science parks" in the world, a figure that does not take into account the thousands of multipurpose "industrial parks" in existence.[29] "Nowhere" turned out to be an interesting place, and MacKaye's dam had turned instead into a sluiceway.

Artisanal Experience vs. Science and Technology

Early Yankee merchants did well out of seaborne trade and they accumulated a lot of capital, some of which they wanted to invest — nearby. Clearly, the way to do it was to invest in something that was capital intensive and had a domestic market focus. The obvious answer in the late eighteenth and early nineteenth centuries was the production of first thread and then cloth.

Their British competitors relied on two things besides capital expenditures for their success in cloth. First, they had a supply of trained artisans who were, on the whole, a fairly stable group of people; to augment this stability, the authorities tried to prevent the emigration of skilled people. Second, the British had a ready-made market in their Empire. The new American Republic could match the latter somewhat in terms of its domestic market, but except for Southern slavery, it could not prevent workers from moving west to new and better farmland. The Yankee entrepreneur therefore had to deal with a temporary labor force that was educated, but relatively unskilled by the artisanal standards of the British workforce.

The key to American success in factory-based manufacturing lay in the intensive use of the division of labor, based on breaking down complex tasks into simple parts and organizing them into a mix of people and machines, so that the end product was much the same as that turned out Britain's artisans. The result was a move away from the practice, experience, and complex actions of the artisan to a simpler person-machine mix of functions that could accomplish the same thing, more or less.

It is the term "more or less" that then becomes important. Early attempts to produce factory-made cloth could not replicate the quality product that came from the talents of good British artisans, so the producers of American textiles opted to supply the rough and cheap down-market segment rather than compete across the whole range of British imports. If American capitalism could not be directed at supplying the rich, as in Europe, it certainly could supply the poor and middle-class mass market. Indeed, this has become one of its hallmarks, down to the global penetration of Wal-Mart, McDonalds, and Microsoft today.

Yet, there were always pressures to improve reliability in production, quality, productivity, and price. Even if the challenge from British and Indian textiles could be met, there were other American competitors to worry about. Making improvements in machinery was a continuous process that could not occur through the slow accretion of artisanal experience. Once the Lowell factories had cut themselves off from artisanal knowledge, they had to replace it with something else.

The answer, to the Yankees, lay in experimentation and scientific discipline.[30] As early as 1829, a Harvard physician began using the word "technology" as a label for the application of scientific principles to improve production techniques and processes.[31] The early textile factories doubled as experimental technology laboratories. The Lowell machine shop, the leader in this process of improvement, was attached to a water producer and development company, the Proprietors of Locks and Canals. It was not split up and spread out among the different factories, all owned by separate, but interlocked, groups of investors.

In its first decade and more, the machine shop focused on the construction of the mill complex itself. The shop also bought innovations from inventors around the region and used trial-and-error methods to improve processes and work flow. The accession of James B. Francis to the post of chief engineer in the late 1830s came as the complex absorbed all of what power could be extracted from the volume of water passing Lowell.[32] At this point, the desire to keep expanding production called for a new approach to improvements.

Francis brought in a new form of division of labor that separated the task of research and technology development from the factory and centered it in a dedicated organization. One of his major tasks was to encourage the more efficient use of water. He quantified into a "mil" the volume and flow needed in a year to produce a fixed amount of machine movement. Each factory was then allocated enough "mils" to manage its existing production. The extra water that might be available was to remain the property of the Proprietors, who could then allocate or sell it.[33] This created a real incentive for the factories to adopt new technology. Science — in this case, hydrology[34] — had been introduced into production. More broadly, science was no longer to be the plaything of gifted amateurs like Benjamin Franklin and Thomas Jefferson, or of thinkers like Isaac Newton, but put to work.[35]

Over the next two decades, Francis led the Lowell machine shop's separate incorporation. The company introduced scientific methods and calculations into the process of extracting energy from water by introducing radical turbine designs to replace water wheels,[36] entered new technologies, such as steampower — including building a number of early railroad locomotives — and sold advanced textile machinery to erstwhile competitors. Even geography and meteorology were used, as the Boston Associates moved to organize and manage the whole Merrimack River watershed in the 1850s.

By the late 1850s, Yankee entrepreneurs were solidly behind the marriage of science and technology to industrial development.[37] The Boston Associates were, as individuals, heavily invested in supporting natural history museums, endowed chairs, and specialized programs in universities, but it was not all culture and giving back to the community that prompted them to do so. It was an easy step from these actions to the use of universities and research labs as the basis for technical improvements.[38] For instance, the Connecticut backers of the exploitation of Pennsylvania "rock oil" (as opposed to whale oil) in the 1850s took their specimens to a scientist at Yale University to be tested for flame and light.

The idea that research and technology could help even the family farm spread quickly. Senator Justin Morrill[39] from Vermont seized upon the example of Yankee Michigan in creating the land-grant Michigan Agricultural College (now Michigan State University) in 1855, and proposed funding a system of land-grant colleges across the country to improve agricultural science and technology.[40] Opposed and defeated by Southern legislators and a presidential veto in 1859, the *Morrill Act* was resurrected, and was one of the first major pieces of legislation approved by Congress after the withdrawal of secessionist States' representatives in 1860–61. The bill, signed by President Abraham Lincoln on July 2, 1862,[41] provided 30,000 acres of federal land for each Congressman and Senator in a given State, to be sold to fund the new land-grant colleges.⁴² Also passing the *Homestead Act* and the *Pacific Railroad Act*, this 37th Congress probably had a greater impact on America's future than any other. Curious how national adversity can produce momentous change.⁴³

In-house research never went away just because of the *Morrill Act*. In fact, it might be said that agricultural colleges, with their extension departments, became, at least at first, nothing more than corporate labs for the family farm. As well, the large corporations created after 1870 tended to have their own research facilities.⁴⁴ For example, Western Union created an Electricians' Department in 1873 that included a research facility; Thomas Edison created his own labs in New Jersey to develop improvements to the electrical grid; Kodak emerged from George Eastman's labs in Rochester, New York; Rensselaer Polytechnic, created as the first American technical institute in 1824⁴⁵ and General Electric, the successor to Edison's company, helped create an applied research nexus around Albany; and Alexander Graham Bell's organizational research progeny, the Bell Telephone system, developed its own facilities.

By the early 1900s, industrial research had become a specialized part of the manufacturing process, with an estimated thirty-nine industrial labs in the country.⁴⁶ Other forms of research also grew alongside the manufacturers. Arthur D. Little founded the world's first consulting firm in Boston in 1886. Its innovation was the idea of contracted technology research. Little, a Mainer, was a chemistry student at MIT when he and a partner opened a company named Griffin and Little, later becoming Arthur D. Little (ADL). They proposed to do "examinations of chemicals, drugs, paints, oil, grease," as well as water and iron and steel (the "D" in R&D). Because of his work over the succeeding years, the chemical industry was the first to embrace industrial research. In 1911, ADL also set up the first materials testing lab. GE became the first to undertake fundamental scientific research (the "R" in R&D) by recruiting an MIT professor, Willis Whitney, to run its lab in Schenectady, New York, near Albany.⁴⁷ By 1929, there were over a thousand research labs.⁴⁸ It would not be a stretch to say that, by the end of the twentieth century, manufacturing had become a specialized part of the industrial applied research process.⁴⁹

The idea of the research university also came into its own in the late nineteenth century. Following the lead of Rensselaer, MIT was created in 1861, then Johns Hopkins (1876), Clark (1889), Stanford (1891), and Chicago (1892). These institutions left the more traditional British style of education to the smaller, liberal arts colleges and adopted a German style⁵⁰ that placed central importance on research as a basis for advanced education.⁵¹ MIT became the most important source for engineers, and by 1920 was producing one-third of the masters and half of the PhDs in engineering.⁵² Much of the demand for engineers came from the railroad industry at first, but trained professionals often migrated to other industries as steel, electricity, and chemicals flourished.⁵³ By the 1890s, there were eighty-five engineering departments in American colleges and universities.[54] Americans tended to dominate the engineering field because they had a more disciplined, formal curriculum than the British and produced many more decent engineers who could go into all kinds of industry than were produced by the French elite system, while the Germans focused on training technicians in what are now the equivalent of community colleges.[55]

The US government made little use of technological research for its own purposes until the Second World War.[56] Government interest in innovation tended to arise out of warfare and, except for the Civil War, where innovations came from company ideas, America's involvement in conflicts prior to the Second World War was of short duration — even its participation in the First World War lasted only eighteen months. By contrast, for the United States the Second World War lasted about forty-six months, and this was followed by the four-decades-long Cold War, both of which incorporated technological progress into the war effort. Involvement in, and subsidization of, scientific research and technological adaptation of this research became a central part of American military strategy.[57]

Vannevar Bush, a Massachusetts Yankee, MIT professor, and later Dean of Engineering, invented radio equipment that led him and others to create what became the Raytheon Corporation in 1922. He helped develop some of the basic concepts underlying the computer.[58] Just as America's involvement in the Second World War began, he convinced President Franklin Roosevelt to create the National Defense Research Council to coordinate American scientific research funding throughout the war years.[59] As part of this effort, Bush led the organization of the Manhattan Project, the development of the atomic bomb.

Commercial spinoffs from this generation of wartime ideas and devices helped lower the average costs of follow-on products that could be used widely within the military, and gave new directions and impulses to the consumer economy. The growth of the Route 128 industrial corridor was an important part of this cycle.

The Universal Machine

From the 1820s and for the rest of his life, British inventor, mathematician, and philosopher Charles Babbage worked on designs for what he called an "analytical engine"[60], but what we today know as a computer.[61] His basic reason for doing so was to try to avoid the problems of individual calculations for tables related to all sorts of engineering, financial, and navigational problems.[62] The British government underwrote the development of a "difference engine", but it was never finished. At a time when a railroad engine could be constructed for under £1,000, the government spent £15,000 for Babbage's partly finished model.[63] His son apparently built some machines based on variations of his father's drawings, and a Swede, Edvard Scheutz, built a small working machine in 1843.[64] But it was not until 1991 that a working model of Babbage's earlier, and more detailed, "difference engine" was constructed according to the tolerances of Babbage's day — and it worked![65]

Babbage's design allowed for programming a sequence of operations. In 1842-43, Ada Lovelace, Lord Byron's daughter and a friend of Babbage, translated from the French a report by an Italian who reviewed lectures Babbage had given in Turin a couple of years earlier; her translation was published in the journal, *Scientific Memoirs*. In her notes expanding on the paper, she included what was essentially a program that the machine, if ever constructed, would use to calculate a sequence of what are called Bernoulli's numbers. It is considered the first computer program, but it is also clear that the programming concept had been Babbage's from the beginning — his "Enchantress of Number" being more a publicist than a computer scientist.[66]

A century later, after the development of adding machines and punch-card sorters,[67] the first real computer, the Z3, was built in 1941 by a German, Konrad Zuse. The Luftwaffe used it to measure wing distortion in fighter designs, but the Nazis denied Zuse's application for funding a better model as "strategically insignificant." Next came a couple of British code-breaking computers, and then, in 1944, a Harvard team under contract to the Navy's Bureau of Ships produced the first digital computer in the United States, the Mark 1.[68] At about the same time, the Air Force contracted with nearby MIT to develop some means of providing a computer-controlled flight simulator for its Whirlwind Project. An analog computer was tried first, but it failed to do the job, so a digital[69] computer was devised in 1947 after team members saw the University of Pennsylvania's ENIAC.[70] Whirlwind expanded into developing a computer-coordinated tracking system for approaching hostile aircraft.

The Harvard project evolved eventually into International Business Machines' (IBM's) "big iron" models of industrial computers, while the MIT project evolved into a computer-controlled radar warning system, the Semi Automatic Ground Environment (SAGE).[71] This latter project gained a decades-long lease on life when the Soviet Union exploded atomic and hydrogen bombs and the Air Force needed to detect hostile bombers early so as to position fighters to destroy them. The local spinoff from SAGE researchers leaving to found their own start-ups led to the Route 128 minicomputer industry in the 1960s and 1970s.[72] But the Route 128 phenomenon was not simply about computers. Other researchers started or went to companies and labs that were involved in other aspects of Cold War research as well as consumer products, such as the Apollo guidance systems, instant Polaroid pictures,[73] microwave ovens,[74] hearing aids, accounting software, biochemicals, and televisions.

Venturing Outward

To take products from the research lab into the market, new sources of finance had to be developed, since neither banks nor other financial bodies were willing or able to absorb the risks of this step. Thus, in Boston in 1946, the first modern venture capital firm was created, called American Research and Development Corporation (ARDC). The brainchild of MIT president Karl Compton and Harvard business professor General Georges Doriot,[75] ARDC's aim was to earn profits for its investors by financing the commercialization of innovations that were coming out of the MIT and Harvard labs after the end of the war. Venture capital was a way to imitate the banks and other investors of nineteenth-century Boston that created America's Industrial Revolution.[76] In 1946, banks could make loans to companies, but were prevented from owning equity in them. Since the debt-to-capital ratio is a common indicator of a company's financial health, making more loans available to undercapitalized entrepreneurs was not helpful, so somebody had to be able to provide risk capital. Since ADRC, as a financial innovator, had difficulty finding institutional investors, for most of its existence it was dependent on individuals' commitments.[77]

Even after start-up capital was found, these new companies discovered that loan funding was difficult to obtain. The Bank of Boston started to lend some money to these new companies in the late 1950s, based on their receivables and some federal contracts.[78] The experience of both ARDC and the Bank of Boston gradually encouraged others to begin financing technology companies. ARDC, in fact, preceded the first California venture capital fund by twelve years.[79] It helped finance between 150 and 200 companies, and got its greatest return out of a $70,000 investment in Digital Equipment Corporation in 1957, which was valued at $100 million when the company went public in 1968 and $355 million by 1972.[80] Doriot, ADRC's co-founder, was recognized as "the father of American venture capital," and after he retired in 1971 ADRC was merged with the Rhode Island conglomerate Textron.[81] Other Boston venture capital companies had begun to form by this time, and eventually their reach would become national, with the firms along Route 128 firms just one part of their considerations.

The business history of the Route 128 area to the start of the twenty-first century can be divided into three phases. From the end of the Second World War to the wind-down of the Vietnam War in the early 1970s, its growth was based on spinoffs from Cold War government research.[82] Then, the rise of companies based on minicomputers acted as the second phase, known at the time as the "Massachusetts Miracle."[83] This lasted until the late 1980s, when the microcomputer, or personal computer (PC), replaced the minicomputer. Again, after faltering, the area recovered with a diverse, technology-based economy that included some biotechnology, computer-aided design software, and sophisticated media products, similar to the diversity that had marked the start of the 128 phenomenon.

Despite these changes, Massachusetts has maintained its rank at or near the top of technology-based rankings. As late as 2008, a Milken Foundation survey ranked it first, ahead of California, in an index of science and technology intensity, similar to the results of a 1979 survey.[84] Part of this continued dominance is in the State's tradition of encouraging higher education and making its graduates welcome. In 1980, the proportion of postsecondary degree holders in the population was a quarter higher in New England than elsewhere, a gap that persisted sixteen years later.[85]

Mainframes to Minis

The computer came out of the military's needs for faster and more accurate computation of the positions of ships and aircraft and vectors of movement during the Second World War. By the time developers came up with methods to calculate these positions automatically, the war was over. IBM gradually adopted Harvard's development of the Mark I as the basis for its mainframe computers in the 1950s, while the MIT version went on to form the basis for the minicomputer industry around Route 128. Both were driven by the military's continuing needs and funds, and just like the marriage of military and commercial interests in Springfield, Massachusetts, in the 1820s and 30s that eventually led to mass production, this relationship produced the computer revolution that has once again transformed how things, including information, get produced and used.

IBM was created as a result of a dizzying variety of mergers and buyouts starting in the 1880s, when the first time clocks were marketed and used to record employee attendance, among other things. These clocks used keypunch equipment that created holes in cards, allowing data to be counted and sorted by category.[86] The rise of IBM is commonly associated with Thomas J. Watson and his son, who led the company through much of the middle decades of the twentieth century. But the company effectively was created by Maine's Charles Ranlett Flint, who gained a reputation as the "Father of Trusts."[87] Trusts, as John D. Rockefeller found, were a useful way to get around state jurisdiction over nearly all corporations. The shares of different companies were entrusted to a holding company that then allocated capital investment, sales, and territories to its various parts. They also served as vehicles for creating monopolies and monopoly pricing, and soon ran afoul of federal legislators with the passage of various anti-trust acts, the most famous of which was the first: the *Sherman Antitrust Act*.

In 1911, Flint brought together a number of companies to form the Computing-Tabulating-Recording Company, which was renamed IBM in 1924. Flint remained a member of the board until 1930, when he was eighty years old. In 1914, he employed a convicted felon, Thomas J. Watson, who had become too aggressive in driving out used-cash-register-reseller companies in order to expand the sales of his employer, National Cash Register, and had been convicted of blackmail and fired. Watson built IBM into one of America's largest businesses and, seeing the computer as just another business machine, took the company into the development, production, and operation of the new tool. The IBM labs produced a number of prestigious researchers, including three Nobel Prize winners. They also inadvertently spawned Microsoft, one of the the world's largest computer products and services companies.

The earliest computers were based on vacuum tubes. They generated lots of heat, were very large, covering the area of large rooms, and were cantankerous and prone to breakdown. Some tube might burn out every few hours and there were always "bugs" to fix.[88] While IBM continued to develop computers for military use, they also sought to introduce them into the country's largest corporations as machines that could do large-scale computation, keep complex accounts, and provide useful sales indicators. The company rented the machines, its technicians operated them, and IBM provided upgrades and new services. It was not unusual to hear of "big iron" being attended by its "high priests" and "acolytes" in the 1950s and 1960s.[89]

Two things changed this scene. The first was the development of the transistor, an alternative to the vacuum tube, at the Bell Labs in New Jersey between 1945 and 1947.[90] Transistors are "solid-state" material, not glass tubes enclosing a vacuum, and they require very small amounts of power to switch currents on and off. Besides this saving of electricity, they also produce very little waste heat. Further, as the technology progressed, they could be made smaller and smaller, and were mounted on wafers of silicon called "chips." One of the inventors[91] of the transistor was William Shockley, who, in one of those scientific feuds, was left off the patent application. Shockley went on to publish a major scientific book on transistor theory in 1950, then applied for a patent for a different and more practical transistor in 1951. He left Bell Labs in 1954, and eventually found himself a financial backer in California's Bay Area. By then, transistors were becoming the workhorses of the whole electronic revolution. They had been applied to a number of products, including hearing aids[92] and other devices where size and weight were important factors. In 1955, MIT, with military funding and IBM help, produced the first transistorized computer, replacing one that used 55,000 tubes.[93] At first, the Boston area, including Route 128, was the leading producer of transistors, but the focus of the companies there was on application products — such as a minicomputer built by Control Data that contained 25,000 transistors — not on production processes, and when the latter changed and the integrated circuit was devised, Massachusetts' lead evaporated.[94]

The second thing to change computing followed from Shockley's lab. A number of his best people, frustrated by his autocratic and erratic management style, left the lab in 1957 and found financing and a home in Fairchild Instruments. One of these, Robert Noyes, developed a transistor embedded in a square, or "chip," of silicon known as an integrated circuit. Noyes was anticipated by six months by Jack Kilby of Texas Industries, but Kilby's chip was made of germanium, and Noyes's silicon-based chip proved to be more practical in terms of ease of manufacture and cost.[95] By 1963, the cost per integrated circuit in a chip had dropped to a fortieth of what it had been three years earlier, and integrated circuit chipboards were being installed in Minuteman missiles and other military hardware. Commercial use was also beginning.

Noyes's eventual partner in founding Intel, the dominant chip manufacturer, was Gordon Moore, who, in a 1965 paper, stated that the number of transistors that could be built into a chip would continue to double every two years.[96] Moore's Law, as it became known, still has not been violated almost five decades later, and the number has gone from 2,300 circuits in 1970 to over a billion today.[97] Depending how you see it, the cost of computing either gets cut by one-half every two years or the capability of a chip to handle data doubles every two years. What this means is that the Commodore 64 I bought in California in 1981 had the same computing power as had been installed in the Apollo moon landers of 1969–70,[98] and its cost then was the same — a bit over $1,000, unadjusted for inflation — as a machine today that is almost 100,000 times more powerful. Where I could entertain myself with "Pong" in 1981, I can now download and play feature-length films. The scary part is that Moore's Law continues to grind along, unbroken, and, when combined with the Internet, continues to cause disruptions in a variety of industries, including banking, insurance, the media, and travel and real estate agencies, to name but a few.[99]

The development of the transistor opened up a niche for computing that was different from that dominated by IBM. Smaller computers that required less maintenance and that could be operated by an in-house technology group were potentially available to companies for less-demanding jobs and at lower cost, and were first used for machinery control and lab instrumentation. The microcomputer did not so much take business away from IBM's "big iron" as it opened up new corporate activities to computerize, including accounting for smaller businesses, word processing, and document management.[100]

Much of the technological fame of Route 128 relates to the development of the minicomputer by Ken Olsen and DEC in the late 1950s and 1960s.[101] DEC's first facility was located in Maynard, about ten miles west of Route 128 and five miles north of US 20. By 1960, it had developed its first minicomputer, priced at $120,000, which included a keyboard and video screen that allowed the operator to interface with the machine.[102] During the early 1960s, only governments, universities, and technical companies could afford minicomputers, but as their prices came down, sales began to expand into more markets. DEC's PDP8, introduced in 1967, was the first computer priced under $20,000. By 1980, DEC was second in computer sales to IBM, and minicomputers accounted for 34 percent of the computer market.[103]

The main growth in the minicomputer industry and other technology businesses continued to locate along Route 128 because it offered good transportation to Boston and New York and allowed employees to live in the pleasant suburbs west of the city. Some of the surrounding towns — in particular, the mill towns — were rehabilitated by new companies looking for near, but less-in-demand locations. For example, Wang,[104] a company that made dedicated word-processing machines, moved to Lowell in 1976 and occupied some of the mill buildings that had been abandoned when the textile industry moved to the South.[105] Other mill buildings were turned into an urban National Park,[106] industrial malls, or shopping areas. Even as military purchases tailed off after 1970, the area continued to prosper from the growth of the minicomputer industry.[107]

By then, the integrated circuit had been developed and the effects of Moore's Law were beginning to make themselves felt.[108] By 1985, the PC was coming into its own and, while "big iron" was largely protected by the complexity of the computational tasks it handled, the minicomputer industry was not.[109] Although a number of minicomputer engineers had come up with designs for PCs, their superiors ignored the opportunities presented, keeping their collective eyes on chasing IBM up the computational ladder.[110] It was a big mistake. DEC faded in 1989,[111] and in 1998 was bought by PC maker Compaq. Wang reached sales of $3 billion in 1988[112] before rapidly declining into bankruptcy four years later. Lotus's spreadsheet/accounting software was hit by Microsoft's Excel and, in 1995, reeling from reverses, it was taken over by IBM.[113]

The spread of the PC thus devastated companies such as DEC, Wang, and even, to some extent, IBM. Instead of a neatly structured market of big, medium, and little machines, the profit potential in ever more powerful integrated circuit chips and related software led to their full exploitation, so that the PC with chip power and complex software could begin to do what a mini or even a mainframe could do. What Moore's Law meant to corporate equipment manufacturers was that, every two years or so, the ability of a PC to do the work of a mini- or mainframe got closer by a leap of magnitude. Although the software took a while longer to develop, this was still too short a time cycle for the makers of bigger machines to develop, sell, and amortize them.[114]

The development of more powerful chips also increased their versatility. They no longer had to be enclosed in a computing machine, but could be used to aim missiles and control complex manufacturing processes. The pressure to innovate using chip technology became widespread in the Western economies. This had unfortunate repercussions in the Soviet Union, which had no large internal market for this technology and no integral connections with Western developers.

Conclusion

The separation of innovation and artisanal knowledge came out of the lack of artisans to fill the growing American factories in the early 1800s. In fact, it could be said that the earliest factories were designed with this shortage in mind. Beyond that separation, the use of formal knowledge and scientific technique, which rested on the definition of a characteristic to be tested and then a search for repeatable results,[115] led to the reproduction of centers of such activity in both corporations and universities. This, in turn, was married to human needs and market realities, so that the process of innovation became subject to the same process of testing as the innovations themselves. Innovation still had some of its random aspect, but a lot of improvements were planned changes.[116] Having a mass market available to amortize the investment in the innovation and to support public providers of expertise in the university system created a society dependent for its prosperity, though uneasily, on reason and science. This system is now being "imposed" on the rest of the world,[117] which shows considerably more unease about it than did Yankee America.[118]

Route 128 was a pioneer in merging the use of the automobile with industrial development. It did not become a huge factory at the end of the trolley line, but the center of a diverse network of businesses whose employees are, in large part, professionals and educated workers. Their products are based on the practical application of research in university and corporate labs, to which these companies have a great stake in staying close. Just as 150 years ago, many of the fundamental innovations have come from a fruitful partnership with a military whose success is based on the ability to produce weapons of war that are superior both technically and in terms of logistics and delivery to those of any enemy. Similar technical-industrial locations have emerged in all parts of the country since the "glory days" of Route 128, especially California's Silicon Valley.

Concord, Massachusetts: Walden Pond and the Idea of Wilderness

In Wildness is the preservation of the World.
— Henry David Thoreau[1]

The eyes do not look as if they read books, but as if they gazed upon a Cause.
— Author Owen Wister, on Gifford Pinchot[2]

Of what good are forty freedoms without a blank spot on the map?
— Aldo Leopold[3]

Walden Pond and Wilderness

Wallace Stegner was a well-known academic and writer on the West. He provides a succinct description of the process behind a Yankee idea that starts at a place just off US 20, a little west and north of Boston, and wends its way across

America to its southwest, where the story takes a new twist away from the subject at hand, sort of like the gopher tracks mentioned below:

> The tracing of ideas is a guessing game. We can't tell who first had an idea; we can only tell who first had it influentially, who formulated it in a striking way and left it in some form, poem or equation or picture, that others could stumble upon with the shock of recognition. The radical ideas that have been changing our attitudes toward our Earth habitat have been around forever. Only if they begin to win substantial public approval and give visible effects do they achieve a plain and even predictable curve of development. But once they reach that stage, they are as easy to trace as a gopher in a spring lawn.

About five miles north of US 20, in the western suburbs of Boston, lies Walden Pond. To those used to living near a lake, the sixty-one-acre Walden is definitely a "pond," though a Westerner might confuse the terms. It is also somewhat of an oddity in that it is a "kettle hole," with no surface source or exit.[4] Walden Pond is otherwise unexceptional. It was used for fishing and swimming as early as the 1600s, and its shores have been logged for firewood and lumber ever since.[5] In 1843, the railroad from Boston though Concord to Fitchburg touched its western shore.[6] Later in that century, the Concord town dump was established nearby, and the railroad opened a picnicking and swimming facility next to the tracks.[7] In the twentieth century, a trailer park was located nearby, and suburban office facilities and subdivisions were proposed on its north and east sides.

Walden Pond thus has not been a promising wilderness retreat at any time in its modern history. It became a symbol of the value of nature, however, because in the mid-1840s, Henry Thoreau, a taciturn, Harvard-educated misfit, lived next to its northern shore for two years, two months, and two days,[8] looking for solitude to concentrate on his writing. While there, he finished a book on his travels on nearby rivers and a draft of *Walden*, a book that loosely touched on his local experiences. The book's influence grew only after its author was dead, and its message was reinterpreted away from transcendental philosophy into a reflection on post–Civil War America's need to be in touch with wilderness and respectful of the natural environment.

The first American literary pilgrimage — a tour to visit a place where a famous writer lived and worked — was to Concord and Walden Pond, and the tours have continued there ever since.[9] A memorial to Thoreau's stay was erected at the site of his cabin in the early 1870s, a decade after he died. Around 1890, a trio of enthusiasts launched a "Thoreau revival" that firmly re-established his name and popularity in America and Europe.[10] In 1922, some local families donated about eighty acres around the Pond to the Commonwealth of Massachusetts as a reserve for "preserving the Walden of Emerson and Thoreau, its shores and nearby woodlands."[11] In 1941, enthusiasts created the Thoreau Society, the oldest professional society devoted to the legacy of an American writer, and in 1945 the site of the cabin was determined and excavated by a local amateur archaeologist.[12] Despite this activity, the Pond was still used for picnicking and organized swimming events into the 1950s. The trustees enlarged the parking lot and cut a road down to the shore to allow more swimming in 1957 — not quite what had been intended by the original donation. This led to a media outcry, and in 1960 the Massachusetts Supreme Court decided that the county commissioners had not acted according to the deed of gift.[13] The fight led to a new Thoreau revival, including support by John F. Kennedy, who had just been elected President. Finally, in early 1975, the Walden Pond reservation was turned over to the State, and work began to restore it in anticipation of bicentennial tourism to the Concord area.

Don Henley, the lead singer of the country-rock group The Eagles, has spent much of the past three decades promoting and supporting the Walden Woods Project, which is devoted to conserving and restoring land in the Walden Pond area.[14] In 1998, it opened a research and educational center in cooperation with the Thoreau Society and the Thoreau Institute at Walden Woods to promote environmental and humanities programs and to house Thoreau memorabilia and archives.[15]

People come to the Pond in such numbers today that access has had to be controlled to assure future generations that something will be left of the milieu. I had to turn back from Walden Pond on my first visit as the parking lot at the state park was full at 11 a.m. Later, I was luckier. Visitors come to pay their respects to a variety of ideas that Thoreau and Walden Pond symbolize. As Wallace and Page Stegner put it,

> Walden is the place for Thoreau's monument as surely as Washington is the place for the temples we have erected to Jefferson and Lincoln. This little lake within sight of the tracks and sound of the train whistle should be part of the American iconography. But we need no marble columns. The pond itself, and the creatures that find life in it, and the woods that surround it, are the most fitting monument for the man who took so much from them, and gave it back in unforgettable terms to his countrymen.[16]

If it weren't a kind of American shrine, Walden Pond today would be, at best, a little country park with holly banks and a covering of forest all around. To the west is a double set of railroad tracks and to the north, hidden by a screen of trees, is a very busy highway. To the east is an old local road that leads past the entrance to the park that surrounds Walden Pond and memorializes Thoreau's time there. Less than a couple of miles distant to the north is the town of Concord, bigger than it was in Thoreau's day, filled with Revolutionary War monuments and other tourist attractions. Surrounding it and Walden Pond are the extended suburbs of Boston.

Henry David Thoreau

Henry David Thoreau was born David Henry Thoreau in 1817, but he changed his name because he didn't like to be called David. The family name and ancestors came from the island of Jersey, in the English Channel. Thoreau grew up in Concord, where his father made pencils and his mother took in boarders,[17] and, after graduating from Harvard, he tried his hand at teaching, worked as a handyman, and made pencils in the family business. He was a reputable land surveyor[18] and a good fisherman. He sat houses and babies for his friend Emerson, who despaired of Thoreau's ever making anything of himself.[19] Emerson had published his famous Transcendentalist essay, *Nature*, in 1836, establishing himself as a leading American writer and philosopher. Thoreau read it at Harvard, and it subsequently affected his life and ideas.

Thoreau was an idler and an eccentric, needing a four-hour walk through the environs of Concord each day to get himself going.[20] He rarely left the area, commenting to those who suggested he travel that he had "traveled extensively in Concord." Instead, his travels took him inward, to contemplation and meditation. He never married[21] or even seemed to have had any kind of love affair. He did not vote or attend church. He was not a joiner, telling Transcendentalists who had urged him to become a part of their communal experiment at Brook Farm that he "would [rather] keep a bachelor's hall in hell than go to board in heaven."[22] Yet he greatly valued friendships, and played with children at every opportunity. He died relatively young in 1861 of tuberculosis, a disease common in his family. In such a life there is nothing particularly exceptional.[23]

What is exceptional is *Walden* and its growing public reception in the decades after Thoreau's death. The idea of a woodland retreat had been in his mind ever since he spent six weeks in the woods with his Harvard roommate in 1837. Thoreau's friend and, later, biographer Ellery Channing pointed out that the woods around Walden Pond might make a good locale for writing, and when Emerson bought a parcel on the Pond in 1844, the possibility of locating there began to make sense.[24] The immediate area around "Walden woods" was home to freed slaves, Irish, loners and drunkards, who were not wanted inside the settled town.[25] John Muir remarked on a visit to the Pond that "we walked through the woods to Walden Pond....It is only about one and a half or two miles from Concord, a mere saunter, and how people should regard Thoreau as a hermit on account of his delightful little stay here I cannot guess."[26]

It was not wilderness.

Thoreau's stay at Walden Pond was in part designed to bring him closer to nature, although, as a Yankee of his time, he measured the size and depth of the

Pond and studied the birds and animals that used it. He wanted solitude to think and to write, not isolation. For the next two years he would write in his cabin, which Channing referred to as "a wooden inkstand." There he wrote the first draft of *Walden*, but his perfectionism forced him to make six revisions, and the book was not published until 1854.[27] Despite his wish for solitude, however, he did not lack for friends dropping in, or for meals in town; the local women's abolitionist group even convened its annual meeting in the cabin in 1846.[28] After Thoreau left, he sold the cabin to a local farmer, who took it away and used it to store corn.[29]

Walden was not a mere outpouring of observations taken from his extensive journals. As early as 1838, he had written in his journal that he would like to read Walden Pond's first page just as it had come fresh from the press of Creation.[30] This was his attempt to write that "page." So, by the time its seventh version was published in 1854, things had crept in that happened both before and after his stay in the woods. As well, he left out of his account many of the people who came to help him work and to discuss his writing. Historical accuracy and social interaction could only distract from the timelessness of his work. Instead, *Walden* is a compilation of pieces of natural history, pond surveys, meditations, and short journeys from the area that are not necessarily continuous in time. The effect is to heighten the senses in the reader's local environment and to instill a kind of wonder at the complexity and beauty of the wild. As Thoreau puts it, one should avoid "the folly of attempting to go away from *here*, when the constant endeavor should be to get nearer and nearer *here*."[31]

Walden, like his *A Week on the Concord and Merrimack Rivers* (1849), was not a great commercial success, even though it gained a number of favorable reviews in America and Europe. Instead, the book gained fame only because Emerson and others who were taken with Nature promoted it and kept its ideas alive.[32] Only after a new edition was published in the 1880s did Thoreau's stature as a philosopher of Nature begin to grow. The quest by many Victorian-era urbanites in both Britain and America to get back to the land might have done a lot to create the aura around Thoreau and Walden.[33] A biologist once wrote of Thoreau that "[a]n amateur naturalist perched in a toy house on the edge of a ravaged woodlot became the founding saint of the conservation movement."[34] Another scholar has noted that "*Walden* is arguably the most widely translated and available book by an American author."[35] The writings of Muir, Leopold, and Edward Abbey all reflect Thoreauvian influence. Others felt they could find inspiration imitating Thoreau's writing experience: Jack Kerouac, thinking himself a modern-day Thoreau, rented a little house in the foothills near Denver in 1949.[36]

Walden and its author spun off a lot of ideas that resonated with later generations. There is the idea of solitary man living off the land. There is the idea of the ecological connection between humanity and the natural order and the Divine. Thoreau's ideas on plant succession led to the concept of ecology. There is the idea of the wilderness as the purifier of the American soul — although this, in truth, has more to do with others, such as James Fenimore Cooper, than with Thoreau, but Thoreau gets the credit. His suggestions about the need to put aside forested land led to ideas about city and state parks, asking, for example, "why should not we…have our national preserves…in which the bear and panther, and some of the hunter race, may still exist?" and stating that each Massachusetts town "should have a park, or rather a primitive forest, of five hundred or a thousand acres."[37]

Wilderness is the theme here. By Thoreau's time, the wilderness had been pushed well to the west and north of New England. His trips to the backwoods of Maine were his primary contact with real wilderness, as opposed to the "gardens of Concord." He found the rugged and windblown slopes of Maine's Mt. Katahdin to be "made out of Chaos and Old Night." He considered that "[I]t is vain to dream of a wildness distant from ourselves. There is none such. It is the bog in our brain and bowels, the primitive vigor of Nature in us, that inspires that dream."[38] Thoreau took to heart Emerson's statement that "the whole of nature is a metaphor of the human mind."[39] He had no desire to leave Concord for the West, even though he noted that his compass needle always settled in that direction.[40] As other outdoorspeople have found since then, the wilderness constitutes a good experience and an antidote to civilization, but it is not a place to be lived in permanently.[41] Some of Thoreau's contemporaries, such as Jedediah Smith and Kit Carson, were "mountain men," comfortable living for months and years in the wilderness, but not Thoreau. Even so, Thoreau began a lecture in 1851 in Concord by saying, "I wish to speak a word for nature, for absolute freedom and wildness.…Let me live where I will, on this side the city, on that the wilderness, and ever I am leaving the city more and more, and withdrawing into the wilderness."[42]

The Yankee relationship to wilderness is a long and complex one.[43] When the first settlers, Pilgrim and Puritan, arrived, their leaders warned people not to go off into the wilderness and be swallowed up by it, if not physically, then spiritually. In truth, the local Indians, decimated by diseases brought a few years before by other Europeans, left a forest whose brush had been regularly burned over and which had numerous clearings and meadows. It was hardly pristine wilderness. Even so, the "wild" was seen as something that harbored dark forces, and therefore had to be overcome. At the same time, because it represented freedom from cultural constraints, the early Puritan preachers and leaders wanted the wilderness pushed back in an orderly fashion through the formation of new and adjoining towns, not by individual pioneers searching out good land on their own.[44] To the hunter and trapper, the wilderness was a place where riches could be found and taken back to civilization. To townspeople and farmers, it was something to be turned into a means for a better life, despite its threatening presence along the frontier. Wilderness was there to be settled and exploited.

Cooper and the Romantics

Thoreau, like Americans generally in the first half of the nineteenth century, was greatly influenced by Romanticism, a European movement that emphasized the uplifting qualities of being connected with nature.[45] Romanticism, in turn, grew out of the ideas of eighteenth-century French novelist and philosopher Jean-Jacques Rousseau, who wrote about how social conventions tended to corrupt the natural goodness of the human character. He derived this approach to civilization from the writings of French missionaries in America, who were asking for donations to convert the "noble savages" around them in New France, since Christianity was the only thing these pure souls lacked.[46]

Romanticism was translated back into American terms by James Fenimore Cooper, especially in his "Leatherstocking" novels, which featured the heroic frontiersman Nathaniel (Natty) Bumppo, nicknamed "Hawkeye." Indeed, George Eliot, in an early review of *Walden*, saw Thoreau through European eyes as Leatherstocking-like — a wise "stoic of the woods."[47] The character Natty supposedly was based loosely on the story of Daniel Boone,[48] who led Scots-Irish pioneers over the Cumberland Gap from Virginia into Kentucky and eventually moved on to Missouri in the early 1800s. Cooper's character, however, was a "Yorker," born near the Hudson River and raised by Indians, and who traversed the country around the Hudson and Mohawk rivers during the second half of the 1700s, helping soldiers and local settlers in various conflicts, then moving on when the frontier got too settled, finally coming to rest in the trans-Mississippi prairies.

Cooper arguably was the first American novelist to win international renown. He was seen in Europe as the Romantic interpreter of the American frontier experience. He was born and spent many years in Cooperstown, New York, the town his father, William, had founded. The elder Cooper was a New Jersey Quaker who had profited from the losses of Loyalist land speculators in the Revolution, and Cooperstown was the center of a land concession that William gradually sold, mostly to Yankees.

As James grew up, he began to associate with the Hudson River elite, married into a wealthy family, and carried on feuds and litigation with Cooperstown locals.[49] His experiences colored his novels, which tend to revolve around his father's development of the Cooperstown area, the values of the wilderness, and the virtues of Natty Bumppo's way of life.[50] He also idealizes the dashing Southern cavalier/frontiersman over the methodical and practical Yankee farm developer, while recognizing that the frontiersman was doomed to be overwhelmed by the farmers — familiar roles played out in later novels and then films about the West.[51] A century later, Western artist Charles Marion Russell expressed a similar sentiment: "I have been called a pioneer. In my book, a pioneer is a man who comes to virgin country, traps all the fur, kills off all the wild meat, cuts down all the trees, grazes off the grass, plows the roots up, and strings ten million miles of wire. A pioneer destroys things and calls it civilization."[52]

Cooper's influence extended beyond prose. His friend, William Cullen Bryant, wrote Romantic poetry. Thomas Cole, an important contributor to the Hudson School of painting, painted at least one scene that figured in Natty's fictional reminiscences about the Catskill Mountains. Historian Francis Parkman's focus and subject matter are heavily influenced by Cooper's tales. But the writer in a garden gave the term "wilderness" a very different implication than did the pioneer in the forest.[53]

Ralph Waldo Emerson adapted some of the Romantics' ideas to Transcendental philosophy, and considered the Romantic poet William Wordsworth to be "a seminal genius of the modern age."[54] Emerson and the Transcendentalists took the Romantic connection between wilderness and freedom and used it to upend the early Puritan idea that the wilderness, and the absolute "freedom" it provided, was a threat to godly people.[55] By the time of the Civil War, however, the Romantic notion of wilderness had largely disappeared from America, although one could still find people living on the edge of "trackless wilderness" farther afield in Canada and Alaska. But that wilderness was left wild simply because there was no other use for it, unlike the wooded but fertile lands south of the Great Lakes.[56] Elsewhere, the discovery of gold in California in 1849 and emigration to Oregon would populate the West Coast, while the completing of the first transcontinental railroad in 1869, followed by others along the northern and southern borders, would soon begin to fill in the middle. By 1890, census officials were declaring the coming end of the frontier, based on population per square mile. The American wilderness had been boxed in, drawn, and quartered.

Even so, the idea that unpopulated and unexploited areas might have some connection to American values continued to resonate. As the country filled in, three ideas came to take the place of the old wilderness: the urban park, national (and later state and county) parks, and designated wilderness areas. Natty Bumppo would not fit well into these, but he was the ideal-type of a day gone by.

Frederick Law Olmstead

The most influential developer of urban parks was Frederick Law Olmstead. Born in 1822 in Hartford, Connecticut, the son of a successful Yankee merchant,[57] he went through a number of jobs until, in 1849, his father helped him to buy a farm on Staten Island, New York. The following year he went to England to look at public gardens, where he was taken by the vision of Sir Joseph Paxton, who designed Liverpool's "People's Park."[58] On his return to New York, Olmstead wrote *Walks and Talks of an American Farmer in England* (1852), which brought him to the attention of the editor of what would become the *New York Times*, who commissioned him to do a series of articles on travels in the antebellum South.[59]

Around the same time, Andrew Jackson Downing, publisher of *The Horticulturist*, introduced Olmstead to Calvert Vaux, a French landscape planner. Downing had proposed plans for the Capitol grounds in Washington, DC, and a large park in New York City, but tragically drowned in a steamboat accident before anything came of them. His work proved invaluable to Vaux and Olmstead later.

In 1857, Olmstead, armed with many letters of recommendation, applied for and got the job as superintendent of the still-undeveloped Central Park in New York City. When Vaux showed him a rough plan of the park, he and Olmstead began to work as partners on it, an association that would last for the next fifteen years.[60] Their proposal was approved by the city's Board of Commissioners in 1858, and Olmstead was appointed architect in chief for the park. The approach of Olmstead and Vaux was to provide a wooded green space where people could escape the stresses of urban life.

Between 1857 and 1895, when he retired, Olmstead, first with Vaux and later with his own firm, was involved in the planning and construction of a large number of parks in such cities as Atlanta, Boston, Brooklyn, Buffalo, Chicago, Detroit, Louisville, Milwaukee, and Montreal.[61] Other commissions included the Vanderbilts' Biltmore estate near Asheville, North Carolina, the Chicago suburb of Riverside, the campuses of Stanford University and the University of Chicago, and the Capitol grounds in Washington, DC. He designed the Niagara Falls Reserve for the New York state government, and served on a commission reviewing the area that would become Yosemite National Park.[62] The climax of his career was to lead the designing of the site for the 1893 Chicago World's Fair, the fabled "White City."[63]

Olmstead expressed his ideas about urban parks as a variation on a landscape painting that helped "in a remedial way to enable men to better resist the harmful influences of ordinary town life and to recover what they lose from them."[64] They were crafted to appear as a kind of local wilderness and to meet the ideal of the pastoral retreat into a kind of Arcadia. Generally, the land city politicians made available was marshy or hilly and unsuitable for building, but these qualities only gave Olmstead more latitude for design. Even the illusion of wilderness had to be constructed, as Olmstead noted in his plan for Biltmore, where there was "not a single circumstance that can be turned to account in gaining any desirable local character, picturesqueness...or geniality." The grounds would have to be fabricated "out of the whole cloth." He wanted guests to feel "the sensation of passing through the remote depths of the natural forest" as they took the three-mile drive from the estate's entrance to the mansion itself, but he left the implementation of this idea to his forester *protégé*, Gifford Pinchot, to execute.[65]

Olmstead's projects were not all focused on the needs and perceptions of the elite; rather, he looked to attract all parts of society, as long as they behaved properly. Thus, Central Park became an outing for large and varied groups of New Yorkers, and his college campuses were appreciated by students, not aristocrats.

The 1893 Chicago Fair attracted a significant proportion of the American population of its day, and Boston's Fenway Park baseball stadium was built at the entrance to his transformed "Fens" swampland. His home and grounds lie at the upper end of Fenway Park in Brookline, Massachusetts, and are a National Monument.

Olmstead brought a constructed version of nature to city dwellers as urban areas exploded with new businesses and immigrant populations. With his interest in designing "natural" environments, he thus represents a middle ground between his contemporary Muir, an advocate of nature as a kind of "temple," and the later Pinchot and Leopold, who grappled with methods to manage the natural environment rationally. The debate is still unresolved.[66]

John Muir

John Muir was born in Dunbar, Scotland, in 1838. When he was eleven, his family emigrated to Wisconsin, settling near Portage, about fifty miles north of Madison. Like many other religious dissidents, including the first Pilgrims and Puritans, Muir's father, a prosperous grain dealer, preferred to go and live along the frontier, where he was free to practice his faith with others of similar belief. Daniel Muir had joined the Disciples of Christ, a church that had evolved out of the great Cane Ridge Revival of 1800 in Kentucky. The church's founders, Barton Stone and Alexander Campbell, much as the Puritans before them, saw the frontier as a place where people could create the Kingdom of God that would predate the Millennial return of Christ. The Disciples of Christ became an influential group in Kentucky and Tennessee,[67] and also had communities in Canada, to which the Muirs originally were attracted. They allowed themselves to be redirected, however, by Disciples of Christ colleagues to the church's Wisconsin centers, where Canadian Scots and Yankees from New York and Ohio were joining the migration to the newly opened lands of what was in 1849 the second-fastest-growing area next to Gold Rush California. The Muir children found themselves in a wilderness unlike their previous home in long-settled Scotland, and they pumped the Yankees whom their father hired for whatever they could learn about the local plants and animals.[68]

Unlike his father, who saw Christian texts as the sole core of necessary learning, Muir also developed an interest in things of the mind. He read of the travels of Mungo Park, the Scots explorer of West Africa, and the German Alexander von Humboldt's explorations in South America. He became fascinated by mathematics, and found he had an aptitude for inventing devices and clockworks. He decided he wanted to become a machinist, and was encouraged to enter some of his devices in the Wisconsin State Fair, where they proved to be one of its hits. Despite his lack of formal education, Muir's talent led him to become admitted to the scientific program at the University of Wisconsin in Madison in 1861. He became interested in botany and earth sciences, but left Wisconsin during the Civil War for Canada to avoid the Union army draft. He spent two years journeying through the Canadian Great Lakes country, working occasionally as an industrial technician.

In 1866, after the end of the war, Muir embarked on a "thousand-mile walk" through the southern Appalachians, studying their botany and enjoying the mountain life and scenery. Then, in 1868, he arrived in San Francisco after illness had forced him to give up his plans to imitate von Humboldt and extend his walk to South America. He made his way to Yosemite and fell in love with the Sierra Nevada range. Due in part to Olmstead's 1865 work on a California commission to delineate a state reserve,[69] Yosemite had been spared serious development, but the land in the valley had been cleared long ago by generations of Indians living there, and the great Sequoia trees had been partially logged by earlier settlers. Olmstead's report to the state legislature advocated that "natural scenes of an impressive character," such as Yosemite, should not become private property.[70] Muir spent some summer seasons in the mountains, first as a sheepherder and then as a sawmill operator, where he made a number of mechanical improvements. Meanwhile he developed theories about how the mountains were shaped, giving preference to glacial action similar to what he had seen around the Great Lakes.

At the behest of an acquaintance who was a former professor of Muir's, Ralph Waldo Emerson sought out Muir on his trip west in 1871.[71] They met at the sawmill where Muir was working. "It was amusing," he recounted, "to see the old philosopher climb the hen-ladder into the 6 × 8 room" that made up his cabin attached to the sawmill, where they had a "fine, clear talk."[72] Emerson, however, turned down Muir's suggestion that the two camp out overnight in Yosemite.[73]

In 1874, a paper of Muir's on the ecology of the mountain range was published by the prestigious American Association for the Advancement of Science, and he began to gain a reputation in California and elsewhere. Muir's books, unlike Thoreau's *Walden*, sold well all his life.[74] His articles in *The Century* magazine on behalf of parkland in the Oregon woods and his reciting the truths that wild nature could bring to Americans resonated with those who felt the last of the American wilderness was being destroyed when it should be preserved. Emerson's vision of scholars-in-action seemed embodied in Muir.

Muir spent the 1880s as a married rancher, though he never stopped writing and speaking on the Sierra Nevada environment. In 1889 he climbed Mt. Rainier in Washington State, and in 1890 he wrote about and campaigned for a National Park at Yosemite. In 1892 he formed the Sierra Club to promote the preservation of mountain wilderness, especially that of Yosemite. The Club's first campaign was to protest a congressional bill to contract the park's acreage.[75] His campaign for a large Yosemite reservation was helped in great measure by the Southern Pacific Railway lobbyists, as a larger park would have greater tourist potential.[76] A similar alliance led to the area's being upgraded to a National Park in 1906.[77]

Muir's facility with the pen, including his ability to express ideas with intensity and enthusiasm, earned him nationwide attention. Emerson was but the first of a number of prominent people to seek out and hear out Muir. Gifford Pinchot enjoyed a short friendship with him in the mid-1890s, until they broke over letting sheep, which Muir had come to detest on ecological grounds, graze on federal forestlands in the mountains.[78] In 1903 President Theodore Roosevelt went camping with Muir, who pressed him to make Yosemite a National Park.[79] Muir also traveled to Alaska with the railroad magnate E.H. Harriman on his yacht. Muir's relationship with such dignitaries helped the effort to expand the National Park system and added momentum to efforts to put aside even more lands for wilderness preservation.[80]

Muir died in 1914, shortly after his defeat in a high-profile fight over damming the Hetch Hetchy Valley, next to Yosemite, to provide a reservoir for the San Francisco area. At his death, he was hailed as "the most magnificent enthusiast about nature in the United States, the most rapt of all prophets of our out-of-door gospel."[81]

With Muir, the public emphasis shifted from pushing back the wilderness to protecting what remained of wild lands from exploitation and ruin by settlers and businesses. Muir only gradually broke with the conservationists, who saw the forestlands of the West as having many uses,[82] and moved toward total protection save for tourists and hikers. Part of this gradual evolution was strategic, as the mixed-use approach of the conservationists was superior to what actually existed in the 1890s on unregulated land.[83] The aim of Muir and others was to preserve some parts of the country, generally those with limited or no economic value, from encroachment, so that others might find some remnant of the freedom that wilderness implies. America was a garden that could rejuvenate the spirit of its inhabitants if it were not plundered for simple material gain.[84] The split between the utilitarian conservationists and the Romantic preservationists was an extension of Cooper's dialogue between characters in his novels of almost a century before.[85]

Muir was influenced first by his childhood religious upbringing,[86] which taught that understanding nature and appreciating its beauty helped one come closer to the spiritual promise of America.[87] He was also influenced by the Romantics and the Transcendentalists. He enjoyed *Walden* greatly, though he seemed more attached to Thoreau's posthumous book, *The Maine Woods*. His notes on a visit to Walden Pond focused on its glacial origins.[88] His Sierra Club, born in what was a "golden age" for organizations of all kinds, soon became one of the most famous naturalist organizations and the first major force for environmental protection in the United States. It was also a regional hiking club, the walking equivalent of the Wheelmen's Clubs that popularized the bicycle about the same time. Unlike the Alpine Club, created by some University of California professors the year before,[89] the Sierra Club admitted women and encouraged them to hike and mountaineer on their own.[90] It was also favored by professionals such as photographer Ansel Adams.

But Muir and those who were like-minded could not keep the remaining wilderness from being exploited. Successive governments attempted to develop trade-offs between wilderness and the extraction of resources located in these areas, such as minerals, timber, and rangeland. The history of the conservation movement in the 1890s can be seen as a struggle between bureaucratic professionals and committed amateurs,[91] although some of the leading participants changed sides over the years. Since the 1920s, furthermore, when the automobile created a new version of freedom for most Americans, the philosophical relation between wilderness and freedom has been reduced to just a part of the equation.

Gifford Pinchot

If John Muir was the quintessential committed "amateur," then Gifford Pinchot was the quintessential committed professional bureaucrat. He was born in Connecticut in 1865 into a well-off New York City family that was half-Yankee. He became taken with forests, partly because his father, like many Americans of his day, had been impressed by the arguments of a polymath named George Perkins Marsh.[92] In 1864, Marsh, a Vermont congressman, linguist, architect, writer, and American ambassador, wrote a book called *Man and Nature*, in which he argued that the collapse of civilizations had been due to environmental degradation, especially of forestlands. He suggested that America could preserve itself through careful management of forests, arable land, and watersheds.[93] Many regard Marsh as America's first environmentalist. Marsh began the long process of introducing nineteenth-century Americans to the idea that there was a connection between their continued prosperity and the ability of the natural world to provide its wherewithal. Young Gifford was presented with a new edition of *Man and Nature* when he was seventeen.

Pinchot went off to Yale and began to consider forestry as a possible profession. The only problem was that, in America, there was no such profession, nor was there an educational program for foresters. He took this as a challenge, writing to his parents: "I shall have not only no competitors, but even a science to found. This surely is as good an opening as a man could have."[94] He continued his education in Europe, especially in Switzerland and Germany, which, he noted, had developed forest-management practices that, if applied to the Adirondacks of northern New York, could be "one under which they will be as picturesque as though left wholly alone, and which will bring a respectable income at the same time."[95]

Pinchot returned to the United States and found work as a consulting forester under Olmstead on the Biltmore estate in North Carolina, which gave him the chance to implement "a management plan without any precedent."[96] His work there led him to create and distribute a pamphlet at the 1893 Columbian Exhibition in Chicago entitled "Biltmore Forest." Over the next decade, he helped to found the Society of American Foresters and taught at the new Yale School of Forestry, created with the help of his family's money.[97] Membership in a federal forest commission brought him into contact with John Muir, but despite their personal affinity, they had two different perspectives on forests and wilderness.[98] Because they needed each other's legitimacy in particular communities — Muir needed professional credibility and Pinchot needed public interest in his concerns — they tended to mute their opposition to each other:.

Pinchot's connections brought him into contact with the Eastern establishment of the time, and he became an associate of Theodore Roosevelt. Believing that the rational management of forests required "the supervision of some imperishable guardian; or, in other words, the state," in 1889 he joined the Division of Forestry in the US Department of Agriculture, and was appointed its head by President William McKinley in 1900. He led the division's separation as the autonomous US Forest Service after Roosevelt assumed the presidency upon the assassination of McKinley in 1901. By then, Muir had come to believe that professional forestry and wilderness preservation were incompatible, if only because he thought Pinchot and the Forest Service were susceptible to bureaucratic "capture" by business interests. Roosevelt relied on Pinchot for policy advice, saying he was his "conscience," yet he also maintained a good relationship with Muir.[99]

Pinchot and Roosevelt shared the idea that public lands, especially in the West, could not be left to those who wanted to use them for free pasture or free resources, and that exploitation had to be kept in sustainable bounds. To that end, they expanded the federal lands that were designated as being reserved for special purposes. It was not enough to have National Parks, but other lands were made into National Forests and National Monuments. Pinchot had a bias for management and scientific principles that seemed to be a variation on his contemporary Fred Taylor's scientific management ideas, and he believed that selective exploitation could go along with the maintenance of these protected areas. Mixing ideas from the old Utilitarian philosophers and his contemporary Progressive colleagues, he wrote that conservation has as its goal "the greatest good, for the greatest number, for the longest run."[100] Under Roosevelt, the federal government fought court cases that established its constitutional right to manage federal lands, took 150 million acres out of the "free" domain, and conducted scientific studies of better watershed and reforestation methods.

After Roosevelt left office in 1909, Pinchot's policy influence declined quickly. He left government in 1910 after a public battle with President William Howard Taft's interior secretary, and went on to form the first of a series of forestry and conservation lobby groups in Washington, DC, that extend down to this day.[101] He worked for Pennsylvania's Forest Commission and served two terms as governor of that state in the 1920s and 1930s. Over time, he became a critic of his creation, the

US Forest Service, claiming it had bent to business, ranching, and tourism interests instead of protecting the broader public interest, just as Muir had feared. Pinchot gradually came to realize that forestry was a part of a broader ecological problem, the need to clean up Pennsylvania's rivers and streams. The revisions of his 1914 memoir, *The Training of a Forester*, over four editions in twenty-three years tend to bear out his partial evolution away from narrow forest-management principles toward a wider view of wilderness and ecology.

During the Depression, another President Roosevelt asked Pinchot to consider whether his pioneering effort to hire unemployed workers to help in reforestation and in constructing outdoor recreation facilities might be applied as a make-work measure for the 1930s. Pinchot thought it would, and the Civilian Conservation Corps (CCC) was created to employ young people to work in the woods.[102] One of these was my father-in-law, who lived in a CCC camp in Michigan's Upper Peninsula for a time. Pinchot also advocated that millions of acres of despoiled crop and forestland be acquired and brought back into a healthy state by the CCC.

Before he died in 1946, he proposed that the Allied governments, then planning what would become the United Nations, hold a conference on the conservation of natural resources after the war's end. In 1949, the UN Scientific Conference on the Conservation and Utilization of Resources was finally convened, though it had lost much of Pinchot's vision.[103] His idea of the need for international agreements to deal with conservation issues can be seen as a precursor to principles underlying such agreements as the Kyoto Protocol, and the connection he made between international peace and the protection of national resources was expressed later in the formation of the Greenpeace organization, though one doubts that he would have subscribed to its activism — he was too much the organization man to go that far.

Aldo Leopold

Aldo Leopold was born in Burlington, Iowa, in 1887, a town that had been settled by Yankee pioneers following the Black Hawk War, which resulted in the opening of the whole of the trans-Mississippi frontier. The first settler in Burlington might have been Jeremiah Smith Jr., the son of a prominent New Hampshire politician and judge, but it got its name from the purchaser of the first lot in the new town, John B. Gray, a native of Burlington, Vermont. For a time, it was the second capital of the Wisconsin Territory, which temporarily extended from Lake Michigan to the upper reaches of the Missouri River; when the state of Iowa was created from this vast domain, the capital moved west to Des Moines. The "Hawkeye" nickname for Iowans, created by the local Burlington newspaper, memorialized not only Chief Black Hawk, who lived out the rest of his life near the town, but also the virtuous character Natty Bumppo in Cooper's novels. The town grew slowly, as it was bypassed by the first two railroads to cross the Mississippi, but after the Irish famine and political upheavals in Germany in 1848, the area's Yankee settlers were overwhelmed by Irish and German immigrants.

Leopold was the grandson of a family of these German immigrants. The year before his birth, his grandfather was involved in the creation of the Northwestern Furniture Company, renamed in 1888 as Rand-Leopold Desk Co., which, under various names, lasted until 1990.[104] The firm, like many other furniture businesses on the Mississippi,[105] prospered, and young Aldo was expected to join it. But his interest in birds and wild animals, which grew from his childhood exploration of the local riverbanks and summer trips to northern Michigan islands in Lake Huron, led him to Yale and into the Forestry School that the Pinchot family had helped create. He went into the woods, instead of into wood products.

Leopold entered the Forestry School just as Pinchot and Roosevelt were adding millions of acres to the federal forest reserves, soon to become National Forests. He worked in the Forest School Camp in the Poconos in northeastern Pennsylvania, near the Pinchot family estate, learning to measure forestland and to determine the proper wood volumes that could be extracted without depleting the forest. Like many foresters, Leopold found himself split between scientific management of the land and the desire to preserve its wildness.

In 1909, Leopold graduated and was posted by the Forest Service to New Mexico. Over the next five years, Aldo began to focus on the need to preserve adequate game for both hunters and predators, and he advocated the creation of "game refuges" to guarantee the continuous supply of this "product,"[106] but Washington rejected his ideas.

After this setback, Leopold was transferred into public relations and tourism. He visited the Grand Canyon, which was being overrun with commercialism. He took Theodore Roosevelt's words about the canyon to heart: "Leave it as it is. You cannot improve on it. The ages have been at work on it and men can only mar it." He and the local supervisor developed a plan to scale back the commercialism, which the Forest Service implemented over the next two years.[107] He then returned to his concern for game protection, writing the *Game and Fish Handbook*, the first such issued by the Forest Service. Having caught the writing bug, he began to turn out a consistent series of articles on wildlife and forestry.

In 1921, Leopold began to argue that parts of federal lands — he pointed specifically to the headwaters of the Gila River in western New Mexico — ought to be reserved as wilderness areas. He defined wilderness as a "continuous stretch of country preserved in its natural state, open to lawful hunting and fishing, big enough to absorb a two-weeks' pack trip, and kept devoid of roads, artificial trails, cottages, or other works of man."[108] After much internal controversy, in 1924 the Forest Service approved the Gila Wilderness plan, which covered 750,000 acres of the Gila National Forest, creating the first formal wilderness area.

Leopold, by now a family man, then suffered a life-threatening illness that cut short his career in the field with the Forest Service, and he was encouraged to move to Wisconsin to head up a game management research program. By the time his wilderness plan had been approved, Leopold had moved on to the Forest Products Laboratory in Madison, Wisconsin. He continued to focus on game management, discovering that controlling both hunting and predators simply led to an exploding game population that was destructive of other parts of the forest. The whole thing would have to be brought into balance. This was an insight that led this scientific manager to ecology and experimentation: "We conservationists are the doctors of the game supply. We have many ideas as to what needs to be done, and these ideas quite naturally conflict. We are in danger of pounding the table about them, instead of going out on the land and giving them a trial…A vote [by the American Game Protective Association] for the adoption of this policy is, in my opinion, a vote for the idea of experimentation, rather than a vote for any one of the particular systems to be experimented with."[109]

In 1931, Leopold's *Report on a Game Survey of the North Central States* was published, in which he reiterated the need to consider wildlife together with its habitat. It led him to scholars who had been studying plant ecology in American and British universities, and he began to see that there were relationships between animal, plant, and soil ecologies on any given piece of land, and that all of these had to be considered together in a kind of "land ethic." As the Depression deepened and the "Dust Bowl" of degraded soils in the Midwest began to shower the East with its dirt, the need for a land ethic was driven home. President Franklin Roosevelt, probably unknowingly, restated George Perkins Marsh's theme when he said "[t]he nation that destroys its soil, destroys itself." When Leopold was asked to return to the Southwest in the summers to help run the CCC's forest and topsoil conservation programs, he was happy to go, even though he had left the Forest Service for a professorship at the University of Wisconsin in 1933.

In 1934, a group of conservationists who were affected by Leopold's arguments met to form the Wilderness Society to press for more wilderness areas. Their leader, forester Robert Marshall,[110] called Leopold "the Commanding General of the Wilderness Battle."[111] Once Marshall became the head of recreation for the Forest Service, he established regulations to create what were known as "primitive areas," and set up a process to better understand them and their natural boundaries.[112] From 1956 until 1964, the Wilderness Society pursued, ultimately successfully, legislation that defined wilderness as "an area where the earth and its community of life are untrammeled by man, where man himself is a visitor who does not remain."[113] Today, more than five decades since the passage of the *Wilderness Act of 1964*, which provided for the hands-off management of designated areas, there are at least 550 wilderness areas in 44 States,[114] including a piece of Alaska half the size of California that President Jimmy Carter reserved in 1980.[115] Wilderness advocates' biggest problem now is not that Americans still see wilderness as something to be thrown back, but as something to be visited and loved — loved to death by millions of people. Ironically, only intensive management can keep this 2 percent or so of America from being overrun by neo-Romantic well-wishers.[116]

In 1936, Leopold visited Germany to look at its forest-management practices, which were based on principles that had so impressed Pinchot fifty years before. Leopold was not impressed, however, feeling that management was not as important as understanding the complex relationships among animals, plants, and soils. There had to be more research into these interdependencies.

For the next dozen years, he continued to write about the need to recognize these interdependencies. His most enduring work, *A Sand County Almanac*, which recorded his thoughts about the natural environment and humanity's role in it, was published posthumously in 1949.[117] The book was written from the perspective of his "Shack," which lay on a piece of deserted farmland not far from the Wisconsin childhood homes of John Muir and Frederick Jackson Turner, the historian of the frontier. After his death in 1948 from a heart attack sustained while trying to help a neighbor at the "Shack" put out a grass fire, farmers in the district surrounding the Leopold land turned over 1,200 acres of their property to a management trust called the Leopold Memorial Reserve, which includes a laboratory study center on site. *A Sand County Almanac* has continued to inspire people to look at "wildness" differently.[118] It has been ranked with *Walden* and some of John Muir's writings as a classic.[119]

Leopold subscribed to the principles of forest management as popularized by Pinchot and Teddy Roosevelt, but soon realized that allowing mixed usage was doing great damage to the habitats of many wild species. He was concerned with game management, but came to realize that habitat, or the ecology, had more to do with the maintenance of wild birds and animals than did restricting hunters and predators. In the 1930s, Leopold finally gave up on the sustainability of Pinchot's ideas about the exploitation of resources in wilderness land, essentially coming to the conclusion that the best management lay in a combination of scientific knowledge, practical management, and a heavy dose of nature's own methods. He saw the ecological discoveries of the twentieth century to be on a par with those of geology and Darwinian natural selection in the nineteenth,[120] yet he merged Yankee practicality with a Romantic ethic. Leopold recognized that his "land ethic" and "ecological conscience" were ideals, but they fired the dreams of the green movements of the next generations.[121]

Edward Abbey

"Cactus Ed" Abbey died a few years before we took the first of our many trips to the Southwest in 1995. We loved what he saw and worried about its sustainability as well, but we were, sort of, part of the problem. Abbey worried about mass tourism, calling it "industrial tourism," where people used the automobile and motels and shopping centers near sites of desert and canyon beauty as a kind of touring — a "been there, saw that, bought the T-shirt" kind of tourism.

Edward Abbey was born in Indiana, Pennsylvania, in the coal country in the western part of that State, in 1927, and grew up near there in the small village of Home, about thirty miles from where my mother was born and spent her early years, leaving with her family for Michigan's automobile plants in 1926. Abbey's father, Paul Revere Abbey, was a farmer, logger, and sometime traveling salesman whose politics were decidedly to the left, supporting the Socialists and the "Wobblies," a radical labor movement in the early 1900s, and later praising Soviet communism. He knew Walt Whitman's poetry by heart and passed on to his son a streak of aggressive non-cooperation with authority. His mother, on the other hand, was religious and a schoolteacher, and she bequeathed him his artistic talent. Abbey early on developed a liking for the woods and for writing. He was also a passable cartoonist.

Like his father, who had traveled as a youth to Montana, Ed went west at an early age, hitchhiking and riding the rails in the summer of 1944 between the eleventh and twelfth grades. He fell in love with the Southwest on that trip and, after graduation and a two-year stint in the postwar Army, ended up at the University of New Mexico on the GI Bill. Even as he took the government's money, he went on an FBI watch list for encouraging young people to turn in their draft cards.[122] Obstinacy, determination, and resistance seemed to be his hallmarks, both in life and in writing.

Over a decade of intermittent study, including a year in Scotland at Edinburgh University, he graduated with an MA in Philosophy. In 1956, Abbey's second novel, *The Brave Cowboy*, was published;[123] it was later made into a movie, "The Lonely and the Brave." In 1957, he attended Wallace Stegner's prestigious summer program for creative writing at Stanford University, preceding writers Ken Kesey and Larry McMurtry and jurist/author Sandra Day O'Connor by a year. During his life, Abbey published over two dozen volumes, mostly non-fiction. His best-known work, *Desert Solitaire* (1968), was his first non-fiction attempt. Its descriptions of desert flora, fauna, and landscape — along with a few tales of exploration — established Abbey, according to McMurtry, as the "Thoreau of the West." Seven years later, he published his most influential novel, *The Monkey Wrench Gang*, a story of eco-activism.[124]

As a person Abbey is hard to define. In private, he tended to be quiet, shy, and reclusive. His public persona was "Cactus Ed," emerging from his study only for money, vice (he was a compulsive philanderer), and the possibility of applause. He was married five times and had five children. His professors labeled him, according to his FBI file, as "rash and immature" and "exceptionally brilliant and an individualist." He favored gun ownership, opposed immigration that caused American overpopulation, detested officialdom, political and bureaucratic, loved classical music, and called himself a "redneck."[125] As late as age forty, the year before *Desert Solitaire* was published and made him famous, he was working as a school bus driver in Death Valley, California, where his then wife was an elementary schoolteacher. The previous summer he had been a fire ranger around Mt. Lassen. In later life, he was a professor of English literature at the University of Arizona. He died of a condition commonly associated with alcoholism. Friends took his body out into the desert and buried him in an unmarked grave, as befitted a Romantic desert rat.

Abbey took to heart the basic American conflicts between the individual and the organization and between the Romantic notion of wilderness as the regenerator of the American spirit and the Yankee utilitarianism of land, wild or otherwise. His father's anti-establishment politics and his mother's religiosity melded into someone who could state that "[s]entiment without action is the ruin of the soul." Even though he regarded the notion of the rugged individualist along the lines of Natty Bumppo and some fictional Western heroes as, basically, a lie,[126] he did advocate for wilderness areas as places to which rebels to an authoritarian government might retreat and organize their resistance, something that others later took up in Idaho and elsewhere.[127] His basic opposition to big business and big government predated the widespread opposition to these institutions as the Vietnam War dragged along. Abbey's action was to write and to live the individualist, anarchist lifestyle. In this, he fit his times.

At a sunrise memorial ceremony at the Arches National Monument, Dave Foreman, a founder of the activist group Earth First!, summed up Abbey's life and role. In primal cultures, he related, there have always been forces like the Zuni mudhead kachinas, tricksters who make fun during the most sacred ceremonies: "In every real society the planet has ever seen, we've had to laugh at our most sacred ideas, at our most honored personages....Ed Abbey was the mudhead kachina of the environmental movement....and it is to our everlasting shame as idealists that more of us didn't understand that Ed was a trickster....Ed was the wise prophet from the desert who tried to keep us on track and [tell us] not to take ourselves too seriously."[128]

Abbey's writings about the natural desert resemble Thoreau's about Walden Pond and its surroundings.[129] Abbey called Thoreau a put-on artist who loved to shock and exasperate, much like Abbey himself.[130] Where Thoreau made occasional trips to the Merrimack River and to Maine's Mt. Katahdin, Abbey rafted the Colorado River through Glen Canyon before it was dammed for electric power generation, and spent weeks alone in the Havasupai Canyons leading into the Grand Canyon. Where Thoreau fantasized about destroying a dam on the Concord River[131] and spent a night in jail protesting the Mexican War, Abbey resorted to anonymous minor sabotage of development projects, but dreamed in print of *The*

Monkey Wrench Gang that would destroy the hated Glen Canyon Dam, liberating the Colorado River.

Though he resembled Kerouac in many ways, Abbey could not help being affected by living in the Southwest after leaving the military. Jobs and travels in the wilderness of southern Utah and northern Arizona gave him time to think and write, and he produced his best work as a park ranger or sitting in a fire watchtower. (Kerouac also spent some time as a fire ranger but, by then, *On the Road* was behind him.)

Both Abbey and Muir opposed the damming of rivers that coursed wilderness areas, and both lost their fights. He was appalled that Muir's creation, the Sierra Club, did not oppose the Glen Canyon Dam at the outset, ostensibly because it would not flood a designated National Park or Wilderness Area.[132] Where Muir saw the mountains as having a spiritual significance, Abbey saw nothing in the desert. Nothing, the opposite of civilization, was just what he was looking for. He was not an atheist, he claimed, but a "nontheist" or an "earthiest": as he advocated, "be true to the earth."[133]

Abbey also bore some resemblance to Aldo Leopold, having considerable experience in the wildernesses of the Southwest and able to see the interconnections between various parts of the desert ecosystem. He shared Leopold's feeling that "wilderness is not a luxury, but a necessity of the human spirit, and is as vital to our lives as water and good bread."[134] Indeed, a number of Abbey's ideas either came from Leopold or paralleled his twenty years on, such as excluding cars from National Parks, preserving wild places intact, reintroducing predators to skewed ecosystems, and demolishing dams that have disrupted riverine ecologies.[135] Today, most of these have come to pass.

Where Abbey differed from Pinchot, Leopold, and Muir was in his willingness to entertain direct action to resist or eliminate publicly sponsored development. There are parallels between the fight over the Hetch Hetchy Reservoir and the Glen Canyon Dam, and Muir and Abbey had similar reactions fifty years apart. Where the earlier case gave particular passion to the Sierra Club, the Glen Canyon tale, *The Monkey Wrench Gang*, gave rise to the direct action of Earth First! and Greenpeace. He had hoped the book would "stir people into action to do things I am too cowardly to do myself."[136] It did.

Abbey saw the effects of large-scale tourism in his stints as a park ranger or fire warden in his fifteen years working in forests, monuments, and parks from New Mexico to California. The growth of the Interstate highway system in the 1950s and 1960s meant that good access was created for large numbers of people, and their use of park lands was harming the natural attractions even in remote southeastern Utah. Access would also lead to large-scale use of more remote areas as well, so that the sheer numbers of hikers, canoeists, rock climbers, and campers were having an effect on ecosystems that could not regenerate fast enough to repair themselves. Were Abbey to undertake a stint in Moab, Utah, today, his irritation would be heightened further by finding that the town had become the "mountain-bike capital of the United States."[137]

Conclusion

Abbey and Thoreau, though writing 120 years apart, were concerned with much the same thing: if one would get way from organized society and live simply, one could enjoy a measure of personal freedom that wilderness of some kind affords. The trade-off between organization and freedom is at the root of the Yankee experience, and in the end Yankees opt for organization while exalting freedom. Today, mass tourism organizes people in their quest to "get away from it all" and be free. They go to be free, however, in a land where it is impossible to be more than twenty-one miles from a road of some kind.[138] They go to be free in a wilderness while carrying a cell phone and GPS homing beacon to keep them in touch with the wider society as they drive their all-terrain vehicles across "remote" lands to do some rock climbing or hang gliding.[139] The Stegners have noted that "an American, insofar as he is new and different at all, is a civilized man who has renewed himself in the wild."[140] Yet, if renewal consists only of the "experience" afforded by an IMAX theatre at the entrance to a National Park, the drive-by viewing of natural wonders, and the hunting of animals and birds introduced into the wild, are Americans really that different from others any more?

Beyond the effects on wilderness of the American talent for organization, another change has come to the way the natural world plays out on the American psyche. As in the past with wilderness preservation, the change emanates from the concerns of urban Americans. Abbey was trying to be provocative when, in his introduction to *Desert Solitaire*, he said, "[t]his is not a travel guide, but an elegy. A memorial. You're holding a tombstone in your hands."[141] He was right, of course, but he was right about the wrong thing.

The American wilderness is still being threatened, centuries after the first Pilgrim stepped ashore. But the truth is that, by the middle 1900s, the notion of the redemptive powers of wilderness had already faded in the American mind. Some still cared, but the stream of concern about what was being exploited in the natural world had started to turn away from wilderness preservation. Americans were beginning to fear they were poisoning themselves, polluting and crippling their whole environment, wilderness or not.[142]

But that is a story for later, near Niagara Falls.

Shrewsbury, Massachusetts: Pincus and the "Pill"

Oh, the sisters of mercy,
They are not departed or gone.
They were waiting for me
When I thought I just can't go on.
— Leonard Cohen, "The Sisters of Mercy"[1]

Life has taught me one supreme lesson. This is that we must — if we are really to live at all — put our convictions into action.
— Margaret Sanger[2]

Katharine Dexter McCormick was "a woman more strange and powerful than fiction could ever invent."
— Loretta McLaughlin[3]

> *You know, Dr. Pincus, your pill does work, but it will never sell!*
> — Dr. Edris Rice-Wray, as she gave Dr. Gregory Pincus
> the data from her field trials in Puerto Rico[4]

Shrewsbury

The environs of Worcester, Massachusetts, have their "science-fiction made real" places like no other part of the country. One day we discovered the monument to Robert Goddard's liquid-fueled rocket flight in the middle of a golf course in Auburn, to the south of Worcester. The same day, going east from the city, we found the place where the oral contraceptive, the "Pill," was developed in the 1950s.

The town of Shrewsbury is effectively a suburb of its larger neighbor. A couple of miles north of US 20 and a little southwest of the center of Shrewsbury is the Hoagland-Pincus Conference Center of the University of Massachusetts Medical School. It is the site of the former[5] Worcester Foundation for Experimental Biology (WFEB), co-founded by Hudson Hoagland and Gregory Pincus, and the location where the Pill was developed. A low brick building in the complex is identified as the laboratory where Pincus and his colleagues did their work.

The Pill was the only device or method of controlling births to be devised in the twentieth century. It has had multifarious impacts on global society since the late 1950s, not only in terms of limiting births, but also on medical practice, biotechnology, popular culture, demographics, and religious practice, to name a few.

Birth Control in History

Limiting births has been a human practice in all cultures for at least as long as recorded history.[6] Until recently, childbirth was fatal to one in eight mothers, on average, so the odds against a mother's surviving increased with her number of pregnancies. Then, those born tended to die as children frequently, creating a social quandary: should society approve risking the life of the mother in multiple births in order to overcome childhood mortality rates, or try to limit pregnancies and risk population decline?

In general, prevention of conception has been the most difficult goal to reach, if only because of the inconvenience or unreliability of the methods. Many mechanical forms of preventing conception, such as condoms and diaphragms, have been known for ages, but received continual improvement only after 1800.[7] There were even precursors to the Pill — different plants, herbs, and concoctions that people claimed would inhibit conception. Nearly all were worthless, and many that were promoted as preventing conception were often more dangerous to the user than they were effective. All methods, dangerous or harmless, shared this problem of being much less than 100 percent effective.

Social methods, including religious ones, worked better, but they carried serious responsibilities with them. These mainly fell on women, and consisted of restricting sexual activity until marriage and making an important social value out of female virginity. As well, restrictions on sexual activity within marriage through taboos related to menstruation and post-childbirth might have helped to space childbirths, but, in many cultures, they also left women to be regarded as "unclean" at certain times of the month. Men were far less affected by social conventions.

The question of population size, therefore, carries a lot of baggage with it. It affected and was affected by health concerns of various kinds, sensitivities about the insertion of devices into women, religious and other social customs, male-female power relations, social roles and expectations, family prosperity and harmony, inheritance rules, and on and on. Compounding the problem was the lack of knowledge of how babies were conceived. The fertilization process itself not widely understood until the beginning of the nineteenth century, many believing that the male "seed" simply grew in the female "garden" until birth.[8] The details and mechanisms of the ovulation cycle were not well understood before the 1930s. The roles played by and the nature of chemicals used to signal egg release and other changes were unknown and unsuspected until then.

In the American context, population size was an issue in the early colonies primarily in terms of encouraging rapid growth. The Puritans' society, for example, was relatively healthy and there was adequate food and land to share among growing families. After the English Civil War, emigration to the New England colonies temporarily declined and expansion came from natural increase. In the early 1700s, New England's decent land began to fill up, and new families were forced northward or up the sides of hills onto less productive land. Meanwhile, in colonial cities, the growth of a prosperous class of people led to concerns about the future for their children.

The result in both cases was a decline in fertility.[9] One attempt at estimating average family size suggests that American women bore 7 children in 1800, and 3.6 in 1900. In 1936, the fertility rate fell below reproduction: 2.1 children.[10] It rose during the baby boom of the 1950s and returned to 1936 levels by the end of the century. It is clear that, at about the time of the Revolution, fertility rates in America, once some of the highest in the world, began a steady decline. This was especially true in New England and in the lands settled by Yankees to the west of the Hudson River.[11] A Congregational minister wrote in 1867: "There is scarcely a young lady in New England — and it is probably so throughout the land — whose marriage can be announced in the paper without her being insulted within a week by receiving…a printed circular offering information…by which the laws of heaven in regard to the increase of the human family may be thwarted."[12]

In response to declining fertility rates, local and state attempts to prevent birth control legally began around 1820; a New York statute restricting abortion was first passed in 1829.[13] Yet the desire to control fertility continued, and was accomplished partly through later marriages.[14] Given a relatively stable period of fertility out to age forty or so, this meant less time in to have children and therefore, potentially less children per fertile woman. For instance, the average marriage age for American women in the 1950s was around nineteen,[15] while today, the average is closer to twenty-seven, with predictable effects on their childbearing. Other chastity approaches were based on the period when the woman was breastfeeding, trying to determine the monthly period of infertility, or the need for men to go away to sea or logging camps or take other long absences.[16]

Another part of this decline came from the age-old use of concoctions of different plants and minerals. These were emmenagogic — that is, they had the effect of restoring menstruation when it was irregular, including the "irregularity" of early pregnancy.[17] In the past, people had considered the crocus, peony roots, nettleseeds, and some minerals as having this potential.[18] Native Americans were also aware of the use of plants to control pregnancy: pioneer names for relevant local plants included "squawbush," "squaw mint," "squawroot," and "squaw vine."[19] Abortions coming later in the term were relatively rare and generally found as asides in court cases.[20] The language used to describe such plants and other devices makes it difficult to understand this side of life. Generally, advertisements such as those for the popular Hooper's Female Pills referred to "feminine hygiene" or medicine for female irregularities or maladies. Frontier women and European immigrants sent letters home asking for seeds of favorite emmenagogic plants that could not be found locally. One such plant was savin, which continued to be imported until it was realized that the American juniper or red cedar was identical chemically.[21] The success rate of these natural drugs was about 70 to 85 percent.[22]

Then, another aspect of population growth — overpopulation in already-settled parts of the world — began to be discussed. This was a problem of population pressure, rather than of the effects of childbirth on women's lives. In 1798, Thomas Malthus, a British thinker, published his "Essay on the Principle of Population," in which he claimed that, in good times, population would grow exponentially until it exceeded society's ability to provide enough food, the supply of which could grow only in a linear fashion. Only war, disease, and starvation could restore the balance, he argued. In an 1803 revision, he advocated moral restraint and the postponement of marriage as a means of limiting population growth. His ideas were criticized at the time, but later served as an intellectual justification for British official inaction in combating famines in Ireland and India. The Irish famine of the late 1840s, in particular, led to new waves of immigration to North America.

At least three factors, besides the waves of foreign immigration, influenced attitudes about family and population growth in mid-nineteenth-century America: the frontier, education, and aspiration. The frontier meant an agricultural society, one where children were of value to the family once they became old enough to be given simple tasks. As a consequence, larger families were not seen as a problem by their progenitors. Once the frontier slowed as an attraction for non-immigrant Americans, as it did in the post–Civil War years, the social value of large families began to be seen in a different light. The growth of cities altered the cost of raising children. The cost of living space and the dependence on cash incomes for long-term survival meant that having children beyond a small number became a problem for most families.

Second, the Puritan/Yankee ethos dominating the northern half of the country after the Civil War placed heavy emphasis on education. This came at base from the Puritan belief that each person had to come to an understanding of the Bible's message, which presumed that each person could read. This, in turn, presumed that they were educated to some degree. Yet, literacy did not restrict one's information sources just to the Bible. Alexis de Tocqueville, in penetrating the Michigan forests in 1831, noted the presence of Eastern newspapers in even the rudest cabin. Female education thereby complicated the role of women in frontier society. Education also extended to urban women and, when combined with town and city life, created pressures on traditional family ways. Mothers could read magazines that, much like our own times, came with all sorts of life skills tips, including carefully worded ones on family planning. Newspaper advertisements for devices and services also existed. There were numerous "little" books for sale that described how methods of birth control worked.[23]

Third, education, whether combined with farm or with city life,[24] brought with it new incentives for aspiration.[25] General advertising, along with the demonstration effects of new products designed to enhance domestic life, led to the desire for additional family income. The new mail catalogues sent to farm families carried the aspiration for a better quality of life to them. Education also meant that better jobs were available to women than farm or mill or sweatshop, including as typists, telegraphers, and store clerks. The jobs were still time consuming and repetitive by today's standards, but they were less physically demanding and paid better. Family values and economic desires collided.[26]

Comstock and Vice

After the Civil War, all of these changes combined with social reaction and a particular variety of antebellum activism. Social reaction took the form of reinforcing the traditional division of labor and the roles of men and women in providing employment income (men) and creating a comfortable household environment (women). This was a middle-class ideal that meant a more desperate aspiration for poorer women, who by necessity had to work, and was largely seen as irrelevant by wealthy women, whose roles were modified by servants, social climbing, and philanthropy.[27]

The activism of the times emerged in a more complex way, and was symbolized by the career of Henry Comstock,[28] a Connecticut Yankee farmboy who returned from Civil War service to live in New York City. Comstock was a holdover from the Second Great Awakening, which generated a variety of social causes before the war, including temperance, anti-slavery, and women's rights.

Comstock's experiences in New York City varied greatly from his religious beliefs. Living as a single man in poorer quarters, he was exposed to the open and easy postwar urban morality, and was especially repelled by its sexual aspects. He began to oppose what he and others called "vice." In particular, he was scandalized by the excesses of the new rich, particularly men who made their fortunes from war profiteering and financial dealings. They carried their financial amorality over into postwar life by openly challenging social mores, indulging in prostitution, public infidelity, and pornography.

Comstock rode on a wave of activism that came from a number of sources. Women's rights groups campaigned for a single standard of morality, asserting that their menfolk should not visit prostitutes, get drunk, or engage in other practices if women could not do likewise. Religious groups opposed infidelity on moral grounds, but noted it also risked introducing disease to the wife who stayed at home. Inspection of prostitutes was opposed as assisting the "profession," and physicians were pressured to declare chastity as being good for men's health. A good workforce had to be one with temperate habits.

By 1872, especially with the publication of his book, *Frauds Exposed*, Comstock had built a powerful political alliance across the country that opposed vice in the same manner as temperance, but because it dealt with family morality, had a more immediate effect. In the lame-duck Congress of 1873,[29] Comstock managed to get some amendments added to a postal bill that included penalties for using the mails, then the most common form of distributing mass communication, to advertise vice or to ship obscene material or devices.[30] Comstock's amendments included as "vice" anything related to birth control, since prostitutes could use this information[31] and related devices to protect themselves against unwanted pregnancies. In reality, the largest group of potential consumers was married women, so a legal prohibition of birth control information also made sense to social reactionaries who believed married women should be at home, expressing themselves through their fertility.

Even though the "Comstock laws" were the core of legal resistance to family planning, they met with mixed success as other purposes and intentions undermined their efficacy.[32] For example, was information or technology related to women's *health* to be considered an *obscenity* if sent through the mails? For sixty-three years, from 1873 until the Supreme Court intervened in 1936,[33] the Comstock laws weighed against efforts to inform women about birth control. But because the key regulations only related to the US Mail, other restrictions on the sale of devices and the provision of services and information were largely left to state law. Connecticut enacted the most restrictive laws in the country, while next door in New York the law was much more liberal. Like local options with respect to alcohol sales, a patchwork of family planning laws and regulations sprang up within and between States.[34] The laws were not often enforced, however, as the Post Office added only four new inspectors to its roster of sixty-three, who had to cover all sorts of trade-related cases, including regulating common carriers, such as the railroads. Comstock himself was made a Special Agent, and he spent his time pursuing miscreants, often unsuccessfully.[35] Legal battles took place over the privacy of the mails, the legality of "sting" efforts by the federal regulators, the power to regulate "health" information, and the rights of families to information. Yet the existence of the laws did have a deterrent effect.

With the First World War, army officials realized that unprotected sex by soldiers was leading to a wave of venereal disease that was having an effect on military performance. Their concerns led to army contracts for condoms, even though, legally, this might have been considered as promoting "vice." Then, in July 1918, Congress created the Division of Venereal Diseases in the US Public Health Service. Health and birth control were slowly coming together. A ruling on a case involving Margaret Sanger by Judge Frederick Crane in 1918 noted that the use of contraceptives for prevention of disease could not be considered as obscene.[36]

War's end saw soldiers coming home from Europe with new attitudes toward sex and birth control. At the same time, social reformers emerging from the Progressive movement, such as Margaret Sanger, were establishing birth control clinics and fighting for them as public health institutions in the courts of legality and public opinion. Stories of immigrant women in city tenements dying from childbirth and botched abortions were an extension of those about sweatshops and the food product and sanitation issues that had mobilized earlier Progressives.

For decades, advocates of increased births by American women had argued the "eugenic" case that the "race" that built the country would be overcome by immigrants — first by Irish hordes and later by the children of immigrant families from southern and central Europe.[37] By the 1930s, this argument had been accepted as having some validity. Yet the fear that the English European "race" would decline in America and on the world stage because of low birth rates resulted in little action in the face of the economic aspirations frustrated by the Depression.[38] Despite government, religious, and social resistance to birth control, American birth rates kept falling, until by the Depression they were below replacement levels.

The most common form of effective family limitation through the nineteenth century and into the twentieth was abortion.[39] Despite attempts by many birth control activists, such as Margaret Sanger, to distance themselves from this practice, it not only was widespread, but mostly performed legally, given the absence of prohibitions in many localities. In 1930, between 25 and 40 percent of all pregnancies were aborted, a third of these illegally, resulting in 30,000 fatalities of women.⁴⁰ The legally prescribed abortions — "for health reasons" — were obtainable mostly by middle- and upper-class women who could afford a family doctor and related hospital charges.

The later 1930s saw the demise of the Comstock laws, but there occurred a new interest in general population problems, prompted by a concern about shoddy products sold as contraceptives⁴¹ and a revival of birth rates. In 1935, the Oregon legislature ordered the State's board of pharmacy to set standards for contraceptives, and in 1937, the American Medical Association recognized contraception as a legitimate service for its members to provide.⁴² A quiet struggle began among doctors over whether the market — in the form of condoms, "feminine hygiene" products, and the like⁴³ — would govern birth control or whether the medical establishment should be the legitimate adviser and dispenser. In 1938, the federal government became more involved when the Food and Drug Administration (FDA) began testing the reliability of condoms.

In the 1940s, however, pro-natal policies saw a toughening of the enforcement of anti-abortion laws. As well, war brought a revival of military concerns about public health in general. Not only was the venereal disease problem revisited, with the official distribution of condoms,⁴⁴ but the military also found it had to deal with the public health challenges of cities it had liberated.⁴⁵ By the end of the Second World War, there was a somewhat more relaxed attitude toward birth control, complicated by the inefficiency and unpleasantness of the methods of achieving it.

Population again became an issue in the postwar years, only this time worry over population decline was replaced by fear of overpopulation, as success in preventing high rates of child mortality in Asia and Africa was not being offset by corresponding decreases in birth rates. Over the next several decades, the world population grew rapidly, reaching over 6 billion around 2000, a growth that became known popularly as "the population bomb," considered by many as potentially devastating as the nuclear bomb.

Family planning, not population concerns, however, was the main driver of birth control usage. In the nineteenth and early twentieth centuries, as people moved into cities, many found themselves unable to earn an income that allowed them the physical space for a large family. Overcrowded tenements and inadequate facilities encouraged the spread of disease and increased death rates. Smaller families could reduce these problems. They would also prevent mental and physical distress to women who found it difficult to bear or care for many children. Among families with higher incomes, aspirations for consumer products and services as well as better education for their children, meant that limitation of childbirths was desirable. For these families, more information and more medical services gradually became available, creating a clear break between well-off families with few children and poorer families with many. The "secret," as access to effective birth control was called, was considered to be a class distinction.

Mounting pressure for women's rights at the start of the twentieth century was focused mainly on acquiring the vote, but it spawned many other causes as well, such as "temperance" (the effort to prohibit alcohol), social and family morality, female education, and labor protections for working women and children. In this multiplicity of causes, birth control played a relatively minor role. Once it became clear, sometime after 1910, that women inevitably would obtain the right to vote, suffrage activists began to become more interested in women's right to control births. As with suffrage and temperance, the promotion of this cause was a long-term effort, one characterized by incremental movement from 1910 through the 1960s. It was the development of the birth control pill by Gregory Pincus and his associates that symbolized the ultimate success of this movement.

"Goody" Pincus

Gregory Goodwin Pincus was born in Woodbine, an inland borough of Cape May County in southern New Jersey, in 1903.[46] Woodbine had been founded a dozen years before as a farming settlement for Eastern European Jews by a millionaire German businessman, Baron Maurice De Hirsch. It became a "model" agricultural community because of its use of modern farming practices and the creation of an agricultural school within its boundaries. Pincus's father had studied at the Storrs Agricultural College in Connecticut, later made the State's land-grant institution and renamed the University of Connecticut. He then taught at the Woodbine school and became the editor of a farm journal. Pincus's mother was also a teacher. One of his uncles became dean of the agriculture college at Rutgers University.[47] The family then moved to New York City, but Gregory grew up aspiring to be a farmer. His father dissuaded him, so Gregory went into biology at Cornell University, graduating in 1924.[48] Three years later he managed to get both an MS and a PhD from Harvard, based on his research on heredity factors in rats.[49] He spent three years in Europe as a National Research Council fellow, returning to Harvard as an assistant professor, a post he held until 1938.

Pincus's interest in genetics led him to the study of procreation in mammals. Beginning in 1932, he began research on hormones and reproductive processes that was to remain his interest for much of the rest of his life. In the late 1930s, Pincus gained notoriety for having succeeded in artificially stimulating the birth of "fatherless" rabbits by *in vitro* fertilization.[50] Although the claim was controversial, so were counterclaims by some detractors that they were unable to repeat his results consistently. The controversy,[51] along with some other academic turmoil — including anti-Semitism — caused Harvard not to renew his contract in 1938.

Pincus, however, had friends. On the academic front, Pincus moved to Clark University in Worcester, Massachusetts, until 1945, then to a professorship at Tufts University in Boston from 1946 to 1950, and then to the Boston University

Graduate School after that.[52] On the research front, at a conference on hormones Pincus attended near Baltimore, the manager of the site at first denied entrance to one of the invitees, an African American, which led to a protest and a later decision not to return that conference site. Pincus and some others were asked to find a new site, and they accepted an invitation from the Montreal Physiological Society to use a facility in the Laurentians, the hilly cottage country north of the city. Pincus became the permanent chairman of the annual Laurentian Hormone Conference until his death in 1967.

The Worcester Foundation for Experimental Biology (WFEB) formed another aspect to Pincus's career and, in the end, probably the most important one. Called by his eulogizer, Dwight J. Ingle, "a bootstrap operation representing scientific free-enterprise," the WFEB was started in 1944 as a private think tank and laboratory facility focused on steroids, including hormonal research, by Pincus and Hudson Hoagland, who had moved to Clark University and arranged for Pincus to be offered an appointment there. Meeting with some funding success in the postwar era,[53] the WFEB moved to a facility in neighboring Shrewsbury. Though Hoagland was its titular head, he continued on with his increasingly distinguished academic career and his work on hormones and schizophrenia. "Goody" Pincus was a respected hormone researcher with a pleasant personality and a gift for finding money and projects. He attracted capable staff and graduate students, becoming a research administrator. In late 1950, he went to New York City and had a fateful meeting with Margaret Sanger, a woman who had devoted her life to improving the methods and legitimizing the practice of birth control.

After the development of the Pill in 1956 and its approval in 1957 for use in menstrual problems, and then in 1960 for birth control, Pincus once again became something of a celebrity. In 1965, he published *The Control of Fertility*, which summed up his research on the subject in thirteen dense chapters and 1,459 citations.[54] He suffered from a rare blood disease in the last three years of his life, and died the day before the annual Laurentian Hormone Conference opened in 1967. His eulogizer, Dwight Ingle, said, "[t]o oversimplify, some scientists become great by making important contributions to knowledge — discovery in the laboratory — and others become great as organizers and by making important applications of knowledge. Gregory Pincus, a scientist-statesman, was one of the latter."[55] Another scientist, Oscar Hechter, who had worked for Pincus at the WFEB, noted: "I feel that...Goody...is too big a man to treat in a ritualistic fashion. Pincus for me represents the prototype of a *new* scientist, whose life and achievements merit critical examination and analysis...because if new Pincuses arise in the future, they will have a powerful impact upon the world."[56] The first full biography of Pincus only came, however, in 2009, forty-two years after his death.[57]

The Sisters of Mercy

By 1950, the WFEB and its research director were known for their work on steroids and their effects on mammals. Some of this work concerned the identification of hormonal substances that might affect the ability of women to become pregnant. The probability that "cures" for this condition might also be used to suppress pregnancy was realized, but in the tenor of pro-natal postwar America, this was not a politically acceptable line of investigation. Enter two "sisters of mercy," who, with Gregory Pincus, made it possible for women to control the conception process with almost virtual certainty and without the involvement of a male partner, sparing women from potential social, psychological, and physical uncertainty and suffering.

Margaret Higgins Sanger and Katharine Dexter McCormick were different in their backgrounds and in the circumstances that led them to seek a common goal. They both wanted to be medical doctors, but were unable to succeed, for different reasons. McCormick was the granddaughter of a Massachusetts-born founder of a Michigan town and daughter of a prosperous Chicago lawyer. She went to the Massachusetts Institute of Technology (MIT), lived in Boston for many years, and was childless as a result of a marriage to a man who soon after proved to be seriously mentally ill; she was a director of organizations and a philanthropist. Sanger was a middle child of eleven born to Irish immigrant parents in Corning, New York.[58] She acquired tuberculosis from her mother, had three children, and trained as a nurse, but became a social activist who gained a global reputation.

As early as 1913, Sanger had wished aloud for a "magic pill"[59] that would allow women to control their own lives by determining when they might become pregnant. In late 1950, then seventy-one years of age, Sanger was asked by Abraham Stone, the director of the Clinical Research Bureau, a combined research and practical clinic operation that she controlled, to meet with Gregory Pincus. At a dinner meeting in New York City on December 7, Stone introduced Sanger to Pincus and encouraged her to explain her dream. Pincus allowed himself an optimistic response. It was enough to generate a partnership, and Sanger began to press various agencies interested in birth control to help fund Pincus's work.

Money came in slow, small amounts, partly because of the prevailing social and political climate, but also because the "magic pill" sounded like science fiction to possible donors, especially when it was to be pursued by an independent research outfit. In 1953, two and a half years after her meeting with Pincus, Sanger brought her long-time associate, McCormick, to the Shrewsbury facility. Four years Sanger's senior and now almost seventy-eight, the millionaire heiress had been prominent in the last years of the suffragette movement and had supported birth control groups since the 1920s. Now, Pincus needed new support to keep the research lab open. As Bernard Asbell relates,

They made a contrasting trio. Pincus' dense bush of graying hair and piercing black, ominously shadowed eyes, almost a caricature of the menacing scientist's played against his gentle and observant look of sympathy....Had Pincus not already met Sanger and McCormick separately, he might easily have mistaken one for the other...taking the one who stood almost six feet tall with the military shoulders, swooping brim hat, and ankle-length matron's skirt to be Sanger, the embattled lifelong radical. But that one was Katharine McCormick....Standing beside her...Margaret Sanger was slight, scarcely five feet tall, with a striking crown of auburn hair...a cautious gaze through wide-apart gray eyes...and a subdued voice.[60]

McCormick, like Sanger earlier, had an immediate positive reaction to Pincus. Experienced in organizing activities, she immediately began questioning Pincus about the operations of the WFEB, its personnel, and the budget needed to make Sanger's pill a reality. "Goody" knew how to satisfy her requirements, and McCormick went on to fund almost single-handedly the millions needed for the "invention" of the birth control pill, with Sanger providing backup support through her influence in the birth control community. McCormick was a constant visitor to the Foundation and followed progress actively. By the time the first version of the Pill was approved for sale by the FDA in 1960, the "sisters of mercy" were eighty-four and eighty years old, respectively. They had turned science fiction into science fact.

Katharine Dexter McCormick

Katharine Dexter was born in Dexter, Michigan, in August 1875.[61] Both sides of Katharine's family were Yankee stock. Her father was a prominent attorney who had moved from the family home in Michigan to handle the Chicago end of their extensive lumber business. Her mother had left Chicago for Dexter late in her pregnancy to be in the cooler, country air. Her father's premature death led to his widow's returning to be closer to her family in Boston.

Katharine grew up in luxury, traveled well, and had a top-flight education. She was determined to become a medical doctor, and entered MIT's biology program, graduating in 1904 as the first woman to be granted a science degree there. Instead of going on to medical school, however, she found herself heavily courted by Stanley McCormick, a son of Cyrus McCormick, the wealthy inventor of the mechanical reaper. Stanley's eccentricities, which at first added to Katharine's initial attraction to him, became worse with marriage, and within two years led to his being committed to an asylum. Because Stanley's sisters and relatives had histories of mental illness as well, Katharine determined that, even if his sexual problems could be overcome, she must remain childless, leading her to support birth control groups.[62] Although she never had a child by Stanley, she recognized from him and his siblings that there must be a hereditary basis for their mental condition. This led her to concerns about population control and the need to restrict births where there was good cause to believe there might be heredity problems. In turn, this led her to the broader question of post-suffrage women's rights, including control over fertility. She helped her personal causes with financial contributions and, in the case of legitimizing birth control, with access to prominent physicians and administrators. She also donated her home in Geneva as the site for the 1927 World Population Conference.

McCormick first met Margaret Sanger at a lecture in 1917, and they remained in contact thereafter.[63] After caring for Stanley until his death in 1947, Katharine spent considerable time untangling the affairs of his estate, although she kept in touch with Sanger and supported her idea of finding a biological way to control fertility. She became interested in Pincus's work after Sanger wrote her about it in early 1952, and around then she met with Hudson Hoagland, whom she had known from her interest in the potential connection between her husband's mental illness and glandular secretions.[64]

After her historic 1953 meeting with Pincus, she moved back to Boston from her then home in California to be able to make weekly visits to the WFEB. She not only supported the project financially, but also spent a lot of time at the Foundation going over chemical tests, first on animals and later on mental patients.[65] She constantly goaded Pincus to keep pursuing the Pill,[66] and, in 1956, after he finally succeeded, she wrote to Sanger: "[n]othing matters to me now that we have oral contraception….Pincus' genius brought us the oral contraceptive we have been seeking and now we must implement it. We must keep testing indefinitely — that goes without saying."[67] She continued to support the Foundation's research until her death in 1968.

McCormick's support for Pincus, estimated at over $2 million, brought the Pill's development to the market at an opportune time for its wide distribution. At that point, the idea of health through drugs was at an optimistic peak, due to the success of antibiotics and the polio vaccine.[68] Only a year after the Pill's approval, the thalidomide crisis that erupted from the propensity of this sedative to cause physical defects in unborn babies led to slower and more restrictive procedures for the approval of new drugs.[69]

Margaret Sanger

Margaret Sanger was born in Corning, New York, in 1879 to a family of Irish immigrants named Higgins.[70] The family lived in poverty due to the propensity of her father to do more talking (and drinking) than working. He was a "sculptor" who carved tombstones for the families of Corning and sometimes elsewhere, but his well-proclaimed politics caused him trouble getting work. The family was supported in the main by the contributions of Margaret's older siblings to the family expenses. Her devoutly Catholic mother died at age fifty, when Margaret was nineteen; ever after, she blamed this early death on the physical exhaustion of bearing eleven children.

Once she was of high school age, Margaret's sisters helped her to go away to the Claverack School in the Hudson Valley. She then did some nursing training at a New York hospital, and had visions of becoming a medical doctor. Like Katharine Dexter McCormick, in 1906 she was pressed by a suitor, aspiring architect William Sanger, to marry him quickly, before, as he explained later, some doctor swept her off her feet. Sanger was Jewish and had socialist leanings, a political stance not unlike that of Margaret's father. He, like her father, also began to show a tendency, aggravated by the professional effects of his politics, toward a more artistic and intellectual life without the means to support it or his family: all talk and no work.

Soon pregnant, Margaret discovered that the tuberculosis she saw in her mother had affected her as well, and that pregnancy had aggravated its effects. She had to go to a sanatorium in the Adirondacks for a time, supported by family and in-laws, where she came to the realization that she would have to space her family if she wanted to remain healthy and not follow her mother into sickness and exhaustion. How to do that was a problem, however, given the knowledge and techniques of the time.

To support her family, she began working as a nurse, visiting women in the New York City tenements. Her experiences helped to radicalize her politics, and she began to write for left-wing publications. She discovered that she could speak effectively in front of an audience, and she became involved with strikes and demonstrations led by the radical union, the International Workers of the World. In 1912, Sanger became the center of a well-publicized evacuation of children from Lowell, Massachusetts, where textile workers were engaged in a lengthy strike. The reaction to the evacuation among Progressive women of upper-crust Eastern society led to the resolution of the strike in favor of the workers, and Sanger received wide attention.

The public suppression of a series of her articles on gynecology and birth control in a radical magazine only increased her profile.[71] More attempts to distribute birth control information led to her being arrested, after which she fled to Europe. She returned in 1915 only when her husband, although now estranged, was arrested for distributing her publications. In 1916, she was arrested again, along with her sister, for starting a birth control clinic in Brooklyn.[72] Their incarceration led to considerable outcry, and society matrons such as Katharine McCormick, Eleanor Roosevelt, and Gertrude Pinchot rallied to their support. By this time, women's rights had expanded beyond mere suffrage to include equality in marriage and the right to divorce and to refuse sex in marriage and reproduction.[73]

America's entrance into the First World War meant the end of the bohemian era and its freewheeling agitation for reform. Sanger meanwhile had derived a philosophy about women and sex and procreation from the anarchist Emma Goldman and the more conservative and thoughtful Havelock Ellis. She began to move more toward organizing and speaking and writing for a nascent popular movement around the single issue of birth control. She dropped, at least publicly, the connection between birth control and social and economic revolution. She continued in this vein for the rest of her long life. In 1921, she divorced William Sanger, and a year later married industrialist Noah Slee, who gave her the wherewithal to continue her work and did not interfere much with her freewheeling personal life. She devoted much of her time to being a semi-independent spokesperson for the birth control/family planning movement, trying to influence the medical profession to add the subject of women's sexual problems to their scope of knowledge and advice.[74] It was a slow process, as the medical profession, having gained a sense of respectability and professionalism around finding cures for illness, was loath to take on something where the patient was not "ill." She also continued to support the creation of birth control clinics, their number rising to fifty-five in 1930 and more than eight hundred by 1942.[75]

Sanger continued to be somewhat active in her sixties, but the Second World War and her declining health meant that, with the exception of her personal interest in the Clinical Research Bureau in New York, she limited herself to high-profile appearances and trips. Her involvement in the development of the contraceptive pill stemmed from her earlier interest in identifying the mechanism that produced pregnancy. Then, the discovery of a research organization, Pincus's, that might pursue a contraceptive pill rallied her energies once more. She died in September 1966, in time to see the end of various state "mini-Comstock" laws and the widespread acceptance of the Pill.

Developing the Pill

Like so many other inventions, the basic building blocks of the Pill were in existence before the work that pulled them together into the right structure.[76] Early in the twentieth century, Austrian physiologist Ludwig Haberlandt claimed it was theoretically possible to create a hormonal contraceptive. In 1917, two Cornell University researchers showed that the stages in the reproductive cycle could be determined by a microscopic examination of the lining of the uterus. Two discoveries in the 1920s, one by the Rockefeller-funded Bureau of Social Hygiene and another by scientists at the University of Rochester, identified the chemical compounds that were used in this process.[77] These were estrogen, a secretion from the ovaries that produced changes in the uterus to assist in reproduction, and progesterone, a substance that made it possible for the uterus to accept a fertilized egg. Concurrently, the existence of plant estrogen was discovered. These compounds, called hormones, are part of a wider group of substances called steroids, different ones of which proved effective in treating various diseases and conditions. This discovery excited biologists and chemists, as well as companies that wished to produce commercial supplies for the market. In some cases, this proved easy to do, but not for estrogen, which might work in treating human reproductive disorders. In the early 1930s, the production of estrogen from female urine or cow organs required enormous quantities of raw material to produce small dosages. Progesterone cost as much as $1,000 a gram, an expense not unlike that of some of today's complex drugs used for treating rare, but severe illnesses.

As an aside, these hormones were found to inhibit ovulation, but at the time this was not seen to be important in itself. Enter Russell Marker,[78] an inventive and somewhat eccentric research chemist who never finished his PhD at the University of Maryland because he considered the requirement that he take some makeup courses after he had finished his thesis to be a waste of time.[79] He went on to the Ethyl Corporation in 1926, where he devised the "octane rating" for gasoline. Marker then spent six years at the Rockefeller Foundation, where he was a prolific producer of scientific papers. He decided that he wanted to do more research into steroids, but the Foundation had other priorities, so he quit and found work with pharmaceutical manufacturer Parke-Davis, which installed him in a lab at Pennsylvania State University. In 1937, he managed to extract 35 grams of progesterone from a batch of steroids the company sent him, the largest production anywhere up to that time.

Marker then began to realize that plant steroids might be used to synthesize estrogen and progesterone. In 1939, he found that he could synthesize progesterone from the sarsaparilla plant. Looking for a cheaper alternative, he became interested in a Mexican yam. After three years in Mexico, Marker discovered that his sponsors were unwilling to continue, so he left Parke-Davis and Penn State and stayed in Mexico to access the plants at their source and to research their properties. In 1944, he co-founded a Mexican company, Syntex, to produce the hormones through a process he had worked out, but shortly thereafter he fell out with his partners and walked away from the company, taking his knowledge with him. Even so, Syntex worked out a method that enabled it to market progesterone at only a fraction of what the then European monopoly charged. The company prospered and eventually became known for its own version of the birth control pill.[80] Marker continued to work in Mexico on his own until he retired from chemistry, and it was only after the popularization of the Pill that he began to be honored for his work. Marker's work was followed by the success of Carl Djerassi and Frank Colton in the early 1950s in synthesizing estrogen and progesterone, thereby lowering their cost still further and enlarging the possibilities for their widespread use.[81]

Another challenge that loomed in the pro-natal late 1940s and early 1950s was the "holy grail" of a drug that could be used to regulate menstrual irregularities so as to ensure conception. Of course, if this regulatory process were to be continued indefinitely, then pregnancy could be postponed indefinitely. This realization led Pincus to begin to work with Margaret Sanger and Abraham Stone in 1951. Planned Parenthood gave them an initial $10,000, partly from funds Sanger had wheedled out of Katharine McCormick, who was still tied up resolving her husband's estate.

Meanwhile, Searle, one of the WFEB's major pharmaceutical clients, told Pincus that it could not support the development of a contraceptive, given the political and social climate. It did, however, agree to keep him on as a consultant and to supply him with the various hormones he needed for his research, even while berating him for not paying enough attention to the work he was supposed to be doing for them.

Pincus assigned the actual research work to M.C. Chang, who had come to WFEB on a fellowship to learn Pincus's techniques for *in vitro* fertilization. Chang was an accomplished researcher in mammalian fertilization. His major contribution to the development of the Pill was to test the effects of different steroids on animals, mostly rabbits, using two hundred compounds supplied by Searle. He eventually narrowed the field to three, two from Syntex and one from Searle, that had already been tested for oral use.[82] Since the process of inhibiting pregnancy was known, it remained only to discover which compound might be more effective when administered orally and at what strength and intervals it might be safely taken.

Once Pincus and Chang had worked out what they thought might be the proper dosage and intervals, it was necessary to do human trials. Although somewhat regulated, such trials were controlled considerably more loosely than is the case today. Pincus encountered a friend, John Rock, a prominent gynecologist, Harvard professor, and researcher into problems of infertility, at a scientific conference in 1952, and they both realized they were going down the same research path toward different ends.[83] Rock wanted to put patients on a "pseudo-pregnancy" regimen, then take them off and see if that would jumpstart an actual pregnancy.[84] Pincus wanted a way to keep "pseudo-pregnancy" working and suggested some variations to Rock.

Rock then tested the three steroids Chang had identified on his patients at his infertility clinic, after which they agreed that the Searle product was more effective with fewer side effects. They then found by accident that this drug worked better if contaminated with a small amount of estrogen. Searle named the resulting compound "Enovid."[85]

Rock first used Enovid as a birth control application on mental patients in Worcester, then on some women at a hospital in Boston, where he worked.[86] The researchers then decided they had to go farther afield for a mass testing at the strength and frequency needed to inhibit pregnancy,[87] as well as to treat menstrual disorders, that would produce considerable publicity, since Massachusetts' laws regarding birth control were very restrictive at the time. Accordingly, in 1956, Dr. Edris Rice-Wray, a faculty member of the Puerto Rico Medical School, began field trials on that island. At the time, Puerto Rico had sixty-seven birth control clinics catering to a population of poorer women who were enthusiastic about limiting family size. Many women did not complete the formal testing procedures, but enough did to provide a scientifically acceptable sample. Over the following year, the Puerto Rico trials proved out the Foundation's approach, and in 1957 the FDA approved Enovid for use for menstrual disorders.[88] When side effects appeared, it was found that these could be mitigated by a lower dosage.[89] When Katharine McCormick heard the news, she exclaimed: "Of course, this use of the oral contraceptive for menstrual disorders is leading inevitably to its use against pregnancy."[90]

In 1959, Searle made an application to the FDA for approval to market a variation on Enovid as a birth control pill, which was granted on June 23, 1960.[91] As an anti-ovulant, it was the first contraceptive developed in the twentieth century.[92] At first, Searle did not market it as such,[93] but applied and received approval for an alternative with half the potency and a lot fewer side effects. In the years thereafter, the FDA approved similar hormonal compounds produced by Syntex and others. The strength of the Pill today is about a third of what it was when first marketed.

The Aftermath of the Pill

The introduction of the Pill into American and then world society has had many effects. It is important to keep in mind, though, that many alternatives existed to control birth. Some groups were pressing for legalization of therapeutic abortions; sterilization of both males and females was also being advocated. The sale of birth control devices had been legal for the most part, and sales were widespread. And methods related to fertility timing ("rhythm") were being promoted as a natural alternative. Compared to most of these alternatives, however, the Pill was divorced from actual sexual activity, since it was taken on a daily basis unrelated to anything else.[94] Unlike the Pill, however, alternatives that shared this characteristic, such as sterilization and abortion, involved irreversible surgery and affected emotions and moral sensitivities.

There was also a kind of romance about the Pill, as it was a way for women to control their own pregnancies. Far from the old notion that the woman was merely the container of the man's "seed," the idea now gained prominence that the embryo/fetus was a part of the woman's body and subject to her exclusive control. The notion of liberation, so long associated with birth control by Margaret Sanger and others, was swept up in the ethos of the 1960s and became an accepted part of life for a lot of women, married or not. There might not have been a sexual revolution in the 1960s,[95] but a revolutionary realization that sexual activity no longer had to carry with it the consequence of pregnancy and parenthood did emerge, which led to the Pill's becoming an icon of the turbulent social change of the 1960s.[96] There was even a feminist reaction to the Pill, with some seeing it as a male physicians' plot to foist responsibility for contraception onto women alone.[97]

The Pill has also affected family planning. The average age at marriage has steadily increased in all developed countries until it is now around twenty-six or -seven for women in America and even higher in some other countries. This has shortened the fertile period of most married women's lives, resulting in fewer children. At the same time the Pill has kept these same women from risking becoming pregnant beforehand, given that sexual activity today, as in the past, begins on average in the teens. Further, married couples now plan their children around career and financial considerations, including increased female participation in the workforce.

The Pill has had an impact on the medical profession as well. It was a massively prescribed medication that treated no illness.[98] Willy-nilly, it carried the medical profession into the broader realm of lifestyles and the use of chemicals and surgery to enhance them. Hormone replacement therapy, botox, facelifts, and performance-enhancing steroids all came out of this shift.

The extensive use of the Pill for both menstrual disorders and as a contraceptive led to its use by over a million women before some serious side effects began to be noted. This led to the now general requirement that pertinent information about potential health risks be appended to every package of a prescription. US Senate hearings on the effects of the Pill in 1970 led to FDA rulings requiring doctors to provide information on health risks.[99] Consumers were expected to become informed about the drugs they are prescribed and to consent to their use, a change from the prevailing idea that professionals knew better and consumers should just follow.[100]

The Pill also came into the debate in the 1960s and 1970s over population control.[101] Today, seven out of eight women in the world using the Pill are not American.[102] Birth rates in dozens of countries have fallen below replacement rates. And the total world population, while continuing to grow, is now predicted to peak in the next forty years before beginning to fall back.

Finally, the Pill has had an impact on some religious groups. It has served to distance many believers from a tendency to obey their clergy. There is little to no evidence that the frequency of use of contraceptives is any different among those with no religion, those who adhere to a church that is favorable to contraception, and those who are members of one that opposes it. The overwhelming number of Americans, and people of other countries, do not seem to connect the decision to have children with what their faith tells them. It is a kind of silent disregard.

Conclusion

In our travel along the Yankee Road, we will encounter many stories like this one, where the building blocks of an invention are created, but it takes inspiration, some financing, and a lot of work to produce something with a global impact. In this case, two elderly women who had long fought for a cause came together in what was the final act of their struggle, and enabled a Jewish scientific entrepreneur, assisted by a Chinese immigrant, to create their dream of a "magic pill." None of them got rich off the Pill, not even the determined Yankee, Katharine McCormick, who put up her own funds and supervised the whole process through to fruition. Instead, the profits went to the company who supplied the chemicals for their experiments and to subsequent manufacturers.

The Pill's impact quickly jumped the US border, affecting population growth in the rest of the world, just as the rest of the world had affected the "woman rebel," Margaret Sanger, who had the initial inspiration. A decade after her death in 1966, birth control had passed from a "private vice" to a "public virtue" in much of the world.[103] The widespread adoption of the Pill emphasized what earlier commentators had seen:

> [T]he single most important variable in the historical process is changing technology. "As I understand it," wrote Henry Adams to his brother, Brooks, in 1903, "the whole social, political, and economical problem is the resultant of the mechanical development of power." By 1933, as exemplified by the Chicago Century of Progress Exposition, this form of modern fatalism had become a popular creed. In the Hall of Science, for example, there was a large sculpture featuring a nearly life-sized man and woman, hands outstretched, groping, and between them stood a mammoth angular robot almost twice their size, its metallic arm thrown "reassuringly around each." The meaning of this icon was made explicit by the official motto, as recorded in the *Guidebook of the Fair:* "Science Finds — Industry Applies — Man Conforms."[104]

Springfield, Massachusetts: The Quest for Interchangeable Parts

[Charles Hall was a] pivotal figure in the annals of American industry....No one...had been able to master the problem of attaining complete interchangeability in firearms....Hall's success in combining men, machines and precision-measurement methods into a practical system of production...represented an important extension of the industrial revolution in America, a mechanical synthesis so different in degree as to constitute a difference in kind.

— Industrial historian Merritt Roe Smith[1]

Propose to an Englishman any principle, or any instrument, however admirable, and you will observe that the whole effort of the English mind is directed to find a difficulty, defect or impossibility in it....Impart the same principle or show the same machine to an American, or to one of our colonists, and you will find that

the whole effort of his mind is to find some new application of the principle, some new use for the instrument.
— Charles Babbage, British inventor of the computer, 1852[2]

Les Yankees, ces premiers mécaniciens du monde, sont ingénieurs, comme les Italiens sont musiciens et les Allemands métaphysiciens, — de naissance. [The Yankees, the best mechanics in the world, are engineers, as the Italians are musicians and the Germans metaphysicians — from birth.]
— Jules Verne, 1866[3]

The Duryeas' Car

In the middle of downtown Springfield, Massachusetts, a historical plaque marks the site where the Duryea brothers, Charles and Frank, two men from the Midwest, built the first series of gasoline-powered automobiles in 1895.[4] They were not the inventors of the automobile — far from it — but they had the idea that they could be created in large numbers using interchangeable parts. They proved it could be done, but, as with so many other inventors, it was left to others to profit from their experience.

The Duryeas, along with another brother, Otto, grew up in Illinois farm country.[5] All three gravitated toward machinery, but Charles and Frank were entranced by bicycles,[6] and by the early 1880s both wanted to get into this rising business. Charles graduated from a seminary in 1882 and left home for the East. He saw his first gasoline engine at a state fair in 1886 and, for the rest of his life, was keen to bring this type of engine and the bicycle together, not as a motorcycle, but as a kind of auto or truck. He asked his younger brother Frank to come to Washington, DC, in 1888, where he was trying to produce a prototype vehicle while making bicycles. The brothers moved on to New Jersey, and by 1890 were in Chicopee, Massachusetts, just north of Springfield.

Charles was the talker and promoter; Frank was a toolmaker and persistent mechanic.[7] In 1892, Charles persuaded a Springfield financier to back his idea for a car to the tune of $1,000 in return for 10 percent of future profits.[8] The Duryeas rented the second floor of a manufacturing plant and brought up a buggy they intended to convert into an automobile.[9] By the early 1890s, they had all the systems in place to be developed into a car: a transmission and drive train, a battery, and a carburetor.

Frank succeeded in designing and developing his own one-cylinder gasoline engine, and in September 1893 produced a machine[10] that he claimed, wrongly, was the first American gasoline-powered auto.[11] Charles, meanwhile, the salesman and dreamer, raised more local capital to keep the enterprise afloat. In the summer of 1895, after an impressive demonstration of their car on an eighteen-mile ride over rutted country roads, he managed to interest a group of investors in the production of a series of two-cylinder automobiles. The audacity of the Duryeas' business concept is evident if one considers that, regardless of propulsion technology, only a few domestic automobiles existed in the whole country, all virtually hand made. The Duryea Motor Wagon Company was formed, and the brothers began to produce a run of thirteen of an improved model, the first commercial production in the United States. As one automotive historian notes, "[i]t was the first time that an automobile company was organized [in the United States] and produced more than one car from a single design."[12]

Their reputation was made when a Duryea auto won the United States' first real auto race, in Chicago on November 28, 1895. By then, a race had already become the standard way to promote cars in Europe.[13] The car the brothers entered was only the second they had made. Dozens of people expressed interest in participating, but eventually only six arrived for the race, including two electric vehicles and three modified imported Benz models. It was a snowy and windy day, and only two vehicles, including the Duryea, managed to finish.[14]

Although the Duryeas entered their car in a British auto race in 1896 and their company's future seemed bright, they could not raise more capital, the brothers quarreled,[15] and production stopped at sixteen vehicles.[16] In the ensuing years, the brothers each went his own way in the auto industry, with Frank cofounding the Stevens-Duryea Co., which stayed in business in a small way until 1927,[17] building larger, deluxe vehicles, while Charles drifted from dream to dream. They did prove that their concept was valid, however, and that the business of making automobiles was about to leave the realm of hobbyists and enter that of manufacturing.

Choosing a Technology

For a century and more, automobiles have been as standardized as personal computers. Where a computer generally has an Intel processor chip surrounded by a plastic box and runs off Microsoft software, an automobile uses a basic Otto-cycle engine to drive a four-wheeled wagon steered by gears and wires. The engine even usually sits in the front, where the horse used to be.

The Otto-cycle engine is named after a German, Nikolas Otto, who began experimenting with gasoline engines in 1866.[18] The first internal combustion engine was patented as early as 1826 by Charles Morey of Orford, New Hampshire, who ran it on turpentine vapors.[19] Others later "reinvented" the engine, including a Belgian, Jean-Joseph Étienne Lenoir, who developed a one-cylinder internal combustion engine in 1863, also fueled by turpentine.[20] Otto's first attempts were too large, heavy, and underpowered to be of much use in mobile equipment, but in 1885, he made an internal combustion engine using pistons that reacted to exploding gasoline to rotate a crankshaft and produce power. It had a number of advantages over steam and electric motors: it had a low weight-to-power ratio, it was cheaper to manufacture and to operate, and it was easy to fix.[21] Fellow Germans Karl Benz[22] and Gottlieb Daimler[23] were the first ones to adapt Otto's engine for automobile use.

Americans learned about the gasoline engine through publications such as *Scientific American*,[24] and by 1889 hobbyist cars began appearing using locally developed engines.[25] One experimenter of Yankee descent, George Selden of Rochester, New York, built an internal combustion engine and applied for a patent on a gasoline-propelled, four-wheeled car. Delays and amendments led to his being issued a patent in 1895, years after other American and European mechanics had built and operated automobiles with gasoline engines. Selden's patent claims complicated auto producers' activities for the next decade and more.

It was the French, however, who were the real pioneers in auto production, with entrepreneurs such as René Panhard, Émile Levassor, the Renault brothers, and Armand Peugeot leading the way in promoting racing and specializing in expensive, custom-built vehicles.[26] From the French we have borrowed such words as "garage" and "chauffeur."[27]

Another technology, steam, was used for this purpose even earlier. In 1769, a Frenchman, Nicolas Cugnot, is said to have created a steam engine on wheels that moved under its own power.[28] Oliver Evans in Philadelphia developed a number of steam-powered vehicles starting in 1805, and steam-driven coaches were apparently tried in England in the 1830s. Over the nineteenth century, enthusiasts in a number of countries worked on machines that began to resemble later automobiles as the size and weight of the engine dropped. For instance, a New Hampshire Yankee, Sylvester Roper, developed a steam-driven car during the Civil War while working at the Springfield Armory.[29] Other steam-powered enthusiasts miniaturized and adapted railroad engine technology to non-rail transportation, such as tractors.[30] The steamers had the virtues of simplicity of operation and smoothness of ride, but they took a long time to work up to operating strength.

A third technology was also in the making. Around 1880, Edison and others realized that electricity could be used to power industrial motors, replacing both the waterpower still widely used in New England and the steampower increasingly used in factories around the country. Adapting an electric motor to power a wagon was a natural extension. Even railroads faced a change to electricity, although this took hold more in Europe than America. Electric trolley cars were in use in America by 1888, and subways and interurbans somewhat later. Batteries could free the vehicle from overhead wires, but the batteries of the time were weak and gave the car little range.[31]

Technological choice happened in automobiles around the end of the nineteenth century.[32] It turned out that battery technology was not sophisticated enough to allow for the performance demands of automobiles — something that is still a problem today. Steampower was efficient, but, as mentioned, it took time to start up. The Otto-cycle gasoline engine started immediately and its range depended only on the size of the fuel tank. It also had another, hidden advantage: as a magazine article in the late 1890s said, "[i]t carries gasoline enough for a 70-mile journey, and nearly any country store can replenish the supply."[33]

Refining oil basically means splitting it into its constituent parts, from "light ends" such as butane and jet fuel, to "heavy ends" such as industrial fuel oil and asphalt. In the late 1800s, few of these "ends" had any particular value. The fortunes of the Standard Oil monopoly of the time depended, rather, on the production of kerosene, which, along with a small production of oils, lubricants and industrial fuel oil, was the only use of a barrel of crude oil. The rest, including the 35 percent that is gasoline, was thrown away. John D. Rockefeller once said: "We used to burn it [gasoline] for fuel in distilling the oil, and thousands and hundreds of thousands of barrels of it floated down creeks and rivers, and the ground was saturated with it, in the constant effort to get rid of it. The noxious runoff made the Cuyahoga River [through Cleveland] so flammable that if steamboat captains shoveled glowing coals overboard, the water erupted in flames."[34] Standard Oil's effective monopoly on refined products at the end of the nineteenth century was exercised through a web of distributorships, so it was no great feat to turn this network from selling kerosene, which was under threat from Edison's electric lighting grids anyway, to providing gasoline to sell to auto enthusiasts everywhere. Conversely, systems to support batteries and steam engines had to be created from scratch.[35]

Interchangeable Parts

The reason the Duryeas brought their ideas on technology and business east had much to do with interchangeable parts. Today, the notion of exact duplicates of parts and whole products is taken for granted, though even as late as 1980, Xerox was complaining that it could not get some parts for its copiers made to their proper specifications.

The quest for interchangeable parts probably began with Neolithic hunters millennia ago. Producing arrow bolts and heads so that all arrows would go to the same place with the same speed and effectiveness clearly was better than sheer, dumb luck when hunting a mammoth. The concept was simple, but the mechanics were not. In early nineteenth-century America, two parallel production processes were used to make yarn and cloth. In New England, Samuel Slater and then Francis Cabot Lowell made textile products using unskilled labor, in part because they could not keep people from moving West and setting up their own farms as soon as they got the money together.[36] Mills paid cash wages, part of which could be used as family savings or, in the case of the mill girls, put aside for a dowry. To use unskilled labor, however, the mill owners had to use technology that lent itself to yarn and cloth that was rough but durable. This forced them to concentrate on the mass market.

In Britain, mills tended to use more skilled craftspeople, who were higher paid but who would remain in their line of work. The technology they used produced a better grade of yarn and cloth. Most immigrants to America after the Revolution, including many skilled textile workers, went to Philadelphia, rather than to either New England, with its Puritan "tribal" overtones, or to New York, where the country upriver was dominated by Dutch landholders who wanted sharecropper immigrants, not freeholders. Pennsylvania makers of textiles for the upscale market had a harder competitive task than New England mills, given the advantages of the British in this market and their relative lack of capital to grow into large establishments like those in Massachusetts. The American advantage thus eventually went to Yankee producers, who focused on the mass market for rough cloth, using unskilled labor and power-intensive machine technology.

The reason machine technology was applied to produce cloth before any other goods is that exact tolerances were not needed either for the product or for the textile machinery. Bolts of cloth were cut to roughly standard sizes on machinery made almost exclusively of wood, which wore as it was used. As long as the shuttles moved back and forth and produced cloth, it did not matter if the pulleys, gears, and looms wore a bit. More likely was that the wooden machinery would crack and break, or warp with humidity. Replacement parts could be hand fitted,[37] and the relatively small number of machines in existence before the Civil War meant that they did not have to be mass produced. But products with a lot of metal parts or complex shapes also began to be manufactured in New England.[38]

Connecticut (River) Yankees

The Connecticut River valley was ripe for the development of manufacturing businesses at the beginning of the nineteenth century. Fertile land was not plentiful, and the families who settled the area could not subdivide the land among their children and still have viable farms. Birth rates were high and infant deaths relatively low, so family members had to find alternatives to farming.[39] An obvious alternative was emigration. Connecticut Yankees, basing their claims on their colony's charter, formed the Susquehanna Company in 1769, and settlers moved west across southern New York to the Wyoming country in what is now northeast Pennsylvania, where they clashed with local authorities until 1799 over land rights in the area.[40] They were also quick to move past the Dutch in the upper Hudson River valley and into western New York as soon as the Iroquois were driven out at the end of the Revolution. Valley people also moved north and settled parts of Vermont and New Hampshire, coming into conflict with the authorities in New Hampshire and detaching Vermont from New York, first as an independent republic and then as a State, an American process that was to repeat itself later in Texas, California, and Hawaii.

A second alternative was to devise non-farming methods of employment. A culture of innovation of sorts dated back to the first governor of Connecticut, John Winthrop the Younger (to distinguish him from his father, the founder of Massachusetts), who established a bog iron foundry in Lynn, Massachusetts and a grist mill in Connecticut.[41] In the 1790s, a Connecticut mechanic, Eli Terry, began to turn out wooden clocks, which were really status ornamentation for farm households that tended to work by the sun, but they were also cheap, and while the Europeans continued to focus on the market for hand-crafted clocks for the upper classes, Terry's products went into farms and frontier settlements.[42] After 1800, he took over an old mill and began to design power machinery[43] that would allow him to mass produce wooden parts for his clock mechanisms. He prospered for forty years and produced new clock-making machinery, clock designs, and trained mechanics.[44] Clockmakers, because they were familiar with complex gear trains, were in demand as machinists.[45]

By the middle 1820s, the Boston Associates had turned their eyes and textile investment funds to the Connecticut River valley, though its attractions lay in the waterpower provided by tributaries such as the Chicopee, rather than the main river. The growing number of larger mills both required and produced large numbers of skilled machinists, who moved about from job to job as well as starting their own shops.[46] Since the technology was applicable, using a certain degree of cleverness, to all sorts of endeavors, this dynamic spread knowledge about techniques and machine tools among all the New England communities.[47] Local merchants also developed smaller mills[48] devoted to making woolens as well as cotton cloth. Others made small quantities of shoes and boots, wooden and brass clocks, brooms, straw hats, wagons and tools, and equipment to service the local economy as well as the larger factories and the Springfield Armory.[49] Even religious communities, like the Shakers, began producing and using washing machines and knitting machines, as well as distilling whiskey and making oak staves for barrels, combs, and buggy whips.[50] A variety of other products and devices were produced by Connecticut entrepreneurs, including steel fishhooks, friction matches, hoopskirts, kerosene oil burners, and even a mechanical player piano. Patents issued per state resident in antebellum years averaged about one for every three thousand residents, but Connecticut was three times more active.[51] Of 143 important inventions patented in the United States before the Civil War, 93 percent came from the free States, with fully half coming from New England alone.[52]

A contrast developed between Yankee machinists and their British counterparts. Because they were physically mobile and their labor was much in demand, machinists in New England especially found their versatility tested by different challenges and different workplaces. The British machinist, in contrast, more often found himself working at the same task throughout his career, generally using the same personalized tools he had developed over the years. In New England, the machinist, moving to a new job in a new shop, was faced with using the tools available on the spot, often made by other shops to a more or less standardized form. The differences in terms of adaptability to a market and to different workplaces called for continuous innovation and job development. Thus, the transition from wood to brass to ironworking as technology moved forward was easier in America than in Britain.[53]

In contrast, farther south, below Pennsylvania, streams of settlers moved up and through the Shenandoah River valley as well as west over the mountains toward Pittsburgh. But these people, when faced with the problems of bad land, enjoyed the prospect of what seemed to be unlimited land farther west into which their families could expand. They also lacked the capital that New England was generating from its maritime commerce, as well as the markets, both local and relatively nearby, where they could sell manufactured goods.[54] Finally, they lacked the cultural norm of providing good, basic education for their children that pervaded New England as a byproduct of its Puritan heritage.

"The great arms factory..."

Enter the early military-industrial complex.

Roswell Lee[55] was an Army officer who became superintendent of the Springfield Armory in June 1815, months after the end of the embarrassment of the War of 1812[56] and weeks after the creation of a new Ordnance Department by Congress.[57] The Ordnance Department, reflecting on the experience of the previous fifteen years, became obsessed with uniformity in its products,[58] and, following the advice of French military officers, looked for manufacturers that could produce uniform parts for muskets. It found Eli Whitney.

Whitney, a Connecticut native and inventor of the cotton gin, had found he could not both protect his patent and produce enough gin machines given the manufacturing methods of the day, and as early as 1798, had successfully lobbied for a government contract to provide a large quantity of muskets with interchangeable parts to be produced in a very short time.[59] He failed to meet his contract date, resorting to legerdemain and political connections to keep and eventually fill his overdue contract over the next decade, although he did so with arms that lacked interchangeable parts.[60] Even so, Whitney pioneered the use of "filing jigs" that allowed a worker to file rough metal parts easily to the proper specifications, which led to "fixtures" to hold parts in place, gauges to measure sizes and tolerances, and water-driven cutting tools.[61] A commentator has noted:

> Among these pathfinders, Whitney ranks first, but not because he originated interchangeability, perfected it, or applied it successfully to mass production; he did none of these. What he did do was to perceive identical parts as a precondition to volume production, particularly in the circumstances of the early United States. He applied his energy and genius to the

problem and influenced his contemporaries and successors who addressed the same challenge. Whitney built the first arch in the engineering bridge that spanned the gulf between the ancient world's handicraft methods of production and the modern world's mass manufacturing, a bridge across which his countrymen and the rest of the so-called developed world marched to prosperity.[62]

Whitney's experience in the pursuit of interchangeable parts led to improvements in a number of other fields. The federal government asked his advice on improvements to its armories, while clockmaker Eli Terry borrowed some concepts and processes, and then, when Samuel Colt began to manufacture his famous repeating pistol, he had the Whitney plant produce the initial orders.[63]

At Springfield — the Armory's 250 employees made it the largest metalworking establishment in the country — Roswell Lee was determined to take up the challenge of producing interchangeable parts. At first, the Armory hired a group of gunsmiths to work together using their traditional skills, but there was little increase in productivity.[64] Then, Lee, still early in his eighteen-year tenure as superintendent,[65] had the Armory produce some of its own arms while continuing to contract out for some of its needs. He also insisted that the Armory's local contractors inform it and its competitors of any product or production innovations that might be of value in producing a better, more uniform weapon.[66] As an example, an armorer and two strikers wielding sledges could make six musket barrels a day, welding them shut inch by inch, but after a tilt hammer was introduced in 1815, a single armorer with one helper could make sixteen barrels in the same time.[67]

Lee was concerned with accounting and quality measures. During his tenure, the division of labor, or tasks, to produce a musket expanded from thirty-six to one hundred people, thus improving the measurement of individual productivity.[68] The Armory also experimented with new machinery and gauges[69] that improved the uniformity of the product.[70] The aim was to have a reasonably uniform gun that would fire reasonably uniform balls.

Lee took a personal interest in new machinery concepts, such as installing a trip-hammer to form gun barrels.[71] When this machine proved to have difficulties, he turned to a local mechanic, Thomas Blanchard, who had a penchant for ingenious problem solving.[72] Blanchard came up with a solution, and Lee asked him to look into other machinery-related problems. The results of this collaboration led to the introduction in 1819 of one of the first machine tools, the waterpowered Blanchard lathe, which allowed operators to turn out uniform pieces of wood with irregular shapes.[73] Used in conjunction with some of Whitney's major tool contributions, jigs that held parts in one place for shaping, the lathe could produce gunstocks from sawn lumber that were of uniform size and shape. By 1823, Blanchard had created a sequence of single-purpose machines that fully mechanized the manufacture of uniform gunstocks.[74]

Industrial archaeologists have noted the attempts throughout the nineteenth century by manufacturers to produce interchangeable parts and products. The introduction of lathes run first by waterpower and then by steam led to the production of quantities of parts made from soft materials such as wood and brass.[75] These techniques spread through the local manufacturing sector. For example, Connecticut-made wooden clocks, produced from uniform parts made by water-powered cutting machines in factories using a division of labor in the assembly process,[76] were sold by Yankee peddlers throughout the West and South.[77] In 1844, an English traveler noted: "Wherever we have been in Kentucky, in Indiana, in Illinois, in Missouri, and…in every dell of Arkansas, and in cabins where there was not a chair to sit on, there was sure to be a Connecticut clock."[78]

Some of the molds and cutting tools of the time, all hand formed by artisans, approached very fine tolerances. But the tools themselves, being made of relatively soft materials, tended to wear quickly and their tolerances faded. By the 1830s, as machine tools were able to cut harder substances than cloth and wood, metal-working began to replace woodworking as the machine-builder's primary skill.[79] It remained for innovators such as Frederick Taylor, working with hardened-steel cutting tools in the 1870s and 1880s, to develop methods of shaping steel parts that were interchangeable and lost little of their shape or cutting edges to wear. Working as an engineering trainee in the heat of a Chevrolet engine plant dynamometer room during my early college days introduced me to the importance of measurement of tolerances and wear and tear on metal parts.

In late 1818, a Maine toolmaker named Charles Hall was given a chance to produce a musket from interchangeable parts when he was sent from the Springfield Armory to Harpers Ferry. The Virginia armory, however, resisted Hall as a Yankee "visionary theorist."[80] Moreover, the effort to produce muskets in large quantities was not cheap: between 1819 and 1835, Hall spent $150,000 on machinery and equipment to produce muskets and rifles with interchangeable parts. Some of his machines were purchased, while others he had to design and have built. They were largely single-purpose machines, laid out in the factory so as to make the task of producing arms relatively straightforward.[81]

By 1834, Hall in Virginia and Simeon North, a gunsmith in Connecticut,[82] had succeeded, using copies of Hall's measuring gauges, in making muskets so alike that their parts were interchangeable, though they were filed to tolerances comparable to guns made decades later.[83] The remaining challenge was to produce them in quantities, which required machine tools. Only in 1841, after expensively re-equipping, could the Harpers Ferry Armory produce muskets in volume from parts whose tolerances were as exact as were needed.[84] When a British commission visited the Armory in the 1850s, they observed a test of muskets made in the previous several years: a workman dissembled ten muskets made in different years, mixed the parts, and assembled ten muskets from the mix.[85]

In the 1830s, Samuel Colt succeeded in developing and patenting his repeating pistol, or "six-shooter," out of relatively uniform parts, but he failed to sell it to the military due to the Panic of 1837, which led to many government cutbacks.[86] Colt's company managed to sell a few thousand of his pistols on the open market, but not enough to escape bankruptcy in the slowdown. Some of these pistols found their way to the Texas frontier, where they were adopted by the Texas Rangers. By the time of the Mexican War in 1846, the Colt pistol's reputation in Texas led to demands by local recruits to be issued the weapon.[87] With the aid of advances in interchangeable parts and a clever chief mechanic, Elisha Root, large quantities of the gun, said to have won the West, began to appear in both military and civilian hands in the 1850s. The Colt 45 Peacemaker, introduced in 1862, the year of Samuel Colt's death, became part of the "Wild West" in the 1870s and 1880s,[88] and Colt's factory in Hartford, Connecticut, was the model for Mark Twain's reference to the great arms factory in his *A Connecticut Yankee in King Arthur's Court*.[89]

Upon the outbreak of the Civil War, soldiers on both side started with smoothbore cannon and muskets, both low-tolerance armaments. As hard, steel-cutting tools and more powerful lathes came into use, it became possible to produce quantities of complex war goods that had both quality and endurance. By the end of the war, the Union army was equipped with rifled artillery and repeating pistols and rifles, the rate of fire and accuracy of which were manyfold greater than the arms generally available to the Confederate forces. Rifling, or creating spiral grooves in the gun barrel, allowed for more powerful and more accurate fire, but it required shells and bullets of constant size, something that was not as necessary with powder cartridges placed behind cannon or musket balls.

A key component of the process for creating large numbers of interchangeable parts was the machine tool. Tools have been part of humanity's stock in trade for millennia, but whether primitive axes or nineteenth-century specialized chisels, almost all were hand tools — that is, the motive power and force behind the tool was provided by a human being. Animals could pull tools such as a plow, but the application of waterpower in mills led to tools powered by an outside force. Machine tools allowed their users to replace human or animal effort in turning or hitting materials with the force of gravity in running water, which was both stronger and more consistently applied, since it was delivered through shafts, gears, and leather belts.[90] Productivity and accuracy both improved. Water, then steam, and then electricity could all deliver power to tools in a consistent and regulated fashion, and the game was on to produce tools to make complex shapes for parts in a way that was consistent and had good tolerances. As David Landes puts it, "[machine tools] must know how far to travel, which is another way of saying that they must make the same pass time and time again. They must be neat, for cleaning entails deformation. And they must be tough, for any wear of the cutting tool will change the result. All these requirements, of course, vary in rigor with the margin of tolerance. The smaller the margin, the more precise and consistent the machines have to be."[91] Machine tools could be programmed so that unskilled labor could be taught easily to use them productively. The computerized, digital machine tools of today are but refinements of the principles of the power lathe of the nineteenth century.

Productivity in American manufacturing made great jumps when these new tools — many copied from British designs and then improved upon — were introduced.[92] One of the earliest machines was a nail-maker that revolutionized the cutting and heading of nails, introduced by Jacob Perkins of Newburyport, Massachusetts, in 1795. It and its successors dropped nail prices from 25 cents per pound in 1795 to 3 cents by the 1840s. The ready availability of nails made practical the adoption of the popular "balloon frame" standardized wood building construction, which used machine-sawn boards and shingles coupled with mass-produced doors and windows.[93] This idea was conceived by a Yankee, George Washington Snow, and first used in two boom towns, Chicago and Rochester, and then used all over the country.[94]

By the 1850s, two great machine-tool manufacturers, the Ames Manufacturing Company in Chicopee, Massachusetts, just north of Springfield and its Armory,[95] and Pratt and Whitney in Hartford, Connecticut, whose founders had been employees of the nearby Colt Arms factory, were leaders worldwide.[96]

Building the Yankee World

Interchangeable parts technology gradually seeped from armaments into consumer products, not unlike how miniaturization in space technology a century and a half later moved to computers and communications.[97] The Remington Company, for instance, moved from rifles to sewing machines[98] to typewriters in the 1860s and 1870s using the same general technology: machined metal parts that were durable but easily replaced.[99] Even the architecture of the factories, long and narrow to take advantage of waterpower, was copied for other purposes until well into the age of steam.[100] The architecture reflected the two ideas that extended into the assembly line — layout and synchrony —that Samuel Slater brought to Rhode Island in 1790.[101]

Not only were interchangeable factories part of the system, so were interchangeable people.[102] Necessity transformed a personnel system that could handle locals who wanted to earn just enough to go west and buy or develop a farm into one that could absorb the millions of immigrants coming from Europe and elsewhere while keeping productivity high and prices affordable.[103] When mass production techniques, such as those developed in the Waltham mills, the Connecticut clock factories, and the Colt Arms factory, were united with precision, interchangeable parts around the Civil War, the Yankee approach to the provision of goods and services to all Americans and, ultimately, to all the world began to take off. As early as 1846, the whole process had become known outside America as "the American System." Bernard deVoto notes it this way:

> If you had spoken the phrase, "The American System," to [President James K. Polk] or any of his supporters or opponents, it would have meant to him the domestic policy fathered by Henry Clay and supported by the Whig Party. That is, strong centralized control, development of the internal market, systematized public works, and the protective tariff. But in England and Europe the phrase had already acquired a different meaning. It meant a kind of factory production new to the world....It meant: the displacement of hand labor by machine labor to an ever-increasing extent, the application of machine labor to successive operations, increased precision, the production of finished objects by such exact duplication of parts that the parts were interchangeable and so independent of the finished object, the progressive rationalization of processes and techniques, and the development of straight-line of processes and techniques, and the development of straight-line manufacture and automatic machine tools. It meant that, by 1846, the American industrial order had so matured that it was manufacturing tools for the manufacture of the goods exhibited at the [National] Fair [held that year].[104]

10

Albany, New York: Democratizing Capitalism

[Corporations] cannot commit treason, nor be outlawed or excommunicated, for they have no souls.
— Sir Edward Coke (1552–1634)[1]

We will say to Martin Van Buren, O.K., you can remain at the White House for four more years.
— Presidential campaign slogan, 1840[2]

Trade planted America.
— Ralph Waldo Emerson[3]

Americans did not invent banking, they democratized it.
— Paul Gilje[4]

The Incongruity of Albany

Albany has tended to grow on me. The city is not as large or as prepossessing as might befit the capital of one of the most populous States of the Union, but then neither is Sacramento, on the other side of the continent. Nelson Rockefeller, when he was governor of New York, tried to give it some heft by having some skyscrapers constructed around a huge public square, but they stand out oddly against a general low-rise profile.[5]

Part of the attraction of the area is the gradually growing-together group of small cities that line the Hudson from Albany north and along the lower reaches of its main tributary, the Mohawk. Troy and Schenectady stand out, the former located across the Hudson from the mouth of the Mohawk and the eastern terminus of the Erie Canal, the latter farther up the Mohawk, where it emerges from the enclosing hills to the northwest. Between this triangle of cities and towns lies a relatively flat plain that is filling with population.

One hundred and eighty years ago, in 1831, Alexis de Tocqueville and his partner Gustave de Beaumont, ostensibly engaged in prison research, took a steamship upriver from New York City to Albany. The reaction of these Frenchmen to the one-time Dutch outpost[6] and rude frontier town, now a gathering point for Yankees and others moving west,[7] was not negative, but the place was small[8] and the presence of government almost non-existent. Paris, with its royal splendor and large population, was more in keeping with a true capital city. Except for some Fourth of July celebrations, there seemed no awareness that the place was a capital of anything. De Beaumont noted in his journal about the approach to the town, "tilled fields [with] trunks in the middle of the corn….Nature vigorous and savage."[9]

Six years later, in 1837, English novelist and traveler Captain Frederick Marryatt felt much the same way, but then he was from an analogue of Paris — London. He noted the effects of Yankee migration into the area:

> Little more than twenty years ago, Albany stood by itself, a large and populous city without a rival, but its population was chiefly Dutch. The Yankees from the eastern states came down and settled themselves at Troy, not five miles distant, in opposition to them. It would be supposed that Albany could have crushed this city in its birth, but it could not, and Troy is now a beautiful city, with its mayor, its corporation, and a population of twenty thousand souls, and divides the commerce with Albany, from which most of the eastern trade has been ravished.[10]

Both travelers commented more on the region's main tourist attraction, the dancing performances of the Shakers at their colony at Niskayuna, in the center of the plain. Today, these Shaker buildings are preserved as a museum hard against the fenced boundary of the Albany airport.

US 20 borders the Empire State Plaza in downtown Albany, at the center of the State's governmental complex. In this place, where government seemed so invisible to early travelers, capitalism was democratized. In 1811, the state legislature passed an act of significant import for the economy of New York, America, and eventually the world. One day, it would be called on to control the forces it let loose — but that is another story.

Back to Albany

The economic growth that the Erie Canal and the railroads brought did have some impact on Albany and its neighbors, but vastly more so on New York City. As the city grew by leaps and bounds, it remained tethered to Albany constitutionally while its interests were national and international. "Upstate" and "NYC" were different in outlook and population. In New York City, there was Yankee influence at the upper end of the political and economic scale, but below was a growing population of foreign immigrants. Upstate remained Yankee in character throughout society much longer. Albany was where their interests were traded off, to the benefit of both, for the most part.

In this trade-off, Albany saw more than its share of scandals and fights. It saw Governor DeWitt Clinton's "Big Ditch" flower with commerce from the West and people going the other way. It saw Martin Van Buren's "Regency," with its political support for financing New York and its growing financial sector. Then came the railroads and "free banking" and more trade with the west. The rails made Albany a hub between New England, New York City, the western part of the State, and later Chicago by centralizing Western produce for shipment to New York City and Boston. The railroad scandals of the late 1860s and '70s left "informal" money in the pockets of legislators. Sweatshops blossomed in New York City with the arrival of millions of immigrants, as did wholesalers, department stores, and financial institutions, all looking to Albany for part of their legal existence.

Locally, the Albany area shared in the commercial boom from the start. By 1826, a normal week would see two hundred canal boats towed into the Albany Basin[11] with goods from the west and from the north (along the Champlain Canal). It was then the seventh-largest city in the country,[12] and its local industries matched the manufacturing prosperity of other upstate cities. There were hatters and textile companies and foundries that supplied railroad parts. Troy produced 85 percent of America's collars and cuffs,[13] and toward the end of the century, the headquarters of General Electric were located in Schenectady. Albany could boast of one of the earliest department stores, which, in turn, introduced a number if innovations to the community, such as electric lights, elevators, and telephones.[14] Gaslight came in 1845, kerosene in 1860, the bicycle in 1869, and a telephone exchange in 1878.[15]

On March 22, 1811, the state legislature, meeting in session in Albany, passed an act of "free incorporation for some kinds of businesses." It was seen as merely a temporary measure as America drifted into war with Great Britain, but it ultimately created the general structure of the modern global economy. A century later, in 1919, corporations, making up less than a third of the total number of businesses, employed 86 percent of the business workforce in America and produced almost 90 percent of its output by value.[16] One hundred years and three days after the passage of the act, on March 25, 1911, a tragic disaster in New York City, caused in part by corporate abuse, led to an investigation that involved a group of people whose experience underlay what was perhaps the defining political event of twentieth-century America, the set of political actions called the New Deal. Capitalism was both democratized and then later controlled as a result of what happened in Albany.

The Logic of Incorporation

The great French historian Fernand Braudel saw capitalism, in its basic form, as the injection of capital between the actions of buyer and seller.[17] This is both simple and profound. It explains the difference between a farmers' market and a supermarket. In the former, the seller and buyer meet face-to-face for the exchange. In the latter, the seller sells to an intermediary, who then may process and resell the good to a supermarket chain that distributes it and resells it once more to the final buyer. To conduct economic relations at a distance requires capital.

Max Weber added another dimension when he defined "a capitalistic economic action as one that rests upon the expectation of profit by the utilization of opportunities for exchange, that is, on a (formally) peaceful chance of profit." Capitalism is identical with the pursuit of profit, *and forever renewed profit*, by means of a rational capitalistic enterprise. Neither greed nor acquisition is relevant to capitalism.[18]

Capitalism is as old as most other economic activities. Seashell ornaments were discovered in the Anasazi ruins in the American southwestern deserts showed there had to have been a capitalist exchange going on. An intermediary with capital must have been between the shell-gatherers of the Pacific coast and the cliff-dwellers hundreds of miles inland. The Romans got pepper from Indonesia the same way. The question in early America was how to provide people all over the wide and distant frontier with the necessities of life. Clearly, a capitalist relationship had to be established. But how could this be organized?[19]

Historically, there have been only four ways to organize people for productive activity. All of today's organizational permutations and combinations are but variations on these four. They are: the sole proprietorship, the partnership,[20] the joint-stock corporation, and the government agency. Most common has been the sole proprietorship, the rarest the joint-stock corporation. Even in Roman times, a shopkeeper might have been a sole proprietor, the owners of a merchant vessel would have formed a partnership, and a government agency would have managed a silver mine. The Romans, indeed, are considered the founders of the components of the modern corporation.[21] Individual tax collectors found they could not personally guarantee the receipts the growing Roman Empire required, so they came together with other rich men to form *societates*, in which shares were distributed. These financial companies used some of their profits to create other large companies to make, among other things, arms for the Legions. The *societates* used professional managers and formal accounting methods, and launched the idea that a group could have a separate identity from that of its individuals.[22]

In medieval times, free cities and towns, monasteries, and universities all involved charters of incorporation provided by royalty.[23] Other forms of organization were not adequate to the task of overseeing common property or to perform other public purposes. A church building and its cemetery could not be "owned" by a sole proprietor or a partnership. Instead, a legal device was created: a "legal" person, as opposed to a real, physical one.[24] The "legal" person would be given a legal "body," or "incorporated," and its behavior would then be controlled by a group of trustees or directors. Much of the medieval use still exists,[25] although its connection to the state is muted. Probably the most obvious remnants lie in government-owned, or state corporations, and in incorporated municipalities.

It does not take a great mental leap to see the potential value of the corporation for economic or business purposes. Control over the legal "body" could be divided in terms of "shares" of its "stock," or assets, that could be assigned or sold to different people or other corporations. Selling shares would then constitute a means for accumulating large amounts of capital from many investors. The business corporation would have the advantage of "outliving" any or all of its owners, something that plagued the heirs of sole proprietorships and partnerships.[26]

Colonization by Corporation

The prevailing view of incorporation from the 1600s until the 1800s was that it was a device that could or should be used only for specific public purposes and benefits. If government were to charter a business corporation, the entity had to have a well-understood public purpose, two of the most common of which were profit from foreign trade and colonization.[27]

In the 1600s, the Dutch, who built their empire in the East through the use of the Dutch East India Company,[28] founded what later became the British colony of New York and developed many of the instruments of modern capitalism.[29] This legacy was passed on in America, and has colored the behavior of New York City since then.[30] The British[31] and the French adopted the Dutch innovation and used incorporation in the same manner. The tendency was to give the company both commercial and civil powers. In the case of the Dutch and the English, the most successful of these companies ruled huge territories in what are now Indonesia, India, and northern Canada. Perhaps the most successful was the British East India Company,[32] which gradually took over and administered most of the Indian subcontinent, to be displaced by the British Crown only following the 1857 Sepoy Mutiny.

In the case of colonization, Yankee origins can be traced to the Virginia Company, whose charter gave it rights to much of the Atlantic coast of North America. The Plymouth colony in today's Massachusetts began as a kind of subsidiary of the Virginia Company, holding what amounted to a subcharter.[33] Shortly thereafter, the Massachusetts Bay Company was chartered as a separate organization to settle the area of Boston and northward.[34] Some other colonies were created by corporations as well, though most were based on Royal grants given to individual "proprietors" such as William Penn.[35] Not only was much of America created as a result of corporate organization,[36] but the first corporate merger in its history came from the absorption of the Plymouth colony into Massachusetts in the mid-1600s.

Though the British government had largely stopped providing charters of incorporation for businesses in 1719 in the aftermath of the trauma of the South Sea Bubble,[37] colonial legislatures were not so inhibited. Thus, the Massachusetts legislature chartered an ironworking company at Saugus in the 1640s to produce iron and iron products locally, rather than importing them. During the 1700s, the most common activities of colonial corporations were transportation, banking, and, to some extent, land speculation and development. At the time, transportation meant turnpikes and canals, where public benefits were fairly clear and whose investors would require relief from competition,[38] not unlike today's utilities.

The reason for chartering some land development companies tends to hark back to the founding of many of the original colonies.[39] The aims were to enable capital to be pooled by either potential settlers who might need a corporate structure to purchase a large tract of land, or by financiers who intended to purchase, survey, and sell land to settlers, and who needed extra capital to provide local improvements as inducements to settle. The Ohio Company, chartered in Massachusetts in 1785 to settle southeastern Ohio, is an example of the former, while the Holland Land Company, chartered in Pennsylvania to purchase and sell land in New York and Pennsylvania, is an example of the latter. Once the Northwest Territory and the lands across the Mississippi came under federal control after 1800, its survey and sales policies made such corporations uneconomic.

"Free" Incorporation

The critical point is that charters were all one-off grants of privilege.[40] Each charter had to be applied for and approved as a separate legislative act. In general, this meant that incorporation was a tool largely reserved for, and used by, the political and economic elites. Entrepreneurs often found it difficult to get their petitions acknowledged and passed.[41] Democratic capitalism required "free" or automatic incorporation for businesses.

In the early 1800s, a rash of charter proposals began to occupy more of the time of the state legislatures. Many were for the formation of banks, especially in the capital-poor Western States. Other proposals arose for large-scale manufacturing, where the size of capital needs and the requirement that the capital remain in place regardless of the personal situations of the various people involved in the formation of the enterprise made the corporate form more appropriate.[42] Canal and turnpike proposals multiplied,[43] especially after the opening of the Erie Canal in 1825. Then the popularity of railroads added to the pressure.

The disruption of trade resulting from the Napoleonic Wars, the American trade embargo of 1809, and an increasingly hostile reaction toward British policy, which led to the War of 1812, encouraged American entrepreneurs to create manufacturing facilities for consumer products that otherwise would have been imported.[44] Legislators in the Northern States especially were faced with charter proponents who argued that, in these circumstances, they met the test for public benefit. This argument, expanded along different lines after the war, led the public and the legislators to identify increases in business activity as an important component of the wider public interest.[45]

Faced with this pressure, in March 1811, the New York State Legislature passed an act that allowed any petition that met the act's list of conditions to proceed.[46] It went a long way toward allowing the "free" incorporation of businesses, but still contained a number of restrictions, including a narrow list of eligible types of businesses, an upper limit on the capitalization of a proposed company at $100,000, and a twenty-year charter term limit.[47] The act was meant to be temporary, but after the War of 1812 its basic principle was never retired, though it was superseded by later legislation. As early as 1817, then-governor of New York and later Andrew Jackson's chosen successor as president, Martin Van Buren, proposed, as a "democratic" reform, the creation of a more generalized "free" incorporation law. Other States largely did not follow New York's lead, however, and it was left to the "Jacksonian revolution" to "free" the rules governing incorporation. One indication of this coming trend was an 1829 article in a legal journal complaining that "millions in capital were being diverted out of Massachusetts by the legislature's refusal to grant adequate charters."[48]

In 1834, the editor of the *New York Post*, William Leggett, wrote a series of articles entitled "What is Monopoly?"[49] in which he called for an end to chartered monopolies, including banks, and advocated equal rights in the use and protection of property: "Their only safeguard against oppression is a system of legislation which leaves to all the free exercise of their talents and industry, within the limits of the GENERAL LAW, and which, on no pretence of public good, bestows on any particular class of industry, or any particular body of men, rights or privileges not equally enjoyed by the great aggregate of the body politic."[50] Leggett, considered a left-wing radical by the New York elite, went on to propose a general incorporation law, with no limits: "Such a law would be the very measure to enable poor men to compete with the rich."[51]

Free banking in New York came in 1838 and general free incorporation, pioneered in Pennsylvania in 1836 and Connecticut in 1837, was provided for in the New York State Constitution of 1846, though the law enacting it came only in 1848.[52] Gradually, other States followed in this direction,[53] though a lot of restrictions on capital size and term of existence remained through the rest of the century.[54] Among other things, free incorporation led to the creation of hundreds of railroad companies as the effects of the Panic of 1837 wore off.[55]

The Value of the Corporation

As it evolved, "free" incorporation divorced the organizational tool from the notion of a public purpose. Instead, a company could incorporate if it simply met a list of rules. Unlike the experience under the legislative charter system, these corporations were not created in an ad hoc, or one-off, system of political action, but according to general rules applied to whole industries, much as we have today.[56]

Incorporation seemed to be the answer to many problems. Prime among these was that, because of the inherently risky proposition that was the frontier, entrepreneurs needed a vehicle that allowed groups of them to syndicate risks and potential losses. Those who took chances and lost needed some protection from debtors' prison if they were to be able to keep trying to make their fortunes. Their liability had to be limited, something that other forms of organization could not offer.[57]

A common practice in early America was the imprisoning of people for nonpayment of their debts.[58] If they were not caught, many debtors simply fled west and started over again.[59] But the threat of imprisonment did act as a deterrent to young entrepreneurs and small businesspeople whose enterprises might fail. During his term as governor of New York, Van Buren was one of the earliest to introduce a bill to eliminate imprisonment for debt, though it did not pass. Later he campaigned for this same measure while in the US Senate.[60] One estimate is that, in 1830, five-sixths of those incarcerated in New England and the Middle Atlantic States were there for reasons of debt,[61] somewhat akin to the proportion incarcerated today who are jailed for drug-related crimes.[62] It took a long time to overcome the influence of the alliance of the large landowners and the new big businessmen of the State, "the Aristocracy," but in 1831, Jacksonian democratic success at the polls meant that imprisonment for debt in New York was finally abolished.

Limited liability was central to the use of the corporation, and by the 1850s, the idea had become widespread.[63] It meant that the investor was liable for the company's obligations to the extent of his or her investment, and no more.[64] The loss in case of the failure of the company was limited to how much the person had invested, and could not extend to the other assets of its investors. Unlimited liability could make sense if a venture were a relatively small project or a specific activity with an agreed end-point, such as a sea voyage. If a project had a life measured in decades,[65] the capital requirements were large, and the liabilities unknown in advance, there was little chance of its ever getting started in the absence of limited liability.

Unlimited liability also impeded the ability of investors to sell their interest, since any new owner of the shares also had to assume the total liability for the actions of previous and existing owners. Limited liability, in contrast, allowed the firm's shares to become financial instruments only, not commitments of one's whole fortune to an enterprise with potentially enormous obligations.[66] The ability to see shares as financial instruments meant they could be sold and resold, which inspired the development of an aftermarket for shares. This led to the rise of a trading community in New York. The New York Stock Exchange, located along Wall Street in lower Manhattan, had come into being in 1817, but its importance as a national financial market took off only with the emergence of limited liability shares and the technological revolution of the telegraph, both coming in the 1840s.[67]

As a result of stock trading, the relationship between management and shareholders changed. They were not normally one and the same, as they were in other forms of organization. Shareholders could be "investors," holding the shares for financial gain only, rather than operators of the company. This led to the rise of professional management, with all its attendant organizational power, specializations, and needs.[68] This became more obvious in 1890 once New Jersey became the first state to change its law on incorporation to allow corporations to own the shares of other corporations.[69] Management then became even more abstracted from ownership.

The attempt to provide for failed businesspeople also affected the fate of failed corporations. For both, the legacy of the "democratic" reforms of the 1830s and 1840s led to a quick and effective method of putting people and corporate assets back to work after they had failed. People were no longer jailed for owing money they could not repay, and corporations were allowed to fail such that their productive assets could be passed efficiently to others who could make better use of them. Today's "Chapter 11" and liquidation procedures do for corporations what personal bankruptcy does for individuals: they give people, both physical and legal persons, the ability to re-enter the work world and try again.[70]

State and Nation in Incorporation

Alongside these evolving provisions, the US Supreme Court made a series of decisions that opened up competition and defined the nature of the corporation relative to the rest of society.[71] Ensuring competition emerged as the most common regulatory means to ensure that prices and services would be in the consumer's interest. The Court first struck down monopolies in interstate trade, then in 1819 it said that state legislatures had to obey the terms of the charters they issued,[72] meaning that it was not their business to meddle periodically and politically in the corporation. In 1837, the Court ended the ability of state legislatures to grant monopolies to intrastate corporations,[73] followed in 1842 and 1844 by definitions of corporations as legal citizens, which allowed governments to treat them under general laws of taxation and regulation.

Outside New York, the use of incorporation as a business tool tended to be more concentrated in the Northeast, including in Massachusetts and Connecticut, than elsewhere. Pennsylvania, as well, saw a gradual increase in incorporations, but farther to the south and southwest, incorporation was little used. Southern and Western state legislatures, with the exception of those areas settled by Yankees, tended to treat even banking incorporation petitions cautiously, with the result that the Northeast continued to dominate national financial and commercial activity. Before the latter part of the nineteenth century, however, federal incorporation was a relative rarity: a national bank was incorporated in 1791[74] and another in 1816; a manufacturing entity promoted by Alexander Hamilton, the Society for Useful Manufactures, was federally incorporated in 1791, and a company created to construct the first segment of the National Road was incorporated in 1806. During and after the Civil War, the transcontinental railroad was incorporated, along with a number of other railroads and utilities.[75]

The fact that States continued to dominate the incorporation landscape before and after the Civil War meant that corporations that did business across state boundaries found them an increasing impediment.[76] In the 1840s and beyond, railroad and telegraph companies found this restriction a nagging inconvenience, which they learned to sidestep by "leasing" a connecting railroad or telegraph line in another State, thus acquiring control over it without affecting the "shell" of the corporation whose assets were leased. After 1870, manufacturing companies such as Standard Oil developed nationwide facilities and distribution networks by turning over the controlling shares of related companies in different States to "trusts" to manage, vote on, and coordinate, thus sidestepping state residency restrictions.[77] This kept the various corporations local, while placing actual control in the hands of national management. In turn, the trust could issue its own shares to replace those turned over to it and to be used as a vehicle for dividend payments. This created a new class of securities that could be traded on Wall Street.

Beginning in the late 1880s and for the next two decades, a variety of industries, including oil production, refining, and distribution, business machines, sugar, and rubber products were organized as trusts.[78] The policy challenge then became how to provide a competitive environment to foster innovation and more efficient production. Breaking up the monopolistic trusts became a central interest of progressive politics in the 1900s.[79]

Financing America

The financial problems on the frontier in the 1830s were not really different than those facing the Pilgrims in 1620, or, for that matter, Greeks colonizing the shores of the Black Sea or Sicily two thousand years before. It is axiomatic that new colonies require significant ongoing capital investment if they are to survive and then later generate a surplus. How much capital they need depends on a number of factors. If the colonists are used to living in a subsistence-type economy, the need for capital might be low. The converse is true if the colonists come from a prosperous society. If there are locals present where the colony is placed, conquest might allow for the appropriation of capital.

In the case of the American colonies, the English chose to have heavier-populated colonies that had a relatively high standard of living and primarily exported foodstuffs and wood, as well as some furs. The colonies always had a problem of how to allocate what capital they possessed. They needed export revenues to buy goods and services that the colonial market could not provide, especially when the British authorities prevented the establishment of local industries that would compete with those on the home island. This, and the requirement that imports had to be paid for in hard (British) currency, could leave little money for a colony's internal development. Even the choice of what to import had to be made in terms of consumption or capital goods for development.[80]

One possible solution was to print money for local use (scrip), and to reserve hard money (specie) for the foreign market.[81] Scrip for local circulation was generally provided by the colonial/state authorities before 1788, but later it was increasingly provided by private banks.[82] As long as there was faith that these new, local notes could be exchanged at any time for hard currency — whether pounds Sterling, Spanish "pieces of eight" (half a quarter), French francs, or German-Austrian thalers (dollars) — all was well: the money in circulation could be used to purchase land, services, and possibly even imports.

Massachusetts issued the first paper money in America in 1691 as a temporary measure,[83] but it proved popular, and further issues were made in 1714 and 1719.[84] The colonial government during this period rejected a plan for a private bank with issuing privileges.[85] Because the notes could be used only for domestic debts, the colony's legal tender was of little interest to the larger merchants, but it did prove popular in the hinterland as a way to settle small debts when no specie was at hand, which was usually the case.[86] Chartering a bank, however, was the most common way to produce money outside direct government means. In general, banks could use a capital base, or capitalization, of hard money as a foundation for making loans to customers at a ratio of say 10:1 — that is, for every British pound, say, that was part of its capital, the bank would print 10 pounds in its own "bank notes" for loan purposes, all of which theoretically could be redeemed at the bank for specie. As long as everybody who held these notes believed in their worth, all was OK.[87]

But there were many threats to the faith in the convertibility of bank notes. Perhaps the bank might "over leverage" — that is, print too many banknotes — until someone began discounting the notes by paying less than the stated face value with another bank's notes.[88] Perhaps the bank would be "undercapitalized": such "wildcat" banks[89] were notorious on the Michigan frontier after that State's 1837 free banking act, where there was little capital or it was fraudulently declared.[90] Perhaps another bank nearby would fail, worrying holders of banknotes in a sound bank.[91] Shaken faith could cause a run on the otherwise sound bank as people demanded "real" money in exchange for their local money.[92] If most of the banks in a jurisdiction experienced such a run at the same time and failed to meet their obligations, there would be a "panic" as the money supply got very tight and business activity, which depended on credit, slowed. The panics of 1819, 1837, 1857, 1873, 1893, 1907, and 1929–32 were not much different than what happened in the global financial crisis of 2007–8.

Captain Frederick Marryatt, who arrived in New York City at the onset of the Panic of 1837, commented on the informal credit systems that had arisen as access to money dried up:

> The distress for change [for large bills] has produced a curious remedy. Every man is now his own banker. Go to the theatres and places of public amusement, and, instead of change, you receive an I.O.U. from the treasury [cashier]. At the hotels and oyster-cellars it is the same thing. Call for a glass of brandy and water and the change is fifteen tickets, each "good for one glass of brandy and water." At an oyster shop, eat a plate of oysters, and you have in return seven tickets, good for one plate of oysters each. It is the same every where.—The barbers give you tickets, good for so many shaves; and were there beggars in the streets, I presume they would give you tickets in change, good for so much philanthropy. Dealers, in general, give out their own bank-notes, or as they are called here, *shin-plasters*, which are good for one dollar, and from that down to two and a half cents, all of which are redeemable, and redeemable only upon a general return to cash payment.[93]

Then there was counterfeiting. With so many banks as well as governments issuing scrip or still having theirs in circulation, though bankrupt, it was not difficult to make reasonable facsimiles of existing bills or, better yet, make bills from a bank that did not exist at all.[94] Small wonder that Boston business people around 1800 petitioned the government to charter a bank that did no banking. The Exchange Office would look at bills and tell the bearer what they were worth in terms of other, more trusted or local bills, thus restoring faith in the whole system.[95] This innovation was later replaced in New England by the Suffolk System[96] and in New York by the Safety Bank System. Keeping faith in the banking system was not unlike the function performed by the US government when it propped up (recapitalized) the major American financial houses in 2008.

"Free" Banking

The first bank chartered in the United States was the Bank of North America, organized by Robert Morris in 1781 and headquartered in Philadelphia, at the time the Capital.[97] In 1784, a group of shipping insurance underwriters wishing to put its profits to work gained a charter for the Massachusetts Bank, which was given the right to print money on behalf of the Commonwealth.[98] In 1791, Secretary of the Treasury Alexander Hamilton, under a federal charter, formed the Bank of the United States.[99] As States began to charter banks, their number grew until more than two thousand were spread across the country.[100]

The very existence of banking companies was controversial. From the beginning of the country's existence, a debate, which could be characterized as between North and South, had raged over the morality of and need for banks. Did banks grease commerce's wheels or were they parasites and frauds?[101] The debate especially played out over the chartering of the Bank of the United States, with Thomas Jefferson, a Southerner, unsuccessfully opposing Hamilton. Twenty years later, Jefferson's successor, James Madison, another Southerner, let the charter and the Bank expire. After the financial debacle of having no federal institution to finance the prosecution of the War of 1812, he was forced to approve the chartering of a Second Bank of the United States in 1816.

Part of the Southerner Andrew Jackson's 1828 election platform was to support the Southern and Western farmers and businesspeople who opposed the Second Bank, and whose federal charter was due for renewal in 1836.[102] Jackson's moral stance against banks was accompanied by a practical problem having to do with the policies of the bank, which was located in Philadelphia and which discounted local bank notes from the weaker frontier banks, constraining income and credit in these areas. Jackson's solution was to cancel the charter and use the federal deposits previously placed in the Second Bank to support the capitalization of local banks in the States.[103]

Jackson's "war" on the bank galvanized New York Democrats, who interpreted it as a crusade against monopoly. This then carried over into an assault on those who gained legislative consent, often through bribery, for state bank charters.[104] When a rising group of New York entrepreneurs found that their ambitions could be thwarted by competitors who were already part of the Establishment through the simple expedient of getting the legislature or governor to deny their petitions for incorporation, they championed free incorporation as their way to democratize capitalism.

To many ambitious businessmen in the growing New York City, access to capital was just as critical a problem as it was for their counterparts on the frontier.[105] Finding they could not acquire bank loans without going through the same rich and influential people, they advocated a logical extension of free incorporation: "free banking." These Eastern businesspeople saw in the Jacksonian Western agenda a way to democratize success and wealth. As well, they were supported by many working people in Eastern cities whose wages were paid in banknotes issued by out-of-state banks that were heavily discounted when spent locally — in effect, robbing them of part of their earnings and raising their ire against banks in general.[106]

In 1834, Connecticut took a step toward a free banking act by allowing banks to incorporate under a general set of rules. Michigan passed a free banking act in 1837;[107] in this case, "free banking" meant that incorporating a bank would not be done through issuing a special charter voted by the legislature, but by meeting general terms for banking, just like free incorporation for manufacturing. New York Whigs, shut out of power by Van Buren's Democrats, saw free banking as a way to win over the business community and regain the legislature. It worked — New York got free banking in 1838.[108]

Free banking wasn't quite what Jackson himself had in mind in terms of finance and business organization. He came from the Southern frontier, where there were few towns or businesses and almost no manufacturing; its people were plantation owners, their slaves,[109] and near-subsistence-level white farmers. At the extreme, these interests were greatly suspicious of city merchants and their banks, so in most Southern States even the creation of local financial institutions was opposed.

Farmers and local manufacturers on the Northern frontier, however, welcomed the prospect of local banks and easier credit terms, and the implicit inflation these local institutions might bring. They could not have anticipated the financial ramifications of their welcome: the Panic of 1837 that followed the winding up of the Second Bank in 1835–6. Overnight, credit went from very loose to extremely tight, causing farmers and businesses to contract their activities and expectations drastically.[110] It took more than four years to emerge from the recession caused by Jackson's "war" and the subsequent Panic. Ironically, Van Buren lost his chance at a second term as president in 1840 because of the fallout from his party's success in executing its platform.

As in Madison's day, the attempt to leave the production of banknotes to State-chartered banks did not survive wartime. Passed during the Civil War, the *National Banking Act of 1864* created a system of federally chartered National Banks and provided for the Treasury to issue a national currency, called "greenbacks" in local parlance due to their common color. These could bear the name of a local bank that was chartered under the federal act, but they could not be discounted regionally.[111] This time the change would be permanent.

A crucial part of American history is concerned with the availability of money.[112] Those who did not want banks to take risks and possibly lose their money on deposit wanted the ratio of local money to hard money kept low. Those who needed money to hire people to clear land or build, say, a fishing boat or a house, wanted a higher ratio. Borrowers — whether New England fishermen in 1696, prairie farmers in 1896, or dot-com startups in 1996 — want money to be available, while savers worried about the safety of their capital.

In the end, the plethora of American banks has shrunk through an ongoing process of consolidation and panics. The continuing nostalgic resonance of Frank Capra's 1946 film *It's a Wonderful Life* — a Christmas story focusing on a helpful small mortgage banker in "Bedford Falls" trying to keep his bank afloat by avoiding a capital shortfall — is testament of sorts to the popular appeal of "free" banking.

Far downriver from Albany, Nicholas Murray Butler, Nobel Peace Prize laureate and president of Columbia University from 1901 to 1945, called the corporation, with a degree of hyperbole, "the greatest single discovery of modern times."[113] There is no question that its use has become a large and continuing feature in the lives of Americans —and of the rest of the world. Today, the notion of free incorporation — the ability to create a corporation by meeting a set of requirements and obtaining licenses — is observed everywhere.[114]

Cooperstown, New York: Heroes or Winners?

Tell them to wait until I make another hit.
— Attributed to Abraham Lincoln in a 19th century fable, when a delegation supposedly appeared at a ball game to tell him of his nomination in 1860[1]

I don't have to tell you that the one constant through all the years has been baseball. America has been erased like a blackboard, only to be rebuilt and erased again. But baseball has marked time while America has rolled by like a procession of steamrollers. It is the same game that Moonlight Graham played in 1905. It is a living part of history, like calico dresses, stone crockery and threshing crews eating at outdoor tables. It continually reminds us of what once was, like an Indian-head penny in a handful of new coins.

— W.P. Kinsella[2]

Just as [George Babbitt] was an Elk, a Booster, and a member of the Chamber of Commerce, just as the priests of the Presbyterian church determined his every religious belief and the senators who controlled the Republican Party decided in smoky rooms what he should think...[about politics], so did the large national advertisers fix the surface of his life, fix what he believed to be his individuality.

— Sinclair Lewis[3]

Cooperstown and the Coopers

Barely eighteen months after the British evacuated New York, a Philadelphia-area Quaker, William Cooper, picked up a patent, or official grant of land, in upstate New York that had been held by a Tory who left with the departing Royal Navy.[4] Soon, he was on his way north from New York City to view and then to begin to develop his property. His significant acreage lay along the west bank of Lake Otsego, which drains into the Susquehanna River. He called the town he founded, located where the waters of Otsego Lake outflow and create the Susquehanna River, Otsego, but its name was changed after William Cooper's death to honor him. He liked to claim that he had penetrated an untrodden wilderness, but this was patently untrue — in the area were considerable Six Nations ruins and village sites, and farmers worked the land just to the north of Cooper's patent.[5] And ten miles away to the east, along the best access to the Mohawk River and therefore to Albany, was the village of Cherry Valley, founded nearly fifty years before Cooper's arrival by families from New Hampshire. The village even boasted an academy, burned in the Cherry Valley massacre of 1778; General George Washington later came to Cherry Valley to honor the victims.

Cooper was a land speculator, economic developer, and political manipulator. He sold his land to Yankees, while the Dutch to the east continued to insist on leases. Though he was the first judge in his district, he was not above lobbying Albany for road contracts and fixing the county ballot boxes to help his party. Judge Cooper died in 1809, apparently from a blow to the head incurred in a fight over politics.[6]

Cooper's son, James Fenimore Cooper, was a more aristocratic type and a popular Romantic author.[7] He drifted away from his father's Federalist politics because he saw in Thomas Jefferson's Democrats greater protection for landed property than he did in the "monarchist" Federalists.[8] He worried about the loss of his property as the area filled with settlers. Though he had more in common socially with the Dutch *patroons* of the Hudson Valley, he saw forming a nation of landowning farmers elsewhere as an appealing policy.[9] In the late 1830s and 1840s, property conflicts led him to condemn the tyranny of public opinion, first when the people in Cooperstown fought him over the ownership of what they considered public land,[10] and then later when an anti-rent movement threatened the property rights of his Dutch landowning friends along the Hudson. Further, he was concerned with the rise of a business elite over the landed gentry and with the corruption of the virtue of the rural population, which was agitating for changes in New York's rental regime.

Cooperstown hasn't changed a lot, only now it is the Clark family, heirs to part of the Singer sewing machine empire of the late 1800s, that dominates the land, institutions, and environs of the village. Some travel accounts call Cooperstown and Cazenovia, its seeming sister-village to the west, "aristocratic towns."[11] They are pleasant in appearance.

The Curious Origins of Baseball

My wife and I have visited Cooperstown at least four times, primarily to view the Baseball Hall of Fame. The "downtown" of today's village of two thousand people is a standard tourist trap:[12] an attraction surrounded by a large number of related souvenir shops and restaurants. In the 1930s, the baseball establishment accepted Cooperstown as the place where Abner Doubleday supposedly devised the rules of the game in 1839.[13] The attribution of Doubleday as baseball's inventor was made on very improbable evidence, however. *The* Abner Doubleday — and there could have been more than one living in upstate New York at the time — was a noted Civil War general, but in 1839 had been a cadet at West Point and unable to leave its grounds.[14]

The National Baseball Hall of Fame and Museum was a child of circumstance. The 1939 decision to locate it in Cooperstown rested on a letter sent more than thirty years earlier to a committee investigating the history of the game.[15] Reacting to baseball critic Henry Chadwick's 1903 claim that baseball was descended from the British game of cricket,[16] A.G. Spalding, who had developed a prosperous baseball-related business, was determined to Americanize the origins of the game.[17] As Zev Chafets puts it, "[w]hat Spalding needed was an alternative creation myth, one backed up by evidence."[18] Spalding accordingly asked Abraham G. Mills, a former president of the National League and a political ally, to lead a hand-picked group of five other prominent former players and sports administrators to investigate the origins of baseball. Spalding produced a letter from a supposed old friend of Doubleday's, a retired mining engineer who claimed to have been there when Doubleday gave the boys of Cooperstown a set of rules that prefigured the eventual game as it was played.[19] Yet Doubleday, who was deceased by the time of the Mills Commission, had made no mention of baseball in his writings. Moreover, his "friend" had been just five years old in 1839. The friend's claim, however, was enough to give Cooperstown its mythic importance, and at the end of 1907, the Mills Commission reported that it was satisfied that Doubleday had created the modern game in Cooperstown in 1839.[20]

Much evidence[21] of the game's standardization in fact points to the son of a relocated Nantucket sea captain, Alexander Cartwright, and a Yankee medical doctor, Daniel L. Adams. In 1845, two years after he and other young Manhattanites had come together to form the New York (City) Knickerbockers ball club,[22] Cartwright devised a constitution for the club and set down the rules for how his club should play the game.[23] These rules then spread across the country until, by 1860, they constituted the standard for play. Cartwright eventually was inducted into the Hall of Fame. More recent research also points to Daniel Adams, the president of the Knickerbocker club, as having had a significant role in the development of the rules and standardization of the game, since Cartwright left for California during the 1849 gold rush.[24] The controversy that began in 1906 over Doubleday and Cooperstown seemed to have been put to rest in 1953, when the US Congress, no less, resolved that Cartwright was the first to put down the rules.

In 1857, the teams that adopted the Knickerbocker rules organized the National Association of Base Ball Players.[25] At first, most of its members were from the New York area, but gradually it expanded to cover most of the area north of the Ohio River and east of the Mississippi. By 1869, the Association, originally devoted to promoting amateur sport, gave in and allowed professional play. The Cincinnati Red Stockings[26] began a string of consecutive victories that gained baseball a large following, and in 1870 the first professional league, the National Association, was formed. Commercialism, indeed, was already a part of the scene. In 1862, Brooklynite William Commeyer built an enclosed field for baseball games and a clubhouse, graded the diamond, and charged admission. With the Civil War raging, he had "The Star Spangled Banner" played before every game,[27] though it would not become the official national anthem until sixty-nine years after baseball had adopted it.

By the time of the Great Depression, and as concerns about mass immigration faded, Spalding's venture to Americanize baseball had been largely ignored for almost three decades.[28] The federal government then was prevailed upon to develop a ball field on the site where Doubleday was supposed to have invented the game.[29] The Works Progress Administration, an agency of Franklin Roosevelt's New Deal, put a number of unemployed locals to work constructing what became known as "Doubleday Field," and in 1935, Baseball Commissioner Kenesaw Mountain Landis laid out a program for a Hall of Fame and museum next to Doubleday Field, along with a process for naming players, coaches, and others to the Hall.[30] In 1937, the Clark family offered to build a permanent Hall and museum for the sport. This would prove to be of commercial help to the village and surrounding area, through not only its construction, but also the continuing benefits of a tourist attraction.[31]

The relationship of the site of the Hall of Fame to baseball's origins might be mythical, but its location in a small town, rather than at, say, the Knickerbockers' old playing field across the Hudson River in urban New Jersey, somehow evokes a sense of appropriateness. Baseball *should* have been invented in a small town only a generation or so past its frontier days, with the accompanying nostalgia[32] of long summer days in a pastoral atmosphere.[33] Let's give the final comment to Frenchman Bernard Henri-Levy, who visited America and the Baseball Hall of Fame 150 years and more after Alexis de Tocqueville passed nearby:

> This is not a museum; it's a church. These are not rooms; they're chapels. The visitors themselves aren't really visitors but devotees, meditative and fervent. I hear one of them asking, in a low voice, if it's true that the greatest champions are buried here — beneath our feet, as if we were at Westminster Abbey or in the Imperial Crypt beneath the Kapuziner Church in Vienna. And every effort is made to sanctify Cooperstown itself, this cradle of the national religion, this new Nazareth, this simple little town that nothing prepared for its election and yet which was present at the birth of the thing....The only problem is that [its] history is a myth, and every year millions of men and women come, like me, to visit a town devoted entirely to the celebration of a myth.[34]

Today, none of this really matters to the stream of pilgrims coming to baseball's historic Hall of Fame.[35] For them, the rules are eternal and the myth is real.

Organizations and the Individual

The modern Hall of Fame seems to have been a German import, as the first ones were developed in that country during the height of the Romantic period. It remained for democratic America to produce specialized Halls of Fame for everything from local businesspeople to sports figures to national heroes. This wide variety of subjects for "fame" might reflect not only democratization, but also the value Americans place on many aspects of their society. It is almost mandatory today for any large group, industry, or association of interests to honor outstanding individual contributions to the activity or interest by creating a Hall of Fame of some kind.[36] Along, or close to, the length of US 20 are Halls of Fame not just for baseball, but also for basketball, boxing, football, rock and roll, and women's rights.

The oddity of such venues is that the individual accomplishment so honored is almost always in the context of an organization or organized activity. Sometimes, as in business or warfare, it is the leader who is so honored; in most celebrity affairs, it is normally a "player" or an "actor," in the broadest sense of the term, who becomes honored. In any case, without the organization, the "star" likely would never "shine." It is harder to get a grasp on organized, systemic efforts. In the six hundred years of Roman Civilization, it is easier to portray a few outstanding or outrageous emperors than the complicated process of Roman rule. In sports, this seems even more difficult. Sports, in classical Greek or Mayan terms, had a religious purpose. In our more secular age, this religious aspect has been translated into a Romantic feeling that ties the sport to the geographical location of teams, thus identifying pride of location with its teams' fortunes.

In *American Exceptionalism*, sociologist Seymour Martin Lipset notes five elements of what he calls the "American Creed."[37] One of these is "individualism." Poking around the concept brings a number of ideas to light. In effect, individualism is about the importance, and perhaps the primacy, of the needs, dreams, and goals of the individual as opposed to those of society as a whole or of some "superior" class, religious dictates, or other group elements. Ironically, however, with America, we are dealing with the promotion of an individualistic ideal in the context of perhaps the most organized society on earth. Organizations, private and public, need skills and actions that probably are not congruent with this aspect of the American Creed.[38]

Carried to an extreme, the primacy of individual goals leads to libertarianism or even to anarchy if people pursue their individual ends regardless of those of others. If one believes in an extended version of Adam Smith's "invisible hand," one might expect that the more than three hundred million Americans can follow their individual dreams peacefully without many rules or serious conflict; otherwise, compromises must be made in terms of the rules to be followed to mitigate conflicts. Individualism in this sense becomes an ideal that can never be met by any one individual.

This conundrum in American society has lasted for more than two centuries. American accomplishments have been due to superior organizational skills, but it is not organizations that have been honored but individual "players." John Rooney, a sports geography professor at Oklahoma State University, argues that the aspect that distinguishes American sports from those in the rest of the world is organization, and that Americans' need for organization can be explained by "our overwhelming need to belong." Pickup baseball has been replaced by Little League and T-ball, complete with uniforms, so the players look like professional athletes.[39]

One hundred and fifty years ago, this organizational bent could have been characterized as the triumph of the Yankees over the Southern Cavalier culture. Since then, a united America has agreed to celebrate the individual while organizing itself and the world to its tastes. Baseball and the other team sports are therefore a useful background against which to look at this aspect of the "reUnited States".

Everyman as the Hero

Squaring individualism with organization is a challenge that is at the heart of American life. James Fenimore Cooper distrusted the aggressive commercialism and organization skills practiced by the Yankees: "What will the axemen do when they have cut their way from sea to sea?" he laments.[40] His main character in many of his popular novels was Hawkeye, a white raised partly by Indians and a resourceful student of the wilderness, the character based loosely on the exploits and description of Daniel Boone of Kentucky fame.[41] Even though Yankee methods doomed the world of Hawkeye in the end, it was still the extraordinary individual who came out morally superior and personally attractive in Cooper's writings.

Wallace Stegner, 150 years after Cooper wrote his best novels, notes the continued appeal of the myth, but disagrees with it:

> Do these figures represent our wistful dream of freedom from the shackles of family and property? Probably they do. It may be important to note that it is the mountain man, logger, and cowboy whom we have made into myths, not the Astors and General Ashleys, the Weyerhausers, or the cattle kings. The lowlier figures, besides being more democratic and so matching the folk image better, may incorporate a dream not only of freedom but of irresponsibility. In any case, any variety of the frontiersman is more attractive to modern Americans than is the responsible, pedestrian, hard-working pioneer farmer breaking his back in a furrow to achieve ownership of his claim and give his children a start in the world.[42]

In *Babbitt*, Sinclair Lewis seems to mock this approach, but he is really underlining the tragedy of the Romantic hero in America: "To [Babbitt and his business friends], the Romantic Hero was no longer the knight, the wandering poet, the cowpuncher, the aviator, nor the brave, young district attorney, but the great salesmanager, who had an Analysis of Merchandizing Problems on his glass-topped desk, whose title of nobility was 'Go-Getter' and who devoted himself and all his young samurai to the cosmic purpose of Selling."[43]

One can see this same conflict repeated over and over again in American history, literature, and song: John Henry and the steam hammer competing to spike rails, the great white whale pursued by the crew of the *Pequod* in *Moby-Dick*, "Johnny Reb" against the well-equipped Yankees in the Civil War, the outlaw against the railroads,[44] the exploits of Bonnie and Clyde in robbing banks, and the hippie anarchists versus the army and police in 1968 Chicago. The individual goes up against the organization — and, though regarded as having a superior morality[45] — generally loses.

Even as American heroes generally lose their quests, or at least are not often paid well for their success, the mythical archetype[46] has moved through the American consciousness in many guises. The Romantic hero of Cooper's novels found his way into popular media through the penny press and the dime novel. Perhaps the most noted of these were Ned Buntline's "Buffalo Bill" tales. Most of these appeared as William Cody took his Wild West shows to towns and cities across the country, but they were about a lone scout and meat supplier for the transcontinental railroad. The incongruity between the lone hero of the Western plains and the more-or-less simultaneous commercial success of Cody's well-organized show was not a problem for his urban fans in America or in Europe.

A related version of the hero in American society is provided by Horatio Alger's popular stories of boys who, through their character and energy, manage to rise out of poverty and achieve fame and fortune. Even as late as the 1950s, stories of youthful heroes fed the dreams of American boys and girls of my generation. We might not have had the advantages of the Hardy Boys or Nancy Drew, but we could imitate their success if we worked hard and exercised initiative.

The lone cowboy hero slid into the urban milieu as the private detective, characterized at first as the "hard-boiled" type, a loner in the urban social wilderness. He was largely anonymous, relentless in pursuing his goal, and as often as not drove off into the sunset still owed for his work, thus never seeing more than enough to keep an office and a receptionist. Sam Spade and Philip Marlowe, both Western (California) detectives, were the base figures for a continual supply of variants, in America and worldwide. In higher-brow literature, the individual/loner/hero took on a different cast. In *Babbitt*, a novel of the 1920s, the individual lives in a conventional household and has a steady life, but dreams of doing something completely different. A fishing trip to the backwoods of Maine only makes the gap between his dreams and reality more obvious. The organized life of urban family and business relationships lacks the Romantic feeling and existence of a Hawkeye, a Buffalo Bill, or even a Sam Spade. Variations on this theme continue to feed novelists with stories to tell.

By the late 1940s, this alienation breaks out in Ayn Rand's characterizations of solitary dreamers whose ideas are right, but who are frustrated, not by the security and entanglements of the wider society, but by the opposition of the larger society, which wants to appropriate the benefits of the hero's work to itself. Rand characterizes American society as sliding into a collectivist mode, where the individual is faced not just with the impediments of normal life or opposed interests, but with a general antipathy. Rand's Cold War theme was quickly followed by the anti-hero of Jack Kerouac's *On the Road*, who actually lives the Babbitt fantasy and crosses the country in search of a different kind of good life, one imitated by thousands over the following couple of decades. Former hippies, like Steve Jobs, became computer nerds and, as the Cold War faded, the nerd as technology entrepreneur continued as a hero, linking the world together through mobile phones and file-sharing programs.

None of this attention can cover the fact that the individual, confronted by even the most primitive of organizations, is generally the loser. Henry David Thoreau went home for dinner most nights. An intricately organized manufacturer made Rambo's machine gun. Rand's architect-hero could not translate a real thought into concrete and steel without a construction company. Nonetheless, the appeal of a Romantic and heroic existence exists as a subterranean force through

American life even though most people pursue a more conventional existence. But, as Stegner points out, like Babbitt, they dream of something more satisfying, but ever so more risky and thus out of reach.

Yankee Organization

Looking back, we can see the role organizing and organizations had in the Yankee frontier. Into the "untrodden wilderness" came Yankee "pathfinders" who identified lands suitable for settlement — lands that had *already* been surveyed by government employees or contractors. They put down money at government land offices and went back to New England or upstate New York to get their families and, in many circumstances, large proportions of their old communities. Once settled on their sections or quarter-sections, they cooperated in staging "bees" to get houses, mills, churches, and schools built and land and roadways cleared.

Even the term "individual" was a misnomer, as it really referred at the time only to the father of the family. He lived in a social environment in which his sons were bound to him by law until the age of twenty-one and all the women were non-voters and non-property owners, their individual goals subservient to a "superior" set of values. Before the Civil War, at least, the idea of "individual self-reliance" applied as loosely in the North as did "democracy" in classical Athens.

The frontier community effort was matched in the growing cities by the organization of large enterprises, first in textiles and railroads and then in manufacturing operations of all kinds. It became possible, through the division of labor, to create tasks in factories that required a certain degree of organizational habit, but not a lot of learned skills. This was especially useful for integrating the millions of immigrants who came after the Civil War.

De Tocqueville, as he traveled the lands through which US 20 now passes, noted the tendency of Americans in this area to organize at the community level as well as in businesses. Pioneers went west in communities, while railroads and factories were incorporated and technology harnessed by companies to provide petroleum products, steel, and machinery. The resulting story of American dominance in the world is one tied to organizational proficiency: one cannot have a society that produces a significant proportion of the world's gross domestic product without having a great deal of organizational skills.

On the side of labor, workers were needed who knew how to work inside an organization. It did not matter what type of organization, only that they had general skills in literacy and numeracy and that they understood how to function in a plant or similar environment. These were skills that each worker could carry along from job to job, which meant that the notion of worker turnover, voluntary or involuntary, was accepted as part of the economic landscape. This is so common in American culture that it is often overlooked, but it began to emerge only in the 1820s in New England. In an environment where machines incorporated skills, managing the machine became the only skill that was required of labor. Turnover is not found or accepted as much in more settled and communal societies.[47] In Japan, for example, "lifetime employment" began early in the Second World War as labor laws restricted mobility in exchange for permanent employment.

In America, the pursuit of both individual goals and organizational goals was enabled by the development of a universal and organized school system. The system taught basic literacy and numeracy, both important for functioning in urban society, but its professional methods of organizing the school day and children's overall behavior were also of fundamental importance. These taught children, as they became adults, to fit into organizations easily and, more important, to enter and leave them as circumstances dictated. It would have been impossible to have a fluid and flexible labor market without these factors. My own experience, working alongside some new Arkansas migrants in a northern factory, shows how hard it was to leave a Southern backwoods farm, with its special rhythms, and adapt to a Michigan assembly line with a time-clock routine. Merle Fainsod, in his 1960s book, *Smolensk under Soviet Rule*,[48] mentions how murderously frustrated Soviet factory managers became in dealing with peasant recruits who did not understand the new rhythms. In a turnover-friendly environment, there is little place for individual exceptionalism except in more senior positions. There, arguments are made for "talent," but they are regarded with suspicion.

The role baseball played in the acculturation of Americans to organizations was to show people how to act in a team, defined as "a small group of people with complementary skills who are committed to a common purpose for which they hold themselves mutually accountable."[49] The authors of this quote note five characteristics of teams:

- They are small
- Members possess complementary skills, such as
  ◊ Technical or functional expertise
  ◊ Problem-solving and decision-making skills
  ◊ Interpersonal skills
- Members share a common purpose and performance goals
- Members develop a common approach
- Members hold themselves mutually accountable.[50]

Baseball fields nine players at any given time, and they play specialized positions. They have a manager who tries to keep them focused on a strategy; during play, they must make a lot of individual decisions, but they win or lose unequivocally as a team. These lessons, if absorbed, contribute to an individual's success off the field in the pursuit of the good life. To me, this points up how individualism must square with the reality of the superior organization, which goes to the heart of the American identity and the Yankee experience within it. In effect, the

Cavalier and individualist culture of the South and the Scots-Irish who adapted to Southern ways amalgamated, though uneasily, with the more organizational culture of the Northern Yankee.

Glory or Victory?

It would be a sensible organizational practice if the "stars" in team sports were single-handedly to make winners of their teams, but this often is not how things work in America. Rod Carew spent many years as the star of the Minnesota Twins baseball team, but his team never got into the World Series.[51] Carew did get into the Hall of Fame, however, because he won batting championship after championship. Likewise, early in his career, basketball star Michael Jordan might score many more points in a game than anyone else, but his Chicago Bulls would still lose the game. Once management provided teammates who were proficient team players, Jordan's greatness began to pay off.

Then there is the opposite experience of winning without stars. In *Moneyball*, Michael Lewis describes the use of team statistics by the Oakland Athletics in the early years of the twenty-first century to build a winning team out of players with certain skills who were not highly paid, because the other teams did not regard their particular talents as worthy of competing for them in the player market. Traditionally, scouts are sent out in all directions to look at high school and college players. They use their experience and wisdom to find the one or two prospects that might be offered contracts. A's management, unable to compete financially, began to rely more heavily on a statistical analysis of these players to find the qualities that matched high-scoring players and effective pitchers. No "look-sees," just analysis.[52] Others in the sport, however, viewed with hostility the A's new understanding of what constituted a winning team, as well as Lewis's praise of this approach,[53] not because it did not work (it did), but because it was alien to baseball tradition and practice: "He [statistician Paul DePodesta] doesn't explain anything because Billy [Beane, the A's general manager] doesn't want him to. Billy was forever telling Paul that when you try to explain probability theory to baseball guys, you just end up confusing them."[54]

Sports performances and records apparently were kept in many countries long before the advent of baseball. Egyptians, Maya, Japanese, medieval English, and Hawaiians all kept records of sporting performances, both as ritual and as part of the religious connection to sport. The critics of the baseball statistics compiled by Sandy Alderson and Bill James[55] and of DePodesta's and Beane's use of them were unhappy because they threw the relevance of other ritually collected statistics into doubt, which in earlier times might have been seen as religious heresy. They called into doubt an aspect of a century and more of tradition.[56] Worse, perhaps, was the implication that a more sophisticated use of technology and mathematics constituted as much a change as moving to aluminum bats or, as was done in the American League, creating the "designated hitter." It smacked of the scientific management that underlay McDonald's restaurants or Monsanto's genetically modified seeds, and threatened to suck the romance out of the game. In the end, though, the big-budget teams also adopted the new metrics and raised competitiveness another notch, though not with much publicity. The approach began to spread into other team sports as well, especially among teams who had smaller player budgets. None of the players who met the team criteria would likely make it into the Halls of Fame, but together, they could constitute winning teams.

This leads into the question of whether, or where, on a spectrum, one should place the individual and the team. It seems to me that the Yankee (the individual, not the team) approach is about winning through a group effort, sustained by a rational approach to team performance, while a more romantic approach would be to find stars and enable them to demonstrate their mastery. Pride in a team can come as a result of either or both of two effects: the identification of individual heroes and the record of the team. Ideally, both should be connected, but often they are not. Potential Hall of Famers might provide drama and bring in fans and revenue without necessarily being effective in team victories. Football coach Vince Lombardi famously might have said that "winning isn't everything; it's the only thing," but Halls of Fame, by focusing on the individual, do not reflect this attitude.[57] Cooperstown has made some effort to provide more balance between individuals and team, over the years, but the achievement of individuals still dominates there.

Americans seem to live comfortably with a number of contradictions at the base of their lives, and compromises are worked out to leave them be. The individual still tends to be the centerpiece of private aspiration and public life, while the organization is the basic framework through which everything gets done. The individual recognizes that it often is through imposing his or her passion on organizations that personal success can be measured, whether as an entrepreneur creating a new business or as a manager making incremental improvements in products or services. Yankee organization, communications, and products and services have become globalized, with many subversive and unintended consequences. Yankee America — the cradle of the complex organization — seems to be more about winning than about glory. Yet enough of Romantic America exists to support a drive to exalt the individual who develops a special ability to excel, so that glory is not far behind. It often masks the more anonymous organizational reality that enables winning.

Ilion, New York: Typewriters and the Second Computer Revolution

[I]t isn't writing at all, it's typing.
— Truman Capote, discussing Jack Kerouac's work in a TV interview[1]

How do you convey the magic of a new dawn?
— Preface, *Fire in the Valley*[2]

My Neighborhood Marvel

In 1981, my wife and I took a trip to San Francisco to see her brother and his family. While we were there, I found a good deal on a Commodore 64 personal computer and decided to take one home. The system consisted of a keyboard/computer housing and two floppy-disk drives, and cost $1,200. I carried it on the plane, where a friendly stewardess suggested that I stow the boxes behind the rearmost seats in First Class and promised she would look after them.

Reflecting on this experience in the age of the Internet is somewhat different, and difficult. In 1981, the practical, popular use of the Internet was still more than a decade in the future. Graphics were crude — a game might have an opening picture with the title, but the rest was abstract dots and rectangles that sort of looked like rocket ships, cannon, or whatnot. But, I recall, that same year my father had retired early as the head of the proofreading department of the *Flint Journal* because the function had been given to writers and editors with new, and expensive, specialized word processors with correcting features. Before our trip to the West Coast, I had read that the Apple and Commodore home computers could do the same word processing. My children were all in high school and my wife was getting her master's degree, so we needed a really good and forgiving "typewriter" that allowed us to do away with the tedious task of "whiting out" typos; as well, teachers by then preferred to receive typed essays rather than handwritten ones.

Word-processing programs were available for our machine, and we could even save our work on big, square floppy disk, instead of having to start over if a paper copy got lost or damaged. All we needed was a monitor and a dot-matrix printer, and we were in business. Our computer needs were simple. Our main concern was to establish with our and the neighbors' kids that typing mom's thesis took precedence over video games — sociology was more important than Missile Command.

Turning Rifles into Typewriters

The little town of Ilion, New York, sits on a hillside above the Mohawk River, about ten miles north of US 20.[3] The Erie Canal and an early expressway, bypassed later by the nearby Thruway, go through town. Ilion was one of a number of manufacturing and distribution centers that grew up along the canal in the early 1800s, encouraged by its relatively low transportation costs. The town is also an odd, yet appropriate, place to start a story about the personal computer.

Ilion's best-known entrepreneur is Eliphalet Remington, a Yankee blacksmith who, in 1799, moved from Suffield, Connecticut, to the Mohawk River valley, just south of what became the town.[4] Remington established a smithy on his farm to make agricultural implements, but the forge also dabbled in other products, including musket barrels made by bending hot sheets of iron around a rod used to keep the barrel hollow. On the early frontier, the wooden parts of muskets, as well as the balls they fired, were made at home, and the metal firing mechanisms and barrels were purchased from blacksmiths and gunsmiths.

Eliphalet II, his son, devised a sporting rifle in 1816, when he was twenty-three, and a demand for his rifled barrels arose from gunsmiths who produced custom equipment for the gentry along the Hudson River. This became an increasingly larger part of the Remington forge's activity. Rifles were preferred because they were more accurate than smooth-bore muskets, but they required standardized bullets.[5]

In 1828, with the death of his father, Eliphalet II moved into town, acquiring acreage on the new Erie Canal, and began to concentrate on providing firearms for those moving west. In 1835, as the business grew, he bought rifle-finishing machinery from Ames & Co. in Springfield, Massachusetts, as well as part of their contract for rifles for the Springfield Armory. By 1840, Remington had brought his three sons into the business. The company prospered as a result of new products, such as a pistol popular with the Army and Navy, and a method to convert muzzle-loading muskets into breech-loading rifles. The outbreak of war with Mexico and then the Civil War meant a good deal of new business.[6]

Remington helped pioneer important innovations toward the end of the Civil War. The traditional way of loading a gun, large or small, was to put powder and shot down the barrel and then set these off, a practice called "muzzle-loading." Toward the end of the war, "breech-loading" pistols and rifles started to appear. These were loaded with cartridges fixed to bullets and placed into firing position from the back of the gun, or its breech. The combination of prepackaged bullets and breech-loading was faster and more efficient than the older technique. The South was unable to produce this technology, and by the end of the war, the Northern preponderance in firepower was not simply in numbers of soldiers, but in their individual firepower, which was many times that of Southern troops. The revolver and repeating rifle required only that the user cock the hammer and pull the trigger, increased firepower still further. Mythically. the guns that won the West were the Colt 45 revolver and the Remington and Winchester rifles, all repeating firearms.

After the Civil War, the Remington firm was incorporated in order to raise capital. Its breech-loading rifle acquired an excellent reputation, and foreign sales offset some of the decline in postwar American demand. Earlier, in 1856, the company had gone into farm implements to supplement its arms income and had succeeded. Under the postwar leadership of Philo Remington, the company turned to products that, like rifles, needed standardized metal parts, something that arms makers had pioneered. The company was approached by a former Singer executive to produce an improved sewing machine[7] — the device had been developed originally by Elias Howe, a Boston mechanic. Remington did so in 1870,[8] but the product never lived up to its potential because of the company's inability to establish a good sales network.[9]

Remington then hit upon an innovative product that seemed promising: the typewriter. Whatever the design, a typewriter needed a lot of precision metal parts.[10] Remington bought the patent for the first practical working typewriter in March 1873, just six months before the Panic of 1873 sent the country into a depression. Two Milwaukee innovators, Christopher Sholes and James Densmore, sold it to the company for $12,000.[11] Remington's first "Sholes and Glidden" typewriter appeared on the market in 1874, but the depression meant it did not sell well. Then, Sholes demonstrated the machine at the 1876 Centennial Exhibition in Philadelphia, which helped stimulate worldwide interest in the product.[12] This time, Remington began to develop its own sales staff for both foreign and domestic markets.

Interestingly, the "Sholes and Glidden" resembled a sewing machine — the Remington executives who championed the purchase and the mechanics who developed it were both on the sewing-machine side of the company.[13] So the sewing machine foot treadle became a carriage return, and the letter arms and platen were inside and under the machine, making it impossible for typists to see what they had typed until they had started the next line or two.[14] The now-traditional QWERTY format came about as Sholes tried to keep the keys from sticking when two that were close on the keyboard were struck in quick succession, causing their arm mechanisms and type heads to get in each other's way.[15] Interspersing commonly used letters with the less common proved a relatively effective solution, although it compromised the speed of the typist. Because of this early mechanical "bug," keyboards today still feature the QWERTY layout — the social investment in the pattern is just too great to be overcome by the quaint notion that it is now completely unnecessary.[16]

Remington had more success selling typewriters than sewing machines,[17] since it controlled the patent and had hired a sales organization to market the machines, Even so, the company continued to suffer from financial problems, not entirely due to general business conditions. In 1883, rights to the typewriter division were assigned to a company called Wycoff, Seamans, and Benedict, the names of the men who were the sales group for the machine. They then bought the division and the Remington trademark outright in 1886,[18] calling their subsidiary the Standard Typewriter Company, renamed in 1902 as the Remington Typewriter Company. By 1893, five hundred people were employed at the Remington typewriter works in Ilion, making twenty thousand machines annually.[19]

The sale was not enough, however, to save the original Remington company, which went bankrupt later in 1886.[20] In 1888, a New York company, Hartley and Graham, bought the arms assets and reorganized the original company as the Remington Arms Co. In 2007, the Remington Arms Co., after passing through a number of other owners, was bought by Cerberus, a hedge fund. In 2011, the world's last typewriter manufacturing company, Godrej and Boyce of Mumbai, India, stopped its production of this line.[21]

Mechanical Writing and Its Impact

Sholes and Glidden were not the first to think of mechanical writing.[22] I would not be surprised if the second idea Gutenberg had in the 1450s as he arranged type in a box and pressed it against paper was about how one might press individual letters. Yet the method for doing this was slow in coming. An Englishman got a patent for a typing machine in 1714, but nothing came of it.[23] An Italian designed a machine in 1808 to help his blind patron continue her voluminous correspondence.[24] Another Englishman designed a machine in 1829, and a Frenchman in 1833 developed individual type keys for each letter. One estimate is that there were at least fifty and perhaps more than a hundred machines designed or built before Sholes and Glidden made their breakthrough into commercial production.[25] They were also lucky: a New Yorker named John Jones had patented a similar kind of typewriter in 1852, and had produced 130 of them before he lost his shop in a fire.[26]

In the end, Sholes was the driving force that made typing feasible on a large scale. He had been apprenticed to a printer as a child, and spent much of his adult life as a newspaper editor. As well, he had been active politically in Milwaukee, becoming a state senator, a postmaster, and collector for the Port of Milwaukee. He and a partner, Samuel W. Soulé, developed a page-numbering machine for offices in 1864.[27] Another inventor, Carlos Glidden, having read a *Scientific American* article about a writing machine devised by an American in Europe, suggested to Sholes in 1866 that something like their page-numbering machine might be modified to make letters instead of numbers.[28] By September 1867,[29] working in a machine shop that was an early "incubator mall" for inventors, they had built their first prototype and put on a public demonstration.[30] Sholes, who coined the term "typewriter" to describe his machine, became fixated on the idea, and spent much of the rest of his life designing and redesigning typewriters.[31]

In 1868, a local investor named James Densmore, a printer and newspaperman, put in $600 in return for a share of the profits and the relinquishment of Soulé's and Glidden's patent rights.[32] He also agreed to finance future development expenses. Then, in 1870, he tried to sell his and Sholes's rights to the Western Union telegraph company for as much as $100,000,[33] but one of Western Union's employees, a certain Thomas Edison, successfully discouraged the purchase by asserting he could build a better one. Edison, fascinated by electricity rather than by writing, designed an electric machine that was more useful as a stock ticker than as a typewriter.[34] He did offer Sholes some useful design improvements.[35]

The typing technology represented a new way of communicating with the world. It was faster than handwriting and definitely more readable, and with carbon paper, an innovation that had also recently come on the market, capable of making exact copies.[36] When added to the mimeograph machine, patented by Edison in 1876, it was now possible to make multiple copies. All of this was a boon to the large corporations that were forming in America and elsewhere. Before the typewriter, the business world included rooms full of copyists sitting at their desks and laboriously copying correspondence and accounts. Dickens memorably pictured these men at work, with bad eyes and poor health from constantly peering at handwritten notes and originals.[37] With typing, consistent, clear communication could be sent up and down the hierarchy.

The typewriter plus the US Postal Service plus the telegraph plus (a bit later) the telephone equaled the ability of organizations that spanned the continent to be managed as single units.[38] The standardized pieces of technology that underpinned the mass-production organizations of the twentieth century were coming together, after their beginnings in Waltham, Massachusetts, a half-century earlier.[39] One of these, scientific management (see Chapter 4) intersected with the typewriter. Frank Gilbreth, one of Frederick Taylor's associates, "helped Remington develop the world's fastest typist by put[ting] little flashing lights on the fingers of the typist and tak[ing] moving pictures and time exposures to see just what motions she employed and how those motions could be reduced."[40] The typewriter allowed for an increased and legible paper stream that contributed to the development of what passed for an information and services economy in the early 1900s.

These innovations even affected government organization. A 1919 biography of Theodore Roosevelt compared his White House with that of Grover Cleveland a decade before:

> Only a few years before, under President Cleveland, a single telephone sufficed for the White House, and as the telephone operator stopped work at six o'clock, the President himself or some member of his family had to answer calls during the evening. A single secretary wrote in long hand most of the Presidential correspondence. Examples of similar primitiveness might be found almost everywhere, and the older generation seemed to imagine that a certain slipshod and dozing quality belonged to the very idea of Democracy.
>
> Nevertheless this was a time of transition, and the vigor which emanated from the young President passed like electricity through all lines and hastened the change....Instead of one telephone, there were many working night and day, and instead of a single longhand secretary, there were a score of stenographers and typists.[41]

Notably, the typewriter re-employed women in factories and offices for the first time since the early factory practices in Massachusetts had been abandoned. The idea of introducing the typewriter to a female audience can be traced to the Remington sewing-machine men who championed the machine. They felt that women would find the typewriter less intimidating than men did, especially if the table on which it sat were decorated with painted flowers. As well, the company used women in demonstrations of the product. Gradually, an "urban myth" assumed shape that a woman's fingers and hands were more suited to the typewriter than were a man's.[42] A third reason for the connection between women and typewriting was that women were hired as stenographers during and after the Civil War, and the typewriter could be used in place of their first copy of the dictation notes they made.

The typewriter thus became a way for women to re-enter the world of the office and factory, easing out the male secretaries and copyists. In 1870, only 4 percent of such jobs were filled by women, but their proportion rose to almost 40 percent by 1880, to 80 percent in 1910, and nearly 96 percent by 1935.[43] No wonder that, when I was in high school in the late 1950s, typing classes were girls' classes. Computer keyboarding thirty years on was not to be so segregated, though the basic skill had been largely dropped from school curricula.[44] Prospective employers want familiarity with software packages more than typing speed today.

In 1881, the Young Women's Christian Association began its first typing class for women. Typewriter manufacturers also began training women, who were then provided as experts to explain the use of the new office machinery bought by their business clients. Part of the instruction of the day was to train typists to use all their fingers instead of "hunt and peck" with two or four. A Mrs. Longley outlined the "All-Finger Method" in 1882 in her manual, *Remington Typewriter Lessons*. At first, typing was by sight, and only in 1889 did techniques for "touch typing" become widely publicized.[45]

The typewriter also affected culture. Mark Twain was an early adopter. In 1874, he and a friend saw one in a Boston store window.[46] Going inside, they were given a demonstration by a woman who, the salesman insisted, could type fifty-seven words a minute. After a series of trials, Twain was convinced, and spent $125 on one of the machines. When he got it home to Hartford, he tried to teach himself to type as fast, but soon realized that he could not do so. He eventually hired a woman to use the typewriter, which could only type in capital letters, to take dictation, but she used it only for correspondence.[47] Though he claimed to be "the first person in the world to apply the type-machine to literature," he merely had someone else type part of an already-published book. It was not until 1883 that gave a typed manuscript, *Life on the Mississippi*, to a publisher.[48]

Then there was Jack Kerouac. More than six decades on from Twain's purchase, Kerouac had become a speed typist in high school — some claimed he could type a hundred words a minute when he was really engaged.[49] It brought him closer to the speed at which his mind worked. In 1941, the nineteen-year-old acquired his own typewriter after having used his father's for years: "Here I am at last with a typewriter...[and] with my new typewriter and a lot of yellow paper, I am grown dead serious about my letters, my work, my stuff, my writing here in an American City [Hartford, site of Twain's home]."[50]

Ten years on, battling writer's block trying to compose *On The Road*, he came upon a possible solution. He taped together a roll of paper into one long sheet, loaded himself up with coffee, cigarettes, and Benzedrine (from inhalers for nasal congestion), and in one mad, two-week frenzy of speed typing, produced a single-paragraph, 120-foot-long novel. He then took it to his editor, who punctured his euphoria with, "but Jack, how can you make corrections on a manuscript like that?" Kerouac, deflated, returned home and turned it into a paragraphed manuscript on standard paper sheets.[51]

The typewriter allowed him to move the flow of his ideas quickly from image to paper.[52] Many of the first drafts of his later books also came from caffeine, nicotine, and "uppers," combined with a long roll of paper, then spooling his subconscious through the typewriter. Kerouac's method opened up a crack into the future. Sixty years on, the distance between Twain and Kerouac, today's word-processing screen functions as a 120-foot-long, or even a 1,200-foot-long, roll of paper. With the addition of an easy way to publish via a web page, the blogosphere has become full of typists channeling thoughts as they flow out of their heads right into the homes of anyone who wants to read them. The digital camera, has even translated the technique into image transfer, avoiding having to "edit" pictures through descriptive narrative. There are lots of Kerouacs out there today, typing and clicking away.

Turning Typewriters into Computers

In 1927, three years after the creation of IBM, Remington Typewriter and the Rand Kardco Power Accounting Machine Company, a competitor of one of the companies folded into IBM, merged, along with a few other companies, to form Remington Rand, under the leadership of James H. Rand Jr.[53] Rand, born near Buffalo of Yankee heritage, graduated from Harvard and, at age twenty-four, took over his father's index card and ledger company when Rand Sr. fell ill.[54] When he recovered four years later, Rand Sr. resumed control and a feud broke out between father and son, leading Rand Jr. to begin a successful competitor firm in 1915. In 1925, after a reconciliation, their companies were merged, with the son in charge.

Rand Jr. led the company into a buying and merger campaign that resulted in its revenues growing from $5 million in 1927 to $500 million in 1954. One can't help feel that Rand and Watson of IBM were always close competitors. Rand's large business machine factory was in Elmira, New York, just down the road from IBM's big plant in the Binghamton area. While Remington claimed to have marketed the first electric typewriter, IBM came to dominate this growing sector of the business market.[55] They were both military contractors in the Second World War and both got into the mainframe computer business, IBM acquired the Harvard-built Mark I and Rand by bought the Eckert-Mauchly Computer Corporation in 1950, thereby gaining the experience of Mauchly and Eckert, the two men who had built the ENIAC at the University of Pennsylvania. The purchase was fateful, as the two went on to develop the Univac computer, a serious competitor for a time to IBM's "big iron."[56]

In 1955, as Rand turned sixty-eight, Remington Rand merged with Sperry Gyroscopes to form Sperry Rand; then, in 1978, the company became Sperry Univac, selling a number of divisions to heighten the company's commitment to mainframe computers. It left its typewriter operations behind to disappear in 1981 in lawsuits between European and American holders of Remington rights. The revamped corporation went on, through a number of mergers, to become Unisys in 1986, which stopped manufacturing computers in the 1990s to concentrate on services to the industry and its customers.

There is still a Remington plant in Ilion and a small museum that contains a brief reference to the Remington typewriter. This is where the story leaves rifles and leads to PCs.

Missing the Micro

It was really no coincidence that the minicomputer was developed about the time the miniskirt came into vogue. Nor was it any leap to name the even smaller personal computer that arose in the 1970s the microcomputer, after the even shorter skirts that followed the mini. Somehow, the connection between cheaper and smaller computers and a trend of the less-structured Sixties implied that they were part of a kind of liberation movement. The symbolism of the large mainframe, with its corporate-only price and the "high priests and acolytes" that tended the machine and scheduled users' time, seemed to those interested in personal empowerment to resemble the railroad, while the PC was like a car that would respond to the user's needs and goals.[57]

The ideas that led to the PC came from a variety of sources, but the most sustained vision seemed to come from the human augmentation project at Stanford Research International led by Doug Engelbart. Human augmentation is a way to describe the use of computing power to assist the mind, much like glasses assist the eyes, or cars assist the legs. In contradistinction to projects on artificial intelligence, augmentation did not seek to replicate or take the place of human thought processes. Over the decade from 1963, Engelbart, funded from the Pentagon office that also helped create the Internet, developed a vision of the computer that was radically at odds with the conventional one of the times, which viewed the computer only as a means to do ever more complex mathematical calculations.

What seems obvious today was not so then. Engelbart's vision rested on cheap and abundant computer power, which undercut the expensive and restricted world of mainframes and minicomputers as it existed in 1968.[58] It eliminated time-sharing, and it proposed a world in which computer users could do more than just undertake complicated data file transfer actions; they could also use the computer to type, communicate, do business, and create.

Yet the minicomputer business around Route 128 in Boston did not understand — or perhaps it feared — this mini/micro connection, though it profited from its cachet.[59] By the early 1970s, with Moore's Law about the increasing power of chips still grinding along, many in the industry could see that the next iteration of computers would be smaller, cheaper, and just as powerful as the minicomputers of the 1960s. The leadership of DEC and its competitors, believed, however, that this development was not practical, and discouraged their employees who wanted to go in this direction.[60] The miniskirt might be a symbol of liberation, but the minicomputer industry had its collective eyes focused on the white shirts and hierarchy over at IBM. DEC and others did not even want to sell to individuals, even when customers demanded they do so[61] — they did not want to deal with a lot of one-off help requests from amateurs.[62] This was similar to the attitude of early upscale car manufacturers before Henry Ford entered the scene: if you couldn't afford your own mechanic/chauffeur, you shouldn't be buying a car.

Even the typewriter guys missed the turn. IBM still sold typewriters, with modifications such as electric power and a rotating ball instead of the old key bars. Xerox tried to sell a typewriter with some memory and font cards and even a small screen so the words might be seen and corrected before put to paper. Today, however, even to collectors of old typewriters, these early attempts to graft computing to traditional typing are of marginal interest.[63] Xerox even went so far as to set up its Palo Alto Research Center (PARC) in 1970 in what is now Silicon Valley[64] to discover some way to overcome IBM's dominance of the mainframe computer market, as well as to develop new tools for improving its copiers, typewriters, and other business equipment.

In 1972, PARC researchers, mostly acquired from the Stanford research groups working on artificial intelligence and human augmentation, developed the first microcomputer, the "Alto," as well as many key features of today's machines, such as the mouse and the Ethernet, which allowed computers to communicate easily with each other. But Xerox was only looking for improvements to its existing product lines, not something that would change the game altogether.[65] Shown the Alto in 1977, Xerox's chairman could only ask how fast the man (not woman) on the keyboard could type.[66] Only in 1981 did Xerox market the Alto, and then it was a $40,000 flop.

In fairness to the companies who missed the PC revolution, one should noted that it is generally difficult to predict the direction a new technology will take. Companies normally stick to the actions that made them successful, and resist a variation that lies outside a simple improvement. As Robert Cringely commented about Xerox and PARC, "Sure, PARC invented the laser printer and the computer network and perfected the graphical-user interface and something that came to be known as what-you-see-is-what-you-get computing on a large computer screen, but the captains of industry at Xerox were making too much money the old way — by making copiers — to remake Xerox into a computer company. They took a couple of halfhearted stabs but generally did little to promote PARC technology."[67] Inventors such as Edison found that the market's use of an innovation generally is not what the inventor first had in mind.[68] Even government help does not work: Massachusetts set up a Massachusetts Microelectronics Center (M2C) in the 1980s in an attempt to regain the area's lead over Silicon Valley, but the effort failed, and M2C turned toward training people for the computer industry.[69] By the late 1980s many of the Boston region's computer companies had gone out of business or been acquired by others.[70]

The area that benefited most from these missteps is located around the southern head of San Francisco Bay, where the Santa Clara Valley meets the water. In the 1930s, a professor of electrical engineering, Fred Terman, was looking for a way to keep graduates in the area, rather than to see them drift to the East. Terman was central in getting two of them, William Hewlett and David Packard, to set up shop nearby to make electronic instruments. In the postwar era, they became legends as Hewlett-Packard (HP) went from instrumentation manufacture into calculators and then PC manufacture and sales. In 1948, HP was followed by other Stanford graduates, who formed Varian Associates. In 1947, the university also created an interdisciplinary research institute to take advantage of military and civilian government research funding.

As the region around San Francisco Bay grew in population after the war, and the GI Bill encouraged more university admissions, Stanford was both expanding and looking for ways to improve its financial situation. It opted in 1950 to designate part of its lands as an industrial park. Varian was the first to sign a lease, in 1951, and the company moved into a new facility there in 1953. Others, including Eastman Kodak, GE, Admiral, and Shockley Labs — created when Terman enticed William Shockley back to his hometown of Palo Alto — followed soon after.[71] Coupled with military research facilities nearby, the Bay Area's technology base began to come together. The industrial park eventually became the most successful of its kind and a model for a host of university-related industrial/research parks across the country and around the world. As well, the growth of spinoff companies, new startups, and branch operations meant that surrounding communities began to fill up with business "campuses," offices, and factories.[72]

The area around Route 128 in Massachusetts and California's Silicon Valley share some characteristics. They are the two largest concentrations of technology-related employment in the United States.[73] Both got their start in the Second World War from research and systems development funded by military and, later, space programs.[74] These led to spinoff companies that substituted, in one, for the decline in textile mill employment and, in the other, replaced fruit orchards with plants and office campuses. Neither was wholly "high tech," but a mixed-product economy. Both benefited from the sponsorship of leading universities and the availability of their graduates.[75]

There are also differences between the two areas. The computer industry developed near Route 128 about fifteen years earlier than in northern California and was more heavily concentrated in equipment production. Silicon Valley, in contrast, led in chip design and production once that industry took off in the late

1960s. The Valley also had a strong instrumentation focus, which led it naturally into the PC-manufacturing business.[76] Boston had a more developed venture capital industry until the 1980s, when the PC business in California expanded rapidly, leading to venture capital development and eventual leadership — indeed, after 1980, computer startups in the Boston area complained that venture capital companies had little interest in client companies not located in Silicon Valley.[77] An important additional difference between the two lay in their styles of management and organizational culture. That of Route 128 was dominated by hierarchical and centralized companies, while Silicon Valley was more fluid, informal, and flexible,[78] making it better able to deal with the shifting fortunes of the PC industry, especially as dominance shifted from hardware to applications.[79]

The Successful Revolution

It took a little while, but some Henry Fords of information technology eventually came along.[80] In January 1975, *Popular Electronics* published a story about a new startup company, MITS, in Albuquerque, New Mexico, that was about to build and sell computer kits for the home market. The computer was called the Altair 8800.[81] The article led Bill Gates and Paul Allen to leave Harvard and move to New Mexico to get in on the action, Allen as MITS's software director and Gates as programmer. They concentrated on developing and improving the BASIC computer-language operating system for the new machine. Eventually, frustrated and spotting opportunity, they left MITS and founded their own company, Micro-Soft.[82] MITS struggled and was sold to a disk and tape-drive manufacturer in 1977 and disappeared from the scene.[83]

Steve Jobs, a dropout from Reed College in Oregon, and Steve Wozniak, a Homebrew Computer Club enthusiast, working together in Cupertino, California, developed a computer kit they called the "Apple,"[84] which they put on the market in 1977 for the symbolic price of $666. Soon, local stores were calling for Apple to provide ready-made computers instead of kits. That year, at least thirty companies formed to produce PCs, but some, such as RadioShack and Heathkit, were not located in the Bay Area.[85]

The revolution accelerated. The editor of *Dr. Dobb's Journal* estimated in 1977 that there were "50,000 or more general-purpose digital computers in private ownership for personal use."[86] Apple began to grow, showing a profit after its first six months and abandoning the kit format in favor of a simple machine that was accessible not just to the hobbyist, but to all kinds of users. Wozniak then designed the Apple II, which was to have an affordable price ($1,195),[87] could be purchased off the rack, and be easy to set up and operate. It supported a color monitor and had expansion slots for peripheral devices.[88] The Apple team developed a nationwide distribution system and got its product into important electronics shows.[89] The company also obtained some venture capital from New York, advertised in

Playboy, and got its product reviewed in *Byte*. The response showed that a large segment of the public was ready to purchase its own computer.[90] By 1981, Apple's success inspired mainframe builder IBM to investigate the market and to offer a machine with "open architecture" composed of standard and accessible parts and software,[91] though its core business remained the minicomputer market.

By 1995, the market for PCs was larger than that for mainframes and minicomputers combined. The minicomputer industry withered and died, and the PC industry became a global one.[92] IBM, using Microsoft software, had overtaken Apple in PC sales, but IBM could not hold its dominance, since the PC's main selling feature turned out not to be the physical machine but the software that ran on it and allowed users to do what they wanted.[93] And in that market Microsoft was and remains supreme. IBM's reputation as a reliable supplier to industry and government gave it immediate credibility once it entered the PC market — indeed, its endorsement gave the PC the credibility it needed to make the jump from the early adopter market to the business mainstream. In doing so, IBM was not so much cannibalizing its own existing mainframe products as destroying the minicomputer market. Because it had been a manufacturer of business equipment for decades, IBM left the provision of software to others, not deeming it important.[94] Soon, however, other, lower-cost producers entered the business machines market as well as the home computer market with machines that were both reliable and ran the same software as the IBM machine.

Microsoft came to dominate operating systems and business function software by providing licenses to both IBM and its competitors. Licensing was dependent on the legal culture of the business, and especially of the corporate world. The hobbyists who had pioneered the PC market in the late 1970s tended as a matter of both principle and cost-effectiveness to advocate the free dissemination of information, including operating system software. Gates and Allen, however, first in Albuquerque and later in Seattle, complained bitterly that the hobbyists were stealing their property, but to little avail.[95] In the business market, Microsoft found it could effectively license its products by defending its rights in court. As the IBM computer took over the market and then the "clones" pushed out IBM, Microsoft moved to another model to defend its dominance. The operating software, and even the office productivity software, was bundled into the machines at the time of manufacture, so that its cost was included in a single purchase price. Microsoft also provided companies and institutions with site licenses that guaranteed they would be given upgrades and repair "patches" as they became available. The company regularly released new versions of its software, leading to more purchases of upgrades. It also aggressively bought or developed competing applications that especially threatened its hold on the business market.[96]

Apple stayed aloof from these developments, continuing to press forward with its own high-priced designs and proprietary software. It was confined by these practices to a small market share, but it survived, especially among those who appreciated its graphics and ease of use. It was the counterculture machine in an era when the microcomputer it had pioneered to the world had become part of normal life.[97]

Cranks and Dreamers

> It was a time when cranks and dreamers saw the power they dreamed of drop into their hands and used it to change the world. It was a turning-point when multinational corporations lost their way and kitchen-table entrepreneurs seized the banner and pioneered the future for everyone....It was a bona fide revolution.[98]

It is impossible to separate the development and spread of the PC from the social ferment of the 1960s and 1970s around San Francisco.[99] Fred Engelbart's vision of human augmentation based on cheap and abundant computer power, which he expressed at the end of 1968,[100] had two main elements. First, he saw the computer as a communications and information-retrieval tool, to place the intellectual resources of humanity at the service of the individual, not just as a way to solve complex problems. Second, he believed the computer should be for *personal* use, rejecting the growing consensus that a system of time-sharing terminals connected to a mainframe or minicomputer would allow more people to access computer power. Inside these principles lay the personal computer, the Internet, email, search engines — in short, the direction that has resulted in today's computer environment.

This approach fit the liberationist temper of the times in the Bay Area.[101] Ken Kesey, the writer and popularizer of LSD, after watching a prototype computer in a lab at Stanford Research International go through its steps in the early 1970s, said, "[i]t's the next thing after acid."[102] Needless to say, Engelbart's lab contained a rather chaotic mix of programmers, political activists, dreamers, and administrators. Though it declined as members left for PARC and other locations around the country, a lot of its internal and external cultural assumptions would prove prophetic in the way Silicon Valley startups in the 1980s and 1990s behaved, with an informal party atmosphere coupled with manic and long work hours and very personalized décor and habits.

For some, such as Stewart Brand of the Portola Institute and publisher of the *Whole Earth Catalog*,[103] the PC represented another tool that the counterculture could use for personal and social liberation.[104] For others, such as members of the 1970s Homebrew Computer Club[105] in Palo Alto, the PC was a means to offset the power of the big corporations and to allow the private citizen better access to databases and government documents, as well as to improve education,[106] a feeling shared with clubs of hobbyists across the country.[107] For still others, programming computers while immersed in the psychedelic and creativity-enhancing subculture of the Bay Area seemed to offer new horizons to explore the mind. They had seen the connection between computers and states of perception in Stanley Kubrick's 1968 film, *2001: A Space Odyssey*.[108] One thing they all shared was the idea that freedom of information was critical to the new society they wanted to build: not only should there be no restraints on what could be said and published, but access to computing should be unrestricted and software costless. The former had political implications, while the latter went against the interests of those who were trying to make a business out of the demand for PCs and would have a huge impact on whole industries as computers were connected through the (free) Internet.

The dichotomy between the computer as a tool for liberation and the computer as "the largest legal accumulation of capital in the twentieth century"[109] is still at the heart of many social tensions of the early twenty-first century. China, among other countries, controls access to political matters on the Internet, while the United States chases illegal downloaders of copyrighted music and films, Microsoft pursues software pirates, and the world's crime-fighting organizations try to control "phishing" (acquiring personal data through Internet fraud), child pornography, and other broadly illegal activities.

Even though this freedom has resulted in a continuing flow of innovations, much of the PC's use since its inception has been in traditional typing of one form or another. Despite attempts to build a useable voice-to-print software package, nothing has worked well so far. Workarounds have been tried with videoconferencing, webcasts, and other non-writing programs, but, at the base, the world still runs on written messages. Typing with the QWERTY system has a lot of social momentum left.

Cazenovia, New York: Organizing the Early Frontier

Massachusetts soon became like a hive overstocked with bees, and many thought of swarming into new plantations.
—Cotton Mather[1]

What is that quest that pulls me on the road?
— Loreena McKennitt, "Caravanserai"

Sell a country! Why not sell the air, the clouds and the great sea, as well as the earth?
— Chief Tecumseh[2]

The Yankee Pioneer Movements

The pioneer experience has been claimed as the basis for many of the social and political values of Americans. Even so, some aspects of the pioneer experience in the forest, as opposed to the trans-Mississippi prairies, might be underplayed.

Some demographic analysis and contemporary social commentary suggests that migrants moving west might have had two or even three distinct experiences. One of these, and the focus here, is the Yankee experience.

Yankee expansion in the mid-1700s and after was directed at the northern reaches of the Connecticut River and toward repopulating the coast of Maine, which had been devastated by the Indian wars of the preceding century. Vermont was the creation of Connecticut River Yankees, who took advantage of dubious New Hampshire land grants to inhabit territory to the west of the river. New York traditionally had claimed both shores of Lake Champlain, and the subsequent struggle over who owned what land dragged on past the Revolutionary War. At the war's end, New York gradually gave up its claim and Vermont became a state in 1791.

Yankee expansion westward came so late that relatively few New Englanders, transplanted or otherwise, actually became involved in Indian wars after the end of the French and Indian War in 1763. Iroquois attacks in the Revolutionary War affected some Connecticut colonists in the Wyoming country of northeastern Pennsylvania, but reprisal raids in 1779 led by General John Sullivan of the Continental Army broke the power of the Iroquois in western New York and drove the Six Nations across the Niagara River. There were enough Yankee soldiers in this army that word spread home that the land beyond the Hudson River was vast, fertile, and soon to be available.

The forcible opening of what were to be the Ohio, Indiana, and Michigan Territories to later settlement was accomplished in large part by Kentucky militia units in battles from 1790 until 1814, though some Yankees settling in Ohio Company lands in southeastern Ohio also took part. Finally, the so-called Black Hawk War in northern Illinois in 1832, which marked the end of Indian resistance in the old Northwest Territory, was fought mostly by militia from southern Illinois, including the Kentucky-born Abraham Lincoln.

The impact of the Erie Canal on American population and settlement cannot be overestimated. Since the railroad did not reach past the Appalachians until just after 1840, water transportation was the cheapest and most practical method of moving people and raw materials into and out of the continent. By that date, a large proportion of Americans lived west of the mountains and a third of the country's wheat and half its corn came from there.[3] The Erie Canal was, with New Orleans, one of the two ways this produce could reach the East.

As well, the canal affected population movement. At a stroke, it shifted immigrant destinations from the southern and middle States, which had predominated since the early 1700s, to New York. By 1825, New England settlers could travel to Albany, and within two weeks find themselves on new land in the Western Reserve in Ohio, or in Michigan. It took even less time to stop off in upstate New York and make a new home there. Costs were reasonable: a mill worker in the

East in the 1830s could move to Ohio via the Erie Canal, buy farmland and some pigs, and build a log house for an outlay of about $200, an amount a frugal worker could save in about six years.[4]

As the great immigrant streams continued, their destination shifted toward New York City, where the Erie Canal and then the railroads could carry them by the millions to the cities and farmlands of what some then called "the universal Yankee nation,"[5] north of the Ohio River. For both Yankees and immigrants, the journey was both quicker and easier than the trip by road from Philadelphia to Pittsburgh or Wheeling.

Time was not the only consideration. First, water transport was only a fraction of the cost of land transportation. Second, much of the land in the Southern States was not part of the federal survey system imposed on the Northwest Territory, and it was subject to land speculators and a different, less precise "metes and bounds" surveying system, unlike the transparent way northern land was priced and sold. Finally, the existence in the South of a plantation society and slavery worked against large-scale immigration either by farmers or those who had urban trade skills.[6] Texas, for example, had been settled to such a degree by 1836 that it could win its independence from Mexico, yet the Chicago area then had perhaps a couple thousand people, while southern Michigan was lightly settled and aspiring to statehood. Twenty-five years later, Michigan iron ore and Chicago-marshaled raw materials were of great importance to the conduct of the Civil War, while Texas was largely irrelevant. By then, though incomprehensible in 1825, Yankee organization and materiel had overcome the Southern cavalier warrior culture that Senator James Webb well describes in *Born Fighting*, his book about the Scots-Irish in America.

In sum, before the second quarter of the nineteenth century, it made sense for immigrants and migrants to use the Shenandoah Valley or the National Road to go west,[7] and to leave the Northern States to depend mostly on natural increase. After that, the situation reversed itself and the Erie Canal/Great Lakes route became the way into the interior,[8] and the South then became dependent on natural increase, while the North experienced an explosion of immigrants.

Cazenovia

In the early nineteenth century, the story of the development of Cazenovia, Cooperstown, and the areas surrounding these small upstate New York towns is a microcosm of all the frontier development from there west and south through New York and into northern Ohio. A group of businessmen would purchase or assemble a large tract of land, survey it into plots, which they would then encourage settlers to purchase, on credit for the most part.[9] The process would be overseen by a managing partner, who would generally be resident in the area. Beyond northern Ohio, the settlers would be subject mainly to federal land agents.

Cazenovia, just northeast of Syracuse, was laid out by John Lincklaen in 1793 as the market town for a 117,000-acre tract of land he was employed to subdivide and sell. The land had been bought for a consortium of Dutch banks by Theophilus Cazenove, who had been sent from Holland to America to invest in American bonds.[10] In addition to Cazenove, the Dutch used the famous French politician Talleyrand, in exile in America during the 1790s, as what we would call a lobbyist.[11] He became excited by the potential of land development and sale, and convinced the Dutch banks to move in that direction. Lincklaen was sent over from Holland to act as Cazenove's field agent.

The purchase, near today's Syracuse (then called Webster's Landing) was the first of a number Cazenove made. The Dutch banks subsequently incorporated the Holland Land Company, which also developed western New York around the town of Batavia. When Lincklaen surveyed the eastern tract, he decided to name its townsite on the lakeshore after his boss in Philadelphia. He found, however, that he had to "hot bed" the area with improvements if he wished settlers to buy his land. At first, he wanted to build a store, mills, potash works, and some roads, but his paymasters reined in his ambitions somewhat, but not so much as to hurt sales. As well, like Cooper to the east, Lincklaen had to help finance his buyers' land purchases. Depending on the tract's distance west of Albany, sales could be slow. For instance, in the area west of Cazenovia, during the ten years that he was the Pulteney Associates' agent (1792–1801), Captain Charles Williamson spent $1 million on their lands and took in only $146,000.[12] Patient money indeed. No wonder landowners from Buffalo to Cooperstown wanted the Erie Canal built.[13]

After twenty-five years, the Dutch informed Lincklaen that they wished to sell their remaining land and facilities; the end of Napoleon's rule in 1815 meant that opportunities were greater in Europe. He agreed to buy them out, but the Crash of 1819 and defaulting purchasers almost wiped him out. He died in 1822, never imagining that his family would continue to lead the area into the next century.[14] After his death, his son-in-law took over the land operations, and by 1843 managed to pay off the debt on the purchase.

Cazenovia was first settled by poor Yankees, and it probably was only the second wave, buying land from the original settlers,[15] who were prosperous enough to develop the area. Lincklaen married into a Yankee family, the Ledyards of Aurora, New York, which succeeded to his estate. In time, the Ledyard connection led through intermarriage to a wider connection between the town and the Yankee merchant and professional families in the region.

Lincklaen's idea of providing infrastructure for Cazenovia led to its having a church, a school, and at least one mill.[16] Ten years after their marriage in 1797, Lincklaen and his wife moved into "Lorenzo," an estate he built on the south side of Cazenovia Lake that was to be the residence of the family for the next 160 years. His sons worked for him in land sales and a daughter married his lawyer,

Jonathan Ledyard, who was adopted into the household, making him an heir. He became the businessman who saved the family fortune. Lincklaen developed two sawmills and a grist mill on the creek as it fell toward the lowland to the north, and in 1816 started a woolen mill. Other settlers set up a number of other small industries in the town, producing paper, chairs, and linseed oil.

When, in the 1840s, a plank road was built between Cazenovia and the railroad at Syracuse, another bout of industrialization took place, which had some family investment, as did the plank road itself. Another member of the family built new woolen mills in 1849 and in 1852. By the Civil War, Cazenovia had about two thousand residents. As the nineteenth century wore on, Irish immigrants came to the town as hired help. Lincklaen had often complained in his time about getting Yankees to work for him, as they preferred to work on their own hook if possible. The Irish did not bother his descendants in this way.

Cazenovia's growth eventually was affected by the extension of a railroad line through the town. Cheaper goods disrupted the small, local industries, but it also brought summer tourists up from the lowlands to the "more salubrious climate" by the lake. By the 1880s, city beautification had become popular in America, and Cazenovia, with the leadership of L. Wolters Ledyard, began to beautify itself. Summer estates, or "cottages," designed by New York and Boston architects, began to appear around the lake, and the rural simplicity of the area, coupled with easy access by rail, increased the number of visitors. As a result of the Ledyard connections in the region, the town became a vacationland for prosperous businesspeople and the wealthy elite of nearby cities and towns, such as the Remingtons from Ilion. A Methodist seminary, started in the town in 1824, produced a number of famous graduates, such as Philip Armour, of meat-packing fame, and Leland Stanford, the California railroad king, who gave his name to Stanford University. The seminary subsequently became Cazenovia College.

Today, Cazenovia still strongly resembles a small New England town, with a small college and a summer population of wealthy businesspeople, who, at least until well into the twentieth century, were Yankees with an anglophile social bent.[17] The town's upland hinterland contains both a village and a state forest named for Lincklaen, while the lowland to the north contains the expanding suburbs of Syracuse and threatens Cazenovia with becoming an exurb on US 20.

Lincklaen and Cooper

Sixty miles to the east of Cazenovia lies Cooperstown, on a tract that had been organized for sale by William Cooper some seven years before the Holland Land Company purchase (see Chapter 11). Lincklaen and Cooper encountered similar problems in the colonizing business, one of which, before the construction of the Erie Canal, was the difficulty of access to their tracts. A rudimentary canal had progressed into the interior of New York using the Mohawk River as a channel, but direct movement west was difficult and sporadic.[18] Even so, between 1790 and 1800, the population of the state almost doubled, from 340,000 to 589,000, and then almost doubled again in the next decade to just short of a million.[19] Much of this growth took place to the west of Albany.

Both Lincklaen and Cooper also faced pressure from Yankee migrants who wanted freehold ownership of their land.[20] Whatever thoughts Cooper might have had of imitating the Dutch *patroons* to the east and leasing his land (something his son James Fenimore Cooper might have desired) were lost in the face of this pressure. The Dutch bankers behind Lincklaen were more practical. They had no intention of leaving Holland and setting themselves up as American aristocrats, but simply wanted a good return on their long-term investments: sell the land and collect the money. Although the Yankee impulse toward freeholding was mainly cultural, the decision to sell rather than lease created value in the local communities. Indeed, the value of property ownership and development to economic development is often unappreciated. Peruvian economist Hernando de Soto has shown how the lack of property rights in less-developed countries makes it impossible for poor families to use their equity as a lease to fund entrepreneurial activity: the capital tied up in "informal housing" is dead, and capital accumulation for entrepreneurship hampered.[21]

The two promoters also found they had to finance buyers. Farmers could buy a tract of land on credit,[22] make improvements, market their produce, and eventually own their farms or sell the land at a profit. A respectable percentage of frontier families, generally the poorer ones, followed a pattern of purchase, develop, and then sell to more conservative settlers before moving westward to do it again, all the while building up an equity stake. It is not uncommon in the records and stories to see families move from New England and, within twenty years, live in three other States, all to the west of one another.

Lincklaen and Cooper had to become "active" landowners as well, providing clear surveys and titles, getting roads built, and opening a store where settlers could buy tools, seed, and other necessities.[23] In Cooper's general store, farmers could buy the things they needed in exchange for wheat[24] and potash, their primary cash crops, which he then transported and sold at a town with water transportation.[25] Cooper also drove cattle on consignment to a New Jersey auction and then credited farmers by reducing their mortgage by the amount earned from their animals.[26]

Potash production illustrates the market nature of the colonizing process. The settler could either cut the wood or hire someone to do it. Converted to ashes, an acre of hardwood could produce most of the money needed to pay someone to cut it, or, if the farmer did it himself, to pay down debt. In either case, there was the bonus of adding to the cleared farmland (if one ignored the stumps).[27] Local "potasheries" would buy the ashes and boil them for the salts. The potashery receipts could be used at a store like Cooper's or exchanged in the community as currency. Cooper, for example, after taking possession of a load of potash, would cancel the debts of the potashery and ship the potash off to Albany or New York for cash or to pay down his own debts to wholesalers. Upstream came new goods in demand by the farmers or cash for Cooper to use for local purposes. In this way, and in many others, settlers were connected to the world market.

Promoters like Lincklaen and Cooper were not "babes in the political woods,"[28] especially when it came to transportation. Cooper, for instance, worked with Albany merchants in 1790 to lobby the state government to build an improved road from the Mohawk River to Cooperstown.[29] Once successful, he then got the contract to construct the road, employing "his" people to do the work, paying them with credit at his store, while the road made his remaining property more valuable.[30]

Both Cooper and Lincklaen then lobbied Albany to fund the Great Western Pike, the predecessor to US 20 in this area. The road was begun in Albany itself in 1798, passing through Cherry Valley and on to Cooperstown, before being extended to Cazenovia.[31] Either Lincklaen or his son-in-law headed the Third Great Western Road Turnpike Company until 1859. The Pulteney Associates, landowners to the west of Cazenovia, extended the turnpike past Cazenovia, connecting it to Syracuse. The state government also assisted in the construction of the Great Genesee Road from Syracuse to the Genesee River, helping to open up the western part of the state and leading to a basic east-west road across New York by 1818. Captain Williamson, the Pulteney agent,[32] also had a road built southerly to connect to Williamsport, Pennsylvania,, allowing commerce to flow to and from western New York and Philadelphia and Baltimore.

On the frontier, the greatest need was for cash money, or specie. Around 1800, Cooper, a Quaker abolitionist, provided some of his people with maple sugaring pots and implements in the hope that maple sugar might replace cane sugar on American tables. This would strike a blow against the American and Caribbean slave owners and their sugar plantations, while providing a new cash product for "his" people. The spring weather proved too unreliable for this scheme to work as it also did in Lincklaen's area,[33] however, and it was not until 1841 that abolitionists in Illinois perfected a viable sugar-beet-processing machine that allowed farmers to compete with sugar cane growers.[34]

Finally, both promoters found the influx of Yankees to be both advantageous and maddening. They came in large numbers, some with money and others without. They also left in large numbers.[35] They wanted cheap money for its inflationary potential, easy credit at the store, little or no regulation, low taxes but good roads, and steady markets for their produce. The Yankees also wanted a church and a school and, if none of them had the wherewithal, for the landowner to build a mill.[36] The settlers soon realized that landowners needed them more than they needed the landowners and agents, which added to their difficult attitudes.

Organizing Farther West

In 1792, the bankers who would later combine to form the Holland Land Company also acquired all of westernmost New York as part of a more massive and complicated land "flip" (see Figure 3). Over the next seven years, their protégé, Thomas Cazenove, assembled and oversaw their widespread American operations, concentrating on the development of assets other than the remote western end of New York, in part because Robert Morris, the seller of the land, had not dealt with the rights of the Indians still left in the area. As well, until there was some kind of road into the area, chances were good that few settlers would go that far when good land was available in more accessible places.

Cazenove left his job in 1799, having hired Joseph Ellicott, the younger brother of a famous surveyor, Andrew Ellicott, to survey the Holland Land Company's 3.3 million acre western holding. Ellicott completed it the same year Cazenove left, and the company hired him to stay on as its area manager[37] and sell property to settlers.[38]

By the time the Erie Canal opened, the day of the large, private land companies had passed, but their example had served to point out both the value of the companies to settlement and their limitations. They brought sensible surveying and titles, and were able to lobby state authorities for public improvements. Yet they were unable to develop basic economic structures such as mills and services because of a lack of sufficient capital. Doing business on the frontier, they tended to attract settlers with few resources, and public authorities could not function well until a decent population size was reached. Cooper and Lincklaen saw roads to Albany or to the Mohawk River as the answer. Ellicott's distant location meant that settlers there had to depend on exporting to Canada via Lake Ontario and the St. Lawrence River, which they did even through the War of 1812.

The next wave of Yankee settlers and foreign immigrants would find a different regime as they pushed farther west.

Syracuse and Utica, New York: The American Religion

The Revolution was effected before the War commenced. The Revolution was in the mind and hearts of the people: and change in their religious sentiments of their duties and obligations.
— John Adams[1]

New-England of the West shall be burnt over...as in some parts of New-England it was done 80 years ago.
— Lyman Beecher, 1828[2]

If we had a hundred Moodys and Sankeys in the country, all the Protestant sects would unite within ten years.
— a Methodist minister attending a Moody revival, 1875[3]

I don't know about God. I know God. He is a nice fellow.
— TV Evangelist, 2005

Syracuse and Utica

A lot of people living outside the United States are puzzled by American religiosity. The numbers that attend church services regularly are multiples of those in Europe, or even Canada. American belief in the literalness of the Bible baffles these outsiders, as does the active relationship Americans feel between themselves and God. Driving through upstate New York is probably the best place to start to understand this difference.

The region from west of Albany to Buffalo, mostly just to the north of US 20, was once known as the Burned-over District,[4] but not from great forest fires or other devastation. Instead, it was where, in the first half of the nineteenth century, continuous explosions of religious fervor sent out shock waves that helped condition religious ideas in America for the next 150 years.

Syracuse sits on the Erie Canal, in the eastern half of the Burned-over District. Today, its metropolitan area contains about 750,000 people. It has been a crossroads for transportation since the swampy area to the south of nearby Onondaga Lake was drained and settled. Before the Erie Canal was built,[5] the area was home to some small settlements, notably Salina, named after salt springs that existed around Lake Onondaga. The area, an early reservation created by the state for the Onondaga tribe, had been gradually acquired by settlers after 1790.[6] Some of these developed saltworks, where the spring water was collected and boiled to extract the salt.

When the canal was opened from west of Salina east to Utica in 1819, the area began to be more intensively settled.[7] Syracuse was incorporated in 1825, the year the entire canal was completed, and growing shipments of wheat from the Genesee valley, followed by those from Lake Erie ports farther westward, created a reciprocal demand for consumer products.[8] Many of these came from New York City, but cities along the canal, such as Syracuse and Utica, saw entrepreneurs come and develop manufacturing enterprises. The Syracuse area had a more mixed economy, as it could sell salt as well as manufactures, plus it was the logical point for farmers to bring their crops for transshipment via the canal.

Utica and Syracuse were the same size in 1830, but Syracuse grew to be the largest in the area.[9] The city absorbed its rival, Salina, in 1847. The saltworks became more valuable as wheat production in the state began to decline in favor of more westerly areas and salt pork production for eastern markets was substituted. Syracuse also became a regional railroad hub, with lines going north to the St. Lawrence communities and south toward Pennsylvania, joining the main line of the New York Central. Syracuse resembled other upstate New York municipalities on the Erie Canal in its settlement pattern, with Yankee settlers making up the majority of its early inhabitants. These were joined by the Irish who worked on the canal and others from downstate and beyond who saw the opportunity to profit from its trade.[10]

Though it was affected by the revivals of the early 1800s, the area was more a place filling with immigrants and factory workers who were either godless or, perhaps worse, Catholic. A Baptist church opened in 1821, followed by a Presbyterian church in 1825. For many years, the radical Unitarian minister Samuel May, uncle to writer Louisa May Alcott, led abolitionist forces in Syracuse. The most famous revivalist of his time, Charles Finney, grew up somewhat to the north and east of Syracuse and began his religious career in the nearby Lake Oneida area. Neither Syracuse nor Buffalo interested Finney at first, but Utica and Rochester did, as they were somewhat more settled, with factories and surrounding farms, rather than being commercial hubs.[11]

Utica had been settled earlier than Syracuse and also prospered from the Erie Canal. It was located closer to the towns and villages whose initial settlement began when the area was just to the east of the British line laid down in the Proclamation of 1763. As the early settlers used leather leggings in the brush, they became known as "leatherstockings," a term James Fenimore Cooper adopted for his tales of the pre-Revolutionary frontier. The fame of these stories led Alexis de Tocqueville and Gustave de Beaumont to explore the area on their trip west from Albany to Auburn.[12] By then, Utica had become one of the centers of the religious revivals of the 1820s and 1830s.[13] It is not a coincidence that de Tocqueville's comments on the nature of American religiosity should seem a lot like a description of his experiences in Utica and beyond.

Head and Heart: Jonathan Edwards

The major American Protestant denominations all began by emphasizing different fragments of Puritan theology and practice. The Puritans in seventeenth-century England tried to emulate how they believed the earliest Christians, infused with the Holy Spirit, might have behaved. The religious trappings that had accumulated in the Roman Catholic church and its Anglican successor were to be scraped away — liturgy, vestments, hierarchy, and all. In the Puritan colony of Massachusetts, the church was to be community based, focused on the Bible, and run by "Saints" who had been touched by the Holy Spirit.[14] Moreover, like the Church of England, the Puritan church was "established" — that is, its costs were borne by the community, and religion and colonial affairs were fused;[15] outside faiths and their adherents were discouraged from settlement. Until the Charter of 1692 enforced official tolerance,[16] missionaries or evangelists for other faiths were forcibly expelled from the Massachusetts colony, or even executed.

Gradually, this vision of a common people joined in faith began to come apart.[17] First, not every person could honestly state that he or she had a valid encounter with the Holy Spirit. Some, like Jonathan Edwards, entertained doubts all their lives about whether their "conversions" were real. Those believers not touched by God's grace were still considered to be church members, but their deficient status meant they could not partake of the Lord's Supper at church meetings and were not allowed to speak or vote on church and public matters until such time as the Spirit might touch them. Eventually, as the numbers of these second-class citizens grew and many were family to the "Saints," pressure grew to admit them to normal privileges despite the Holy Spirit's having passed over them. This "Halfway Covenant" (1662) split the church in more settled areas into factions, though it did not seem to improve the minority proportion of the population affiliated with it.[18]

As well, almost unnoticed, the number of indentured servants — those who had signed an undertaking, or indenture, to work for a settler for a fixed period in return for passage across the Atlantic and room and board — was growing.[19] These mostly single people could not become church members — only "free" people could. As early as 1649, the majority of men in Boston were not affiliated with the church. In 1683, 83 percent of taxpayers in Salem, Massachusetts, failed to claim church membership. Some Connecticut communities allowed as few as 15 percent of their population in church.[20]

There were other debates as well. Could children too young to know about God and His ways be baptized?[21] The question arose as early as the 1500s in Europe, and it, among other disputes with the Puritan leadership, led Roger Williams to abandon the Massachusetts colony and create the colony of Providence and Rhode Island Plantations, that haven for dissenters leading eventually to the Baptist church. By the 1690s, Baptists were organized in New Jersey, and were active in Pennsylvania and Delaware in the early 1700s.[22]

A third problem arose with the expansion of Puritan settlements past the Connecticut River west and north into real hill country.[23] These isolated communities were supposed to be formed with a meetinghouse (that is, a church) and school like the original Puritan towns, but these regions lacked the resources of the richer bottomlands. Further, these communities' exposure to Indian raids and the legal claims of other colonies (New York, for example, claimed Vermont) meant they were relatively lawless[24] and their people probably illiterate — fair game for missionaries of other denominations and home-grown religious movements. As well, Puritans moving onto Long Island and into southern New York found the rights of their church restricted, first by the Dutch, then by the Anglicans who replaced them.

There were fears, too, that the desire for prosperity was overcoming the communal spiritual sense of New England and encouraging even more settlement beyond the pale of the "Shining City upon a Hill." In 1676, William Hubbard noted that the founders of Massachusetts "could not well tell what to doe with more Land than a small number of acres, yet now men more easily swallow down so many hundreds and are not satisfied....Many of these settlements [on the frontier] were contented to live without, yea, desirous to shake off all Yoake of Government, both sacred and civil."[25] The struggle to keep New Englanders within the Puritan flock proved futile. In 1679, Increase Mather, echoing Hubbard, complained about the expansion of New England settlement: "yea, so as to forsake Churches and Ordinances, and to live like heathen, only so that they might have Elbow-Room enough in the world."[26]

A century after the colonization at Plymouth in 1620, the Puritan church faced schism over the Halfway Covenant and found itself having to put resources into missionary work among its own members. One family heavily involved in both of these problems was the Stoddards. The Reverend Solomon Stoddard was the pastor at Northampton, Massachusetts, equidistant on the Connecticut River between Connecticut and Vermont.[27] An articulate and learned pastor, with great credibility in the puritans' Congregational church, he was sometimes referred to as "Pope" Stoddard.[28]

Stoddard was a champion of a more open and "experimental" (experiential) church, and encouraged all who attended his services to experience God's grace and join the congregation. The more conservative Boston-area ministers who dominated the church barely tolerated the Halfway Covenant; they were much less enthusiastic about Stoddard's further opening the church doors. Yet the need to evangelize the sons and daughters of the New England church who had moved north and west to find farms for themselves required new approaches. Stoddard was also the region's premier revivalist, holding five such sessions between 1679 and 1718, which led to his church's becoming the largest in the Connecticut River valley.[29]

Stoddard's grandson, Jonathan Edwards, was educated at Yale and also became a minister. When Stoddard died in 1730, Jonathan was asked to take over the church in Northampton, a post he held for the next twenty-one years.[30] He was a prodigious reader and writer, considered by many to be one of America's most original thinkers, and famous in his time second only to Benjamin Franklin. He was more a preacher than a practical pastor to his people, and continued his grandfather's revivals. His emphasis on learning was not uncommon in pre-Revolutionary New England, where the church insisted on the duty of its members to be literate and to read and interpret the Bible for themselves,[31] along with technical assistance from a highly trained clergy.

As an intellectual, Edwards found his early Puritan roots increasingly attractive. For a number of years, he did not change his grandfather's liberality toward those in the church who had never experienced the Holy Spirit. Yet he became increasingly convinced that the early Puritans were correct. Finally, he did two things. He stopped the practice of allowing everyone who attended to participate in the full rights of the service, and he began preaching old Puritan Calvinist "hellfire and damnation" to "sinners," calling on them to receive the Holy Spirit into their hearts and revive their faith.

By the mid-1730s, Edwards's revivals among church members had become so successful that they also began to attract frontier people, who had little churching or education.[32] And there were plenty to attract, as religious affiliation in New England at the time was less than 20 percent — one odd estimate was that sexually active single women in the region outnumbered church adherents of both sexes, and this under an established church.[33]

Edwards's activity was followed by that of the British Methodist George Whitefield, a spectacular preacher who had been banned from many churches in Britain for his enthusiastic approach to preaching.[34] In 1738 and '39, Whitefield ranged through the middle colonies and the South, attracting large crowds wherever he went.[35] He preached in New England in 1740, wherever someone would open a church for him or, if not, in an open field. His style was democratic in that he tried to simplify the theological essentials to generate mass appeal, an approach that has been imitated ever since.[36] He encouraged "promiscuous" crowds — meaning men and women together — and his detractors referred to him as a "peddler of divinity."[37] He had a voice that could be heard by thousands, and for many the emotional experience was so intense that they later deserted Congregationalism for the Baptists, where they could experience something like the Methodist Whitefield's intensity on a more regular basis.[38] What became known as the First Great Awakening[39] took hold, primarily in New England but all along the coast down to Virginia, south of which itinerant preachers were discouraged by the established Anglicans.[40]

Whitefield preached twice in Edwards's church, and Edwards remarked in his journal: "I have heard of one raised up in the Church of England to revive…the gospel, and full of a spirit of zeal for the promotion of real vital piety."[41] The problem Edwards faced was threefold. Here he was, a learned scholar[42] whose most successful device to restore the old Puritan spirit was to revive people by using an emotional route. Many Puritan ministers felt it made their scholarship redundant, though Edwards did not see it that way.[43] They also opposed the notion of lay people undertaking their own revivals without clerical involvement.[44] Since frontier communities often could not afford to support a settled pastor, even if they could find one, might their people acquire a faith based on emotion, rather than on individual learning from the Bible? In the interpretation of Gary Wills and others, American religion then divided between "head" and "heart."[45]

But this led to a second problem. Whitefield and other Methodists, seeing colonists spread out beyond eastern Pennsylvania, thought they might use talented but relatively untrained preachers to spread the Gospel into the West using the revival technique. Baptists, who were locally organized, had already embraced the emergence of farmer-preachers who provided for themselves and their families while answering the "call" to preach the Gospel to their neighbors. At first Edwards publicly recognized the logic of this approach, but in the horrified reaction of the conservative clergy in the more settled areas around Boston, he found himself at odds with his erstwhile allies in the task of restoring Puritanism. Retracting his support of the idea of uneducated preachers, he then alienated the revivalists.[46] The Puritan church began to split between the "Old Lights" and the "New Lights."

Edwards then managed to alienate his own congregation as the fervor of the Awakening died down, and in 1751 they dismissed him.[47] After a stint at an Indian school in Stockbridge, Massachusetts, in the center of the Berkshires, he was asked to become the president of what is now Princeton University in New Jersey. Shortly after arriving to take up the post, however, he and a couple of family members died when a smallpox vaccination program went wrong. At the end, he was mystified why God would call him from his work on the Berkshire frontier just to let him die suddenly in New Jersey.

Edwards's preaching and writings provide much of the tone of American Protestantism.[48] The revival would become a great tool, and an accepted one, for many denominations.[49] The emotional connection to the Almighty would become an important part of American identity. Reliance on the truth of the Bible, a Puritan fundamental, would be retained, but severed from the notion that an educated clergy was needed to explain the meaning of the text. For many Americans, God's grace, through personal revelatory experiences, would provide a truer understanding than scholarly wisdom.[50]

Another enduring aspect of Edwards's Puritan belief was that the settlement of America by God-fearing people was a sign of the coming Millennium, when a long period of utopian life, brought about by the good citizens of "the City upon a Hill" would lead to Christ's Second Coming and the end of the world.[51] The notion stuck with Americans, but later preachers argued that Christ would come *before* the Millennium and lead it. These pre-Millennialists have tended to dominate the imminent-end-of-the-world strain of thinking in the past century, arguing, logically, but in the extreme, that progress and non-spiritual social reform on earth are irrelevant, given the pending arrival of Christ.

Disestablishment and Competition

As the tensions leading to the Revolution increased in the decade after Edwards's death, Yankee preachers, including his son, Jonathan Edwards Jr., began to merge the Puritan religion with patriotism. During the French and Indian War, Edwards Jr. praised the struggle as part of God's plan. Upon the outbreak of the Revolution, he transferred the label of Antichrist from the French to the British, now ensconced in Quebec and allied with the Indians. This alliance of religion with war[52] helped give later Revolutionary rhetoric a religious tinge that has stuck with Americans ever since.[53] Indeed, the nature of American religion itself began to change as a result of the Revolution. For one thing, it led to the decline of the Anglican church in America, as many of its leaders and supporters had resisted the Revolution and gone into exile, taking their hierarchical establishment with them.[54] In the colonies where it had been the established church, this had the effect of placing all faiths on the same plane as far as government policy was concerned. It also led to antipathy toward state-supported religion of any kind. The Founding Fathers were strongly influenced by Deist ideas that all religions are equally right — and equally wrong.[55] For them, a democratic state could not be trusted to choose the "true" religion. In any case, by 1787 Congregationalists, Presbyterians, Methodists, and Baptists were all growing in numbers, followed by Quakers, Catholics, and Jews. America did not need some future war of religion to complicate its existence.

The answer at the federal level was to disestablish all religions through the First Amendment to the Constitution. This meant that, although the federal government was prohibited from passing laws pertaining to religion, the States were free to keep established churches.[56] Few did, though Connecticut abolished its establishment law only in 1818, New Hampshire in 1819, and Massachusetts as late as 1833.[57] The Massachusetts Constitution had not established any particular denomination, but had stated that every taxpayer had to pay taxes to support the church of his or her choice, with the taxes of the indifferent going to the largest denomination — the Congregational heirs to traditional Puritanism. After the Civil War, the Fourteenth Amendment was interpreted to mean that no one could be forced to support any church anywhere without their consent, effectively disestablishing churches at all governmental levels.

The case of New York state presents a useful example of the evolution of American disestablishment.[58] Clearly, the complicity of the established Anglican church against the Revolution meant that state support for it was impossible to maintain, even though the Anglican remnant reorganized itself as an American Episcopal church in 1784.[59] As well, there were scores to settle: Anglicans had worked hard to prevent the incorporation of Presbyterian churches in the colony, thus hampering their fundraising and building efforts. In 1769, the Presbyterians and Baptists created the Society of Dissenters, which seven years later produced the nucleus of the local Sons of Liberty organization.

Disestablishment was written into the state constitution of 1777, but this did not mean secularism, as the delegates still believed the Protestant ethos should condition the public affairs of the State. Today, this idea is more controversial, as we see in opposition to the placing of a sculpture of the Ten Commandments on courthouse lawns and the inclusion of the phrase "under God" in the Pledge of Allegiance. Indeed, over the decades after the Revolution, that ethos was gradually eroded by a number of legal cases.

In the early 1800s a Catholic was elected as an assemblyman. The oath of office, however, called for members to reject any allegiance to foreign crowns in matters both civil and ecclesiastical. This meant that a Catholic had to reject the structure of his religion to take office. The member refused, and was not seated. Eventually, the Assembly changed its rules. A second case concerned a man who decided to join the Shakers. His wife did not want to do so, but he took their children with him as per the legal doctrine of *coverture*. His wife, whose family had considerable political influence, took legal measures to get custody, and had the Assembly pass a law that considered a person who joined the Shakers legally "dead," with the children to be placed with the "surviving" spouse. Nothing was done in the case, however, and the law became a dead letter. By the 1820s, court cases relating to prosecution for blasphemy had disappeared from view and the requirement that witnesses in court be believers had faded — it was abolished in the Constitution of 1846. Sabbatarianism, or legislating Sundays as days of rest, began to be preached in the 1820s, but it made little headway in the state before the Civil War.[60]

Disestablishment had many effects nationally.[61] The most notable was to make denominations competitive, since they could no longer count on the state help to gain an advantage. Further, competitiveness meant that people could change their affiliation if they wished. Churches began to avail themselves of marketing tools, such as incorporation, to foster capital accumulation and develop facilities and personnel to carry their versions of the Gospel to potential converts. As well, although churches could influence the ideas of their congregations, they also had to be aware of prevailing opinion if they were to retain allegiance. Preventing the splintering of denominations became a concern. The separation of church and state also meant that the state could not be used to repress organizational divisions and new faiths.[62] Dislike of unpopular faith groups such as Catholics and Jews tended to be local, social, or illegitimate in nature — though not in the case of harassed Mormons (see Chapter 16). The overwhelming result of these three factors — churches with shrunken hierarchies, voluntary contributions, and official neutrality among churches — meant that the Protestant churches of the day had to absorb democratic values. This in turn led to growing anti-Catholicism as Catholic immigrants increased as a proportion of the population after 1845, in part because of the Catholic church's undemocratic, hierarchical nature.

"Converting the West"

By 1800, all denominations were turning their eyes westward. All, it seems, reflected different facets of the Puritan church. Some were evangelists, concerned with bringing people to an appreciation of spiritual truth and, more practically, through gaining church members. Others were fundamentalists trying to strip Christianity to its core practices and insisting on an individual interpretation of the Bible. Still others were Pentecostals, looking for direct contact with God, again and again. These concepts overlapped among individual churchgoers, and the emphasis placed on one or another led them to different churches

Americans are often thought of as being at least as religious during and after the Revolution as they are today, but that is false. Today, some 60 percent or so claim to be regular churchgoers, but it is estimated the proportion in post-

Revolutionary America was below 20 percent.[63] In part, this was due to a lack of churches and ministers— though most of the unchurched did absorb the basic Protestant morality and its ties to republican forms.[64] But these were also less pious times: the Continental Congress and the Constituent Assembly were dominated psychologically, if not numerically, by Deists and secularists, not Christian faithful. Only one of the educated and upper-class Founding Fathers identified himself as "born again."[65]

Four main denominations contested for adherents: Congregationalists (the Old Light Puritan church), Presbyterians, Methodists, and Baptists. Without going into theological distinctions very far, all shared the need for a personal experience with the Holy Spirit to seal membership, and reliance on the Bible as the basis for religious truth. Differences ranged along a spectrum of ecclesiastical authority from almost complete local control by churches to a hierarchy, including persons of authority who could speak for the denomination. They also ranged along a spectrum from trained, professional, and settled clergy to itinerant or part-time local clergy with little or no formal religious training or even general education.[66]

The Congregationalists had split, after the firing of Edwards, into New Light churches that favored local control and emphasized the personal experience, and Old Light churches that accepted a greater degree of organization and relied on learned, settled preachers to provide guidance to church members. The New Light Congregationalists, those whom Edwards had reacted against, managed to dominate the westward movement of New Englanders, while the Old Light faction held on for a while in New England before giving way in part to Unitarianism. Having the benefits of being settled in a community and deriving a salary from the taxpayer, both groups were loath to undertake missionary work themselves.[67]

The New Lights, along with the Methodists and Baptists, were seen as Arminian, rather than Calvinist — that is, they believed that, instead of being predestined by God to be saved or damned, individuals could, by their actions, save themselves. This belief was in accord with the ethos of settling the frontier: after all, what was the frontier but a challenge to people to "save" themselves economically? Salvation as settler could have its setbacks and despairing moments, but the New Light approach fit this condition on a theological scale — people could fail to improve themselves spiritually as well as economically, and they could be redeemed in both senses as well.

Presbyterians tended to fall along organizational lines in ways similar to those of the Congregationalists, which led the two denominations, for the most part, to accept a Plan of Union in 1801 that called for interdenominational cooperation in evangelizing the West, while maintaining their own organizations in the East.[68] Even so, the attachment of the Congregationalists and, to some degree, the Presbyterians to an educated and settled clergy hurt their efforts in the West, particularly where they should have been strongest: in upstate New York, filling up with Yankee migrants out of New England. As well, the Presbyterians were wracked with internal divisions, growing antipathy toward camp meetings, and mixed feelings toward revivals. Their initial advantages were lost.[69]

Methodists tended to favor a hierarchical style reminiscent of its Church of England antecedents, with bishops and an evolving church organization. After Whitefield's successful participation in the First Great Awakening, Methodist ministers adopted his populist and emotional style of preaching. Its first post-Revolutionary bishop, Francis Asbury, converted a British Methodist idea of enthusiastic but unsophisticated lay preachers into a cadre of lightly trained circuit preachers as a means of evangelizing the frontier.[70] Until his death in 1816, Asbury led this effort by personal example, traveling an estimated 300,000 miles on horseback and foot, preaching and organizing the Appalachian West. Methodist circuit riders would go from one small community to another, conducting Sunday services in whatever facilities they were allowed to use. They had a more-or-less set route, and as the demand grew for religious services, they gradually were replaced by settled preachers and reassigned to areas of new settlement farther west. The circuit riders endured an exhausting routine and were discouraged from marrying, as they would be on the road too much to be effective parents.[71] Many died young, from sheer exhaustion.[72]

The Baptists, inheritors of Roger Williams's break with the Puritans, also accepted the importance of a personal encounter with God, but differed in that they required adult baptism to become a member; other denominations were more willing to accept children as members. The Baptists were given a boost by Whitefield's preaching, which emphasized the salvation experience, rather than doctrine.[73] They spread into the Chesapeake Bay area and then into Appalachia, depending on preachers who "rose up" out of the farm communities in the valleys and hollows, following a "call" from God to preach the Bible. Baptists countered the Congregational insistence on an educated clergy by pointing out that the Apostles, many of whom were clearly too poor to have received much, if any, education, were the ones Christ had chosen and had been infused with preaching skills by the Holy Spirit.[74]

The Baptists' loose organization and both the Baptists' and the Methodists' willingness to accept preachers motivated by the Spirit,[75] whether educated or not, helped these two denominations grow among the small farmers on the frontier. While their organizational structures differed, they commonly exchanged ministers,[76] and their simpler methods of developing clergy meant they could generate surplus ministers, which helped them expand further. Though all of the denominations grew along the frontier, the Methodists and Baptists grew fastest in the South and in Appalachia. Their preaching also reached the slave population, where their reliance on the Spirit to move and convert souls, irrespective of literacy, found a receptive audience. Both denominations also had some female preachers at the start of their spread into the frontier, but this practice gradually declined.[77]

The first time the country as a whole recognized that something was happening in the West came with reports of a massive interdenominational camp meeting held at Cane Ridge, outside Lexington, Kentucky, in August 1801 that drew an estimated 20,000 participants, ten times the population of Kentucky's largest town.[78] The settlers came to hear a number of preachers, each established in a separate booth, and to conduct all sorts of secular and commercial activities as well. The reports of various "shakes" and "jerks" and emotional outbursts added to the picture presented by newspapers to the coastal state populations. Cane Ridge established the mass revival style, begun in late Puritan New England, as the core evangelical instrument, regardless of denomination.[79]

The results were impressive. As one historian notes, "only land could compete with Christianity as the pulse of a new democratic society."[80] Between 1776 and 1850, church adherents increased from 17 percent to 34 percent, even as the country's population surged.[81] Congregationalists dropped, however, from 20 percent to 4 percent of those who claimed to be religious adherents, Episcopalians dropped from 16 percent to 4 percent, and Presbyterians declined from 19 percent to 12 percent. The Methodists, in contrast, grew from 3 percent to 34 percent and the Baptists grew from 17 percent to 21 percent.[82] One commentator has put it succinctly, if in an exaggerated form: "[T]he American Jesus was born at Cane Ridge, and is with us still."[83]

The Burned-over District

The Burned-over District, the area from west of Albany to Buffalo, was the great competitive evangelical battleground for souls in the first thirty years of the nineteenth century. It was a region rocked by an outpouring of religious fervor in existing churches that transcended their organizations. Also, it led to the creation of a number of authentic American religions, their doctrines devised in post-Revolutionary America and not imported from Europe.[84] Finally, the Burned-over District led in the creation of a multitude of voluntary charitable and social reform societies.

Yankee settlers began moving past Albany into and south of the Mohawk Valley as soon as the Revolutionary War had ended. As segments of the Erie Canal were opened in the 1820s, the movement increased even more, welcome or not. One New Yorker called these settlers "a set of fierce republicans, if anything sneaking and drawling may be so called, whom litigious contention [Shay's Rebellion] had banished from their native province….Among this motley crew there was no regular place of worship, nor any likely prospect there should, for their religions had as many shades of difference as the leaves of autumn, and every man of substance who arrived, was a preacher and magistrate to his own little colony."[85]

Once Yankee migration west began in earnest,[86] so did the formation of Congregational missionary societies.[87] Because the Congregationalists were committed to educated and settled preachers, the task of providing these to frontier churches, as well as to foreign places such as Hawaii, was a daunting one. Charitable societies were formed in New England to provide support, including subsidies, to ministers who would agree to settle in western New York, Ohio, and beyond to tend frontier congregations that were unable to support them.[88] The Congregationalists placed little emphasis on areas of the country where Yankee stock did not predominate; their aim seems to have been to have the church follow its adherents' descendants.[89] The problem of attracting and training potential frontier ministers proved too great to overcome, however, and the Congregational and Presbyterian churches turned instead to the rising classes in the new towns, leaving potential converts elsewhere to other denominations. Thus, Congregationalists in the West were largely swallowed up by the more organized and active Presbyterian churches.[90]

Other denominations also focused their attention on the same area. The Methodists came up the Susquehanna Valley with settlers from Pennsylvania, the Presbyterians followed their adherents west from New Hampshire, and the Baptists followed theirs from Rhode Island and elsewhere. Even the needs of small Quaker colonies in the Genesee Valley were supported from Philadelphia. In what became the Burned-over District, all the denominations made intense evangelization efforts in the decades before the Civil War, with all kinds of unintended consequences.

Revivals began in 1799, and more followed in 1807 and 1815.[91] Then, in 1824, just as the Erie Canal was reaching completion, Charles Finney, the Great Evangelist, began preaching in the towns along its route. Alexis de Tocqueville came through the same area in 1831, the year Finney concluded his greatest revival in Rochester.[92] Not surprisingly, de Tocqueville was impelled to comment: "[t]here is no country in the world where the Christian religion retains greater influence over the souls of men than in America." Yet, three groups resisted the call to revival: the settlers who came north from Pennsylvania and stopped at the southern edges of the Finger Lakes, the older lake villages with a large proportion of Episcopal residents, and the substantial farmers who had found Universalists too liberal for their Calvinism, let alone the revival preachers.[93]

Experiments in belief and action boiled out of this Second Great Awakening.[94] One historian notes that a lot of the social causes and religious zeal could not have happened without "the veritable host of evangelists who swarmed over Yankeedom, old and new, preaching every shade of gospel, heresy, and reform to a generation of people who had been saturated with spiritual and moral intensity."[95] This led to a number of complications.

First, there was the danger of going too far with the New Light idea of self-assisted salvation and toward Perfectionism, or antinomianism.[96] This doctrine held that people who were truly converted or saved would be given enough strength or grace to resist all conscious sinning, thereby becoming spiritually perfect in God's eye. The Oneida Colony led by John Phillip Noyes was Perfectionist, and Mormons and Shakers also looked somewhat to Perfectionism, though New Lights found this extension of their theological approach barely tolerable.

The area also produced the Millennialist Adventists called Millerites, followers of a Massachusetts-born Baptist minister-turned-Deist-turned-Adventist, William Miller.[97] He confidently predicted the date of the Second Coming of Christ to take place in 1843, when his adherents would be taken into Heaven.[98] When the date passed without incident, Miller returned to his calculations and, finding an error, predicted another date in 1844, which also passed quietly. His disappointed followers regrouped to wait a bit longer, the largest group becoming the Seventh-Day Adventists.[99] The Millerites' particular focus on the end of the world had continuing impact. When it did not end as Miller predicted, the movement fell apart, but the idea of pre-Millennial apocalyptic preparation, including a belief in the Rapture — when the faithful would be assumed into Heaven — entered the mainstream of American belief.[100] It largely supplanted the older post-Millennial idea that the world first had to be converted, leading to a religious utopia, when Christ would return for a thousand-year reign. The Shakers advocated the equality of masculine and feminine principles and added a kind of passive apocalyptic vision. Shaker colonies were dedicated to perpetual chastity, a formula for extinction if the end did not happen soon. Millennialism has continued to shape American myths of the country's origins and destiny.[101]

The Burned-over District was full of socio-religious experimentation that reflected an emphasis by preachers on the need to couple faith with social action — in effect, personal perfectibility led to attempts at social perfectibility.[102] Utopias were a persistent feature of these movements. Notable among them were the Shaker colonies throughout New England and as far west as Kentucky, the prosperous Oneida Colony just east of Syracuse, and the Mormon colonial migration out of New York to Ohio, then to the Illinois shore of the Mississippi River, and finally into the Western deserts and mountains to a refuge in the Great Salt Lake country.

In the more social and less religious sphere, a wave of new ideas also came out of the Burned-over District that tended to extend the equalitarian principles of the frontier into new parts of society. Three of these — code-named abolition, temperance, and female emancipation — created huge national crises, and their implications are still being dealt with today. In the Burned-over District, these and other causes found religious adherents lining up on different sides, even within some denominations, which only added to the intensity of feeling in the area.[103]

Abolition referred to the ending of the institution of slavery in America. The French Revolution had attacked slavery in the 1790s, and the British gradually undertook its elimination in the Empire after 1815. The moral beacon of the "City upon a Hill" projected by the Puritans meant that their Yankee descendants were increasingly frustrated seeing Europeans do what other Americans so fiercely resisted.[104] The controversy over slavery had more to do with the notion of what ought to be considered "property" and about the nature of freedom than it did about concerns over human civil rights.[105] Northern communities, while abolitionist, were unwelcoming to free blacks and escaped slaves alike, denying them the vote and restricting their living, working, and educational rights. Southerners argued that there was little difference in condition between their slaves and that of "wage slaves" who existed in dire poverty in Northern city slums. Northerners saw this as irrelevant. They also ignored Southern claims that Northern legal restrictions on free blacks living there were at least as onerous as those of slavery. The moral question of *owning* people was paramount, however, to abolitionists. In religious terms, they saw slavery as a threat to the idea that everyone must have the opportunity to use their religious connection to the Lord and to exercise their ability to offer their services to the highest bidder in order to work out their physical and spiritual salvation. Churches tended to split on the issue along regional lines, leading, for instance, to the creation of the Southern Baptists.

Temperance, another code-word, usually implies the willingness to act in moderation, but when applied to the consumption of alcohol, it soon began to mean total abstention and, eventually, abolition. In the early 1800s, alcohol consumption by Americans was, on a per capita basis, many multiples of what it is today, perhaps reflecting the pressures and insecurities of life on the frontier. When communities began to take on a more settled existence of businesses and factories, alcohol consumption as a way to deal with personal problems became less tolerable. It created inefficiencies and lost productivity and accidents, as well as brutality, abuse, and neglect within families. As one historian has noted, "[i]ntemperance was made the symbol of the sinner's slow descent into hell, and 'the pledge' became the token of the new life."[106] The result was a movement allied to religion that opposed the excessive use of alcohol. Because many reformed drinkers slipped back into the vicious cycle of alcohol abuse, the Temperance movement quickly changed into one that opposed its use outright. Upstate New York became a center of the movement, and many of the churches affected by the Second Great Awakening adopted a ban on alcohol.[107] The movement continued to grow through the next century and eventually led to the legal experiment of Prohibition in the 1920s. Its logic is still found today, live and well, in the "War on Drugs."

A third code-word, female emancipation, extended into the area of women's rights. "Emancipation" originally was used to refer to the abolition of slavery, but women also found a use for the term. Evangelism that required social reform enlarged women's experience — as early feminist Lydia Maria Child said, "[t]he sects called evangelist were the first agitators of the woman question."[108] As well, in the Burned-over District, the shift of female labor from the farm to the mill and the office, which happened somewhat later and under different conditions than was the case in New England, might have contributed to feminist action.[109] Other reform movements, such as dietary and prison reform, also played a role.

Camp Meetings and Evangelist Preachers

People in western New York were continually "revived" in their enthusiasm, and attracted to open-air and camp meetings, church revivals, and mass preaching in halls and auditoriums. They were also enrolled in Bible study groups, charitable organizations, and reformist movements. Pamphlets, hymnbooks, and lecturers on all sorts of subjects proliferated, while charismatic preachers led some people into new variations of Christianity.[110]

One of the most popular revival methods was the camp meeting.[111] These were derived from the Scottish Presbyterian practice of outdoor "communion sessions," which engendered a degree of emotional religious feeling. Apparently, the first American camp meeting was held in 1797 in upstate New York by Presbyterian and Methodist ministers.[112] Farther south, the Scots-Irish Presbyterians who flooded through the Cumberland Gap into Kentucky brought the practice with them. It was especially useful where the population was light and dispersed.[113] The 1801 camp meeting at Cane Ridge, Kentucky, became a model for other evangelists on the frontier.

As a result of the popularity of these meetings, itinerant preachers from one or another of the denominations, but who were generally estranged from them, dominated the revival scene, whether rural or urban, preaching to thousands and becoming celebrities in their own right. The best-known were Yankees and, not surprisingly, their activities over time became more organized in terms of tours, advance publicity, fundraising, and popular anticipation. The denominational churches often closed their doors to these preachers, but local businesspeople and officials welcomed them, or at least gave them the run of public spaces and facilities as long as they compensated the town and brought in business.

The Sensation: Lorenzo Dow

One of these itinerant charismatic preachers was Lorenzo Dow, born in Coventry, Connecticut, in 1777. As a child, he was sickly and subject to religious visions. He became a Methodist as a young man, and applied for and was given a preaching circuit in New York State. After a year of preaching there and in western Massachusetts and Vermont, he left on a trip to England in 1799, returning there again in 1805. He was an eloquent and eccentric character who introduced the English to camp meetings and felt he had a message from God that he should preach to the Catholic Irish, which did not endear him to the Irish or, at the time, to the Methodist church in England.[114] Dow remained a nominal Methodist, but, once back in America, he struck out on his own as a preacher. He would follow his own circuit, preaching and standing up to local drunks and bullies, normally earning him a few choice thrown eggs and other sundries, which did not add to his otherwise dirty and disheveled appearance. Even so, his message appealed to those who had left relatives and settled lands behind for the uncertainties of the frontier.[115]

Through much of the first decade of the 1800s, Dow preached in the South, opposing Catholics and Calvinists alike. An abolitionist and temperance advocate, his Southern tours sparked controversy as he tied liberty to religion: "But if all men are born equal, endowed with unalienable rights by their Creator in the blessings of life, liberty and the pursuit of happiness, then there be no just reason, as a cause, why he may or should not think, and judge, and act for himself in matters of religion, opinion and private judgment."[116] His sermons were like nothing people had seen before, and the emotions he generated in them ran the gamut from laughter to fury, with a mix of jokes and Biblical references thrown in.[117] He spoke to the Georgia Legislature and to camp meeting throngs of ten thousand people. He also traveled to Canada and the West Indies. He cared little about the religious preferences of those to whom he preached: he would challenge and appeal to them all. Dow encouraged his audiences to trust in dreams that seemed to be revelations, and claimed he could find lost objects and raise the Devil, leading people to call him "crazy" Lorenzo Dow.[118] He also was not averse to doing a little business: he would carry along and sell his own medicines to help heal the sick, and was involved in some land speculation in the then Northwest Territories.

Dow outlived his wife[119] by fourteen years and died in Washington, DC, in 1834. It was said that parents named more children after him than anyone except George Washington. Twenty years after his death, "Lorenzo" was still a popular given name for male babies.

The Entrepreneur: *Charles Grandison Finney*

Another famous preacher, Charles Grandison Finney, was, like Dow, a Yankee, born in Warren, Connecticut, in 1792.[120] Shortly after, he moved with his parents to western New York, first to just north of Lake Oneida and then farther north along the eastern shore of Lake Ontario. He noted in his autobiography: "When I was about two years old, my father removed to Oneida County, New York....The new settlers, being mostly from New England, almost immediately established common schools; but they had among them very little intelligent preaching of the Gospel....My parents were neither of them professors of religion and, I believe, among our neighbors, there were very few religious people."[121]

After teaching school for a few years, Finney returned to the town of Adams, on the eastern shore of Lake Ontario, and began articling and then practicing as a lawyer. He became interested in studying Scripture, decided to follow a calling into the ministry, and found he had enough talent at it to be licensed as a Presbyterian minister. His mentor urged him to pursue his studies, but he felt he had a gift from the Holy Spirit and did not need an educational straitjacket.[122] Instead, in 1821, he began to preach and revive the faith among rural farm families.

Finney's "language was based on the Bible, Shakespeare, and Blackstone [the legal commentator]."[123] His ability to convert rural folk quickly led to his being called to a variety of places in the area. Because of the mix of churches and faiths, he soon learned not to discriminate between those who needed saving, but to focus on a Biblical message that could resonate with all of them. Finney was the first itinerant preacher who held extended sessions in one locality, finding that this increased zeal in the community.[124]

Then, having married a woman from Whitestown, near Utica, he gravitated southward and began preaching in 1825 in the villages to the north of Lake Oneida, conducting a successful revival in the town of Western. Soon, a pastor in nearby Rome invited him to preach, and Finney's career in the towns that would come to make up the Burned-over District began. At least three times in the next thirty years, he went through the towns of the area from Troy to Utica to Buffalo on extended revivals. In one revival in Rochester, he claimed that a hundred thousand souls had been converted over a number of months. The reports of his 1831 revivals in the District set off a wave of revivals from Maine to Ohio.[125] Lyman Beecher, who, as late as 1827, was a critic of Finney's, called the Rochester revival "the greatest work of God and the greatest revival of religion the world had ever seen in so short a time."[126]

Finney's success did not go unnoticed elsewhere. He continually found himself embroiled in controversy with the organized churches, especially in New England.[127] His theological grounding tended toward an Arminian view of salvation through faith and action that fit frontier thinking, but was too loose for the educated and settled ministers of the East.[128] He made social reform a religious obligation: Christians were no longer to wait for the Devil to push them but were to push the Devil first.[129] Congregational opposition did not stop him from conducting revivals in Hartford, Albany, Providence, Boston, Philadelphia, and New York City. He also went more than once to England and Scotland for revivals. Many of his sermons and lectures were transcribed and published. He claimed he never wrote a sermon beforehand, but uttered whatever the Holy Spirit inspired in him.

In 1835, Finney was asked to go west to Ohio and become a professor of theology at a new non-denominational college in the Western Reserve town of Oberlin. For the next forty years, until his death at age eighty-three, he spent part of the year at Oberlin, part in the East and summers in upstate New York, continuing, in the 1840s and 1850s, to conduct revivals. The appeal of Oberlin was, at least in part, due to its adoption of a combination of education and social activism.

Finney's priorities in conversion did not include bringing people into the Presbyterian church. Instead, like Dow and the Methodist itinerant preachers, he focused on a message of direct contact with God, mass evangelization, salvation for all who came forward (no Calvinist "elect" for him), reliance on the Bible for truth, and little concern for the intellectual side of religion. He brought the rough egalitarianism of the frontier to the city, organized his processes, and set out to convert the country. He made his mark in the heyday of Jacksonian democracy, in the late 1820s and 1830s. After that, in his fifties, he became a part of the growing upper classes as professor and then president of Oberlin College and a famous author and churchman. He gave his last sermon at Oberlin only a few days before a heart attack killed him in 1875.

The Industrialist: Dwight Moody

Dwight Moody[130] came on the scene after the Civil War, when the focus of evangelism was on the cities and their burgeoning immigrant population.[131] With the exception of the South, which was left to its own devices for decades after its defeat, the rest of the country began to turn its interests from agriculture to industry. The churches in this period largely caught up with the spread of the native-born population and became more settled and institutionalized.[132] Even so, by 1900, there were 650 active evangelist preachers and 2,200 part-time ones. Between 1914 and 1918, this "industry" would hold some thirty-five thousand revivals.[133]

The upper and middle classes in the cities were characterized by a split between the work world and the home, which encouraged a distinction between the "women's sphere" of domesticity and the men's sphere of paid work and public activity, including rather loose morality, both in terms of money and sex. The poor, including immigrants, had to work for pay regardless of gender, lacked basic amenities or good education, and morals that were dictated ethnically, if at all. Together, these conditions generated a demand for the evangelization of city dwellers from rich to poor.

Moody was born in 1837, three years after Lorenzo Dow died and the same year Charles Finney left New York for Oberlin College. In effect, he was a third-generation American evangelist.[134] Like Dow and Finney, he came from New England — Northfield, Massachusetts, a town on the banks of the Connecticut River just south of the border with New Hampshire. His father, a local farmer and stonemason, died when Dwight was five, and the children were sent off to neighboring families to work as soon as they were old enough to be useful. In 1854, seventeen-year old Dwight found work in his uncle's shoe store in Boston. In the city, he was "converted," and left his family's Unitarianism for the Congregational church.

In 1856, Moody decided to go west and seek his future in Chicago. He had a letter of reference from his uncle and soon had a job as a shoe salesman. Moody had come to his conversion through Sunday schools and he began to work with them in Chicago, relating especially to the Methodists. His focus was always on the practical method of attracting people, especially poor children, who could be given both spiritual and practical education. Moody's own childhood poverty and lack of education led him to the practicalities of organizing; he left the theological points to others. A visitor relates a story about him:

> When I came to the little old shanty and entered the door,...the first thing I saw by the light of the few candles, was a man standing up, holding in his arms a Negro boy, to whom he was trying to read the story of the Prodigal Son. A great many words the reader [Moody] could not make out and was obliged to skip. My thought was, If the Lord can ever use such an instrument as that for His honor and glory it will astonish me! When the meeting was over, Mr. Moody said to me, "I have got only one talent. I have no education, but I love the Lord Jesus Christ, and I want to do something for Him."[135]

Moody's success at Sunday schools was such that he was allowed to use a large hall at Chicago's North Market, where attendance expanded to an occasional thousand or more. One of Moody's supporters was influential in Abraham Lincoln's presidential election campaign, and got the president-elect to visit the school when he visited Chicago.

After a great revival in Chicago in 1857, Moody became involved with the Young Men's Christian Association (YMCA). During the Civil War, Moody worked with the YMCA and met with men on the battlefields in the West, such as Shiloh. He also entered Richmond with Grant's army at the end of the conflict. Meanwhile, he honed his speaking skills at YMCA conventions and rallies all over the country and in Canada. During the war, he also managed to raise $20,000 for a new mission building near the North Market called the Illinois Street Church. After the end of the war, he was elected president of the YMCA and immediately set up a joint-stock corporation to raise funds for a central YMCA building, subsequently opened at a cost of $199,000.

In 1870, Moody met Ira D. Sankey at a service in New Castle, Pennsylvania. Sankey was a talented singer, and Moody was so impressed that he asked Sankey to give up his government job and work with him. "You do the singing," he said, "and I'll do the talking." In revivals all over the country and in Britain, Sankey would play on a portable organ and sing or lead the singing, while Moody preached to the crowds.

Through the 1870s, Moody traveled in much the same manner as had Finney and Dow before him, except that his venues were large cities like Boston, Montreal, London, and Dublin. These revivals took a lot of organizational ability and funding.

The logistics were complex, though the relatively mature urban railroad systems did mitigate much of the burden of travel. Moody planned his revivals with military bureaucratic precision. In preparation for an 1876 revival in New York City, his advance team leased Gilmore's Concert Garden, formerly known as Barnum's Hippodrome. The cost of the rental, for a month of refurbishing work ($15,000) and three months of revival meetings approached $25,000. Five hundred ushers were found, and Sankey had a choir of 1,200 to work with. Two hundred and fifty people worked seven inquiry rooms around the city to question the veracity of those claiming to have been saved. Moody employed "decision cards" to keep track of the numbers of those converted, and used squads of volunteers organized into districts to get people to come to the revival. Seven thousand attended the first meeting in New York City, with another four thousand in overflow rooms and thousands more left to their own devices outside. On the fourth day, Moody held five meetings with a total of twenty thousand in attendance, and the first Sunday saw twenty-five thousand at the meetings. And on it went, from early February until mid-April, when a YMCA convention closed out the revival. By then, $135,000 had been raised from donations to support future projects.

Despite his lack of theological training,[136] in 1886 he established the Moody Bible Institute in Chicago to train young people for the ministry. Like his predecessors, he preached the literal truth of the Bible, but added a rejection of evolution, a theory that had not existed formally when Dow and Finney preached. And like these two, Moody was no respecter of denominational distinctions, holding some feeling for Congregational predestination with a touch of Methodist Arminianism. Though he acknowledged the Second Coming, he did not seem to see it as imminent as did the Adventists and Mormons. As Bernard Weisberger puts it, "[a]s theology grew simpler, technique became predominant."[137]

Moody continued his revivals until he collapsed at one in Kansas City in 1899 at age sixty-two. He returned to Northfield, where he had summered for years and where he had worked to create a number of institutions devoted to training and education. His funeral in Northfield was attended by high and low alike, and he was buried on a hill behind his birthplace.

Conclusion: A Democratic Christianity

The most influential of the Revolutionary leaders tended to incline toward a Deistic understanding of God as being relatively passive in the affairs of humanity. The wheels of the universe were set in motion and people were placed within it and given free will to work out their own destinies. This notion, however, was not acceptable at the time to the vast majority of believers in the American countryside. Instead, they tended to follow a logic whereby an act of faith or acceptance opened one up to God's influence, in which a personal relationship could develop between God and the individual. Personal behavior thus worked to promote or frustrate Divine plans for all of society. Evangelism freed people to find their version of how to relate to God, first by throwing up a different kind of leadership other than the learned and the certified to point the way, and, second, by giving a choice of religious organizations for community support in this quest.[138]

This belief was both democratic and personal — democratic because it did not admit the authority of trained and/or ordained others as necessary to understand Divine Will.[139] Each person who made an act of faith opened himself or herself up to direct contact. As early as the 1840s, many would say, as their descendants do today, "No creed but the Bible!"[140] The church provided a social context for the furtherance of the Divine Plan. Harold Bloom extends this to an American version of freedom: "Freedom, in the context of the American Religion, means being alone with God or with Jesus....In social reality, this translates as solitude....The soul stands apart, and something deeper than the soul, the Real Me or self or spark, thus is made free to be utterly alone with a God who is also separate and solitary, that is, a free God."[141] Thoreau's rejection of the Brook Farm communal experiment grew out of his assertion of his being "a community of one."[142] He spoke for generations to come.

The relationship was *personal* along the lines of Christ's appearance after his crucifixion and resurrection to some disciples on their way to the country town of Emmaus, and to some of the Apostles as they were fishing on the Sea of Galilee. In the first case, He engaged the disciples in a conversation and had a meal with them. In the other, he suggested they should cast their nets on the other side of the boat if they wanted to catch a lot of fish. Later he had some dinner with them as well.

The essence of these stories to Americans 1,800 years later was that individuals could have a direct, personal relationship with God, and that relationship could be of benefit to the individual. Listen to Divine advice, and you will be guided to a full net of fish or some alternative form of prosperity. Individual decisions in daily life required this direct advisory contact. Of course, the cautionary tale of Job was always in the background, but, in the context of attributing American expansion and general prosperity to Divine advice, it was not taken too seriously.

Because this approach, taken in conjunction with loosely organized church structures, suited American sensibilities so well, it has endured through the past two centuries, and forms the basis for much religious feeling in the country today. This is a long way from Jonathan Edwards. Nathan O. Hatch asks, "What are most churchgoers to do with a man [Edwards] who, according to the best recent analysis, 'explicitly denied the efficacy of petitionary prayer to bring about external change in the world'?"[143] H. Richard Niebuhr put it another way in a 1958 sermon given in Edwards's Northampton, Massachusetts, church on the two hundredth anniversary of his death: "What hearing could he gain, if he stood in this pulpit today, or in any pulpit in America and spoke to us about our depravity and corruption, about our unfreedom and the determinism of our lives[,]...about the awfulness of God's wrath?"[144]

All things considered, the American religion seems to be one of assured salvation, irrespective of deeds or faith. It is a religion where the walk taken by Christ and His disciples to Emmaus takes place often, in many individuals' lives. Christ is a friend and advisor on a personal level. There is no mediator, as in the Catholic church, and even there mediation is more tolerated than needed by the American faithful. Most other faiths, with the possible exception of the Mormons, have kept the unmediated, but Biblically guided, connection with God, and have prospered mightily. As Bloom notes, "the American Christ is more American than He is Christ."[145]

Seneca Falls, New York: Women's Rights

[C]reatures of the same species and rank, promiscuously [we would say, randomly] born to all the same advantages of nature, and the use of the same faculties, should also be equal one amongst another without subordination or subjection.

— John Locke[1]

By marriage, the husband and wife are one person in law, that is, the very being or legal existence of the woman is suspended during the marriage, or at least is incorporated or consolidated into that of the husband; under whose wing, protection and cover, she performs everything.

— Sir William Blackstone[2]

This is Mrs. Lucretia Mott and her daughters....I thought you should know one another. Mrs. Mott's a great abolitionist, but she's a fine cook too.

— Daniel Neall[3]

> *When, in the course of human events, it becomes necessary for one portion of the family of man to assume among the people of the earth a position different from that which they have hitherto occupied, but which the laws of nature and nature's God entitle them....We hold these truths to be self-evident: that all men and women are created equal.*
>
> — Elizabeth Cady Stanton[4]

Seneca Falls and the 1848 Convention

In upstate New York, US 20 runs past the northern edges of a series of glacial gouges collectively known as the Finger Lakes. Narrow and long, they extend almost to Lake Ontario, leaving only a narrow corridor for east-west travel. The Erie Canal, connecting Albany with Buffalo, had to be constructed along this same corridor.[5] A string of towns grew up beside it to provide service to the canal and take advantage of its freight capacity.

Where the largest of these lakes, Seneca Lake, drains northerly over a waterfall, the small mill town of Seneca Falls was created, somewhat to the south of the path of the canal but with easy access to it. A small and pleasant town of about 7,500 people, it is noteworthy as the site of the first national Women's Rights Convention in 1848. US 20 is the town's main street, and on it is the Women's Hall of Fame, fairly modest as halls of fame go: no crowds wait to get inside as in Cooperstown.

If one had to choose a starting point for the convention and the organized movement that followed, it would be the World Anti-Slavery Convention, held in London in 1840.[6] While there, prominent abolitionist Lucretia Mott met the new wife of a New York delegate, Elizabeth Cady Stanton. Stanton's husband, also an ardent abolitionist, had joined the majority in opposing the recognition of women delegates, and the organizers had insisted they sit off to one side behind a screen and not participate directly in the proceedings. The new Mrs. Stanton was not a delegate, so she was seated in the "ladies' section" as an observer. There, among the disallowed female delegates, she was quite impressed by Mott, and the two women spent a lot of time in conversation about women's deficient social and legal status. Henry Stanton, in the meantime, reversed his stand on female participation when confronted on this issue by the powerful personality of his wife.

By 1847, Stanton had relocated his law practice from Boston to Seneca Falls. His wife found herself in a small town with small children and a husband who was away a lot of the time. Strong willed and impulsive, she was keen to become involved in social action, and the following year, 1848, was a heady one for political change. It was a time of upheaval in Europe. France changed its regime and the German states had riots, strikes, and rebellions. There was a brief communist regime in Hungary, a legacy of the French Revolution, and, regardless of the success or failure of European revolutions, the example of 1789 was resurrected. Karl Marx and Friedrich Engels published *The Communist Manifesto* as a kind of call to arms to European workers.

In America, 1848 marked the end of the Mexican War. Thoreau had been briefly imprisoned in Concord, Massachusetts, for refusing to pay his taxes because he saw the money as partly funding the war. He and many other Northerners, including Abraham Lincoln,[7] had reservations about the war, seeing it as enlarging the number of slave States, especially if its successful conclusion were to lead to other conflicts to annex the sugar islands in the Caribbean. Abolitionists, including Henry Stanton, were creating a new political party, the Free Soil Party, to stop the geographic spread of slavery into the West. Closer to home, the New York Legislature had passed a Women's Property Rights Act that spring.

Seneca Falls seems an odd place for the founding of a women's rights movement that had such important effects on American life and whose claims still resonate. It came almost as an afterthought, but its timing sparked a large number of local and annual national conventions and the creation of lobbyist and activist groups that persisted until the achievement of national suffrage in 1920, more than seventy years later. Suffrage, however, was not the main item on the feminist agenda — Lucretia Mott had insisted that its inclusion in the Seneca Falls Resolutions would hold the movement up to ridicule, but she deferred to the determined Stanton. Instead, a thicket of economic and social laws and practices had to be overcome as well that affected women in the antebellum decades in a real fashion.

The challenge was immense. Beyond generally hostile public opinion, including that of most clergymen, the women's rights movement faced considerable competition from other causes clamoring for attention and resolution, including abolition of slavery, temperance, labor activism, peace activism, healthy living advocacy, utopian communities, school reform, and revived religious fervor.

Legality and Equality

The American descendants of the English Puritans were a significant force in the American Revolution, supporting Jefferson's text of the Declaration of Independence, in which he noted that "all men are created equal." Coming from an aristocrat and slaveholder, one might interpret this claim as an abstract philosophical statement on Jefferson's part,[8] or as an extreme reaction by an aristocratic commoner trying to hold on to his privileges, which were under threat from the King in London and the King's (more legitimate) cronies. At any rate, Jefferson did set the stage for a limited kind of equality in America.[9]

Meanwhile, Abigail Adams had been virtually alone in championing female equality. She asked her husband, John, in a private letter, not to forget the cause of women in Congress, but nothing was done.[10] In France, women attempted to get the 1789 "Declaration of the Rights of Man and of the Citizen" expanded beyond men to include women. Their petition was denied, but the public nature of their attempt set off a 125-year struggle in Europe and America to turn male rights into basic human rights.

A close English observer of the Revolution, Mary Wollstonecraft, was prompted by this failure and by the attempts of various French and English writers to justify the unequal status of men and women to write her *A Vindication of the Rights of Woman*.[11] In it, she made the case that women were not given the same advantages as men, especially in terms of education, and that they were socialized and even pressured by their circumstances to de-emphasize reason for emotion in order to live decently. Given the same legal and social advantages as men, women could demonstrate their ability to meet the conditions for Lockean equality and therefore for claiming the same rights as men.

The core argument around universal rights was all about "nature" or "nurture." Opponents of universal rights claimed there were inherent differences between the sexes that could not be bridged. Women and men functioned best in their respective "spheres" of influence, with the woman's "domestic sphere" focused on such activities as bearing and raising children, while men went out in the "public sphere" to earn a living and manage the country.[12] Those who insisted on universal rights pointed, in contrast, to the effects of culture and socialization that created the very behaviors that were then used to justify continuing these effects. Give women equal rights, and they would show themselves to be every bit as "rational" as men. The ascendancy of these "nurturers" came only over the next two centuries in Europe and America, albeit imperfectly, and is a work in progress in the rest of the world.

In the nineteenth century, the legal and customary relationships between men and women had changed little from the times of the Greeks and Romans. Roman law,[13] which formed the basis for the legal systems of western Europe, gave the *pater familias* the right to govern and dispose of his family and property as he deemed fit. Sons, at a certain age, could become independent and start families of their own. Daughters passed, for the most part, from the control of fathers to the control of husbands. Inheritance laws provided for the widow and children, but even then, they were subject to guardianship by relatives. English common law came out of customary sources in the 1300s and 1400s, and it reflected strong "Roman" influence.[14] Indeed, the concept of *coverture*[15] summed up legal family relations that were even more restrictive of the behavior of women than Roman law had been. Women and children were under the "cover" of the father, who owned the property, managed the family's finances, disciplined everyone, and decided the children's futures. In turn, the father owed allegiance and obedience to his feudal lord.[16] In the feudal societies that arose between the Roman era and the Renaissance, such laws applied only to the small number of elite families. Peasant farm families essentially had no rights, except what their lords would offer. Male or female, they lived in a world of custom and obligation. Even so, custom on this level did not differ significantly from many elements of Roman law, buttressed by the teachings of the Roman Catholic Church and later its Protestant successors.[17]

In early America, colonial development and the common law seemed to work well together, and the system functioned at least from the 1600s until after the Civil War.[18] The legal system supporting the family unit provided the labor necessary to clear land and plant and harvest crops. Keeping children under a legal inhibition meant that parents could count on having their labor until the age of majority, in the meantime enabling production increases on the farm as the parents grew older. By the time sons reached the age of majority, their parents might have accumulated sufficient savings to allow their sons to acquire and settle their own land. The system also allowed fathers to indenture children as apprentices,[19] thus finding them a trade to pursue as they grew and often deriving some profit from their children's labor.

As well, until the 1830s, relatively few men had political rights since they were not sizable property owners, so the idea of rights was little more than a pious declaration even for most men. The main struggle during this time was to achieve male equality — Alexis de Tocqueville was almost alone in seeing "that one consequence of democratic principles [would be] the subversion of marital power."[20] Then the influence of *Blackstone's Commentaries*, published in America between 1771 and 1773, was pervasive in the legal thinking and practice of the new States. Blackstone's style was accessible to the "do-it-yourself," semi-apprenticeship method of learning and practicing law of the time. His interpretation of the common law was conservative,[21] but the radical ideas of the Revolution never penetrated deeply into law in any case, precisely because the existing law was accepted all across the new country.[22] The Revolution was really all about preserving American customs and economics, not about changing them. Relative to what the King and his advisors might wish to impose on the colonies, the Revolution was revolutionary, but not so to most of those already on American soil.

Things Come Apart

A number of factors led to changes, beginning about 1830. First, male suffrage was expanded in the 1820s and 1830s to include most free, white males over the age of twenty-one, irrespective of their wealth.[23] All of a sudden the political world, which had been somewhat distant from most families, began to involve them all. In some families, women became interested in affairs of the world, and they became even more so once universal manhood suffrage began to include Irish and other immigrant men, who were generally looked down upon by native-born men and women alike. Why should these low and ignorant men be allowed to vote when educated and upright women could not?

As well, the expansion of public educational opportunities for girls meant that more women were able to read and understand aspects of the world around them.[24] For many, this was of small interest, but there was a growing number who did care. In 1847, an Englishwoman, Elizabeth Blackwell, was admitted to the Geneva Medical College in New York, where, two years later, she graduated at the head of her class.[25] Her example led others to take a chance on admitting female medical students, and to the creation of institutions specifically for them including the Female Medical School of Philadelphia (1850), and the Boston Medical School for Women (1852).[26] The first female minister of a "mainstream" church was Antoinette Brown, whom the Congregational church ordained in 1853.[27] During the Civil War, the *Morrill Act* providing for land-grant colleges also encouraged the admission of women, which resulted in increased educational opportunities for them beyond traditional colleges such as Oberlin, Antioch, and Vassar.[28] Education for women also led to the rise of a legion of female authors and publishers. *Godey's Ladies Book* (1820) and *Peterson's Ladies Magazine* (1840s) appealed to middle-class interests. Margaret Fuller, a Transcendentalist, became the editor of *The Dial* in 1840, a literary critic for Horace Greeley's *New York Tribune* in 1844 and an author in 1845. Amelia Bloomer began her temperance paper, *The Lily*, in Seneca Falls in 1849, but soon mixed its message with that of women's rights.[29] By 1855, Nathaniel Hawthorne was complaining that "America is now given over to a d----d mob of scribbling women, and I should have no chance of success while the taste is preoccupied with their trash."[30]

Then, the rise of manufacturing meant that women who otherwise might have remained at home to do piecework went off to a factory to make money to support their families.[31] At first, the owners and managers of New England textile mills exercised a kind of *in loco parentis* over the women and children who worked for them and lived nearby, but that disappeared as the century wore on.[32] Between 1800 and 1860, regardless of population increases, the proportion of women in the labor force *and* the proportion employed in non-farm labor both more than doubled.[33] By 1850, one-quarter of the workforce in manufacturing was female.[34] Moreover, the experience of Yankee farm girls going off to mill towns and earning their own money gave them a sense of personal independence that was not lost after they left for a new life in the West. Pious noise about the home as the woman's "sphere" began to fade into the background as increasing numbers went off to work.[35] It continued, however, in relation to women in the professions, where their admittance was strongly resisted.[36]

Despite such resistance, a widening number of female occupations began to emerge. The Massachusetts educational reformer Horace Mann, for instance, felt strongly that the teaching profession had many ties to women's domestic sphere, and encouraged women to regard it as a profession.[37] Of course, when women discovered that they would not be paid as much as men for the same work, political sensitivities were aroused. Soon, other jobs opened up, due in part to the telegraph, the sewing machine, and, later, the typewriter.[38] Also, in the late 1840s and early 1850s, the disappearance of thousands of men to the Mexican War and then to the goldfields of California left women to mind the store or farm back East.[39]

Urbanization, too, allowed women not only to find work, but also to be put in close proximity with other women of like mind, which enabled them to organize more easily. Feminists in the cities began to learn political techniques such as petitioning to push changes to the law.[40] As well, the separation of work and home meant greater physical separation between men and women on a daily basis, one effect of which was the contradictory creation of the "Cult of Domesticity" after the Civil War while also freeing up women to become more politically active in social causes. A study of women's groups in Rochester, for example, suggests that about 10 percent of local women were actively involved in voluntary organizations, including the Female Missionary Society (1818), the Rochester Female Charitable Society (1822), the Rochester Female Anti-Slavery Society (1835), the Female Moral Reform Society (1836), the Female Association for the Relief of Orphan and Destitute Children (1837), and the Ladies' Washingtonian Total Abstinence Society (1845).[41] Reading and discussion circles also became an urban fixture in the North after 1815.[42] As well, Anti-Slavery "Fairs," where women provided items for sale to raise money for abolitionist causes, thrust them into organizing positions. The first such fair was organized in Boston in 1834, with Lydia Maria Child, later a prominent feminist, as its president.[43] During the election of 1840, women's involvement was such that the Massachusetts radical, Edward Quincy, was prompted to exclaim that, "Truly the 'Sphere of Woman' must be made of [rubber] of the most elastic description, it so well accommodates the never-ending inconsistencies of this…generation."[44]

Finally, religion also provided a kind of training place for women's rights activists. The personal tie to God that came out of the Second Great Awakening in the late 1820s did not differentiate between men and women. Presumably, everyone had equal access to the Redeemer, so why did they not have equal rights in a Christian society?[45] The Quakers already accepted equality between the sexes, and had female preachers. Local egalitarian communal colonies in New England and upstate New York also had an influence by example. The Oneida Colony in New York, just beginning in 1848, was based on equality of the sexes, as were the Shaker communities in New England and the Albany area.

The effect of the religious Second Great Awakening in the late 1820s raised a number of social causes that required participation. Revivalist preachers encouraged "promiscuous," or mixed-gender, prayer meetings. Middle-class men were preoccupied with their jobs, so it fell to middle-class women to be recruited to these causes.[46] They pioneered the development of voluntary associations and, in doing so, began to learn the rudiments of political organization.[47] Women activists had been gaining experience from the two controversial, but increasingly popular causes that had come out of the Burned-over District of western New York — abolition and temperance — and the organizations promoting these goals were the two main channels for women to enter political life.[48]

As *coverture* began to come apart, those who wanted to retain the system, whether male or female, encountered contradictions.[49] The notion of "spheres," whereby women were said to be best suited in roles that related to their function of motherhood — the domestic and the educational — began to be stated only as more people settled in cities, where there could be a distinction between home and office or factory. This was also a class- and race-based argument that excluded working poor women, female slaves, and black women in general from being put on a pedestal of domesticity. As Sojourner Truth, a former slave, famously told the 1851 Women's Rights Convention in Akron, Ohio: "That man over there says that women need to be helped into carriages, and lifted over ditches, and to have the best place everywhere. Nobody helps *me* [to] any best place. *And ain't I a woman?*"[50]

So, one has to keep in mind that the agitation for women's rights in the nineteenth century was a complex process, including a changing society, powerful personalities, and considerable resistance from clergy, some male activists in related movements, and most important, women themselves. The title and the subject of Debra Gold Hansen's *Strained Sisterhood: Gender and Class in the Boston Female Anti-Slavery Society* points to this problem of women's conception of their role.[51] Susan B. Anthony, in frustration, said, "[w]oman is the greatest enemy of her own sex. She spurns the betrayed, but feels flattered by the betrayer."[52]

As the century wore on, a lot of the early resistance to women's rights wore down, but it would be folly to see this movement as simply taking off from the Seneca Falls Convention toward ultimate success. Some of the issues raised 150 years ago still have not been resolved. A gender relationship that has existed in different forms for millennia does not fade that fast.

The *Married Women's Property Rights Act*

The events leading up to the Seneca Falls Convention had somewhat to do with the overall status of women, but a lot to do with another long-term issue. Since the Revolution, there had been a conflict between men whose prosperity came from the ownership of large tracts of productive land and men whose prosperity came from raw materials and manufactured products.[53] In effect, the legal system inherited from the British was full of feudal protections for those who owned or controlled land, but had little to offer for the kind of capitalist country that was emerging in nineteenth-century America.

After the War of 1812, Democratic Party dominance of state legislatures and the federal government was based on offering more rights to their supporters. The attack on privilege centered on the existence of a dual legal system, one based on the "normal" set of laws passed by legislatures and the other based on decisions of a court of chancery, a medieval institution that emerged out of the practice of appealing cases to the King if a written law produced an unjust, or inequitable, result. In a democracy, however, a chancery review of cases was seen as leading to "judge-made" law, not "people-made" law, and thus "undemocratic."[54]

Democratic or not, one concern of both landowners and other wealthy men was that abolishing "equity" legal decisions in favor of legal codes would lead to what to them was a serious complication. Holders of large estates in land or cash, which went in part or in total to a daughter, would be affected. Under feudal marriage laws, the chancery had provided that fathers could use trusts to keep their estates out of the hands of undeserving or dissolute sons-in-law. Abolishing equity law meant abolishing this protection. The choice, fought over for a dozen years, was either to restore trusts or pass new legislation allowing daughters to maintain personal control over the estate.[55] But this would fly in the face of *coverture*, under which a daughter's civil existence was "covered" by that of her husband and where her property became theirs — in effect, his. Thus, since the attack on feudal law was concerned with property and commerce at this point, some undermining of feudal marriage law could be tolerated.[56]

The old rules could also stunt commercial activity. For example, if a woman could not be the beneficiary of a life insurance policy on her husband, then there would be less demand for the product.[57] In like manner, the increase of savings accounts at banks was inhibited by not allowing women to have them in their own name.[58]

In 1836, the use of trusts was severely restricted, and the following year a *Married Women's Property Rights Act* was first proposed in New York State, but did not pass.[59] Further attempts at such legislation were made in 1840 and again in 1846, when it was thoroughly debated as a clause in the State's Constitutional Convention, but defeated.[60] Meanwhile, the Constitutional Convention had finally abolished the chancery court, and finally, in 1848, the right of women to own property became law with almost no opposition.[61] It was, some claimed, "the death-blow to the old Blackstone code for married women in this country."[62] This was an exaggeration, as the conservative nature of the judiciary tended to rein in the actual use of these laws for decades to come.[63] The law allowed women to own property in their own name and to shelter it from their husbands if they so desired. It also protected a woman's property from being seized by her husband's creditors. It was a start, at least for women in wealthy families,[64] but did not address, for example, who controlled wages paid to the wife or a woman's ability to sue and be sued.[65]

Until after the Civil War, the essence of the women's rights movement focused on changing the legal status of middle-class women in ways that would undermine feudal marriage law. A women's rights convention held in Rochester in 1853, for example, called for equal pay for equal work; wives' entitlement to their earnings; guardianship of widows for their children; and the abolition of gender distinctions in the administration of estates, the ownership of property, or inheritances.[66]

In 1860, New York passed the most significant reform of women's rights up to the Civil War in the *Earnings Act*, which allowed women to engage in lawsuits and for a wife's earnings to be included in her separate estate.[67] It was only after the postwar constitutional amendment that gave freed male slaves the right to vote that the women's rights movement began to focus on their own right to exercise the franchise.[68]

Complex and deep-seated issues such as these cannot be resolved in anything under a few generations unless there is some force that propels a solution — a civil war or some other crisis — because the practices associated with the status quo are so solidly entrenched in social practice, the economy, and the law. That being said, there are points in time when a movement to correct social dysfunctions suddenly jells.[69] For women's rights, that happened in Seneca Falls in 1848.

Lucretia Mott

The beginnings of the women's rights movement can be best seen through the lives of three women: Lucretia Mott, the "soul" of the movement; Elizabeth Cady Stanton, the "brains" of the movement; and Susan B. Anthony, its "arm."[70] Both Mott and Anthony were born in Massachusetts, while Stanton was born near Albany of Yankee and Dutch parentage. Mott and Stanton were married and had seven children each; Anthony remained single. Mott and Anthony were Quakers; Stanton did not support any organized religion. Mott was a generation older than the other two, and before the Civil War the best known of the three. Having all died by the early 1900s, none saw the culmination of their movement in the Suffrage Amendment of 1920.

Lucretia Mott was born to a Quaker family named Coffin on the Massachusetts island of Nantucket in 1793.[71] Largely unknown today, she nevertheless has been called "the greatest American woman."[72] Her father, Thomas Coffin, was a whaling ship owner and captain, often gone for years on expeditions around Cape Horn and beyond. Like other women of the island, Lucretia's mother was accustomed to handling the family's affairs, including shopkeeping and financial and legal matters, while her husband was away.[73]

In 1804, the family moved to Boston, where Lucretia's father went into business, primarily as a wholesaler and ship outfitter. Lucretia, having started school on Nantucket, went to a public school in Boston for a year, after which her parents sent her to the Quaker Nine Partners Boarding School in Dutchess County, New York, one of the few schools of its type to admit girls.[74] She graduated from the school at age seventeen and was offered a job teaching there. She took it, but was irritated that the women teachers were paid only half what the men teachers received. By this time, the Coffins had moved to Philadelphia, and when Lucretia decided to visit in late1810, another teacher, James Mott, followed her there and asked to her to marry him. They were married in a Quaker ceremony in April 1811. James went into the Philadelphia wholesale business with the Coffins, and remained in that business for the rest of his life. Lucretia maintained their household, bore seven children, six of whom survived childhood, and taught school for a while — an unremarkable story for a woman born into an American merchant family of the time.

As Mott grew older, however, she became active in causes such as abolition, temperance, the promotion of peace, and, later, women's rights.[75] James followed her lead and supported her in these causes until his death in 1868. Mott did not write much;[76] instead, she preached, beginning at Quaker meetings in her early twenties, and by age twenty-five she was certified as a minister, a position largely unheard of for a woman outside the Quaker sect. An early listener said "he went home feeling as he supposed they must have felt in the old time who thought they had heard an angel."[77]

Soon, she transferred this facility to speak to public meetings. At the time, participants in such meetings were assumed to be male and the audience partly female. At an abolitionist meeting in Philadelphia, for instance, a declaration was drawn up to be signed by the participants, but Mott spoke up from the audience and prodded her hesitant husband, anxious about the effect on his business, to sign it, saying, "James, put down thy name."[78] Having thus spoken up, she then was asked her opinion on other matters and took an active part in the rest of the meeting.

As well as speaking to various groups in the Philadelphia area, including giving sermons at a black church, she started traveling to meetings in other cities. She also learned how to organize a meeting by getting a black churchgoer, James McConnell, to walk her through the formalities. In the early 1800s, adult women had little understanding of these extra-domestic actions, and there was no *Robert's Rules of Order* from which to learn. From a friendly politician, Mott also learned to use the most common political tool of the day, the petition.[79]

In 1833, a convention of sixty-two delegates from regional abolitionist groups met in Philadelphia to form a national society. Women were not invited on the first day, but were admitted on the second. Mott went to the site with alacrity, and rose from the audience to offer advice on a topic of debate, was recognized by the chair, and saw her advice adopted. She and other women were then asked to create auxiliaries to the National Anti-Slavery Society, but were still not allowed to become members of the main body.

Nonetheless, Lucretia Mott's talents on the podium and her prominence as a leader soon led her addition to the boards of various abolitionist organizations, often in the face of male opposition.[80] Indeed, in 1839, the decision to admit women to full membership in the National Anti-Slavery Society led to its splitting into two organizations. The most controversial move to limit women's involvement came in 1840, at a World Anti-Slavery Convention in London, where Mott — one of seven women delegates selected by the American national body along with some other female delegates from New York and Massachusetts — had their credentials rejected and were relegated to an audience area screened off from the podium.[81] William Lloyd Garrison, the most prominent American abolitionist attending, elected to sit with the women as a sign of protest. In his paper, *The Liberator*, he wrote: "With a young woman placed on the throne of Great Britain [Queen Victoria in 1837], will the philanthropists of that country presume to object to the female delegates of the United States, as members of the Convention, on the ground of their sex? In what assembly, however august or select, is that almost peerless woman, Lucretia Mott, not qualified to take a part?"[82] The controversy led to Mott's becoming a public figure in England for a time, and she was invited to aristocratic parties and receptions. She was also asked to visit Ireland, in the hope that she would support the cause of Irish independence. Perhaps more important, she had met and befriended Elizabeth Cady Stanton, then twenty-five and a generation younger than Mott, while they were seated in the convention's observer section.

The controversy in England served to raise Mott's profile in America, too. Her speaking engagements, mostly in churches, were invariably crowded with the famous and the ordinary alike. Ralph Waldo Emerson wrote of hearing her in Washington, DC, that the sensation "was like the rumble of an earthquake" and "no man have [sic] said as much and come away alive." Elsewhere, Emerson referred to her as "a noblewoman" and "the flower of Quakerism."[83] While in Washington, she met with President John Tyler, a slaveholder, and former president John Quincy Adams.[84] Henry David Thoreau went to hear her in New York City, and reported: "After a long silence, Mrs. Mott rose, took off her bonnet, and began to utter very deliberately what the spirit suggested. Her self-possession was something to see, if all else failed, but it did not."[85] Mott's speaking engagements took her from New England to the Western States and into Upper Canada (now Ontario), in tours through the 1840s. The Motts' home became a regular stopping place for prominent visitors. Gerrit Smith, the husband of a cousin of Elizabeth Cady Stanton, called one day, and while there met Frederick Douglass, the former slave and now Rochester anti-slavery activist, and the actress Fanny Kemble, as they discussed issues while Lucretia did some darning and dried glassware.[86]

In the summer of 1848, Lucretia, now fifty-five, and her husband went to Genesee County in New York to attend some Quaker religious meetings,[87] and to visit her sister, who was living in Waterloo, a few miles west of Seneca Falls. While at her sister's home, Mott, her sister, Martha Wright, and two local Quakers, Mary Ann McClintock and Jane Hunt, met socially with Elizabeth Cady Stanton, and the conversation turned to Stanton's long-held idea of holding a convention on women's rights.[88] to begin the project of creating the women's rights movement she and Mott had discussed, first in London eight years before, and then in Boston a year later.

Mott had misgivings about the time of year, mid-July, but they decided to make the attempt, in part because the others wanted to take advantage of Mott's presence as the "star attraction" at a founding convention to promote women's rights. "Conventions" were a popular political device of the time to bring together sympathizers and activists for various causes. These local women had no real experience in organizing one, but Mott and, in particular, her husband, who was along for the trip, did.

Looking for an adequate venue led them to the Wesleyan Church in Seneca Falls. In a rush, the women scattered to complete different tasks, such as publicity, logistics, an agenda, and some kind of declaration before the announced dates of July 19–20, 1848. As Mott had largely shunned writing in favor of the pulpit, Stanton took on the task of preparing the declaration. In the end, it went off with only minor problems. One embarrassing complication was that none of the women except the featured speaker, Mott, had ever run a formal meeting like this, but as she was the guest speaker, it would have been inappropriate to make her the chair as well. Accordingly, her husband James was pressed into service, which left Lucretia free to intervene in, comment on, and influence the debate. She also closed the meeting with a moving speech.

Hurriedly, they put together a convention. It was publicized through Frederick Douglass's abolitionist paper, published in Rochester, and news of it was also sent out along the very new telegraph system that connected western New York to the East. Three hundred people attended it, and Elizabeth Cady Stanton authored and presented a Declaration, with Resolutions, based on the Declaration of Independence, that affirmed the legal equality of women. About two-thirds of the delegates signed the document, though many retracted in the face of later criticism.

Two weeks later, another convention was held in Rochester, again featuring Lucretia Mott.[89] This was followed by a spate of similar meetings throughout the "West" (Ohio, Michigan, and Illinois). Annual national women's rights conventions were held, beginning in Worcester, Massachusetts, in 1850, which brought together activists from all parts of the Northern States.

For the rest of her life, Lucretia maintained her 'star' status, dividing her time among a number of causes, primarily abolition and women's rights, and playing the role of elder stateswoman at meetings and conferences to her death in 1880.

Elizabeth Cady Stanton

Elizabeth Cady was the daughter of a well-known judge and congressman, Daniel Cady, whose family had come from Connecticut and were now living in Johnstown, New York, northwest of Albany.[90] She and her sisters were educated at Emma Willard's Academy in Troy.[91] She then found that she could not follow the boys, with whom she had studied mathematics and the classics in her local school, on to college, as girls were not admitted.[92]

In 1840, at the age of twenty-five, she married Henry Stanton, a gifted Yankee orator and abolitionist she had met at her cousin's house in Peterboro, New York, just off what is now US 20. Her cousin was married to Gerrit Smith, one of the richest men in the State, an abolitionist, and generally radical in his politics.[93] The marriage took place in a hurry, as the couple was going off to London to the World Anti-Slavery Convention at which Elizabeth would meet and fall under the spell of Lucretia Mott, the first woman she had ever heard give a lecture and who became her role model.

After their European working honeymoon, the Stantons returned to Johnstown, where Henry read law for a time under Elizabeth's father. Elizabeth, who had already learned a lot about the legal profession from her father, continued her informal study.[94] The couple then set off for Boston, where Henry planned to make his career. There, in 1841, Elizabeth Cady Stanton met Lucretia Mott once again. They discussed the possibility of setting up an organization to promote women's rights, but nothing happened.[95] Stanton also attended debates in the state senate and got involved in political discussion circles.

By 1847, Henry Stanton had decided that his health and his fortune would be better served by relocating once again to western New York, so the family moved to a house on the outskirts of Seneca Falls purchased for them by Elizabeth's father. Henry's legal and political work — he was a founder of the Free Soil Party in 1848 and the Republican Party in 1856 — took him away often on the new railroad that passed close by, and Elizabeth found herself thrown upon her own resources in managing a growing family — it eventually would include seven children — and trying to find new friends. She was energetic, restless, and lonely.[96]

Elizabeth had a broad education, having gone to the best secondary school of the time and improved it on her own, as well as learning about the law from her father. She would recall hearing women emerge from her father's home office crying because there was nothing he could do for them within the law about their abusive, drunken, or wastrel husbands. Young Elizabeth had tried to tear out the offending pages from his law books, only to be told that that would not work: she must change the law first.[97] She also saw how grieved her father was when her brother, his only son, died young, as he "had no boy" to whom to pass on the family estate. She was aware of the long process through the 1840s to get a Married

Women's Property Act passed in New York, and began to work for that goal as soon as she arrived in Seneca Falls from Boston.[98]

Her political ideas were based on Thomas Jefferson's egalitarianism, which she, of necessity, carried on past its originator to include slaves and women.[99] Indeed, the problem that worried many Democrats in the 1840s was how far equality should be stretched.[100] Stanton wished to change contract law to make marriage simply another contract between two equals. Equal pay for equal work was also a natural outcome for her, as was the right of women to sue and be sued, to have equal access to education, and to join the professions of law and medicine. A *laissez-faire* capitalist, she felt that women should have the right to own and manage businesses on their own account.

Elizabeth Cady Stanton also had a flair for the dramatic; as one commentator has noted, "[p]rudence was foreign to her; half-measures impossible."[101] Whether true or not, the story of her trying to tear out the pages of her father's law book shows this trait. In later life, she would issue some dramatic statement or article on an issue such as divorce or birth control that would serve for some as a talking point, while it scandalized the rest. William Leach characterizes her as "the most brilliant, witty, sarcastic, and dynamic of feminists, and she had a romantic side."[102]

For the declaration of the 1848 Women's Rights Convention, Stanton fixed on the Declaration of Independence that Jefferson had written almost exactly seventy-two years before.[103] Stanton's was a "Declaration of Sentiments," written in the Jeffersonian style, followed by eleven resolutions, After clause-by-clause debate, the convention unanimously approved all of Stanton's resolutions but one: women's suffrage.[104] Indeed, Lucretia Mott herself had opposed the resolution before the convention began, thinking it would bring ridicule and retard the nascent movement,[105] although Frederick Douglass, also attending the convention, had lent his support for it. When Stanton learned that the publisher of the *New York Herald* had reprinted her declaration in full, meaning it to be taken derisively, she exclaimed: "Just what I wanted!"[106] The dramatics in the declaration had publicized the nascent movement across the country, and a seven- decades-long political action campaign got under way.

At an Anti-Slavery Society meeting in Boston in the spring of 1850, a group of eight women, including Lucy Stone, Abby Kelley Foster, and Paulina Wright Davis, met to discuss the idea of another national convention on women's rights. Davis became the chief organizer of the Worcester convention of that year, the first to be adequately planned, financed, and publicized.[107] Because there was no formal women's rights organization before the Civil War, these annual conventions took on an important role in keeping the movement coherent.

Stanton, however, was not particularly talented at organization — indeed, her flair for the radical and dramatic gesture tended to fracture organizations. Her talent lay, rather, in functioning as the movement's thinker or ideologue.[108] This is not to say that others had no ideas, but Elizabeth Cady Stanton was the most articulate writer and had the broadest vision in the movement. Writing meant she could stay at home and be with her young family. As another early feminist, Caroline Dall, wrote in 1858, "I love this new wine, but I do not wish to break all the old bottles."[109]

After the Civil War, the controversy over granting suffrage to male ex-slaves but not to women resuscitated the women's rights movement. It also divided it into two new associations, with Stanton and Anthony opposed to extending the franchise to ex-slaves until women also got the vote, which caused many former abolitionists to desert them. A less radical group, led by Lucy Stone, accepted suffrage for male ex-slaves and some assurances that women's demands would be met "next." But "next" never seemed to come, and in 1880, after a decade of division, the two groups agreed to merge.

Susan B. Anthony

Susan B. Anthony was not involved in the 1848 conventions, though her mother and sister did attend the convention in Rochester and reported favorably on it to her. She was born in 1820 in Adams, in northwestern Massachusetts, to a Quaker family.[110] Her father owned a small mill there, but soon moved on to New York, to a site just over the Vermont border. He did well until the Panic of 1837 and the subsequent depression wiped him out. Susan, who had had a good education from her father, a former teacher, and a year at a Quaker high school in Philadelphia, took up teaching to help pay the family's debts. In 1846, she took a job at the Canajoharie Academy, located along the Erie Canal east of Albany. While there, she found to her aggravation that female teachers were paid only a quarter of what male teachers received.[111]

By 1850, her father's fortunes had rebounded and he moved the family west to Rochester, so Susan quit teaching and moved to her parents' farm. She became the secretary for a local women's group, the Daughters of Temperance, but a year later, having been denied admission to a state temperance society meeting in Albany because of her gender, she resigned.[112] That same year she went to Seneca Falls to meet with a famous temperance editor, Amelia Bloomer, of subsequent "bloomer" clothing fame. On a street corner in the small town, the two ran into Elizabeth Cady Stanton. Anthony and Stanton became close friends and allies in the women's rights movement. Anthony was a nervous public speaker, but Stanton helped by providing her with notes and speeches. Anthony spoke at the third National Women's Rights Convention in Rochester in 1852, and her life as the most prominent woman activist began.

Throughout the 1850s, Anthony traveled to speak to groups about women's rights, giving up to a hundred speeches a year. She directed petition-gathering campaigns to pressure state legislators to expand legal protections and rights for women. She organized conventions, put out publicity, arranged venues, and lined up speakers. Before the Civil War, she managed speaking tours for abolitionists William Lloyd Garrison and Wendell Phillips, among others.

After the Civil War, Anthony and Stanton formed the more radical of two national feminist associations. Anthony always depended on her friend Stanton for ideas, speeches, controversies, articles, and intellectual direction, and the two collaborated with others to produce a multivolume *History of the Woman Rights Movement* in 1881. Both Anthony and Stanton lived into the twentieth century, Stanton dying in 1902 and Anthony in 1906, but not long enough to see the fruition of their struggle to overcome what they viewed as the key legal barrier to women's rights, their inability to vote.

Conclusion

The Seneca Falls Women's Rights Convention marked the crystallization of yet another social movement to come out of the Burned-over District of upstate New York. In the late 1820s, the Second Great Awakening inspired religious interpretations of society that led Yankee immigrants in the area to splinter into Mormons, Millerites, Hicksite Quakers, Shakers, spiritualists, and perfectionists. Reform groups cut across this new religious landscape devoted to temperance, abolition of slavery, health food promotion, and women's rights, including co-education, female property ownership, equal marriage, and equal pay.

At the same time, the expansion of male suffrage, democratic changes to business creation and funding, as well as the growth of the factory system in the cities led to an awareness of the virtues of equality for women. Family relations that seemed to work on the New England farm came apart in the big city.

The first movement away from tradition was by wealthy businessmen who wanted to protect their estates from the depredations of their daughters' husbands. The solution was for lawmakers eventually to "democratize" the right of women to own property in their own name. It was intended to be a small exception to the common law interpretation of marriage rights, but it was large enough for women's advocates to expand into a longer list of demands and a different conception of social institutions and relationships. After the Civil War, the movement turned its focus to women's suffrage as a result of the constitutional amendment giving it to male ex-slaves.[113]

Ironically, the rearranging of the feminist agenda around suffrage might have delayed other changes to the legal status of women, in areas such as the right to bank accounts, their own name, and own domicile as late as the 1960s and beyond.[114] In this way, the struggle for suffrage had mixed results. It is interesting to speculate where the women's movement might have gone had the fight over suffrage for male ex-slaves not been so traumatic for women's rights advocates.

The three women who began this slow revolution, Mott, Stanton and Anthony, were Yankee by birth or background, from Nantucket, western Massachusetts, and western Connecticut. They were all active in the abolitionist and temperance movements before promoting women's rights. Their efforts were supported by men who were of mostly similar background and experience. Their initial success on different issues came often from the frontier States — Wyoming Territory was the first to grant women's suffrage, in 1869.[115]

16

Palmyra, New York: Saving the World, One Utopia at a Time

We are indebted to the Shakers more than to any or all other Social Architects of modern times.

— John Humphrey Noyes[1]

We are a little wild here with numberless projects for social reform. Not a reading man but has a draft of a new community in his waistcoat pocket.

— Ralph Waldo Emerson,
writing to Thomas Carlyle in England, 1840[2]

This is the place.

— attributed to Brigham Young as he first viewed
the Great Salt Lake valley, 1847[3]

The Separatist Tradition

Separatism is an integral part of Yankee, and American, heritage. The people who founded the Plymouth colony came to the shores of North America precisely because they wanted to remove themselves first from societies, both in England and the Netherlands, they saw as religiously "impure" — hence the title for them of Puritans. They were joined by others, less separatist, nearly a decade after the first settlement at Plymouth.

For the first two years after they landed, the Plymouth colonists — or Saints, as they called themselves[4] — tried to build a new and different society. They assumed that their isolation from other societies would allow them room to try to recreate the communist life that existed before the Fall of Adam — communist not in the mode of Marxist-Leninism, but in its older meaning of life in a society where all things and relationships are held in common. They tried to make a system of communal labor work, but inequalities of family size, different tasks, and the different abilities of the colonists led to dissension and the system's collapse.[5] In the long run, however, the feelings of equality that came from owning equal shares in the corporate venture led to an American democratic ethos.[6]

The second and larger group of Puritans who followed a decade later and created the Massachusetts Bay colony did not share the separatist aspirations of the Plymouth colonists. Many of their leaders were prosperous at home and saw their lives continuing in the same vein in the New World. They were keen to keep close connections with England, especially after the Puritans there became the government after the English Civil War. They were community-minded, but did not aspire to some kind of "communist" ideal.

As these colonies expanded in number and extent, the separatist urge largely gave way to the need to integrate their religious brethren who had penetrated the Maine coast and the Connecticut River valley into the Puritan promised land. It was not until the 1770s that a new communal religious group, the Shakers, appeared, first in New York City and then resettling in the upper Hudson River frontier in Niskayuna, between Albany and Schenectady. The uncertainties of the Revolutionary War at first prevented their looking farther west, so they proselytized in Yankee New England. Only as the frontier became more settled did the Shakers move into western New York, Ohio, and Kentucky.

Another frontier-oriented community, the Mormons, came into existence after 1830 and its members, originally Yankee emigrants to western New York, began a slow move westwards, through Ohio to Missouri, back to Illinois, and then, in an overland trek that resembled the voyage the Puritans made over the sea, settling in the basin around Great Salt Lake.

The 1840s were characterized by the growth of separatist groups, all trying to show new ways to organize society, presumably hoping that their small-scale successes could be translated into larger groupings that would affect national policies and life. In 1840, a group of Transcendentalists in the Boston area failed, with Brook Farm, in their attempt to revive the communal, separatist spirit of the Plymouth Pilgrims. Another Yankee example, the Perfectionists, left Vermont in the 1840s but went only as far west as Oneida, New York; the active religious environment of the Burned-over District seemed to them a good one in which to develop their settlement.

Most of the many other attempts at separatist colonies were made by groups of German or other European immigrants, who found tolerance for their beliefs in America but who desired little contact with or appeal to the broader American society. The most successful of these refugee groups were the Anabaptists, made up of Mennonites, Amish, and Hutterites. Another group, German Pietists, settled first near Buffalo in the 1850s and then moved farther west to Iowa, where they created the Amana colony.[7]

The Shakers

The United Society of Believers, or Shakers, a shortening of the description of the "shaking Quakers," were an extreme offshoot of the Quakers and, unusually, led by a woman, Ann Lee. A distraught mother of four, all of whom died in infancy and disappointed by a failure of a husband, Lee had joined the group in England 1758, when she was twenty-two. The English Shakers had been heavily influenced by a group of radical Calvinists in exile from France, whose worship was punctuated by trances, agitation of the limbs, prophecies of the end of the world,[8] hearing heavenly voices, seeing lights in the sky, and testimonies of miracles. They had no clearly formulated doctrine, at first.

In 1770, Lee was arrested and imprisoned in Manchester, England, for breaking the peace, having caused a near-riot when she insisted on preaching her ideas in the city's streets. In jail, she had a vision that God was both the father and mother of humanity and that the sin that had caused the Fall from Paradise was the sexual act. She became "Ann of the Word," the "Mother of the new creation," who preached perfectionism, celibacy, and a church led equally by men and women.[9] After her release from jail, she communicated her vision to her confederates, many of whom decided to regard her as their leader.

Both Ann Lee and a male confederate dreamed that God pointed them to America — specifically, New England — as the place that awaited her words,[10] and a sympathizer arranged the passage of Ann and eight of her few followers. They landed in New York in 1774 and stopped there to consider their options.[11] Instead of remaining in the city, the Shakers opted to build a separatist colony on what was then the frontier. Two of the men were sent up the Hudson River to Albany, where they found some marginal land some eight miles northwest of the town, in Niskayuna, later called Watervliet, about five miles north of today's US

20. There, they leased some land from the local Dutch owner and built a rough house in a clearing, with one floor for the "sisters" and one for the "brethren."[12]

Their timing and location was both fortunate and unfortunate. The Shaker community kept to itself, but the outbreak of the Revolution meant a demand for the men's skills in Albany while their religious practices were ignored by a distracted public.[13] On the other hand, the area was made insecure by the threat of Indian raids. The invasion of a British army from Montreal in 1777 stopped only at Saratoga, some thirty miles to the north. As well, the group was suspected of being British spies or sympathizers, since, consistent with their Quaker origins, the Shaker men refused to bear arms. By 1779, they had managed but one convert.[14] "Mother" Ann counseled patience, and despite her handicap as a female leader[15] and her insistence on celibacy, things did begin to change.

As active warfare moved away from New York, the Shaker faith began to expand. In 1779, a local revival broke out among New Light Baptists in New Lebanon, New York, just across the border from Massachusetts, and its leadership came to Niskayuna, attracted by what they had heard about the Shakers.[16] Their acceptance of Ann Lee led many other revivalists to make their way to the Shaker land, and a kind of rolling camp meeting grew up. The activity attracted the attention of the authorities, who investigated and briefly imprisoned some Shakers under suspicion of being anti-revolutionaries.[17] This marked the beginning of a process that was to last through the 1780s at least: the Shakers would have some evangelical success, which would generate fierce local opposition, a riot or demonstration would break out. and the Shaker evangelists would be forced to leave.

In 1780, Ann Lee, following her dream of a decade before, decided to proselytize in New England. She found some sympathy in parts of Connecticut and in areas around Boston — in particular, the town of Harvard, about thirty miles northwest of the city. There, the appeal of a deceased local "prophet," the New Light preacher Shadrack Ireland, promised a good response for Shaker evangelism. Lee remained in the area for about two years, but was eventually expelled by the locals.[18] The Shakers also found adherents among the Free Will Baptists in New Hampshire and Maine who rejected the prevailing Calvinist decorum and morés. Among many adherents, then and later, the emotionalism and mysticism of the Shakers were regarded as the epitome of devotion.

When Ann Lee returned to Niskayuna in 1783, she had reason to feel successful, having made converts and set the stage for a number of Shaker communities in New England. By the time Mother Ann died in September 1784, the Shakers, realizing that their doctrines could not survive social pressure, decided to take themselves out of "the world" and create colonies to house believers.[19] Their new leader, James Whittaker, raised the first Shaker meeting house in New Lebanon — later renamed Mount Lebanon — in 1785.[20] Whittaker, however, died in 1787 and with him the British connection to Shakerism. The group's leadership passed to Yankees: "Father" Joseph Meacham, a former Baptist minister from Enfield, Connecticut,[21] and his co-leader, "Mother" Lucy Wright, from Pittsfield, Massachusetts.[22] The Shaker leadership was somewhat authoritarian, but it was based on the notion that those less perfect ought to submit to those more perfect. The leaders were considered the successors to Ann Lee, just as the Apostles were to Christ.

The new leadership carried forward the emerging consensus that the Shakers had to withdraw from the world, and on Christmas Day, 1787, they held their first communal meal at New Lebanon. Then they got down to work to provide for the colony. As a chronicler describes, "[b]y 1789 a tannery, fulling mill, clothing shop, chair factory, blacksmith shop, and cobbler's shop were in full operation....The Shaker garden-seed industry was also stared in 1789, and the farm produced, among other crops, three thousand bushels of potatoes. Meanwhile the sisterhood was occupied with various household arts, chief among which were the spinning and dyeing of yarn and the weaving and dressing of cloth."[23] Classic Shaker design was also conceived. As a Shaker document explained, "[a]ll work done, or things made in the Church for their own use ought to be faithfully and well done, but plain and without superfluity. All things ought to be made according to their order and use; and all things kept decent and in good order, according to their order and use. All things made for sale ought to be well done, and suitable for their use."[24]

The Shakers became known for their inventions and redesign of tools, which raised their productivity and provided new products to sell, and they used science and careful experimentation to improve seed quality and uncover the uses of various herbs — activity that places the Shaker leaders among the Americans who were making the transition from household to mass production.[25] Gradually, the sect acquired land and built facilities in some twenty sites from Maine to Kentucky.[26] Communal living meant that, in the Shaker communities, there would be no poor and no rich, no bosses and no slaves. Everyone, from the leadership on down, with the exception of the sick and the aged, was expected to do some manual labor.[27] The membership of Shaker communities overwhelmingly came from laboring people, and they felt a kinship with Christ and the Apostles in this regard, at least.[28] The property of new adherents was added to the capital stock of the communities, while the development of crafts and manufactures diversified the colonies away from selling only farm produce.[29] Conflict with the outside world was discouraged and a new emphasis placed on internal improvement and perfection of the faithful.[30] Some Shakers were trained as salespeople and sent "outside" to bargain with businesspeople for markets. The colonies on the whole were well run; tasks were separated by sex, not out of some notion of inequality but to keep the sexes apart. Shaker products were of the best quality and their manufactures had clean, functional designs.

Shaker thought began to crystallize with three publications in the 1810s, more than thirty years after their arrival in America. The first was Benjamin Seth Youngs's *The Testimony of Christ's Second Appearing* (1810),[31] which explained how Christ and Ann Lee were the two earthly founders of God's kingdom on earth, God being both Mother and Father creator. The faithful were uplifted by manual labor and were helped by continuous revelation, religious communalism, the practice of the confession of sins, and the quenching of carnal desire in their quest to complete the unfinished work of human salvation. The second was *The Manifesto* (1818), by John Dunlavy,[32] and the third *Summary Views of the Millennial Church* (1823), by Calvin Green and Seth Youngs Wells, both of whom elaborated on *The Testimony*, but particularly in justifying the reason religious communism should lie at the heart of salvation.[33]

What relations between the sexes existed were carefully regulated by having "union meetings" as often as four times per week where men and women, in chosen small groups, could have conversations about topics of mutual interest while sitting in rows about five feet apart.[34] The colonies had schools, which, after 1821, were under the supervision of a Shaker who had been a professional school administrator. They were somewhat open to the neighboring children, so they could receive school money from the town,[35] and they were open to outside inspection. The focus of instruction was on practical training.[36]

Shaker communities prospered and attracted adherents until after the Civil War. At its height, the corporation owned hundreds of buildings and over 100,000 acres of land.[37] Their existence, however, because of the rule of celibacy, depended on the continual attraction of new adherents. Obeying Ann Lee's insistence on hospitality, the Shakers welcomed visitors, who also bought things and occasionally became adherents.[38] The French travelers Alexis de Tocqueville and Gustave de Beaumont visited Niskayuna in 1831, but left baffled by the group's communism and sexual segregation.[39]

In the changing society of post–Civil War America, the communal nature of Shaker colonies began to have less appeal. Shaker discipline was also undermined by the colonies' prosperity — there was conflict between those who wished to stay and those who wished to leave and gain their "share" of the assets.[40] As well, attracting new members became more difficult. The Shakers depended on the arrival of unhappily married or deserted or widowed wives and their children,[41] and men who had fallen on hard times, who saw their colonies as a refuge. Now, such people tended to live in the growing cities, and for them the promise of a communal refuge in a rural area was not as attractive as it had been for rural New Englanders. Shaker life was too ordered to cope with the social change implicit in urban, industrial society.

Toward the end of the 1800s, Shaker leaders attempted to bring their communities closer to the world by involving them in the suffragette movement and holding a conference on universal peace; its resolutions were presented to President Theodore Roosevelt in 1905.[42] Despite these actions, membership continued to decline and the colonies gradually emptied. From a high of about six thousand adherents in 1850, by 2010 only three Shakers remained.[43]

The Reluctant State of Deseret

Northwest of Seneca Falls and just east of the village of Palmyra, about a dozen miles north of US 20, is a ridge topped by a large monument. Here, in 1823, Joseph Smith claimed to have had an encounter with an angel called Moroni.

Smith's parents and their family had emigrated from Vermont to better farmlands in western New York in 1816.[44] He was unschooled, and his parents had a difficult time making ends meet even in this better country. As a teenager, Smith was affected by the Second Great Awakening that swept the Burned-over District and had his first vision in 1820 while trying to puzzle out which of the creeds of the time was correct. In his vision, he was told that none was true and that he should await further contacts.

His 1823 meeting with the angel led, three years later, to his locating golden tablets on the side of what is now the Hill Cumorah.[45] Smith spent a year and a half translating the tablets, which others claimed also to have seen and which he returned to their original repository when he was done. The finished work was the Book of Mormon, which related the story of how some tribes from Israel had found their way to the New World and their visitation by Christ after His resurrection.[46] In 1830, Smith and others organized the Church of Jesus Christ of Latter-day Saints to promote the ideals of these early Americans and to recreate Zion in the Americas.[47]

A number of elements of the Mormon church are recognizably part of the "American religion."[48] The hierarchy is reasonably democratic, at least at the lower levels, and its clergy are not employed as such, but have "day jobs."[49] There are resemblances to the Puritan belief that every member is a "saint" and, ultimately, a priest, and given a certain equality of grace.[50] The Millennial "Latter-day" suggests that adherents believe the time of Christ's return is soon. Other Mormon ideas reflect the nature of the communal or communistic groups that sprang up alongside the new church.[51] But where other communal groups tended to abandon some of the fundamental tenets of orthodox Christianity, the Mormons tied themselves to Christian belief and added to it the revelation that Christ had visited peoples in North America about the same time as He appeared in Palestine. In effect, the Book of Mormon, as translated by Smith, was offered as an addition to the Christian corpus — the title of the Mormon scripture reads *The Book of Mormon: Another Testament of Jesus Christ* — giving a place to the Americas and, in particular, to the United States, where the plates containing the Book of Mormon were found. As well, the Mormons believe that revelation and prophets did not stop with the New Testament, and that the church's leaders are the prophets of today.

Had Smith opted to create a community like that of Noyes in Oneida or that of the Shakers, who at their peak had only a few thousand adherents in self-contained colonies, his church might have faltered and dissolved after his death.[52] Moreover, he and his followers would have been left to their devices as long as they did not evangelize the surrounding neighborhood. But that is what he did not do. He did not offer life in a small community, but instead began to compete with the established churches, though with a twist. The twist was that Mormonism offered a different Christian faith, one that had American historical roots. American churches could accept people shifting from Catholic to Episcopal to Presbyterian to Baptist churches — after all, these were debates first and foremost over church organization, control, and legitimacy. The Mormons, however, offered a different alternative: a church that essentially drew inspiration from the migration of Old Testament Jews to their Promised Land, coupled with an American visitation by Christ, neither of which any mainstream church could accept. Moreover, the Mormons created a variety of behavioral rules that promoted a form of exclusivity that extended even into business dealings. Instead of celibacy (like the Shakers) or eugenic free love (as the Perfectionists professed), in 1842 the Mormons formally adopted the Old Testament custom of polygamy.[53] Finally, reflecting its birth amid the Second Great Awakening, the Mormon church was aggressively evangelical.[54]

Smith decided shortly after its founding that the church should shift its location from western New York to Ohio, and in 1830, members established themselves in Kirtland, just east of Cleveland. At the end of the first year there, they had grown to about five hundred.[55] Smith then announced that the new Zion was to be built near Independence, Missouri, and some of his followers moved on to that relatively unpopulated land. By 1835, about twelve hundred Mormons were living in Missouri. Their relations with local non-Mormons were strained by their aggressive Millennialism, evangelization, and virulent verbal attacks on other churches. The exclusivity of Mormon business dealings alienated many local non-Mormon leaders, and fears mounted that the growth of the Mormon community would give it voting power and political dominance. Then, the migration of a few Mormon free blacks into slaveholding Missouri in 1833 led to violence and to the abandonment and destruction of the Mormon colony.

A second attempt to create a Mormon colony in Missouri was an outgrowth of that State's politicians efforts to provide for the refugees from the first attempt. The Mormons were given a new county in land northwards across the Missouri River, where they were expected to keep to themselves. Yet, by 1838, Mormon business competition and aggressive expansion outside "their" county had led to considerable bloodshed on both sides and the destruction of still another colony. Smith and others were jailed on charges ranging from murder to treason, but he escaped across the Mississippi River to a new colony in Nauvoo, Illinois, on the bank opposite the Missouri-Iowa border.

In Nauvoo, Smith managed to get legislation passed that gave him great political control of the town and surrounding county. The influx of Mormons from the east, from Missouri, and from England, where a mission had been sent in 1837, led Nauvoo to become the largest town in Illinois after Chicago by 1840.[56] To protect the colony, Smith formed a body called the Nauvoo Legion, established as part of the Illinois militia, though under his control. It had two thousand volunteers in 1840 and four thousand by 1844 — the regular US Army totaled just 8,500 soldiers. Smith began to have illusions of grandeur, appointing himself "lieutenant-general," a rank that, at the time, only George Washington had ever held. Mormon leaders apparently accepted him as the divine delegate on Earth until the millennium occurred and, as such, ruler over all its countries by divine right, which the US government considered treasonous. Finally, in 1844, he declared himself to be a candidate for president of the United States. By that time, however, Mormon aggressiveness and growth and local antipathy to their faith again led to violence.[57] Smith and his brother were killed by a lynch mob in a neighboring town, and the church was thrown into crisis. One of Smith's loyal followers, Brigham Young[58] — like Smith, born in Vermont — assumed the leadership, and a number of dissidents left, many peacefully returning to Independence, Missouri, and settling down. The Illinois government then told Young that he and his Mormons had had until 1846 to leave Nauvoo and the state — another Mormon colony would have to be abandoned.

Young and other Mormon leaders decided the solution was to leave the settled part of the United States altogether, and they organized a massive trek of fifteen thousand believers westward to the valley of the Great Salt Lake,[59] in the borderland between the United States and Mexico, a move to escape their persecutors that in some ways imitated the Puritan flight to the New World some 225 years earlier. As one historian has noted,

> [n]othing in American history — not the ephemeral towns of mining rushes nor the hardier ones of real-estate booms — is like Winter Quarters [Iowa]. An entire people had uprooted itself and, on the way to the mountains, paused here and put down roots. The endless church government went on. Not only the other camps [to the east, toward Nauvoo] had to be managed from Winter Quarters but all the missions too, in the United States and overseas. Supplies had to be kept moving; hundreds of teams were out all the time, freighting grain, flour, beef, pork, hardware, dry goods.[60]

The new colony of Deseret — a name referring to the local honeybees and signifying the colony's work ethic and communalism — was founded in 1847, just after the outbreak of war between Mexico and the United States over the American annexation of Texas.[61] The Mormons offered to help, and a "Mormon

Battalion" marched with US troops to occupy New Mexico and then went on to California to help in its occupation. The battalion's wages helped relieve the currency problem facing the new colony, and its efforts on behalf of the United States bought the Mormons peace with its tormentors, for a time.

Problems then emerged with the discovery of gold in California. Overland trekkers stopped in the Mormon colony of Deseret for food and supplies, which brought valuable currency into a colony dependent on imports for many vital necessities. It also brought trouble, as the Forty-Niners encountered Deseret's solid phalanx of Mormon church, State, and business — a highly "un-American" way of doing things, to use a phrase coined a century later.

In 1849, Young declared the creation of the state of Deseret and appointed its first legislature and senate. In 1850, not recognizing the Mormon State, the government in Washington created the Utah Territory, which included large stretches of the intermountain region of the West, and sent out officials — non-Mormons — to administer the area, while appointing Brigham Young territorial governor.[62] The unity of the Mormon population intimidated the new officials, however, and they were gone within a year, once again reaffirming Mormon control over their own society.

Converts to the faith — many from Britain, where Mormon missionaries were active — were encouraged to migrate to Deseret, and were helped to move west from the Missouri River alongside pioneers on the Oregon Trail. As well, the colony sent out exploration parties in all directions, as far north as Montana and south into Mexico. Settlers occupied strategic sites and useful farm and ranch land as well as searching for minerals. These smaller settlements — 135 communities were built in the first decade after the Mormons arrived in Utah — were maintained from Salt Lake City, and the reach of the Deseret colony was vast.

Mormon communal methods proved most successful. Mormonism combined Jewish and Christian mysticism with Yankee organization in the face of the desert's challenges.[63] Historian Ray Billington relates that, in the fall of 1848,

> Young revealed his program to his followers. No longer, he told them, could each person live where he chose or farm as he wished. Instead they would dwell together and work together under the leadership of their church. Their homes would be in a great city that would follow a plan he and the Twelve Apostles had devised....He also laid down principles that would govern farming....There would be no sale of land, he told them, and no private ownership of streams or timber or anything else essential to the social welfare. "These," he said, "belong to the people: all the people." Instead each person would be given just the amount of land that he could till most effectively.[64]

That Brigham Young could devise a workable irrigation system was truly remarkable, for the practice was completely foreign to the Anglo-American tradition, yet that put into operation during the autumn of 1848 was so efficient it might have been planned by a modern engineer.[65]

Billington connects the Mormon effort to the past: "Only once before in the history of the nation's expansion had a similar co-operative social order developed: among the Puritans of seventeenth-century Massachusetts."[66] The Mormons, however, cut across accepted American ways of relating politics, business, and religion. Mormons were neither secular, diverse, nor competitive — or, one might say, they were too competitive. Another historian, Harold Bloom, compares Mormonism to Judaism and the Jews in that it is a religion that became a people.[67]

At first, the authorities in Washington were content to let Brigham Young act as territorial governor, particularly as he had proclaimed the state of Deseret along American lines and requested admittance to the Union. But the lack of separation of church and state and the inability of other entrepreneurs to gain a foothold in the Salt Lake trade in the face of the church-business axis led to charges by the federal government in 1856 that Young's state was separatist and lacked duly constituted order. That same year, Congress ignored a formal bid by Utah for statehood for its 30,000 inhabitants, and the new Republican Party called for the prohibition not of only slavery, but also of polygamy in US territories, which President Lincoln signed into law in 1862. Attempts to separate the territorial government from the church, given that the vast majority of voters were Mormons, could not be accomplished democratically. Other means had to be used.[68]

Tensions were such that, in 1857, 2,500 US troops were dispatched to the Great Salt Lake to back up a newly appointed non-Mormon governor. The soldiers — led by future Confederate general Albert Sidney Johnston — had to travel overland as there was no railroad west of the Missouri River. Mormon irregulars harassed them on the way, delaying their arrival through the winter. Young, meanwhile, could boast of a force of 7,000 ready to meet the US soldiers, who now numbered 6,500. In the winter of 1857–58, Johnston wrote to Washington: "[The Mormons] have placed themselves with premeditation in rebellion against the Union, and entertain the insane design of establishing a form of government thoroughly despotic, and utterly repugnant to our institutions"[69] — an ironic statement given that Johnston would be killed while commanding the Confederate army four years later at the battle of Shiloh in 1862. By the time the Army arrived in 1858, however, the Mormon authorities had arrived at a compromise with Washington whereby the Mormons would be granted pardons in return for their accepting outside political rule.

The main effect of the troops' arrival, in fact, was to bring new money and supplies into the Deseret colony. As in every new settlement, the Mormon colony was continually short of hard money for economic development, particularly to buy irrigation machinery, iron and steel products, and some consumer goods.[70] The high cost of transporting goods and people before the arrival of the transcontinental railroad added to the cash shortage. The Deseret government stressed the need for local business activity to produce substitutes for expensive imports. As well, every effort was made to relieve soldiers and travelers to California of their spare cash and to explore for minerals that could be sold.[71]

With the coming of the Civil War, the troops departed, leaving their equipment for local use, and Deseret reverted to the *status quo ante*. The Mormons saw the Civil War as part of their Millennialist prophecies and did not get involved. As the war progressed, however, the federal government again sent a detachment of Union troops to Salt Lake City, this time to protect the overland mail route to California. Wartime also saw the admission of Kansas and Nevada to the Union, but not Utah, even though almost 80,000 people lived in the Mormon West.[72]

After the Civil War, President Ulysses S. Grant made it clear that the federal government intended to impress conformity on institutions and practices in both the ex-slaveholding South and polygamous and theocratic Utah.[73] Federal agents were sent into the territory to enforce the law, but met with unrelenting, though mostly passive, resistance. The completion of the first transcontinental railway in 1869, which ran north of the Great Salt Lake, meant increased communication with the outside world and renewed tensions. The ostensible cause again was the Mormon acceptance of polygamy, which was offensive to abolitionists and women's rights activists as well as to the other Christian churches. More fundamental was the tangled and interwoven relationships among the church, State, and business that were not tolerated elsewhere in the country and that impeded outside business interests.

In 1887, the Mormons seemed to surrender on the issue of polygamy after the federal government disenfranchised Utah's voters and seized church property, but the reality hardly changed.[74] Finally, in exasperation, the federal government again sent in troops. The Mormon leaders were hunted down and arrested, and the communal basis of Mormon society destroyed.[75] By 1890, 1,300 Utah Mormons were in jail for violating US laws, and in that year the US Supreme Court allowed the federal government to seize property and close churches to enforce its laws. The Mormon leadership was forced to accede to the "Americanization" of the territory. Polygamy was dropped as church practice, ties between church and business were weakened, and the party system was reorganized around Republicans and Democrats, rather than church members and non-members.[76] Utah was finally granted statehood in 1896.[77]

It is questionable, however, how much real change has taken place since then. Business in Utah is more diverse, given the rise of national corporations and the erosion of Mormon exclusivity, yet the church remains wealthy through its business contacts.[78] Mormon religious doctrine has become more compatible with general American thought, especially after the church began to downplay its spiritual connection with the Old Testament world of Abraham in favor of the New Testament world of Christ. Yet some of the communal heritage of the Mormon colony of Deseret remains.

Brook Farm

Brook Farm differed from other attempts to create an American utopia in that it was not religious at its base. In fact, for the times, one might call it naïve. The idea, propounded by George Ripley in Massachusetts in late 1840, was that, by sharing the workload of a farm, participants, both men and women, could have a decent life and yet save enough time to pursue their intellectual and social interests. The goal, as expressed by Ripley, was "to insure a more natural union between intellectual and manual labor than now exists; to combine the thinker and the worker, as far as possible, in the same individual." Elizabeth Palmer Peabody, the editor of the Transcendentalist journal, *The Dial*, added: "none will be engaged merely in bodily labor….This community aims to be rich, not in the metallic representative of wealth, but in…leisure to live in all the faculties of the soul."[79] In effect, since participants were not rich enough to lead a "normal" life of leisure, a proper organization might allow them to do so with a minimum of time spent on physical labor. In exchange for three hundred days of labor, participants were to be given free room and board, and all got equal wages. Brook Farm was one of perhaps eighty attempts at communal living that blossomed in the 1840s.[80]

All of this was dressed up in Transcendentalist thinking, a commitment to personal equality, and to demonstrate the superiority of communal living over individual striving. The experiment was started with the creation of a joint-stock company, The Brook Farm Institute of Agriculture and Education, wherein the investors, including those who were to live on the farm, were required to purchase one share apiece for $500.[81] Although hard-headed Boston businessmen might have treated the experiment with disdain, Brook Farm's premise of the equality of the sexes meant that the women of Boston's First Families did not hesitate to support it.[82] The capital was to be used to acquire a dairy farm and construct buildings and infrastructure. The share price effectively kept out those of more modest means, including, unfortunately, those who might have had the appropriate farming skills.

The farm, in West Roxbury, about ten miles west of Boston and eight miles south of today's US 20, was bought in late 1841, though the shareholders had begun to move in earlier. A second, neighboring farm was also acquired and merged with the original. Brook Farm's main asset was a functional farmhouse where meals were served and the group met for discussions. Gradually, other buildings were built, one to act as a school and to house paying guests, while the rest fulfilled other functions, including draining the treasury of needed cash. There did not appear to be too much concern about either the quality of the land, which was poor, or the general management of the enterprise.[83] The farm itself did not produce much except hay, a low-priced product that required considerable hand labor. Since one of the principles of the Farm was that both men and women could choose what work they undertook, managing the distribution of labor was sometimes a problem, and standards had to be instituted to ensure that all worked an equal amount. The best income sources proved to be charging visitors to come to the Farm and assessing tuition for attending the Farm school — in other words, tourism and education, for which a farm would seem unnecessary. Years later, in *The Blithdale Romance* (1852), Nathaniel Hawthorne had one of his characters at a fictional Brook Farm observe that farm labor is not compatible with creativity.[84]

Even so, Brook Farm was a considerable experiment in social change. Outsiders saw the idea of equal, communal work as a threat to the nuclear family. The tolerance of different religious persuasions among the participants also was seen as novel and threatening. Moreover, the mixture of work roles between men and women violated social conventions, while the Brook Farm's approach to early childhood education and to gender-equal adult education predated similar, broader education reforms by many decades. The ambition of Brook Farm, indeed, reflected a kind of secular Millennialism, and was intended "[t]o indoctrinate the whole people of the United States with the principles of associative unity [and] [t]o prepare for the time when the nation, like one man, shall reorganize its townships on the basis of perfect justice."[85]

Gradually, the leadership of the enterprise became enamored with the notions of a practical order for communal life based on the detailed description by French philosopher Charles Fourier of how a "Phalanstery," or "grand hotel" — his ideal version of a utopian community's center — should be organized.[86] In 1844, Brook Farm's shareholders moved away from the idea of sharing farm tasks to free up time for study and reflection — which they were accused of not really doing anyway — toward Fourier's ideas. The adoption of Fourierist discipline led, however, to some disaffection and desertions,[87] and to a call for new construction on the Farm to fit Fourier's model.

Moving to the Fourierist model coincided with worsening conditions at the Farm as revenues fell short of expenses, forcing amenities such as coffee and meat to be cut back. Then, in March 1846, a fire destroyed the just-finished main new Phalanstery building, which had been designed as new and better quarters for the participants. The building was uninsured, however, and the enterprise was left all but bankrupt. Ripley left two months later, when only the school was left functioning, and the last participant closed the Farm in 1847. The farm and buildings were sold to the city of Roxbury at auction,[88] and George Ripley spent the next thirteen years of his life slowly paying back the loans on Brook Farm.

One lesson that can be derived from the Brook Farm experiment is the importance of good management — especially financial — to the success of a communal society. Brook Farm, perpetually short of funds, had to be mortgaged four times between 1841 and 1845. One participant commented: "I think, here lies the difficulty — we have not had business men to conduct our affairs....[T]hose among us who have business talents, see this error."[89]

Brook Farm's flirtation with Fourierism was also found elsewhere. Frederick Law Olmstead contributed an article to the *New York Daily Tribune*[90] in July 1852 concerning his visit to a Fourierist Phalanstery in Colt's Neck Township, New Jersey. That farm, which was started in 1841 and which seemed to him to be dominated by New Englanders, was a secular communal experiment with little production beyond market gardening and related farm goods and included a mill driven by a steam engine. The shares of stock that capitalized the Phalanstery had been taken up by sympathetic capitalists in New York City.[91]

The Perfectionists

John Humphrey Noyes, the founder of the Oneida colony, is said to have noted of such experiments that "right reason" was not a working substitute for the grace of God, and one could not defeat industrialism with plow and scythe.[92] Noyes, born in 1811, came from an upper-class family in Vermont.[93] He was attracted to religion by a revival service preached by the Reverend Charles Finney in Putney, Vermont, in 1831, studied philosophy at Andover Theological Seminary and then Yale, and decided to become a minister. His reading of the Bible and religious history convinced him that Christ had reappeared in AD 70, at the fall of Jerusalem, and that the world was working its way toward the Millennium, which would come soon. He also believed that salvation came from faith, and that whatever a person did in life was understandable as long as it was done with the intent of moving closer to God. Sin arose if the action was done for any other reason. After much thought and mental anxiety, Noyes declared that, in this context, he was henceforth "perfect" — that is, without sin — since he had dedicated his life and his actions to bringing people closer to God by following His plan. In times past, this Perfectionist declaration had been considered "antinomian"[94] and attacked as heresy. The officials at Yale were none too happy either, and in 1834 they revoked his license to preach.[95] Noyes instead went off to preach wherever people might invite him.

Applied to sexual matters, the Perfectionist attitude got him in trouble from the beginning. Since outward acts done with the right attitude toward God were, like other acts, deemed to be following God's plan, they were "good." On one occasion, Noyes left Brimfield, Massachusetts, and crossed the state line into Vermont just before another visiting evangelist was discovered in bed with two local girls who were testing his and their resolve to sleep together as Perfectionists. The incident, dubbed the "Brimfield bundling," cast public doubt on Perfectionist concepts.[96]

From 1836 until 1848, Noyes remained in Putney developing his ideas and putting them into practice. In 1838, he married one of his followers and the couple, along with other followers, formed a community near the town, Noyes having inherited enough wealth from his father to incorporate the group.[97] There, Noyes continued to elaborate on his ideas about "Perfectionism," including economic communism and what he called "complex marriage." Noyes noted that Christ had said there would be no marriage in Heaven, so as the community of believers on Earth approached the perfection of Heaven, they would not need this concept either, but the "free love" and "spousal sharing" that grew out of such a notion outraged the sensibilities of neighbors.[98]

Word spread about Noyes and the community in Putney devoted to his principles, including the idea of a "marriage" that included all of the "family" of believers. Faced with legal prosecution, Noyes skipped bail and fled to upstate New York,[99] where a supporter had a farm near Oneida. Noyes and his followers managed to raise enough capital to set up a separatist colony on the farm in 1848; at first, the colony had eighty people, but the number doubled the next year.[100]

For many years, the colony continued to exist only through the infusion of fresh capital by new adherents and the sale of farm produce to the local community.[101] Then, in the late 1850s, a recent adherent, a metalworker, suggested they begin making metal animal traps. His design caught on and sales took off.[102] Over the next decade, the colony became profitable, leading it to explore other fields of manufacture. Eventually, it hit upon cutlery, flatware, and silverware, and this too became such a success that it replaced the declining market for traps.[103] Over the long term, these businesses sustained both the Oneida colony and another in Wallingford, Connecticut, northeast of New Haven. The Connecticut colony had been started in 1851 by a local farmer who had converted to Noyes's ideas. It grew to more than forty people, developed a publishing business, and began making plated spoons, which soon developed into a full-fledged business, just as in Oneida. As the demand for the colony's products grew, locals began to be hired to work in the plant.[104] Eventually, production was moved to Niagara Falls, New York, where there was ample power for Oneida Community Ltd. to produce its popular "Community silver."

The core value of the Oneida colony was "communism," which then meant living as a group in all aspects of life. The colony also believed in equality and the use of group criticism as a means of discipline. The colony's management was delegated to a large number of committees and departments.[105] Like the Shakers — and unlike the Mormons — there were no schisms in the colony. Life centered around a large, rambling "Mansion House" that provided rooms for everyone. The colony had a crèche and pre-school where all the children, once weaned, were brought up by "the family" — that is, the whole community. Tasks were allocated in part by preference and in part by talent or the needs of the group, and there was no tendency to distinguish work by gender; traveler and writer Charles Nordhoff, who visited the colony in 1875, noted that women were taught to be machinists.[106] Meals, entertainment, learning, and work were all group projects. When a crop had to be weeded or brought in, a "bee" would form of as many people as were needed to do the job.

The Perfectionists' idea of complex marriage was a compromise between the idea of two people having sex for what Noyes called "amiability" or, alternatively, for procreation. The need was to keep everything in a group context, since monogamy was seen as a selfish act, contrary to God's plan. Amiability could be practiced between two people and included "male continence," or breaking off coitus before orgasm. Procreation, however, was a group decision, as the colony attempted to transfer what was known at the time about plant and animal breeding to a rough form of eugenics.[107] Noyes was a respecter of scientific knowledge, to which he attributed in part the exceptional health of the Oneida "family."

The colony did not try to cut itself off from visitors, as it depended on tourism for some revenues and on attracting new followers, although it did no proselytizing.[108] Singing and dancing were encouraged as group activities, as were team games such as ball and croquet.[109] Female equality was at least as advanced as that proposed by the first Women's Rights Conference, held ninety miles west of Oneida the year the colony began. Women took part in running the colony, and they could work in production and crafts because their children were all in the crèche. Provision was made for adult education for all, and a planned institution of higher education, never realized, would have admitted both sexes.[110]

The colony had no special religious ceremonies, as Noyes believed their whole existence ought to be a prayer to God. They studied the Bible and other religious tracts, however, so the colony was not secular. Its population varied between one hundred and two hundred, including children, but Noyes discovered how to develop strong group cohesion, as he welded the large group of adults and children into a "family." Members had children as various and changing couples, so permanent bonds of affection between parents and between parents and children were censured. Everyone lived under one roof and took their meals together. Children were raised as a group of brothers and sisters. Yet, despite its size, the "family" managed to inspire a "we" rather than an "I" approach to life. A century later, around 1970, one researcher encountered descendants of the Perfectionists still returning to the colony grounds: "We all love the old place. Many of our folks lived there, and most of us played there as kids. We know the building down to the last brick and board. It's odd, but so many of the people who moved away come back to the Mansion House as they get older. It's because they had such good times and such happy memories."[111]

The colony began to unravel in the late 1870s in the face of criticism of its complex marriage policies and an upwelling of political sentiment against Perfectionists, Mormons, and other non-conformist groups. Besides, Noyes himself was getting too old to lead the colony. In 1877, he resigned, and in 1879 secretly moved to Niagara Falls — the Canadian side — perhaps to avoid further prosecution for his beliefs. The colony formally disbanded at the beginning of 1881,[112] though most of its members preferred to remain in the area. On a visit to the colony's grounds around 2000, we met a number of older people still living in the area who had inherited many fond recollections of the colony from their grandparents and others.

The Oneida Community Ltd. was formed in 1881, and the experiment in economic communism ended with the issue of shares of stock in the company to colony members.[113] Under Noyes's son, the company flourished into the twentieth century — the company went public in 1967 — and descendants continued to run it until 1981.[114]

Conclusion

The Shakers, Mormons, Perfectionists, and Brook Farm — all Yankee-based experiments in communal living — disappeared or were altered by the intrusion of larger "Americanizing" forces into their midst. The basic Yankee approach to organizing life is generally one of individualism, but a minor strain has sought to return to the principles, real or imagined, of early Christian life, in imitation of the early Puritans. Sociologist William Kephart has identified a number of reasons communal and separatist groups fail: the inability to put the group on a sustainable economic base; a failure of leadership; aberrant membership, where members cannot discipline themselves to work for the group or follow its orders; and the lack of a lasting commitment or sustainable social organization.[115]

Then there is the attempt, in three of the four cases, to remodel the family and sexual relations, the main focus of outsiders' complaints about the communal organizing practices of these colonies. It was not enough to enter a community where assets and income were pooled and used in a radically egalitarian fashion; each group interpreted the same passage of the New Testament — that, in Heaven, there would be no marriage — to fit its needs when it came to sex and the family.[116] The Shakers interpreted Christ's statement to mean that there would be no sex in Heaven, so, in anticipation of Heaven, there should be no sex on Earth. The Perfectionists interpreted the passage to mean that marriage would not be a part of the heavenly experience, but sex was part of human experience. Marriage should be expanded away from the monogamous couple so that the "family" arising from it would be coterminous with the colony and they would be saved as a common group. The Mormon interpretation, through its first sixty years, at least, was that no marriages would be performed in Heaven, but, in imitation of the Old Testament figures, those most able, spiritually or otherwise, ought to produce more offspring, which justified polygamy.

The impulse toward communal organization," as American educator Constance Rourke has pointed out, "[has] formed a conspicuous strain in our history." The "single-minded community" — whether of Pilgrims, Mennonites, Dunkers, Shakers, Fourierists, or people in frontier hamlets — was a common type of settlement, representing "wide-spread popular concerns" and, therefore, reflecting a major aspect of America's early social character.[117]

Rochester, New York: Telegraphy and the Five Internets

By the early 1870s, the Victorian internet had taken shape: a patchwork of telegraph networks, submarine cables, pneumatic tube systems and messengers combined to deliver messages within hours over a vast area of the globe.
— Tom Standage, The Victorian Internet[1]

My God! It talks!
— Emperor Don Pedro do Alcontara of Brazil, upon hearing Alexander Graham Bell's voice through a telephone at the Centennial Exposition, Philadelphia, 1876[2]

I got there first, so I got to choose any punctuation I wanted. I chose the @ sign.
— Ray Tomlinson, sender of the first email between two computers[3]

The Reynolds Arcade

In 1803, a land speculator from Maryland, Nathaniel Rochester, purchased a hundred-acre tract near the mouth of the Genesee River in western New York,[4] thinking it would make a good townsite. In 1811, having moved his family first to land he owned upriver, Rochester moved again to what became known as Rochesterville, soon its official village title, where he was soon joined by a few families, including that of Abelard Reynolds, a Yankee from Pittsfield, Massachusetts.[5]

Rochesterville flourished as a way station for military transport in the War of 1812 and because of mill sites on the river. By 1823, when the Erie Canal passed over the Genesee River on an aqueduct, the "ville" had been dropped from Rochester. Five years later, with Rochester booming from the canal trade and approaching nine thousand in population,[6] Abelard, now the postmaster, moved his house off the Main Street lot he had purchased and began construction of the equivalent of a present-day shopping mall.[7] The Reynolds Arcade took a year to build. It was four stories high, with shops opening onto a covered walkway and balcony, and had eighty-six rooms and fourteen cellars. No one knew how he would ever pay off the mortgage for what was then the largest building west of Albany. No one needed to worry, as it required three expansions over its 103-year history.

Over time, the Arcade held a post office, a library, the first site of the Bausch and Lomb optical business, and a large seed distributor begun by Abelard's son William. George Selden conceived and patented the first practical gasoline engine in an office there. As well, the father of the creator of the Kodak camera, George Eastman, moved his commercial college into the top floor in 1854.[8] Most important of all, the Arcade became the first headquarters of Western Union, the United States' first commercial monopoly, an inspiration to other Yankee businessmen, such as J.P. Morgan and John D. Rockefeller, who would follow. Today, Western Union is the company that transfers money from migrants to their families living abroad, but that is only a remnant of what it was 150 years ago. Then, it was the Internet of the nineteenth-century industrial age.[9] Indeed, the telegraph, once it had developed into a network with common standards and reasonable prices, was a more disruptive technology to those who had never been able to communicate faster than a horse could gallop than is the Internet today. There is not a human alive on the planet who has not grown up with some form of instant electrically based communication. Today's Internet is but an expanded, more complex, and more diverse version of the telegraph.[10]

Professor Morse

Samuel F.B. Morse was a complicated character.[11] He was born in Boston in 1791, the son of a clergyman[12] and later famous Yale professor Jedediah Morse, the author of the first geography of the United States. Samuel went to Yale when he was just fourteen, became a friend of Professor Benjamin Silliman's family, and attended some of his lectures on chemistry and "galvanic electricity."[13] Yet he decided that he wanted to be an artist, and after graduation in 1810, went off to Europe with the Romantic painter George Washington Allston and Allston's new wife. He gained some artistic fame in England as he waited out the War of 1812, but on his return to America in 1815 he found that few of the wealthy had any interest in Romantic pictures, preferring instead to have portraits done of themselves in the mode of the European aristocracy.

Morse married, and his wife bore him three children in rapid order, dying while giving birth to the third.[14] Unable or unwilling to support them, Morse farmed the children out to relatives and continued his artistic career in New York City.[15] In 1829, he went back to Europe, where his art was respected.

While in Europe, he was appointed an art professor at what is now New York University. He was on the way home in 1832 on a packet from Le Havre to New York[16] when his life took a radical turn. At dinner aboard the ship, a fellow passenger described some lectures he had attended in Paris given by Professor André-Marie Ampère and others on the subject of electricity.[17] When his fellow passenger mentioned that electricity could pass through wire almost instantaneously, Morse had a flash of inspiration. "If this be so, and the presence of electricity can be made visible in any desired part of the circuit, I see no reason why intelligence might not be instantaneously transmitted by electricity to any distance."[18]

Morse, consumed by this not-so-original insight,[19] went out on deck and developed some diagrams of how interruptions might be transmitted in a current and, perhaps more important, a method of turning these interruptions into a simple series of dots and dashes on a piece of paper, which could represent numbers. He then felt that combinations of numbers could be used to represent words — say, yes = 56 and no = 222.

Professor Morse came crashing back to reality when he arrived home with no money and few prospects. For the next few years, he had to live off his artwork, as the professorship paid him only whatever fees he could get out of the students. He did not get a classroom until a building was finished on Washington Square, later the center of the artists' colony of Greenwich Village. There he taught, lived, and built his first telegraph.

In September 1837, Morse managed to send a message through 1,700 feet of wire in his classroom, and the next month he filed notice with the patent office. Morse's colleague, Professor Leonard Gale, who taught chemistry and geology, became involved,[20] as did Alfred Vail, a mechanic and nephew of a man who owned a New Jersey ironworks. Vail's father agreed to fund work on the telegraph, and Alfred gained 25 percent of the patent rights.[21]

Gale and Vail, and an assistant, George Baxter, set up shop in the ironworks and got down to improving the equipment. By early January 1838, they had increased the range to three miles and redesigned the sending and receiving equipment. Vail also advocated abandoning Morse's fascination with relating words to numbers in favor of a simpler and more direct system of relating dots and dashes to actual letters that he developed in cooperation with a local printer.[22] The next month, they demonstrated the equipment at the Franklin Institute in Philadelphia, and Morse's connections then got a demonstration scheduled before the House Committee on Commerce in Washington, DC.[23] F.O.J. "Fog" Smith, a Maine congressman and the committee chairman, convinced Morse that he could get a federal subsidy through Congress and the president in return for his receiving 25 percent of the patent rights. In April 1838, Smith put forward a bill to subsidize the construction of a line between Washington and Baltimore for $30,000, but it would be almost four years before he could deliver. First, the financial fallout from the Panic of 1837 intervened. Then Smith and Morse went to Europe, on Smith's money, to promote his telegraph there and to get European patents. The trip was unsuccessful, as many governments perceived the military use of such a system and would not allow private investors to take part in its development.[24]

When Smith finally did get the experiment going with help from Washington,[25] the results at first were a disaster. Morse had chosen to bury the line in the ground, which not only ate up most of the budgeted funding for construction, but the insulation available at the time could not prevent the line from "grounding out." Ezra Cornell, a Yankee plow salesman and friend of "Fog" Smith's, who had the contract to bury the line, saved the day by getting Morse to adopt a British idea that Vail had read about: stringing the wires on poles above-ground.[26] The solution proved workable and relatively cheap.

The section from Washington to Baltimore was still being constructed when Vail, sitting in Annapolis, Maryland, heard from a Washington-bound train passenger the results of the Whig Party's presidential convention. He then sent the first telegraph message to Morse at the Post Office headquarters in the capital. Long before the train arrived, Morse could tell people there that Henry Clay had been nominated. Three weeks later, at the official launch on May 24, 1844, Morse sent the famous "original" message — "What hath God wrought?"[27] — to Vail in Baltimore. Vail returned it quickly to Morse and the assembled politicians, including Clay.[28] Six weeks later, the Democratic Party used the telegraph at its convention in Baltimore to conduct negotiations between the delegates and a reluctant draftee for vice president sitting in Washington. The electric telegraph had been validated.

Morse did not invent the telegraph.[29] What Morse did was file the earliest American patent and organize others to develop his dreams. He was at his best in using his connections to get the scientific community and the politicians to support his claim to be the inventor. Even though he had initially offered his patents to the Republic of Texas for free and later unsuccessfully offered them to the US government for $100,000, he got the credit for making the telegraph into a potentially successful commercial product. Ever the artist, the fame seemed to be what he was after. Fame, however, would come only if the telegraph revolutionized communication. Morse used up the last of his federal subsidy providing free messages, including intercity chess matches, to the public until February 15, 1845. Congress then provided $8,000 to operate the line until it figured out what to do with it. As in European countries, it became a part of the Post Office, and Morse and his associates were hired to run it.

Shortly thereafter, Morse and his partners engaged Amos Kendall, a Massachusetts Yankee considered by John Quincy Adams the "ruling mind" of the Van Buren and Jackson Administrations, as their agent to develop the telegraph business. Kendall would get 10 percent of the resulting revenues. Morse and Kendall went on to incorporate the Magnetic Telegraph Company on May 15, 1845, to extend the line to New York. Their concept resembled that used in the financing and construction of the railroads, which were being built at the same time. They incorporated a short line and then sold shares to those who might directly benefit along the proposed route. After the Philadelphia investment community had been tapped, "Fog" Smith and Ezra Cornell went on to New York to raise money for that end of the project. The line from Philadelphia to New York opened in January 1846. The Hudson River could not be crossed by a wire at first, so messages were taken across from New Jersey to New York by courier.[30]

It was not long before others became involved. Two weeks after the Magnetic Telegraph Company was formed, two entrepreneurs from Utica, New York, stagecoach owners Theodore Faxton and John Butterfield,[31] obtained a license from the company to build a line from Buffalo to Springfield, Massachusetts. On the way back upriver to Albany, Butterfield met Henry O'Reilly, a businessman from Rochester, and interested him in the potential of the telegraph. O'Reilly then got a contract from Kendall to build a line across Pennsylvania and down the Ohio River to Cincinnati, then the largest city in the old Northwest. A quarter of the shares of this new company went to the patent holders. O'Reilly promptly brought into his group a number of investors from Rochester and the surrounding area, and was off on a turbulent and relatively short career as the most aggressive of the telegraph line developers.[32]

In May 1846, the Mexican War broke out, and by November Kendall and "Fog" Smith had negotiated terms to create a company to build a line to New Orleans to hasten communications with the Army headquarters there.[33] Again, similar to the financing of the early railroads, agents of the Washington and New Orleans Telegraph Company encouraged citizens of towns along the projected line to subscribe to shares of stock. Such was the enthusiasm for the war and news of the battles that nearly $300,000 was pledged quickly. The line was not completed until July 1848, too late for the biggest battles of the war. Even so, Northern newspapers quickly realized the telegraph's commercial potential, and sent quick teams of messengers out from New Orleans to meet the approaching line and post their stories onward. This expense led eventually to the formation of the Associated Press syndicate.[34] Morse was now truly famous. He was also lucky. At the end of 1846, Congress, encouraged by a disinterested, and perhaps distracted, postmaster general, opted not to buy the patent rights, and turned over the Washington, DC, line to the patent holders.[35]

Back to the Reynolds Arcade

By 1848, there were two thousand miles of telegraph line in America and by 1850, twelve thousand miles operated by twenty companies; in Britain, in contrast, there were just over two thousand miles of line at the same time.[36] In 1852, the superintendent of the census reported that there were over twenty-three thousand miles of line in operation connecting nearly five hundred cities and towns, with ten thousand more miles planned.[37] A little more than eight years after the first message was sent, an author reviewing this industry could note that "[t]elegraphing, in this country, has reached that point, by its great stretch of wires and great facilities for transmission of communications, to almost rival the mail in the quantity of matter sent over it."[38]

Yet, through the early 1850s, various short lines all over the country were troubled or slipping into bankruptcy. Part of the problem with financing was that investors were torn between the challenges and opportunities offered by both the telegraph and the railroad, and the same basic capital-intensive structures and problems existed for both technologies, including the need to consolidate large regional and national networks. In general, however, the public was "railroad mad," and the railroads gained the upper hand in competing for investors' dollars. As Ralph Waldo Emerson noted, "[t]he Railroads is [sic] the only sure topic for conversation in these days. That is the only one which interests farmers, merchants, boys, women, saints, philosophers, and fools….The Railroad is that work of art which agitates and drives the whole people; as music, sculpture, and pictures have done on their great days respectively."[39]

Another problem was the growing use of different transmission technologies. By the late 1840s, two other methods of sending and transcribing messages on a telegraph line,[40] the patentable core of the business, had been devised, and rival companies began stringing lines using different systems that could not interface.[41] Then there was the problem of charging for the service. From the outset, the standard charge was per word, but the definition of "word" varied as did the rate by distance, so that a message of ten words sent from Washington could be sent to Dubuque, Iowa, for less than one sent to Frankfort, Kentucky, less than half as far away.[42] Competition over heavily used routes meant rate wars and unstable pricing. Pricing also was based on the quality of some local lines, which, of course compromised reliability, as low-quality line equipment was prone to outages — but the rate was cheap.

These challenges complicated an already commercially risky situation. So, some Rochester businessmen set out to reduce the odds. The Rochester group had stood patiently behind O'Reilly in his attempts to link all the major places in the West and South, but by 1850 their patience was exhausted and money was tight, and O'Reilly, his life complicated by his own financial arrangements, sank into bankruptcy the next year.[43] Then, a new group of local businessmen, mostly Yankees and mostly already involved in telegraph businesses, met in Hiram Sibley's office in the Reynolds Arcade in 1851.[44] Sibley, a local businessman and a political operative in Washington, had been involved in getting Morse's subsidy passed by Congress. He thought the telegraph could be one of the "wonders of the age."[45] The group subscribed $90,000 for shares in what became the New York and Mississippi Valley Printing Telegraph Company (NYMV).[46]

Because the Faxton and Butterfield company already had a line based on the Morse patents operating through western New York, this new NYMV venture had to acquire the rights to a competing technology — in this case, the House printing machine.[47] From 1851 until early 1854, the company built a line between Buffalo and Cleveland that connected to a House line that competed with Faxton and Butterfield across New York State. Things did not go easily. As some Rochester stockholders began to fear that a collection of failures would lead only to a greater failure, Sibley and a few others bought their stock off them. Sibley, by 1854, had most of his fortune tied up in the NYMV.

In February 1854, Sibley led a reorganization and refinancing of the NYMV.[48] It came on the heels of a successful arrangement that gave the company access from Cleveland to Detroit and Chicago by persuading some railroads servicing these cities to give him exclusive permission to use their rights-of-way and to help finance the construction of the telegraph lines. After the Erie Railroad had shown that the telegraph could be of great value in coordinating trains and avoiding delays and accidents, both sides embraced the marriage of the two technologies:[49] the railroads got priority access to the telegraph and shares in the NYMV, while Sibley, in effect, got routes for his lines from Cleveland to Detroit and Chicago for free.[50]

Over the next two years, Sibley managed to acquire various other lines using either the Morse or the House technology for a fraction of its worth. He also leased the lines and gradually bought out the owners and he traded NYMV stock for that of others. By 1856, through diplomacy and maneuvering, he gained a monopoly on all telegraph traffic in the Midwest. In a typical deal, he "sold" the House line across New York to the competing Morse line company owned by Faxton and Butterfield in exchange for gaining majority ownership in the combined new company. This new company then joined the NYMV system, a move agreed to by Faxton and Butterfield, who recognized the threat that all the Western business would be sent through the NYMV line if they did not merge. The NYMV now had a trunk line from Chicago to New York.

Using such tactics, Sibley bought or pressured most of the fifty-odd disconnected Western lines into the NYMV fold, including those controlled by Ezra Cornell, who had struck out on his own.[51] When Cornell joined Sibley and his associates, he insisted that the cumbersome name New York and Mississippi Valley Printing Telegraph Company be changed to the Western Union Telegraph Company, to signify its control over the West.[52] Eventually, the company's control would spread along the East Coast and, after the Civil War, to the Southern States, until it had more than 90 percent of the commercial lines in the country.[53]

The Western Union system was gradually rationalized by using standardized technology — the House technology was abandoned in favor of Morse's — centralizing purchasing, and common standards for poles, wires, and language and codes. The spread of this network provided communication at a reasonable price and without the complications of different equipment standards and practices. It was the first commercial monopoly to cover the American market,[54] a model for others who tried to build similar networks for railroads, kerosene distribution, the telephone, and electricity, followed around the turn of the century by the growth of "trusts" in a variety of goods-producing industries.[55] By then, the battle to restore competition in the American economy had far overshadowed the achievements of Sibley, Cornell, and others.[56]

Western Union also provided management, accounting, and pricing structures that paralleled those of the railroads, and probably preceded the railroads' more high-profile use of them.[57] The capital required to string a telegraph line was far less than that for a rail line, and the need for coordination was not as acute. Western Union also started the practice of supporting professional organizations and technical centers devoted to corporate product development. In 1873, the company created an Electricians' Department as its corporate research arm, becoming the seed for the profession of electrical engineering as well.[58] The organizational relationship between the railroads and the telegraph was evident in the final connection that gave the world its globe-spanning connection: the design and laying of a seven-thousand-mile telegraph line between Australia and North America. The project was carried out by a Scottish-born Canadian railroad man, Sir Sandford Fleming, who had surveyed the Canadian Pacific line across the Rockies and coordinated the creation of railroad time zones. In late 1902, Fleming would send the first "round the world" telegraph message to Canada's governor general, Lord Minto, in Ottawa, the message going from the telegraph office in Ottawa, via Australia and South Africa, and back to the Canadian capital.[59]

The telegraph also had a transformative effect on other industries. It made papers, often called "journals" because of the dated quality of their information, into "newspapers" — publications that were up-to-date and, at least in larger cities, put out in several editions every day, especially when events were fast moving.[60] The telegraph led to the creation of a national financial network, as bankers and stockholders were now able to send and receive information on business opportunities and risks from all across the country.[61] Between 1845 and 1871, eleven commodity exchanges were opened around the country, all based on the telegraph for information.[62] The telegraph business was also early — though after the textile mills — to employ women in large numbers. Its first female telegrapher was Sarah Bagley, of Lowell, Massachusetts, a former mill worker, who was hired in 1846. The Civil War led to increased hiring of women, though there was no promotion track for them. In 1869, Western Union opened a school of telegraphy for women at Cooper Union in New York City; in the 1870s a third of the operators in its main New York office were women, and by 1900 so were 12 percent of its telegraphers across the country.[63]

The headquarters of Western Union remained in the Reynolds Arcade in Rochester until the end of the Civil War. The company moved in 1866 to larger headquarters in New York City under new owners, and the days of Rochester as the "communications hub" of the country passed.[64]

From the Telegraph to the Internet: A Redefinition

The "Internet" is a useful way to describe any communications network based on electrical connections. We tend to use the term for the digital method devised after computers became powerful enough to enable the packaging and unpackaging of different types of communications, but it can be just as useful to describe the impact of the telegraph, the radio, and the telephone.

The first "Internet," the telegraph, is the fundamental invention underlying modern telecommunications.[65] In its initial form, it depended upon the interruption of an electric current sent along a wire. These interruptions created two current states, on and off, which also became the basis for today's digital communications that use the binary numerical system.

The second form of Internet was the telephone system. The first usable telephone was patented almost forty years after the telegraph as a way of converting electrical pulses into mechanically produced sound.[66] Western Union had no real interest in this invention, and agreed to an arrangement with the Bell company that provided each its own space in the communications field.[67] This lack of interest in developments outside a company's core technology is not uncommon in business history.[68] The American Telephone and Telegraph Company (AT&T) ignored pressures to move away from its leased black phones, IBM ignored software issues when it entered the computer field, focusing on the equipment side of the industry, and Digital Equipment and others ignored the PC in favor of hanging on to the declining minicomputer market. The list goes on and on.

By the first decade after its invention, there were a quarter of a million telephone subscribers worldwide. Twenty years later, in the first decade of the twentieth century, the telephone was dominant in the local communications business, with 10 percent of American households having one,[69] and its long-distance capabilities were growing so as to threaten Western Union's telegraphy. A relative of Alfred Vail's, Theodore Vail, began to build AT&T into "Ma Bell," the telephone monopoly that existed from the 1880s until technological changes and the federal courts broke it up in 1982.

The strength of the telephone lay in the feeling that it was more personal than the telegraph, which was dominated by business and government traffic. It was with a business purpose in mind, however, that the first long-distance line was built, only three years after the formation of AT&T, between Boston and the mill town of Lowell.[70] And it was only about the turn of the century, after the invention of the automatic exchange, that the company began to court the average user. As a 1914 magazine article noted, "[u]ntil [1900] the telephone was a luxury — the privilege of a social and commercial aristocracy. About 1900, however, the Bell Company started a campaign, unparalleled in its energy, persistence and success, to democratize this instrument — to make it part of the daily life of every man, woman, and child."[71]

In response to the dominance of the telephone, Western Union attempted to innovate — for example, it introduced the first charge card in 1914, and by 1960 it was replacing telegraph lines with microwave towers. The first private American communications satellite, Westar, was launched in 1974 for Western Union.[72] Along the way, the company also experimented with computer system testing and services and home shopping.[73] But the use of Morse code was halted worldwide in February 1999,[74] and by 2000 Western Union was into financial and other services and virtually out of communications — its last telegram was delivered in 2006. Today, Western Union handles a significant part of the remittances from migrants back to their home countries.

Telephony, having trumped telegraphy, was in turn followed by the invention of wireless communications — radio. The new technology was based on the understanding that an electric current generates an electromagnetic field perpendicular to the direction of the electrical current in a wire.[75] In 1900, Guglielmo Marconi set up a system that allowed him to use interruptions in the electric current made at a kind of telegraph station to "send" in the electromagnetic field that went along with the current, to be picked up at another station, which reversed this process. The signal at the receiving end would be the same as the one sent.[76] Early radio operators used the same Morse code as the early telegraph operators. Radio, of course, was improved by telephone-inspired inventors who added sound to the signal. One aspect that radio, or wireless, had that wired methods did not was its ability to be broadcast. If the radio wave was strong enough, not only could the intended receiver get the message, but so could anyone with the proper equipment set at the broadcasting wave frequency. It proved easy to reach a wide audience, unlike the switching requirements necessary to allow a wire-based service to do the same.[77] Think of a poor telephone operator of 1900 manually trying to plug ten people into a "conference call" — it simply couldn't be done without today's switching system. Broadcasting allowed communication on a one-to-many basis rather than the one-to-one basis of telegraphy or telephony.

Adding pictures to the sound broadcast by radio led to the development of television, whereby powders inside a cathode ray tube were stimulated, and the separate pictures were coordinated, changing many times a second, to produce the appearance of motion, similar to the technique used in films. Until recently, the technology was so expensive that it could be used only to broadcast in the one-to-many format. Edward Bellamy, in his 1888 idealistic *Looking Backward* could dream of a one-to-many telegraphy system in place in 2000, while in the 1930s movie serials Buck Rogers could use a video phone, but the date was 2542; Dick Tracy had a wristwatch phone, but no one really knew how these might be made or when. It was not until after the Second World War that such things appeared to be possible. The key was the application of the new field of quantum mechanics,[78] which led in 1947 to the development of the transistor and to new combinations of wired and wireless communications. A decade later, the transistor, which acts like a radio vacuum tube, was incorporated into a "chip" or integrated circuit. With a lot of transistors, information could be processed in digital form and eventually displayed on a television screen.

The Fifth Internet

It is perhaps jejune to speak of Internets when discussing the history of electrical communication technologies, but it at least touches on the connections between them. Building on Standage's notion of telegraphy as the first Internet, one could see telephony as the second Internet and radio as the third. Television was as much an improvement on radio as the telephone was on telegraphy, so perhaps it was the fourth iteration. The fifth is a further improvement, using wires, radio, telephony, and television all together, allowing one-to-one interaction as well as broadcasting. It is both mobile and fixed; it is highly standardized, allowing a global network; and it is incorporating the information industries into its web.

I was sixteen when the Soviet Union launched Sputnik, the first man-made satellite, in October 1957. It was wonderful, even if it was Soviet, because I was a science fiction fan, with a subscription to *Astounding* and dreams of walking on Mars someday. The next year, because the little Catholic school I attended offered chemistry and physics only in alternating years, I enrolled in physics in night school at the local technical high school. Like a number of my contemporaries, I was going to go into "rocket science" via engineering physics, and I began applying to universities with that in mind.

It never worked out. Science and engineering lost their allure when I realized there were more interesting things in life, to me, than differential equations and chemical formulas. Neither did my hopes of going to Mars remain when I found that astronauts had to be of about average height and I was tall enough to be a basketball center. The space race didn't last, either. Once the United States ended its moon landings and the odd probe had been sent to Venus and Mars, the Soviet Union seemed to lose interest in space, content with knowing its military could deliver bombs by rocket anywhere on Earth. The American effort continued, but primarily in near-Earth orbit, occupied with scientific projects and communications satellites. I switched to political science and economic development and went back to reading sci-fi novels.

In the wake of Sputnik and the start of the space race, the Pentagon created the Advanced Research Projects Agency (ARPA). ARPA was designed, in part, to develop technology to assist the space race, as well as other technologies that might have significance. Much of the funding went to university researchers and to non-profit and corporate-sponsored research institutes. And therein lies an unintended consequence. The staff at ARPA liked to keep in touch with the people working on the projects they funded. Neither teletype nor the telephone was satisfactory, and constant site visits, especially to the West Coast from the Pentagon back East, were difficult and time consuming.[79] So, in 1969, ARPA contracted for a communications system that would bypass the normal phone network and connect its computers in the Pentagon and those of the various research teams.[80] AT&T, the monopoly telephone company, was not yet weaned off analog switches, so the system would have to be built on dedicated lines that used digital switches capable of matching computer speeds.[81]

Then another problem arose: the different research teams used different computers with different operating systems and could not talk to the Pentagon or to each other. Not only did this impede contact between ARPA and the researchers, it also prevented many research projects from being conducted jointly by different institutions.[82] Someone then had the idea that, if a standard machine could "translate" incoming messages into something that a computer, of whatever make or operating system, could understand, then communication would be possible. Today we call these machines "routers," but the one that hides under my desk bears little resemblance to the first, refrigerator-sized ones.[83] A company called Bolt, Beranek and Newman, or BBN, in Cambridge, Massachusetts, won the contract to build them, and came up with a way to break messages into electronic packages that were reconstituted at the receiving end.[84] This distributed method of sending packets was inspired by the realization that such a system might prove durable and of military use in the event of a Soviet nuclear attack.[85] By the early 1970s, the fifth Internet was on its way.[86]

Of course, if researchers could use this system to share their work, it could be used for more mundane purposes as well. As time-sharing capability grew and computer users multiplied, they began to communicate with each other through the ARPA system, which served to expand demand, especially after the first public demonstration of ARPAnet in October 1972.[87] Yet, capability lagged demand, and it was not until 1985 that as many as a thousand computers were connected to the system.[88]

Then came another unintended consequence. Like a stealth program, email took over the ARPAnet. Its creators had not planned it as a tool for person-to-person communication; it was meant only to transfer research data. Yet the desire to communicate through computers was there. A program developed in the 1960s at the Massachusetts Institute of Technology called "MAILBOX" allowed users on the same time-share computer to send each other messages. In 1972, Ray Tomlinson of BBN devised a program that allowed him to send a text message from one computer in his lab to another — the first "real" email.[89] Others adapted his program — including the use of the @ symbol,[90] which allowed for standardized addresses for email "boxes" — to the ARPAnet file transfer protocol.[91] By 1973, ARPA discovered that three-quarters of ARPAnet traffic was email.[92] Within a short time, email traffic was going back and forth between America and Britain; it was the beginning of the World Wide Web. Soon, defense contractors as well as foreign researchers wanted access, but the problem was how to tie their computers to the ARPAnet. The answer was to create a software protocol — the Transmission Control Protocol/Internet Protocol (TCP/IP) — that would control message packets. route them to their proper addresses, and put them into their original order. TCP/IP was adopted for common use in 1983, after a decade of development and trial. It provided a standard that allows computers in any country, using any human language, to communicate with any other computer.[93]

The "father of the Internet" is Vint Cerf, who moved from ARPA — now renamed the Defense Advanced Research Projects Agency (DARPA) — to MCI, the company that broke AT&T's monopoly over telephony. Cerf created the first commercial email product and later headed up ICANN, the not-for-profit company that assigns Internet addresses worldwide. ICANN assures every Internet user has a unique address, avoiding the confusion and misdelivery that would be caused by identical names, streets, and towns in traditional addresses worldwide.

By 1980, DARPAnet had grown far beyond DARPA's needs, budgets, and control, and it approached AT&T about taking over the system. AT&T, however, felt that packet switching was not compatible with its existing telephone switching system. Over the 1980s, government involvement declined, and the system's main users then set down standard user charges and methods of connection and use, including address domains, such as "dot-com" and "dot-edu."[94] One of the protocols adopted was the American Standard Code for Information Interchange

(ASCII), which could be considered a grandchild of Morse code and which is at the core of digital information communication today.[95]

As computers and peripheral devices kept getting more powerful, smaller, and cheaper, their cost became affordable for ever-larger numbers of consumers, who also wanted access to the system. By 1991, commercial Internet Service Providers were offering to connect individuals to the Internet via telephone lines.[96] Some, like Yahoo and AOL, bundled services that they then offered as a kind of mini-Internet, and encouraged their subscribers not to leave their friendly sites for the Internet wilderness outside. Others, such as the "Gopher" service at the University of Minnesota, offered connections to different databases available around the Internet, but using these comfortably required a lot of skill. It was a lot like the early days of the automobile, when drivers needed their own mechanics to keep their cars going.

In 1991, things changed again. Tim Berners-Lee at the CERN — a French acronym originally standing for the *Conseil européen pour la Recherche nucléaire* — research facility in Switzerland devised software that allowed users to connect easily to Internet databases and was a standard way of controlling the addressing of sites and protocols for handling information.[97] His World Wide Web did for global communications what Morse code had done for national communications a century and a half earlier.[98] In short order, graduate students at the University of Illinois, including Marc Andreessen, came up with Mosaic — a graphics-based, rather than text-based, interface, dubbed a "browser" — that allowed users to access sites on the Internet easily. To use the automobile analogy again, it resembled the introduction of the electric starter in the late 1920s. Andreessen left university to found Netscape Communications, which in 1995 offered a version of Mosaic called Netscape Navigator that was instantly popular with consumers. Microsoft, in the competitive tradition going back to Hiram Sibley and others, acquired a version of Mosaic, redeveloped it, and offered Internet Explorer as a new part of its Windows operating system; by 2002, it had effectively driven Netscape from the market.

The Internet, with its evolving graphics and text, went from 600,000 users in 1991 to more than 40 million in 1996 and to a billion in the next decade, literally becoming a household, research, and corporate tool.[99] After the turn of the century, with processing power doubling and doubling again, the telephone went wireless and cellphone radio towers began to dot the globe — indeed, some areas not served by wired phones, particularly in developing countries, skipped right to wireless. Married to the Internet, the cellphone became a hand-held version of a desktop or laptop computer. Whole industries, such as travel, banking, and news distribution, began to migrate to the computer. Just as railroads pushed stagecoaches to the periphery in the 1850s, so the Internet began to cripple advertising agencies, radio broadcasting, news and music distributors, retailers, and real estate and insurance agencies.

One Internet, One World

The question naturally arises as to where the "natural monopoly" is located in the fifth Internet. This gets complicated. In essence, there is a global monopoly called the "Internet." It is not made up of its physical infrastructure, which is dispersed among servers that hold the files identified by a single standard of addresses managed by a not-for-profit corporation. It does not lie in the standards, or protocols, that have been developed to allow access both to the system and to individual electronic files, though this rulebook is global. Nobody really runs it. One could create an alternative Internet, such as a corporate Intranet, but it would have little power in the world if it were not attached somehow to the larger one.

Here "Metcalfe's Law" comes into play. Bob Metcalfe is one of the founders of the Route 128-area Ethernet and router company 3Com, which helped to enable Internet access in homes, offices, and even on cellphones.[100] He mused that the cost of Internet connectivity is inverse to the number of adopters and that the "power" of a network increases by the square of its participants. The combination is potent. The power involved is psychological: the need to be a participant. The mail service has potentially billions of participants, but it is slow and therefore not attractive. The telegraph never penetrated much past central stations to companies and the homes of a few wealthy people, so the number of participants was restricted. The telephone's participants were largely local; until quite recently, for most people, long distance and international calls were costly and rarely made. Radio and television were used mostly for broadcasting, a one-to-many approach that limited participation. The Internet, however, allows virtually all forms of interaction, on an immediate basis, to a large proportion of the world's population. It incorporates mail, telephone, radio, and television (via podcasts, video clips and streaming) as well as the mobility of wireless and continual contact, globally, and, after paying an access fee, free. This is real power, in Metcalfe's terminology. You cannot fail to be there. Global and online are starting to become synonyms. The Internet is poised, at this writing, to consume other forms of entertainment, such as movies and television, and, over time, is likely to pull education and much of health care into itself.

In the end, the evolution of the telegraph into a communications system that encompasses all forms of information and is potentially accessible to nearly everyone on the planet constitutes a form of democratization of information that promises both change and opportunity to people in all countries. In the short run, the attractive power of a network that already encompasses billions people can be overcome only by threatened authoritarian governments. Morse and Sibley, two Yankees from opposite ends of Massachusetts, set it all in motion.

18

Batavia, New York: The World's Greatest Land Sale

By early youth, R.C. had become the kind of mean, ornery cuss his neighbors had associated with years of Maturity....'Twas his profession did it to him. As a young Surveyor, from the rude shocks of attending his first boundary dispute, he understood that he must exercise his Art among the most litigious people on Earth.
— Thomas Pynchon, *Mason and Dixon*[1]

Come all ye Yankee farmers who wish to change your lot,
Who've spunk enough to travel beyond your native spot,
And leave behind the village where ma and pa do stay,
Come follow me and settle in Michigania, —
Yea, Yea, Yea in Michigania.
— From an 1830s popular promotional song[2]

If a man is disappointed in politics or love, he goes and buys land. If he disgraces himself, he betakes himself to the west...If a citizen's neighbors rise above him in towns, he betakes himself where he can be monarch of all he surveys.
— Harriet Martineau, Society in America, 1837[3]

Batavia

Batavia is a town of about twenty thousand, about halfway between Rochester and Buffalo. It lies some six miles or so north of US 20 and the Thruway touches its northern edge. Unlike many similar-sized towns dotting the New York landscape, Batavia was not named for some place in classical antiquity (Troy, Palmyra, Rome, Ithaca, Utica, Syracuse) or for ones in England (Rochester, Auburn, Albany). It is, instead, named for the Batavians, the ancestors of the Dutch, who both cooperated with and fought against the Romans. Batavia was also the Dutch name for their main *entrepôt* in Indonesia, now renamed as the capital of that country, Jakarta. We visited the rundown museum of Batavia in portside Jakarta once, and a decade later visited the Holland Land Company Museum in Batavia, New York. They have nothing in common.

It might be suspected that the naming of the American Batavia had something to do with the Dutch origins of the city and state of New York, but that too would be erroneous. Batavia came into existence nearly 150 years after the Dutch were relieved of their North American holdings by the English. Instead, the town was named by its founder, surveyor Joseph Ellicott, to recognize the nationality of a foreign company engaged in large-scale land speculation on the American frontier in the 1790s. The Dutch company managed to acquire, among other holdings, the entire western end of the state of New York for the purpose of subdividing it and selling off parcels to settlers coming in from the East — in particular, Yankees. Batavia was the location of the land office.

The ancient Batavians would have appreciated the events that led up to their namesake town. They were under constant pressure from the Romans, including invasion, political control, and colonization. Batavia, the present-day town, also emerged from war and the dispossession of the native inhabitants, but by American columns under General John Sullivan that destroyed Iroquois villages and farms and drove most of them across the Niagara River into what is now Canada.[4] After the Revolution, even those Iroquois who had supported the winning side were dispossessed and reduced to dependency on small reservations, a practice that many other tribes were to encounter over the next century.

The Iroquois lands extended roughly from Oneida Lake westward to the Niagara River, comprising more than half the good land west of Albany. New York claimed the territory, but Massachusetts contested this, arguing that to the west of the Dutch *patroons*' holdings, other grants, and a military reserve, the land was theirs by pre-existing Royal Charter. The claim must have had legitimacy of sorts, as a compromise worked out by the two States in 1786 gave Massachusetts ownership of the land, but New York acquired legislative sovereignty. In effect, one state became a major landowner in another State. This was not unlike the solution to Connecticut's claims in the Susquehanna River lands in northeast Pennsylvania, where settlers from Connecticut were allowed to keep their grants from that State, but became Pennsylvania citizens.

Once New York and Massachusetts had settled their claims,[5] both States moved to replenish their treasuries by selling off land. In 1787, Massachusetts sold its Susquehanna River lands in the Southern Tier of New York to a group of speculators for 12.5 cents an acre.[6] New York began selling land to the west of the Dutch *patroons*' estates in 1789 in a series of auctions that lasted for a number of years.[7] Most went to large buyers, though some tracts did go to small farmers. A military tract reserved for soldiers was made available in 1791, but since the warrants were transferable, most had been acquired by speculators in the years following the Revolution.[8]

The Massachusetts-owned tract to the west of Lake Ontario's Sodus Bay was sold to two politically connected speculators in 1788 for about 3 cents an acre, but they were unable to generate cash flow fast enough to meet their obligations and went bankrupt after a year.[9] In 1790, Robert Morris, a US Senator from Pennsylvania, a signer of both the Declaration of Independence and the Constitution, "financier of the Revolution," and arguably then the richest man in the United States,[10] bought the whole tract from Massachusetts for £75,000. Morris subdivided it into three tracts. The eastern one — running from the Genesee River, which flows north into Lake Ontario where Rochester is now, eastward to the end of the tract — was the easiest to resell to settlers, so he, in turn, flipped this to British speculators, Pulteney Associates, for what he had paid for the whole property.[11] Having covered his obligations to his creditors, he kept the Genesee River valley for himself.

Then Morris ran into Theophile Cazenove. A consortium of Dutch bankers had originally planned to invest in American bonds, but had shifted to frontier land development. They had hired Cazenove, a protégé of one of them, to assemble and oversee their widespread American operations. He had just made a large purchase farther to the east in New York (see Chapter 13.) and was looking to buy more. In late 1792, Morris sold the westernmost part of the state to this partnership, a 3.3-million-acre tract. This deal left Morris with both a tidy profit and the Genesee River valley lands free and clear. He had, however, to get the Indians to agree to sell their rights, which took a number of years. Morris's operations make today's land speculations look like small potatoes.[12]

From 1793 on, Cazenove concentrated on assets other than the remote western end of the state of New York, in part because Morris was slow to deal with the land rights of the Indians still left in the area. He went on to secure a large tract in northwestern Pennsylvania, bringing the partnership's holdings to over 4 million acres. The Dutch bankers then incorporated the Holland Land Company (HLC) in 1795 as the overall controlling body for their land sales.

Once the land in western New York was unencumbered, Cazenove hired Joseph Ellicott,[13] the younger brother of the famous surveyor, Andrew Ellicott, to survey the Company's huge holding in western New York, a task that was completed in 1799. See Figure 3. Ellicott was then hired as the local HLC manager[14] and tasked to sell this large property to individual settlers.[15] Ellicott laid out the town of Batavia at a bend of the Tonawanda River, which acted as the regional headquarters for this real estate activity, as well as being designated as the county seat for Genesee County. Ellicott then laid out the town of Buffalo in 1801 as the HLC's lake port.[16]

Figure 3: The Western New York HLC Lands

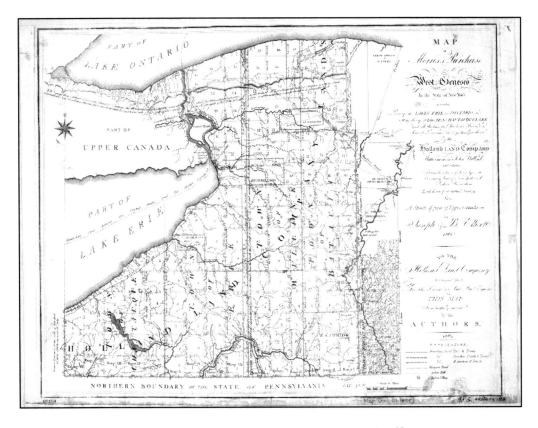

Source: Adapted from map copy provided by the Holland Land Office Museum, Batavia, New York. A variation may be found at www.wikipedia.org/wiki/Holland_Land_Company

The problem with selling the land was that it was relatively inaccessible to potential settlers, who would first pass through comparable lands farther east and presumably stop there. The problem of good roads in the area plagued the Company through all its existence,[17] because Ellicott did not feel that large expenses on improvements would make the land sell any faster.[18] He began by offering land at $3.75 an acre, but soon found he had to drop his price to $1.50 to get any takers. He began to advertise in New England and also began taking goods "in kind" for payments on the land. Sales began to pick up after the state extended the Great Genesee Road westward to Buffalo in 1803.

So, as the more easterly New York territory became better organized, Yankees began to buy their way directly westward. Stopping in the river lands of eastern New York was not an option as the large landowners there wanted tenant farmers, and Yankees wanted to own land freehold.[19] The destruction of the Iroquois farms and villages in the Revolutionary War and the gradual development of roads west from Albany signaled the way west for Yankee farmers — and off they went. "On to the Genesee!" was the byword at their taverns, Sabbath houses, and kitchen gatherings. From practically every village in eastern New York and the New England States, adventurers and farm folk in their thousands went to plant Congregational meetinghouses and village greens through western New York all the way from Utica to Buffalo. As one historian notes, "[o]ut to the wilderness by way of the Mohawk from Albany, up the valleys from the Susquehanna, settlers poured into every western county by single families, by twos and threes, and by whole colonies....Connecticut must have been beggared of inhabitants, so fast did hundreds of her families make their way into New York."

Joseph Ellicott was to gradually develop the territory over the next two decades and more.[20] By 1812, 200,000 people lived in western New York and 25,000 in the HLC lands.[21] Two-thirds of those in the area were from New England.[22] Later, Ellicott promoted the idea of the Erie Canal to the State's politicians.[23] During the long years between the conception of the canal and its completion, Ellicott became a close advisor to Governor Clinton, the chief politician driving its construction. He had the HLC's contacts in Washington lobby for federal assistance, and had the company influence the sale of construction bonds.[24] The canal route he favored, which ran directly through the HLC lands, proved unworkable,[25] but the canal eventually ran through at least part of the Holland Land Company's holdings. Ellicott and his family lived in Batavia for most of the rest of his life, retiring in 1821.[26] He went to New York City along the mostly-completed canal in 1824 to be treated for mental illness, but committed suicide there in 1826. The HLC benefited from the opening of the canal in 1825,[27] eventually sold the rest of its holdings, winding up its business between 1835 and 1837.[28]

Land as an Asset

We take for granted the historical notion that European countries could claim, and fight over claims, to sovereignty over North American lands, as though the inhabitants of those lands were incapable of defending them and would have to submit to an overseas power.[29] Moreover, European monarchs could grant their subjects dominion over portions of these lands on much the same principle as they gave lands to feudal nobles. As well, giving such grants, or charters, over territories in North America implied that, in some fashion, the recipients could extract any resources they found.[30] In Central and South America, the most obvious of these resources was gold; in North America, it was furs, valued by luxury markets in Europe and Asia.[31]

At first, resident Europeans in North America were limited to military bases and trading compounds, just as they would be in Africa and Asia; there was no thought of the wholesale replacement of indigenous populations by colonists.[32] In the beginning, then, land claims had little to do with occupying the land itself, but with controlling a distinct area to get at its desired resources. England, however, created an exception to this. After 1600, various of its governments began to look with favor on the idea that emigration to America, whether voluntary or forced, could be a pressure relief valve for the country's surplus population, as well as rid it of political undesirables and convicts. One such undesirable group, before the outbreak of the English Civil War in 1642, was the Puritans.

The combination of emigration and Royal grants of land led to the plantation model, which worked well where black slaves were used, as they could be found relatively easily if they escaped. Elsewhere, however, the plantation model could work only if there were effective boundaries to keep enough people inside to do the work. But in the more northerly colonies, the existence of wilderness to the west meant that the disaffected and the landless could simply disappear into the hills and adopt the Indian custom of ownership through squatting. Needless to say, after the Restoration of the monarchy in 1660, English aristocratic musings about the desirability of creating or expanding the proprietor model in America by dispossessing freehold small farmers or preventing their expansion westward was a contributor to the grievances and fears that eventually would prompt the Revolution.[33] Only in what is now Canada were large landholdings acquired in this manner, especially on Prince Edward Island.

A second model arose in the Puritan colonies and was broadly copied elsewhere, especially inland: the corporations that created the colonies of Plymouth and Massachusetts Bay were taken over by the colonists themselves. Think of it as an early version of a leveraged buyout. The English government's intent in chartering these corporations was never to give them self-government, but the two colonies managed to rule themselves and, though only a minority may have been shareholders, no one person or small group was able to become a large plantation owner. Instead, the New England colonies' corporate bodies, whose shares were held widely, turned themselves into governments whose "shares" were likewise held widely. As they opened up new land, the number of shareholder-citizens grew as well. The colonies spread out, largely town-by-town in a kind of incremental manner. The colonies of Plymouth and Massachusetts Bay eventually merged, but religious dissidents left to create Rhode Island, and economic and geopolitical pressures led to the establishment of Connecticut. New Hampshire maintained its autonomy from Massachusetts by virtue of a separate Royal charter. Vermont emerged later, as did Maine, which was a part of Massachusetts.

The founders of the Massachusetts Bay colony established three principles that carried over into later American land policy. The first was adherence to the Mosaic law concerning inheritance, whereby the eldest son received a double portion and the rest was divided equally among other children. This effectively meant that large landholdings could not be passed along intact to one son. Second, Governor John Winthrop successfully opposed the granting of large portions of land to "deserving" individuals, preventing the growth of an American aristocracy in favor of land grants to form communities of small farmers.[34] Third, these grants became a means of government financing, as settlers were expected to improve the land over time and sell its produce, thus generating tax revenue, or pay government for it outright.[35] New England colonists assumed that land could be sold like any other personal property, and that the land was held in "freehold" — that is, with the owner having the rights to whatever resources it might contain. These principles also carried with them some of the seeds for revolt, however, as the revocation of a corporate or Royal charter technically voided all derivative land titles and the land would revert to the Crown.[36]

With the end of the Revolutionary War, movement westward, which never really halted, began to pick up speed. As they moved, settlers encountered a number of different land regimes. Until the federal government took control of the process, America, as we have seen, was sold off by financially strapped States or well-lobbied legislators to land speculation companies, big and small, and to wealthy individuals. Some of these speculators were successful, but most were embarrassing failures.[37] The new federal government found early on that it needed to sell land directly as well as to control the survey process — no one would buy land that had an uncertain title.

Surveying was a respectable skill in both pre- and post-Revolutionary America. Washington himself practiced surveying, as did Lincoln and Thoreau, who had a notice posted around Concord, Massachusetts, advertising his services.[38] Land surveys go back to earliest times as a necessary adjunct to tax collection. Surveying was a recognized profession under the Romans, who laid out lands on straight lines and right angles. This practice largely died out over the centuries as surveyors leaned more to irregular measurements based on land and water forms, as well as other natural objects.[39] This was called the metes and bounds system.[40] In England, the Domesday book, which listed William the Conqueror's lands after 1066, adopted the existing Anglo-Saxon method for organizing territory. This entailed the idea of shires, or what we would today call counties,[41] divided into towns or townships. This method of organizing land was carried over into early America in the northern colonies; the southern ones had a slightly different structure based on parishes, a term still used in Louisiana.

As early as 1643, Massachusetts was divided into four shires, and the representatives of the towns gathered to govern each shire. The New England colonies grew by the addition of towns, and the process for creating new towns was similar in each colony. A group of people, usually about a hundred, would petition the government for a grant of land for a new town. The land had to be surveyed and partially cleared before the land was turned over. The grant was often fairly large, but to discourage speculation, conditions were imposed as to the number of houses that had to be built, the creation of a church congregation, a meeting hall, and a school. Land also had to be set aside to support these purposes. The acreage was not entirely used for these purposes directly, but could be sold or rented and the proceeds used. This practice became part of the Northwest land survey in later years.

After the survey, which reintroduced the Roman method of regular, straight lines, the initial settlers became the "proprietors" — the first government of the town — and could allocate land to each person. They also made sale and purchase decisions and decided boundary disputes. In keeping with the corporate undertones of New England settlement, taxes were called "subscriptions," and the proprietors had the duty to collect them.[42] In the Southern colonies, the system was reversed, in that one got a warrant from the landowner, perhaps the government, laid out the desired plot, and then had the land surveyed, creating a patchwork quilt of settlement.[43] As well, these colonies generally used the older metes and bounds method whereby property lines were determined by reference to topographical features such as rivers, hills, rocks, and even trees. This led to all kinds of disputes, especially after a tree fell, a rock moved, or a river changed course. More complications might ensue if the survey was made with reference to magnetic north rather than true north, as the shifting of the North Magnetic Pole over two or three centuries could throw off old surveys by yards and maybe a mile. Such a system was more useful for large plantations, where the gain or loss of a few feet was not important.

The Seven Ranges as a Model for the West

In January 1785, a group of American commissioners succeeded in getting some Indian tribes to cede lands in Ohio. The intent was to lessen the tension between squatters and the Indians.[44] Then, in May, Congress passed the Land Ordinance, which dealt with the sale of federal lands in the Northwest Territory.[45] Besides confirming the land grants made to soldiers and those made by Virginia to settlers on the north bank of the Ohio River, the Ordinance called for a survey of an initial part of the territory in what is now southeastern Ohio, preparatory to its sale. The land was to be surveyed in the New England style,[46] making the English mile the official standard and using Gunter's chain, a measuring device developed around 1620 by an English mathematician, Edmund Gunter, that is 66 feet long and allows for the regular measurement of straight lines across land; a mile is 80 chains, and 10 chains on a side of a piece of property measures 1 acre.[47] The land was to be surveyed in regular squares called ranges, made up of townships, each made up of 36 sections of 1 square mile, or 640 acres, and largely ignored the topography. Later, this became known as the Public Land Survey System (PLSS), and is the reference grid for most property identification in the United States today. See Figure 4. Under this scheme, shown in Figure 5, a baseline runs east-west and a meridian north-south; range lines parallel these at regular intervals.

Figure 4: States included in the PLSS

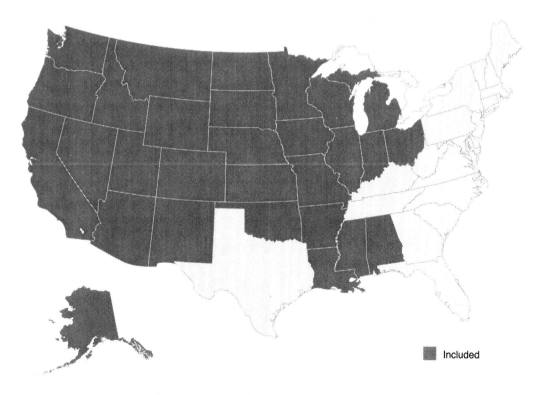

Source: www.nationalatlas.gov/articles/boundaries/a_plss.html

Figure 5: The PLSS System

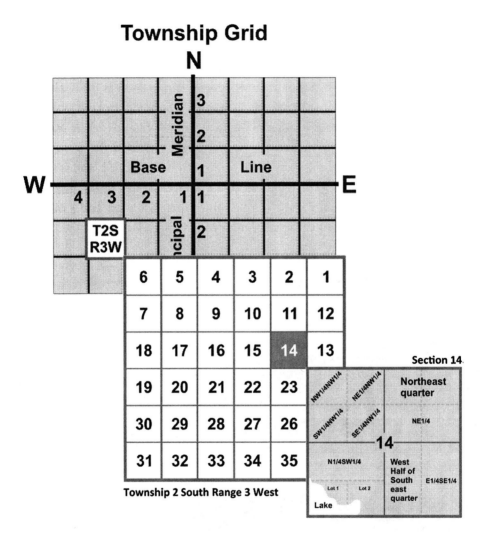

Source: www.nationalatlas.gov/articles/boundaries/a_plss.html

The eventual process of surveying the more than 500 million acres of the old Northwest Territory was begun by Thomas Hutchins, an American career officer in the British Army who had mapped a considerable portion of the interior of the North American continent before the Revolution. Accused of treason by the British, he resigned his commission in 1780 and fled to the American lines, where, in 1781, Congress declared him "Geographer of the United States." As the war wound down, he was called upon to develop maps and lead survey teams.

In Ohio, the federal government moved quickly. In late September 1785, just four months after the Land Ordinance was passed, Hutchins led the first survey team to the Ohio River area just downstream from Pittsburgh and just inside the Territorial boundary.[48] Hutchins began laying out a baseline extending forty-two miles to the west, while his team began surveying an area to its north and south, called the Seven Ranges, into geographically uniform square units, generally ignoring topographical features and interruptions by hostile Indians.[49]

Meanwhile, in 1786, Massachusetts investors along with some of the land surveyors who were active in the Seven Ranges, copied the broad intent of the Plymouth Company model of the 1600s and that of the Susquehanna Company formed by Connecticut in the 1760s, and created the Ohio Company, even though only a few settlers had crossed over the Ohio River from Kentucky.[50] Its founders offered shares at $125 in gold or $1,000 in "Continentals." Shareholders would get a "city lot" and other land in proportion to the amount purchased by the Company from the federal government.

In July 1787, Congress passed the Northwest Ordinance laying out the governance of the area,[51] and in October of that year granted the Ohio Company the right to buy 1.5 million acres just to the south of the Seven Ranges at a price of a dollar an acre, with a price reduction of one-third to compensate for "bad" land. Because the Company was allowed to pay with federal warrants for land originally given to soldiers and others and with government securities that speculators had bought at steep discounts, the land actually sold for about 8 cents an acre.[52] In general, such land development companies attempted to sell off affordable forty- or eighty-acre pieces at around a dollar an acre, providing a tidy profit if settlers could be induced into the country.[53]

Despite tensions between whites and Indians and sporadic attempts by the British to expand their influence in the area, the Ohio Company sent a group of forty-eight men to the Ohio river border of the Northwest Territory. There, they established Marietta, the first permanent American settlement in the Territory and its first capital. The families of the men arrived later in 1788, as did the Territory's first governor. Other groups of settlers from New England also decided to make the long journey across Pennsylvania to the Marietta area. The Ohio Company sold them land surveyed in New England township style, and in keeping with the Seven Ranges survey, and promoted schools, roads and townsites.[54]

The Associates intended their Company to settle the area in "a regular and judicious manner" and that it would "serve as a wise model for the future settlement of all the federal lands."[55] Time would prove this sentiment right, as the experience of Kentucky, with its absentee landlords and metes and bounds surveys, generated unhappy marginal squatters and tenant farmers. Many, like Abraham Lincoln's father, were pulled across the Ohio River by the promise of better opportunity. They either bypassed southern Indiana or remained there only a relatively short time, as its landholdings bore some resemblance to those in Kentucky.[56]

The Western Reserve

When Yankees flooded into northern Ohio and Michigan after the opening of the Erie Canal in 1825, they found two different land regimes waiting for them. With the exception of the Western Reserve and the Firelands, the rest of the land back from Lake Erie was owned by the federal government, which was also the land vendor.

The Western Reserve has a checkered history.[57] Connecticut claimed that under its Charter, it owned all the land in a sixty-mile swath westwards to the Pacific Ocean. Of course, they were unable to contest the right of New York to the whole of the Hudson River valley, but they did create the Connecticut Land company to settle the portion of the Wyoming country to the west of New York and fought with Pennsylvania for jurisdiction until the Revolution, when a settlement was made giving the Connecticut landholders deeds to their property, but placing them under the jurisdiction of Pennsylvania.

After the Revolution, Congress allocated part of the lands in the then Northwest Territory along the bounds of the Connecticut Charter claim for the state to sell and settle as its property. The land could not be settled until 1795, when Indian title was extinguished through a treaty. The Connecticut Land Company then immediately bought it from the state of Connecticut for about 36 cents an acre,[58] and in 1796 sent a team led by one of its shareholders, Moses Cleaveland, a state assemblyman and general of the state militia, to conduct a survey.[59] The land had already been surveyed by federal surveyors, but the Company claimed it had the right as part of its purchase, to carve the land up in five-square-mile blocks, not the six-square-mile standard used by the federal government. Presumably, this would give them more pieces to sell, and the settlers would not know the difference.

After going across the rivers of western New York and around Niagara Falls, Cleaveland reached the Cuyahoga River and found an area that would make a good townsite.[60] The survey party found much of the land to be swampy and unhealthful, however, and a number of the party died of fevers, including malaria. The party surveyed its township blocks, and the land was offered in these large parcels, the owners subdividing them as they pleased, though all the towns were begun around a typical New England square. Cleaveland returned home to Connecticut after his survey and never returned. His only other legacy was the name of his town, which was shortened to Cleveland in the 1830s when a newspaper editor found his masthead had one too many letters and eliminated an "a."

The Western Reserve population grew slowly until the opening of the Erie Canal. When the state sold the land to the Connecticut Land Company, the funds had been payable in part from the expected receipts from land sales, but by 1800, only just over a thousand people were living in the Reserve and receipts amounted to only 6 percent of the sale price.[61] The Company failed in 1809,[62] and the uncertainties that led to the War of 1812 further hampered settlement. Until after 1825, the population of Ohio was heaviest in the southern part, as it was in Indiana and Illinois.

Federal Land Sales

The era of giving Northwest Territory land over to large-scale speculators wound down quickly. After the failure of an early land speculator in southern Ohio and the turmoil this caused for settlers, the federal government decided to stop granting tracts of land to speculators and developers and to get into the retail side of land sales.[63] At the urging of the Ohio Company, which was concerned for its own stability and reputation, and with the assistance of Thomas Jefferson, among others, Congress moved in 1787 to open up land sale offices of its own, holding auctions in cities in the eastern States. Because these were awkward places for small buyers to go, land offices began to be located after 1800 near the land for sale, and a fixed price system replaced auctions.[64]

The Seven Ranges, as noted above, was the initial survey of what was to include most of the land area of the United States.[65] Each township was to have thirty-six sections of land, for sale by local land offices.[66] At first, settlers could purchase a section, or 640 acres (one square mile) from the government: it was not until 1800 that a minimum of 320 acres of land could be bought on credit.[67] In 1804, the minimum purchase was lowered to 160 acres, finally falling to 80 acres in 1817.[68] Federal sales started at $2 cash an acre, but while pressure to lower the price was ongoing, cash sales (and eventually credit) remained a constant feature. Private landowners, in contrast, had long been offering land for cash, but also in kind, taking hogs, cattle, and potash as partial payment.[69] The proceeds were divided, with some going for local roads and schools and the rest given over to the federal treasury.[70] Land sales became a big part of federal revenues after that, and political debate raged around whether the land was to be filled quickly or used as a revenue base for government. It was possible for speculators to acquire acreage,[71] but their proceeds would be low until all the neighboring land had been sold, so they tended instead to concentrate on laying out towns and selling lots in them.

Once the issues of price, size of minimum purchase, and the location of sales transactions were dealt with, land sales rose from fifty thousand acres a year in the late 1790s to between three hundred thousand and a million acres a year until the 1830s. After the opening of the Erie Canal, sales jumped to a yearly average of seven million acres.[72] John Quincy Adams complained in the late 1820s that his dream of using land sales as a way to finance federal internal improvements had failed and that Congress was determined to populate the land rather than generate cash.[73] A decade of prosperity and easy money contributed to a boom that ended only with the Panic of 1837. During that time, through the proceeds of these land sales, President Andrew Jackson managed to eliminate the federal debt, for the first and only time in American history.

Michigan was a classic case of the American land process at work.[74] The War of 1812 was hardly over when federal surveyors began laying out a prime meridian (1815) and a baseline (1816) in the territory.[75] As Indian titles were extinguished, the surveying task expanded.[76] The surveyor general reported to Congress that the territory was mostly swamps, so although Congress had cut the minimum purchase back from 160 acres to 80 in 1817, and land was being offered in 1818 for $2 an acre,[77] the price had to be cut to $1.25 in 1820 for lack of demand.[78] This changed when the Erie Canal opened in 1825. In 1826, steamships advertised passage from Buffalo to Detroit, with stops, in thirty-seven hours, and the first real wave of settlement arrived from western New York.[79] In the boom of the 1830s, land sold at a fast clip, with 4 million acres being taken up in 1836 alone, the last year before the Panic and subsequent depression slowed migration. By that time, however, the population had grown sufficiently for Michigan to gain statehood.

Federal land policy was always in the process of change. Congress was under constant pressure to shrink the size of the minimal acreage allotment to individuals and to alter the terms of payment, either to accept specie ("hard cash") or bank scrip for payment. Even the destinations of the proceeds changed, as Congress passed laws such as the *Morrill Act* subsidizing land grant colleges in each State. States such as New York, where the public land was gone, received lands in a western territory, or at least the revenues gained from their sale.

As settlers moved farther west into drier lands, the acreage needed to sustain a farm family increased, and this was reflected in larger acreage allotments. Congress gave land free to people who would emigrate all the way out to Oregon, which created a land rush to the Far West.[80] Water rights entered the picture as well, since the surveyor's straight lines could make some land extremely valuable and surrounding acreage worthless. The livestock "wars" that are a feature of Western novels and movies came out of this clash of rectangular surveys conflicting with the location of water.

A large complication arose as the railroads, whose initial capital needs were heavy and whose credit before construction was shaky, appealed to state and federal governments for assistance. One way of helping them without affecting the short-run prospects of government treasuries was to provide land grants of acreage that could be pledged or sold to raise more capital.[81] Sometimes the land would come from alternating blocks on either side of the proposed route, but land was also granted far from the railroad line, especially where it passed through deserts and over mountains. Once the railway had its grant, it could opt to sell the land or keep it and borrow against it. Selling railroad land became an interesting business in the Midwest farmlands as well as across the border in the Canadian prairies. Land-rich railroads developed promotional campaigns in Europe to attract immigrants, and subsidized their moving costs to the new country and the new land. In this way, they not only sold land but also filled it up with customers. Scams also occurred. Sometimes, directors of a railway would sell all or parts of the land grant to a shell company they themselves owned at a low price. The shell company would then lay out town sites along the proposed railroad route every sixty miles or so. The directors would make large profits selling town lots and local farmland, while cutting railroad shareholders out of the benefit. If a town began to boom in anticipation of the railroad's arrival, the expected line could be shifted a mile or two onto railroad land, and the townsfolk left with their bubble pricked.[82]

In the end, most of the usable land in America west of the original States was sold off rectangular piece by rectangular piece.[83] In the North and much of the west, the surveyors came first and the settlers followed. The lonely wagon carrying a family to lands they marked out for themselves was, outside the South, a myth.

Niagara Falls, New York: Mr. Love's Model City

This town does not actually exist, but it might easily have.
— Rachel Carson, *Silent Spring*, 1962[1]

I saw entire families in inexplicably poor health. When I walked on [the filled-in] Love Canal, I gasped for air....My eyes burned. There was a sour taste in my mouth.
— Michael Brown, *Niagara Falls Gazette*, 1977[2]

Niagara Falls

The environmental movement both grew up alongside the conservation movement and, in the 1960s and 1970s, absorbed it. Instead of a focus on wilderness and preservation, the movement developed out of a growing suspicion that Americans were beginning to foul the human "nest" — and for that matter so was everyone else. The growth of that suspicion can be seen through three stories, that of utopia and hell at Niagara Falls, Rachel Carson and the "silent spring," and

Robert Goddard and the "big blue marble." The three do not follow a neat progression through time, but tend to intersect and weave around each other.

Like many others over the past two centuries, my first impression of Niagara Falls was a mixed one. The Falls are impressive in themselves, especially the Horseshoe Falls on the Canadian side, but the scenery around them was a collection of tourist traps and unsightly factories and houses. It was 1960, and I had been invited to come to the area by a college friend who lived on Grand Island, upriver from the Falls.

The area was first settled on the Canadian side in the 1780s by those whom Canadians call "Loyalists" and Americans call "Tories." These were people who had to flee their homes and jobs as the Revolution generated political extremes. At the beginning of the Revolution, there were two political parties: Whigs and Tories. By the end of it, there were those loyal to the new government and those loyal to the British Crown. Some of the Loyalists made their difficult peace with the new government and stayed on, but many others left as refugees for Nova Scotia and Lower and Upper Canada (now Quebec and Ontario). Some, like Thomas Edison's ancestors, returned in later years to the United States, while others made their careers in the British Empire.

Though there were "unofficial" settlers on the American side of the falls, the land along the river was put up for sale by New York state only in 1805. A surveyor, Augustus Porter, established a number of small businesses along the riverbank. Then, with the completion of the Great Western Pike and especially after the opening of the Erie Canal in 1825, tourists came in increasing numbers. Their presence encouraged the construction of accommodations, restaurants, and added attractions to see and do when the impression of the natural wonder of the falling water started to lose its edge. Since then, the attraction has generated millions of images: pictures, paintings, stories, plays, and movies.[3]

The defense of the Niagara frontier by the British in the War of 1812 was one of the central military events of that inconclusive war, though overshadowed by Oliver Hazard Perry's victory on Lake Erie and Andrew Jackson's defense of New Orleans. Porter's enterprises and a dozen or so homes were destroyed by British raiders, who also leveled the villages of Buffalo and Black Rock farther upstream.[4] After the war, Porter continued to add to his lands, acquiring Goat Island, which separates the Horseshoe Falls from the American Falls. Porter refused to develop the lands adjoining the Falls, however, desiring to keep it in its natural state. His family continued to respect his wishes until the 1880s, when the land was taken over by New York State.[5]

The success of the mills at Waltham and Lowell, Massachusetts, set entrepreneurs to searching for falling water to power their mills. Niagara was seen as the mother lode — too much so, in fact, as little of its power could be harnessed effectively. The village on the American side was named Manchester, in the hope that it would emulate the industrial growth of its British namesake, but it was not to be.[6] An attempt was made in 1853 to get the state government to build a power canal to Lewiston, near the lower Falls, and a ship canal to connect Lakes Erie and Ontario to compete with the Welland Canal on the British side, but nothing ever came of it.[7] As late as 1860, when Toronto to the north had 55,000 people and nearby Buffalo 80,000, there were only 3,000 permanent residents in the village on the American side.[8]

Thinking Big: Tourism and Industry

The railroad age, including the completion of the transcontinental line, opened up a variety of tourist alternatives to Niagara. A number of attempts were made to reestablish its prominence as the century wore on. In 1861, Pope Pius IX declared it to be a pilgrims' destination.[9] More mundane attractions were also tried. In 1872, a local promoter staged a reenactment of an Indian burial and buffalo hunt. He hired Wild Bill Hickok and Buffalo Bill Cody to transport Indians and live buffaloes from the Midwest and to act in the extravaganza. P.T. Barnum also recognized the possibilities of the place and tried to purchase Goat Island as a permanent site for his big top. Although Barnum was unsuccessful in his enterprise, others succeeded in institutionalizing the circus at Niagara.[10]

Even so, the prospect of factories lining the shores of the river, the high fees charged for viewing the Falls, and the persistent hawkers detracted from its appeal. Americans and Canadians interested in preserving the romantic views of the Falls were united in trying to get their respective governments to clean up the place. A young Englishwoman, Isabella Lucy Bird, who visited Niagara Falls in 1854, directed her comments first to the American side, where she saw "an enormous wooden many-windowed fabric, said to be the largest paper-mill in the United States. A whole collection of mills disfigures this romantic spot." On the Canadian side, she found something equally unattractive: "a great fungus growth of museums, curiosity-shops, taverns, and pagodas with shining tin cupolas."[11]

After the Civil War, New York state made an attempt to restore the site by removing industry from the banks of the river on the American side at the behest of a group of businesspeople and conservationists led by Frederick Law Olmstead.[12] Progress was slow until 1883, when Governor (and soon President) Grover Cleveland signed a bill allowing the state to expropriate the land at the Falls.[13] The Niagara Reservation was created in 1885,[14] and Olmstead and others began to create a natural park along the river and on Goat Island. The design of the Reservation resembled the parks, such as those in New York City and Buffalo, that had made Olmstead famous. The aim was to make the Reservation free of charge, the movement pressuring for the Reservation being called "Free Niagara." Conservation was linked to democratic cultural improvement, as anyone from the cities of the East could come and enjoy the Falls in their natural surroundings. Natural, that is, according to the designs of Olmstead and his partner Calvert Vaux, the two having reunited for what Olmstead called the landscape artists' greatest professional challenge.[15] For the working class, instead of drinking and carousing, there would be the uplifting and reforming experience of a grand natural attraction in an ideal setting.[16] The free park would also serve to raise adjoining property values, as Central Park had done in New York.

The opening of the free park created a tourist boom in Niagara Falls (no longer called Manchester). It restored the image of the Falls as a natural wonder and established the town as a leader in landscape planning and civic policy reform. It set a precedent as the first use of "eminent domain" to claim lands for natural scenery. But it didn't last, and a century later the "reservation" had statuary, concession stands, and even a helicopter tour business located on Goat Island.[17]

Concurrent with the establishment of the Reservation came a resurgence of interest in the late 1880s in the Falls as a location for industry, this time using the falling water to produce electric power to operate factory machinery. The riverbanks would not be needed for this development, as power would be generated in a tunnel bypassing the Falls, while still using the drop between Lakes Erie and Ontario to power dynamos to generate electricity. It was estimated optimistically that the power potential of the Falls equaled all the power produced and used in all American factories in 1890[18] — thinking big was the order of the day. The tunnel was to be driven through the rock of the escarpment from above the Falls to below the lower rapids, and the electricity thereby generated would be sent to local homes and new factories.

Problems, however, abounded. The technology of placing turbines underground inside a tunnel full of flowing water was little tested. The amount of power created would be far in excess of local needs and would have to be transmitted to Buffalo, then considered a long distance away for electricity transmission, to make the project viable.[19] And consumer demand for the power had to be jump-started by cooperative city officials substituting electric streetlights for gaslights. Mere details: the lure was that Niagara could become the manufacturing hub of the greatest manufacturing region on the planet.

It all got done, of course.[20] Through the 1890s, turbine technology improved so that the stress of huge volumes of water would not cause them to buckle and disintegrate, in 1895 Nikola Tesla's demonstrations of the utility of alternating current sent at high voltages overcame Edison's reputation and opposition to alternating current,[21] and the electric lights went on in Buffalo to greet those who journeyed to the 1901 Pan American Exposition there.[22]

There were less practical dreamers, too. In the late 1880s, a mysterious stranger, said to have come from the West,[23] appeared in Niagara Falls. His name was Henry P. Love, and he was taken with the power potential and the possibilities for land speculation. In 1893, Love proposed to build, to the east and north of Niagara Falls, a new "model city" of 700,000 inhabitants[24] along a seven-mile navigable ship canal passing by a downstream generating station that would produce clean power for both factories and households. It was to be a planned community, possibly resembling the one built in the 1880s by George Pullman, the sleeping-car magnate, for his employees in the south of Chicago. It would boast worker-owned cooperative industries and a university.[25] Love claimed to have a $25 million fund established to finance the venture.[26] He was a skillful promoter: one of his gimmicks was to rewrite "Yankee Doodle Dandy":

"Everybody's come to town,
Those left we all do pity,
For we'll have a jolly time
At Love's new Model City.[27]

They're building now a great big ditch
Through dirt and rock so gritty
They say 'twill make all very rich
Who live in Model City.[28]

Love got a charter of incorporation for his "Modeltown Corporation" from New York state in 1893, which allowed him to construct and operate just about anything he wanted, and to access water from the Niagara River. He began to travel to raise capital for the project and to acquire options on as much as 30,000 acres of farmland east of Niagara Falls. He made regular announcements of companies moving to Model City to take advantage of the cheap power soon to be available. His company began laying out streets and roads for Model City, with all the appropriate publicity, and began to cut the canal from the river toward the escarpment. Though only a few houses were built, Love, a teetotaler, refused all requests to permit the establishment of a saloon inside the city limits.

The Panic of 1893 then hit, and over the next three years Love's credit sources dried up, along with his dreams, though he continued to publish a newsletter about the project until 1895. His canal lay unfinished, extending some 3,225 feet in from the river and was as much as 30 feet deep.[29] Love denied that the credit squeeze had affected his plans, but by late 1895 New York City investors had taken over the project and it ground to a halt. The project was killed for good with the signing of an international protocol in 1906 that regulated how much water both sides could extract from the river,[30] as the existing power company was already using most of the American share.

Eventually, during the Second World War, a considerable portion of the land Love had optioned was taken over by the military for a giant munitions plant and a dumpsite for chemicals, radioactive materials, and discarded ammunition and shells. All that remains of Love's dream today are Model City Road, which runs to the east of Niagara Falls, past a few commercial establishments that use its name, and, miles to the south, the site of the unfinished canal that, ironically, now bears the name of its utopian promoter.

Another dreamer about the potential of Niagara was King Gillette, the descendant of Yankees who came through New York, then Michigan, and then on to central Wisconsin. Gillette grew up in Chicago and was a bit of a failure at a lot of things. One day in 1895, so it was said, he was looking at himself shaving in a hotel room when an inspiration hit him: there must be a market for a razor whose blades could be replaced easily. This would avoid the need for men to be stropping their razors continually to keep them sharp. Make the blade separate from the razor, and when the old blade lost its edge, just replace it with a new one. Much like Eastman's idea about cameras, pictures, and replacement film that caught the popular imagination about the same time, the razor would be sold cheaply and the blades would provide most of the profit. Gillette pursued his idea and prospered mightily from it. He became, as one writer noted, the architect of the disposable society.[31]

Even before his name became synonymous with shaving, Gillette, like others at the time, was concerned about the state of the future, given the labor and other unrest he saw about him, as well as the waste he observed in competitive capitalism.[32] Utopian Yankee Edward Bellamy's classic novel, *Looking Backward*, had been published in the late 1880s, inspiring Bellamy Clubs all over America. As well, other thinkers, from H.G. Wells to hack science-fiction writers, were all trying to suggest social formulas that might produce utopian harmony and peace to the developed countries of the world.[33] In this context, and as Love's dream faded in the credit crisis, Gillette wrote his *Human Drift* (1894), in which he posited a North American utopia that would gather all the people north of the Rio Grande into one enormous, electrically powered city called "Metropolis," which would cover a rectangle 140 miles across and 35 miles wide centered on Niagara Falls. The Falls would be dried up as all the water flowing through Metropolis would be used to power and provide water for the thousands of great towers that would make up the city. The rest of the continent, except for farms, forests, and mines, would be given over to wilderness for the recreation needs of the citizens of Metropolis.

All business functions would be combined into one giant "trust,"[34] which would be owned by everybody equally, cooperation and centralization being the keys to a fair and efficient society. Formation of the World Corporation, the title and subject of a second book in 1910, would "determine the great dividing line between the reign of brute and the reign of soul. It is the triumph of mind over matter and the birth of divinity in man….Life will be worth living. Heaven will be on earth….It is not a vision of the future, it is a vision of now. It is at our very door, and the door is open. The dawn of a new era streams across the threshold and lights the pathway to the future."[35] Gillette's inspiration seems to have been the electrically powered[36] "White City" of the 1893 Columbian Exhibition in Chicago, since at one point he refers to Metropolis as a fully developed White City and a perpetual World's Fair.[37]

Nobody took him seriously. Even Theodore Roosevelt, after he left office in 1909, turned down a huge salary from Gillette, who wanted him to become the first president of the World Corporation that would be destined to take over all businesses and also run Metropolis.[38] In the end, the name of his proposed city was Gillette's only legacy, becoming the title of a famous 1927 film by Fritz Lang about an urban revolt against a dictatorial government as well as the title used since 1939 in Superman comic books for the large city where most of the action takes place.

Behaving Badly

As the twentieth century wore on, the demands of war, the growth of local industry, and the increasing electricity needs of distant populations meant the continual expansion of the internationally agreed limits on how much water could be diverted from the Falls.[39] The most egregious development came about in the 1950s, under the leadership of New York public works czar, Robert Moses, when just about every interest other than power production, including an Indian reservation, was pushed aside.[40] But power was not the only product of the region.

The companies attracted to the Niagara area by cheap electric power created a pernicious legacy. First to come, in 1892, was an aluminum smelter. Next came chemical industries, developed out of discoveries in German and American labs, which provided products based on the transformation of petroleum, wood, and minerals through electric power.[41] By 1914, 11,000 jobs in Niagara Falls were based on electricity, and by 1940 the city had become the world's largest producer of electrochemicals. Eventually, Niagara Falls electric power was also being used to make the uranium products to feed America's atomic-bomb-making efforts.[42]

But, just as J.D. Rockefeller's refineries did in Cleveland thirty or forty years earlier, the industries of Niagara Falls just dumped their waste into the river or burned or buried it.[43] Another handy site, not just for solid waste, but also some radioactive materials and liquid industrial waste, was Henry Love's uncompleted canal. As early as 1920, the canal site was condemned at auction and used for a municipal garbage and industrial waste dump. During and after the Second World War,[44] some two hundred kinds of chemical waste, including twelve carcinogens and two hundred tons of a chemical containing dioxin, were dumped there.[45] Where the canal meets the Niagara River, a large landfill dump was created, and waste dumping along the upper river even helped to build up the bank for potential real estate use.[46] Yet postwar developers of an expanding Niagara Falls region either did not know or did not care what went into the old canal site.[47] Once urban development reached the area in 1953, its owners, Hooker Chemical, covered the site with the same kind of clay liner that Love had used sixty years earlier, then donated it to the city for a dollar, in return for which the company was given absolution for any responsibility for personal injury on the site.[48] Other, more northerly parts of the old Model City also became dumping grounds for nuclear waste, deadly in the immediate term, as well as for munitions, casings, and other noxious castoffs. The whole area became a mockery of Olmstead's vision of a natural reserve and Love's dream of a utopian city for people to enjoy happy, healthy lives. A disaster was waiting to happen.

An American Valkyrie

In Norse mythology, the Valkyries were spirit-women who chose those who were to die in battle and afterward took them off to greet the god Odin. Rachel Carson, an ex-bureaucrat and writer, pointed Valkyrie-like at those at risk in her 1962 book, *Silent Spring*.

The essence of Carson's thesis was that American chemical producers, government officials, and consumers were spreading around chemical agents, particularly dichloro-diphenyl-trichloroethane (DDT), that were poisoning, first, the birds (whence the "silent") and animals of the land and, more slowly and subtly, people themselves. DDT was synthesized in 1874, but its power as an insecticide was not known until the Second World War. Its first mass trial took place in Naples, Italy, in 1944, shortly after its liberation by Allied troops, when, during a typhus epidemic caused by lice, more than a million Neapolitans were deloused in a mass spraying campaign.[49] Since then, agents such as DDT had become a quick, technological fix to rid areas of unwanted bugs and weeds, but they were harming other parts of the biosphere. There was no subtlety in Carson's evidence, her approach, her cases, or her conclusions: we either stop using chemicals such as DDT or we eventually will shorten our lives and create a toxic nightmare that we might not be able to clean up.

Carson was born in 1907 and raised northeast of Pittsburgh.[50] She became interested in nature from her life on the family farm and through the example of her mother, who taught her three children to take an interest in the world around them. Carson was a loner, preferring to read and write, and her first published story came at the age of ten. She went to the Pennsylvania College for Women in Pittsburgh, majoring first in English and then biology, probably switching into the latter because of the influence of a biology professor, Mary Scott Skinker. After graduation, she spent a summer at the Marine biology Laboratory in Woods Hole, Massachusetts, then went on to Johns Hopkins University in Baltimore for graduate work, where, in 1932, she received a master's degree in zoology; her thesis was on an aspect of fish biology. She could not afford to continue for a PhD, however, as her family's fortunes had worsened with the Depression. Professor Skinker helped her get a job as a writer with the US Bureau of Fisheries, and by 1936 she had become only the second woman to be given a permanent job in the Bureau, as an aquatic biologist.

During the Second World War, she published a book about the undersea world and numerous articles for national magazines. By 1949, she was the chief editor for publications in the Bureau of Fisheries' successor agency, the Fish and Wildlife Service. Her 1952 book, *The Sea Around Us*, won a number of awards and was turned into a documentary, which won an Oscar, although Carson objected that the script was false to her message. The book's success, however, enabled Carson to leave government and devote herself full time to writing. By 1956, after another successful book, *The Edge of the Sea*, Carson began to take an interest in the environment. Sparked by the federal government's use of aerial spraying to eliminate fire ants, despite opposition from landowners and groups such as the Audubon Society, Carson became concerned about the environmental threat from the overuse of pesticides such as DDT. An alliance with the Audubon Society led to a four-year project that resulted in *Silent Spring*.

The book was meant to question the idea, common after the war, that scientific innovation could be pursued without negative effect. Carson's arguments met with considerable criticism from chemical producers and scientists,[51] but the almost simultaneous thalidomide tragedy, where deformed children were born to mothers who had taken the drug, seemed to vindicate her. The selection of *Silent Spring* for the Book of the Month Club and a CBS television broadcast featuring Carson took her message to millions and led both Congress and the president's Science Advisory Committee to investigate and report to the public. Ironically, even as she was researching links between synthetic pesticides and cancer, she was diagnosed with breast cancer and died from the disease in 1964.

Carson's facility with words and her technical background gave her credibility when it came to explaining concepts in ways the average person could understand. In particular, she drove home the realization that nobody was safe from the impact of sprayed chemicals; anyone, rich or poor, black or white, urban or rural, could be affected simply by the wind currents and the impreciseness of the technique. Further, it was not just corporations that were regarded with suspicion: governments looking for maximum environmental impact for minimum budget impact were seen as complicit as well.

The Big Blue Marble

And now for something completely different....Robert Goddard was the quintessential Yankee inventor. Born in 1882, he was raised and lived much of his life in Worcester, Massachusetts.[52] Goddard was a sickly boy who fell behind in school and did not graduate until he was twenty-two. Spending lots of time home in bed, he became a voracious reader, and was highly taken with H.G. Wells's *War of the Worlds*, which was published when he was sixteen. Already experimenting with kites and balloons, at seventeen he discovered his life's work while staring at the sky as he pruned trees around his parents' house. He would try to devise a way to escape Earth's gravity and travel through space.

Goddard went through Worcester Polytechnic as a physics major, served as a professor's research assistant, and in 1907, still an undergraduate, had a paper on airplane stability published in *Scientific American*. He graduated in 1908, and after working at the Polytechnic as a lecturer for a year, entered Clark University, also in Worcester, to do graduate work. He was bright and disciplined, completing his MA in 1910 and his PhD in 1911. He went to Princeton University on a research fellowship, though two years later he returned to Worcester to convalesce from tuberculosis.

A year later, he took a position at Clark as a physics instructor, and began a productive career that combined academic research with practical inventions. During his convalescence, he patented concepts for a multistage rocket and a liquid-fueled rocket motor. In 1916, his second year at Clark, he got a grant from the Smithsonian Institution to undertake rocket research and testing, and on the side he developed the basic ideas for the antitank weapon later known as the bazooka. He acquired quite a record for applying physics to various problems, accumulating hundreds of patents over his lifetime and posthumously gaining more as a result of the mining of his journals after his death by his wife and others.

But rocketry was where Goddard's heart was. In 1919, the Smithsonian published his landmark work, *On Reaching Extreme Altitudes*, which had an effect on those interested in the subject elsewhere. Hermann Oberth, the father of German rocketry, referenced Goddard's work in his own 1923 book, *By Rockets into Interplanetary Space*. In turn, Oberth influenced Russians such as Konstantin Tsiolkovski, who then republished his own speculations and calculations on rocketry that he had written at the turn of the century. The US domestic press, however, had a field day making fun of Goddard's work, claiming he did not understand basic physics.[53] The experience convinced him to keep his work unpublicized.

Throughout the 1920s, Goddard worked on his engine and launched his liquid-fueled rockets on test flights, the first in 1926 from a site on his aunt's farm in Auburn, south of Worcester, and now a historic landmark.[54] My wife and I had fun finding it: GPS and electronic maps showed it to be in the middle of a suburb, but it's a four-foot obelisk in the left-side rough at the dogleg of the ninth hole of the Pakachoag Golf Course, the successor to his aunt's farm. There was no landmark at all until Werner von Braun spoke in Worcester in 1957 and asked Goddard's widow to show him the site. Upset that there was no recognition, von Braun lobbied Washington to get this modest and ignored marker.

Goddard's rockets were the first real advance over the gunpowder rockets invented by the Chinese centuries ago. By 1929, his work had come to the attention of Charles Lindbergh, who supported it and put Goddard in touch with the Guggenheim family, which, through the Guggenheim Foundation, became heavy supporters of his experiments, helping him to hire assistants and to continue his work at a research facility in Roswell, New Mexico, throughout the 1930s.

Goddard spent the war years helping the military develop a rocket-assisted takeoff propulsion system for carrier planes. There was little other interest in the use of rockets until German guided bombs began to hit Allied warships off Italy in 1943 and V-2 rockets began to rain down on Britain in 1944. The Germans, prohibited by the Versailles Treaty from developing a specific list of weapons, had turned to rocketry as an alternative because it was not on the proscribed list. The first liquid-fueled rocket in Germany was launched in 1930, and by the beginning of the war, the Germans were well ahead of everyone, not that anyone outside Germany at the time cared about this exotic device.

At the end of the war, both the United States and the Soviet Union each managed to capture their share of German rocket scientists. Goddard had died in 1945 and his contribution was neglected in the United States in favor of the German team led by von Braun. Even so, the United States made little progress in developing rockets until the Russians put a small satellite called Sputnik into orbit in 1957. The implications soon became obvious: if a rocket was powerful enough to launch a satellite, then a nuclear V-2 could be fired at the United States using the same equipment. Though the main aim of the "space race" in the 1960s was to build better nuclear-tipped missiles, President John F. Kennedy also used the effort to inspire Americans and build national prestige by promising to put a man on the moon by the end of that decade. Goddard was posthumously recognized for his pioneering work in rocketry, and one of NASA's space flight facilities was named after him.

The success of the American space effort led to a number of spinoffs, including artificial foods, miniaturization, and some very famous photographs. One of these last was taken on December 7, 1972, as the Apollo XVII mission, the last to head to the moon, caught Earth in full light, showing the planet as a "big blue marble" floating alone against the black background of space. With this picture, one of the most famous ever taken, it suddenly became clear that the terrestrial ecosystem was finite and comprehensible by humanity. Americans only grudgingly had began to accept that they might be poisoning their own nest, so to speak; now it was graphically evident that there were no alternative nests, or "new Wests." In succeeding years, concern grew about the potential effects of the thinning of the ozone layer in the upper atmosphere, including a potential rise in ultraviolet radiation, leading to increases in cancer rates. It took only thirteen years from the publication of the first scientific paper linking a thinning ozone layer to chemical compounds called chlorofluorocarbons, common in refrigeration, firefighting, and aerosol sprays, to the implementation of an international ban on the manufacture and use of these substances. We are learning, it seems, to appreciate the symbolism of the big, blue marble.

Return to Love Canal

The Apollo XVII picture was illustrative of a lot of changes of perceptions for a lot of people in the 1970s.[55] As environmental threats began to take their place alongside the threat of nuclear annihilation, people began to see that humanity's stewardship of the big, blue marble had to become more rational and responsible. And that meant dealing with the legacy of yesterday as well.

Where everything came together was at the site of Henry Love's unfinished canal. When it transferred title to the canal to the city of Niagara Falls in 1953, Hooker Chemical had warned against subdividing the adjoining land for housing. But the city went ahead anyway, approving development and building a school along the eastern edge of the canal in 1955. The school, in fact, was built without a basement because of the chemical wastes the contractors uncovered there. As well, a drainage system was constructed around the building that carried off rainwater and chemical leachates into the storm drain system and then into the Niagara River.[56] Even more inexplicably, the school playground was laid out right on top of the canal. The area gradually filled with small houses popular with younger families even as chemicals leached into the surrounding land and migrated along underground watercourses. The resulting sicknesses, especially of children, were seen as individual cases, not symptoms of environmental degradation, and treated accordingly.

By 1976, after some years of heavy precipitation, problems began to intensify as barrels of chemicals actually began to rise out of the cap placed over the canal twenty years earlier and noxious substances began to seep through basement walls and back into residents' drains. An alarming consultant's report to the New York State Health Department was ignored, as were local health concerns.[57] Then, in August, 1977, the *Niagara Falls Gazette* began to publish a series of articles by reporter Michael Brown, who noted that something was wrong in the Love Canal area and that public health was at risk.[58] Prodded by the articles and subsequent publicity, the state government finally decided to investigate, and discovered high incidences of miscarriages, stillborns, and birth defects in the area. But it now found itself in a bind: residents — who knew how many? — would have to leave, but who would pay for their relocation?

As the controversy dragged on, area residents began to organize for action and heighten the profile of the problem.[59] The mayor then criticized the activists for hurting tourism with their negative publicity. The state health department held a public meeting in June 1978 and then another in August, inexplicably in Albany, on the other side of the State. It did not make a decision about the area's safety, but it nevertheless ordered residents not to eat anything from their gardens and pregnant mothers and children under age two to be "temporarily relocated" — a half-measure that only made the uncertainty worse, since the order was not accompanied by any offer to pay for the cost of relocation. The resulting embarrassing uproar, coming just before the 1978 elections, led President Jimmy Carter to declare a state of emergency at Love Canal and the governor, only a week after the temporary relocation order was issued, to announce a permanent relocation of those in the most affected area.

As a result of continuing pressure by residents, others were moved in early 1979, and in May 1980 President Carter declared another "health emergency" in the Love Canal area and provided funds to relocate 810 families "temporarily," if they wished to move. In October, noting the residents' "mental anguish," Carter ordered the permanent relocation of all families who wished to go. Part of the anguish came from the residents' association discovering that elaborate precautions had been made for the safety of the crews who were contracted to clean up the canal site, while residents meanwhile were simply expected to stay inside their homes. By 1988, after the area had been cleaned up, some of the less-affected houses were deemed "habitable" and put on the market once more.

The controversy over Love Canal forced a recognition among Americans that they and their children were at risk from environmental problems of which they were barely aware. In 1980, the Environmental Protection Agency noted the existence of 30,000 such industrial dump sites across the country.[60] Of 336 facilities where former and existing employees were to be given compensation for exposure to nuclear and radioactive material, 13 were in the Niagara area, more than in all of New Mexico, where the atomic bomb had been developed and tested. The New York Department of Environmental Protection listed 649 sites of concern in Erie and Niagara counties alone.[61]

Niagara Falls has been the focus of great optimism about human progress and, indeed, a site of real success in the exploitation of nature's power — one of the earliest examples of electric power transmission.[62] Yet it is also the site of the environmental disaster of Love Canal. Like the Apollo photographs of Earth from the immensity of space, Love Canal brought home to us the fragility of our biological existence. It is ironic that the disaster should happen on the ruins of a utopian project.[63] In the end, Niagara still shimmers between visions of heaven and of hell.

The story of Niagara Falls teaches us two lessons. The first is that, under pressure, industry can become more efficient, substitute materials can be found, and waste reduction and recycling can be factored into business economics. The second lesson is that industrial pollution is not just a matter of the dumping of chemicals and other noxious wastes into the ground and water table; it also involves, perhaps more importantly so, the emission of air pollutants that many scientists argue are having an effect on our climate. In a sense, this evolution has merged the conservation/efficiency ethic of a Gifford Pinchot with the environmental health issue of the residents of the Love Canal area. The challenge now is to find a way to accommodate the rising economic expectations of countries such as China and India without poisoning the planet beyond remediation.

Buffalo, New York: Expanding the Yankee World through Air Conditioning

Buffalo Gals, won't you come out tonight,
Come out tonight, come out tonight.
Buffalo Gals, won't you come out tonight
And dance by the light of the moon.

— Cool White, 1844[1]

Young man, you can't stay around here if you don't apply yourself.
— Henry Wendt, managing director of the
Buffalo Forge Company, as Willis Carrier stared at the
ceiling while devising the first air conditioner.[2]

I hate air conditioning: it's a damnfool invention of the Yankees.
If they don't like it hot, they can move back up north where they
belong.
— A Florida woman to historian Raymond Arsenault.[3]

Without Freon, we'd be dead.
— A Houston resident after a 2007 heat wave.[4]

Buffalo was laid out by Joseph Ellicott, resident manager of the Holland Land Company in 1801 as the port for its western New York holdings.[5] Located near another settlement in the area, Black Rock, built on land purchased from the state by an enterprising Yankee merchant. Buffalo was exposed to open Lake Erie and would require the eventual construction of an artificial harbor. Black Rock lay somewhat to the north, at a creek mouth sheltered from the lake by the land mass of the Niagara Peninsula to the west across the Niagara River. The two settlements soon became fierce competitors.[6] The British burned both places in a raid during the War of 1812, but both were rebuilt and once again began competing for the Lake Erie trade at the end of the war. At first, Black Rock's more protected location made it the base for the first steamship on the Great Lakes, the *Walk-in-the-Water*.[7] Black Rock also lobbied hard to be the terminus of the Erie Canal, but the influence of Ellicott with the politicians in Albany led to the canal's by-passing Black Rock for Buffalo and its newly constructed harbor.

As the West opened up, Buffalo became a transshipment point for the Erie Canal and the embarkation point for the thousands of settlers moving to Ohio, Michigan, and beyond. The city grew from 2,600 people in 1824 to 42,000 in 1850.[8] By 1842, the volume of Western wheat flowing into Buffalo had become so great that it required the building of the first grain elevator. Later, to save transportation costs of bulk wheat, steam-powered flour mills were added. Manufacturing plants serving the Western markets also appeared. After the Civil War, Buffalo's rapid growth was slowed somewhat by the extension of the eastern railroads to Chicago,[9] and after 1900 growth slowed further as the city was bypassed in favor of Cleveland, Detroit, and Chicago. Even so, its manufacturers, such as the Buffalo Forge Company, found they could use Buffalo's excellent rail connections to generate work all over markets between Chicago and New York City for their products. In more recent times, Buffalo has fallen on the same hard times as the rest of the "Rust Belt."[10] In the early 2000s, the already closed Buffalo Forge facilities were demolished; leaving only a large graded empty field where the plant once stood — a sad testament to the spot where modern globalization began.

Willis Carrier

Little is readily known about Willis Haviland Carrier, "the father of air conditioning."[11] We do know he was born in 1876 on a farm near Angola, New York, just south of Buffalo on the Lake Erie shoreline. Both his mother and father were of Yankee parentage, and he was descended through his father from one of the women said to have been killed in the Puritan witchcraft trials in the late 1600s. She was married to the original Carrier ancestor who had left England at the time of the Restoration in 1663, presumably because he had been one of the more radical Puritan leaders. Carrier was supposedly an assumed name.

Later Carriers moved from Massachusetts into New York in 1799, and finally settled in Angola, just south of Buffalo, after the opening of the Erie Canal. Willis's father, Duane, attended Oberlin College for a time, was a school teacher, a merchant, and then a dairy farmer. His mother, who greatly influenced him, had been a school teacher as well, and fostered his interest in mathematics and problem solving, but she died when Willis was just eleven. He became known for his absent-mindedness as he thought about problems related to machinery. He is said to have assembled a thresher from a kit his father ordered from a mail order house, and to have disassembled and reassembled other mechanical devices around the farm.

The family struggled under the effects of the Panic of 1893, and Willis, who wanted to go to Cornell, found he could neither afford the tuition nor qualify because of the poor educational level of the local high school. Instead, he taught in a local school for a couple of years; then, his father having remarried, he was sent to live with his stepmother's relatives in Buffalo, where it was hoped he would get his secondary schooling up to Cornell's entry standards. At this, he succeeded, even getting a tuition scholarship, and entered Cornell in 1897. He completed his degree in mechanical engineering there in 1901. Carrier had also taken a number of electrical engineering courses, and after graduation hoped to get a job with General Electric, whose facilities in Schenectady were the high-tech attraction of the day. But when he was approached by a representative of the Buffalo Forge Company, he took the offer.

In 1902, Carrier patented an air-conditioning apparatus and married Edith Claire Seymour, of Gloversville, New York, and a classmate of his at Cornell. They lived in Buffalo for the next ten years, but she died at age thirty-four. Her obituary in the Cornell alumni news gives no cause of death and the couple had no children. Thereafter, the personal picture of this budding engineer gets cloudy. His only biography was written a couple of years after his death in 1950 by a female engineer, Margaret Ingels,[12] who had been an employee of his company since 1917, and then reworked by a company publicist.[13] The biography quickly veers away from his private life and into his engineering work and the corporate changes that took place over the decades.

So, what other clues can we trace about Willis Carrier? In 1914, two years after the death of his first wife, he married Jenny Tifft Martin — like him, from Angola. Tifft is an odd Rhode Island Yankee name, but also prominent in Buffalo. It is also the family name of Willis's father's second wife, Eugenia Adelaide Tifft. Willis and Jenny were technically stepbrother and stepsister. Jenny was Eugenia's daughter,[14] and was twenty-four at the time of her mother's marriage to Willis's father in 1891.[15] She had graduated from Cornell in 1889 in science and worked in Chicago for the Young Women's Christian Association between 1892 and 1895.[16] She might have been a schoolteacher for a while, and it also appears she had two children. By the time she and Willis were married, in 1914, she was forty-seven and he thirty-eight. Willis adopted the two boys, who took his last name. Sometime around 1921, with the formation of the Carrier Engineering Corporation, the family moved nearer to the company's new headquarters in New Jersey, and lived there until the company's headquarters were moved to Syracuse in 1937. Jennie died there in 1939, age seventy-two. Willis married again, in 1941, to fifty-year-old Elizabeth Marsh Wise, who survived her husband by fourteen years.

Carrier appears to have been a model citizen, a Presbyterian, and a Republican. His business associates were loyal to him all their lives and even in later life, when he no longer ran the Carrier Company directly, continued to refer to him as "the Chief." Besides his stepsons, he took in the children of late associates and, during the Second World War, provided the children of his British associates with a home and a good education. He received two honorary doctorates, from Lehigh University in 1935 and Alfred University in 1942. He and all three wives are buried in the family plot in a Buffalo cemetery.

Climate and Humanity

Part of humanity's success in expanding to nearly all corners of the globe has been our ability to mediate the effects of climate. At the simplest level, anywhere other than the subtropics and tropics, this meant clothing to keep warm and shelter to ward off bad weather. Keeping cool was a different matter. In warm climates, people tended to live in the open and to seek relief from the heat in shade and bathing in cool streams or lakes. Digging into the ground or living in a cave also could serve to keep temperatures down. In early civilizations in hot climates, servants were used to fan air over a cake of ice;[17] one Mesopotamian ruler was said to have had snow and ice transported from distant climes and packed into the double walls of his palace. If the climate was also dry, it was discovered that hanging up wet cloths and allowing the water to evaporate also would cool a room.[18] Yet, for most people, rich and poor, heat was to be tolerated, not mediated.

In the 1800s, ice became a popular method to preserve fish caught in the western Atlantic for transport to the population centers of Europe.[19] It dawned on some Yankee entrepreneurs that they could develop a profitable business cutting ice from lakes in the winter, insulating it under wood chips and sawdust from sawmills, and then transporting it to serve the cooling needs of the Caribbean as well as the growing British middle and upper classes. Cold drinks and cool air from fans blowing over ice became interesting luxuries.

With the growth of factories and urban tenements, controlling temperature began to take on increasing importance. Bakers, for example, could keep cooler by making bread in basements, but these were also places where humidity and warmth might combine to host a variety of diseases and fungi that could affect the product. Besides keeping warm and/or cool, there were other problems relating to an artificial climate. First was the problem of spreading the air around so that a wider area might enjoy the benefits. Many of the fireplaces used before Ben Franklin's innovative stove did not circulate heat far from the flames. Some kind of ventilation system was needed.

Gradually, the nature of heat and cold began to be discovered. The first text on heating and ventilating was published in Scotland in 1815. As the nineteenth century progressed, powered fans, boilers, and radiators designed for temperature control were derived from technical advances in steam engines. In 1850, Dr. John Gorrie, a Florida physician, succeeded in making artificial ice, which he used to cool yellow fever patients.[20] Artificial refrigeration became an engineering function in that decade, when a Yale University professor, building on earlier attempts by others, succeeded in using a compressor system to produce nearly 2,000 pounds of ice per day. In the late 1800s, British engineers succeeded in humidifying air in textile plants.[21] In 1889, a prominent heating and ventilating designer, Alfred Wolff, designed an air-cooling system for Carnegie Hall in New York City that has been called the first attempt at air conditioning, because Wolff consciously tried to control temperature, humidity, cleanliness, and ventilation.[22] In 1901, he convinced the New York Stock Exchange that its busy premises needed to be cooled in the summer, much as they were warmed in the winter. He succeeded in installing cooling equipment, with the refrigeration process itself powered by the waste steam from the building's electric power plant.[23]

Many of the problems facing manufacturers and urban populations around 1900 could be alleviated by fans, ducts, and other ventilation, especially as the use of electricity became widespread. Even so, simple factory ventilation might be overcome by outside heat and humidity, and ambient temperatures could rise so high inside as to affect workers' productivity or even cause them to collapse. Some factories sent workers home on especially hot days — indeed, many cities had laws requiring such closures. In the summer of 1960, I worked in a room with exposed, operating auto engines that could raise the inside temperature to 130°F. I thought taking salt pills while working was for sissies until I awoke one day in the plant infirmary, having passed out on the job.

Residents of urban tenements faced a similar situation on hot summer days, and many slept on fire escapes or rooftop to escape the heat. During heat waves, mortality among the poor and the elderly would skyrocket. Such crowded urban construction was generally out of the question in the South, where very hot temperatures were normal in summertime, and the Northern-developed manufacturing model also could not be exported successfully to Southern States.

Nonetheless, progress was slowly being made. As the twentieth century began, all of the components of air conditioning were in place — the public was first exposed to room cooling in 1904, at the St. Louis World's Fair, where the state of Missouri Building, including a thousand-seat auditorium, was cooled by refrigeration. Now, all it would take was a Willis Carrier to put the components together into one system, and the science-fiction notion of a civilization independent of climate could become a reality.[24]

Devising an Artificial Climate

Later in life, Edna Littlehaes, one of the children Willis Carrier took in after the death of her father, a Carrier associate,[25] described an incident that took place in the Carrier household in the 1930s. They were visited by a shaggy, pleasant, gray-haired man who became heavily engaged in conversation with Carrier and was invited to stay for dinner. The man was Albert Einstein. Much earlier, he and Carrier had had a similar insight about the world at about the same time — namely, that conditions were heavily affected by the relative interactions of forces. Einstein, of course, was concerned about the basic forces of energy, mass, and speed, while Carrier was focused on the more mundane concepts of atmospheric heat, humidity, and airflow.

In joining Buffalo Forge after graduation from Cornell,[26] the young Carrier was called to solve a number of knotty problems facing a company that had once produced movable forges for blacksmith shops and was now manufacturing heating and ventilating equipment for industrial facilities. Generally, this involved working on a stream of custom projects, many requiring the retrofitting of an existing facility. The day Carrier began work at Buffalo Forge, he also began a friendship that was to last a lifetime. On the Broadway Avenue streetcar out toward the Buffalo Forge plant, Carrier asked a fellow passenger if he knew at what stop he should get off. The man, J. Irvine Lyle, told Carrier to follow him, as he was going there as well. Lyle was a salesman for the company, based in New York City. At first, Carrier too was assigned to sales, but Lyle, his sales mentor, soon sent him back to Buffalo with the recommendation that he be used in product research and design.[27] Already, Carrier had begun to acquire a reputation of being absent-minded in the extreme — not a virtue in sales.

Carrier began work designing steam-heating systems for drying products such as lumber and coffee. A year later, after he had become known for his research and development of solutions that gave the company an edge on its competitors, Lyle presented him with a challenge. A Brooklyn printing firm had approached a consulting engineer in the New York area to solve a problem: the company did multicolor printing jobs, but these were complicated by the fact that, as the heat and humidity changed through the day, paper expanded or shrank, causing the colors to mix and blur. No one, including, as it turned out, the consultant, knew how to fix the problem. But Carrier came up with an insight that led to what was later called the "invention" of air conditioning.[28] In fact, like the "invention" of the telegraph, the telephone, and the electric light, the technological elements of air conditioning were already in place, but it took someone to pull them into one technique or process and then formulate a company with a vision to exploit it. And unlike Edison, Morse, and Bell, Carrier spent the rest of his life focusing on all aspects of the industry he largely created.

Carrier knew that changing the temperature of a volume of air affects how much water vapor the air can hold — cooling the air increases its relative humidity because cooler air holds less water. But keeping the amount of water vapor held by the volume of air constant and warming the air causes its relative humidity to go down because the air can now hold more water vapor. Carrier established the regularity of this relationship between humidity and temperature in a more precise fashion than even the US Weather Service did at the time. He then devised a way to manipulate the relationship to meet the conditions needed by a given industrial production process. The engineering trick lies in controlling relative humidity at different temperatures. For instance, a textile plant might require 55 percent relative humidity at 75°F, regardless of the outside temperature. To solve the problem, it would be necessary to calculate the heat produced in the plant from a variety of sources, the normal humidity in the factory when the windows are closed, and how to adjust refrigeration, the addition or extraction of water vapor, air cleaning, and proper ventilation continuously so that the proper conditions exist everywhere in the factory, all at once. Refrigeration experts already knew that compressing a fluid, either a gas or a liquid, heats it up; then, having dissipated the excess heat, allowing the compressed fluid to expand cools it down. Fanning an air supply past coils with an expanding fluid in them then cools the air (and heats the liquid). The reverse works as well. If the air is cooled until there is more water vapor in the air than it can hold, then the dew point is reached and water condenses out of the air onto a nearby surface. Draining the condensate away before rewarming the air produces dry air.

Characteristically, Carrier devised an important part of his process daydreaming while waiting on a train station platform in Pittsburgh. The heavy fog inspired him to see that, if he could saturate air with a fine spray, creating an artificial fog, he could manipulate the heat and humidity of the air. An agriculturist friend suggested that a useful attachment might be a fine nozzle used to spray flowers. Spraying would also wash particulates, pollen, and dust out of the air.[29] Adding automatic feedback controls would "condition" the air to whatever state was desired, leaving the design of ductwork to distribute the air properly as another challenge.

Working out this process in a mechanical system led to Carrier's 1904 patent application for "an apparatus for treating air" and its issuance sixteen months later.[30]

In a way, Carrier did for climate what the pioneers of interchangeable parts did for materials: develop a method to produce a consistent product — in this case, conditioned air — in fine detail.

It's not the Heat, It's the Humidity

Temperature control was an important consideration for nineteenth-century businesspeople and consumers, but the creation of more complex machinery and parts also generated a need to control humidity. Printing plants and spinning and textile mills, among other factories, were plagued with problems of product quality associated with temperature variations and humidity: too much relative humidity and the fibers got too sticky: too dry and the spun threads might separate.

Once Carrier had successfully patented his air-conditioning process, Lyle began to sell Buffalo Forge's custom engineering projects to overcome the problems of fluctuating heat and humidity — not, it should be stressed, to improve the comfort of the workers, but to improve product quality. Gradually, Buffalo Forge's systems made inroads into a diverse lot of businesses, such as printing, textiles, pharmaceuticals, film production, and even pasta drying. In all cases, the idea was to produce a product under its required ideal conditions of temperature and humidity, regardless of the season or weather outside. Buffalo Forge's owners, the Wendt brothers, were concerned, however, about the possible negative effects of this new technology on their core business of manufacturing and selling blowers and ventilation equipment. Accordingly, in late 1907 and at Willis Carrier's suggestion, Buffalo Forge created a subsidiary named the Carrier Air Conditioning Company of America (CACC).[31] From 1908 to 1915, CACC managed to get a stream of contracts for engineering, design, and fabrication activities, with Buffalo Forge providing the component products whenever possible. While continuing to work on Buffalo Forge problems and products, Carrier performed two critical roles for CACC: that of system designer for the projects and of inventor of equipment that kept the company in the forefront of the industry.[32] Then, at the annual meeting of the American Society of Mechanical Engineers in 1911, he presented a paper and a set of tables — his "Rational Psychrometric [sic] Formulae" — that set out the principles governing the relationship between relative humidity, absolute humidity, and dew-point temperature, making it possible to design air-conditioning systems that, in theory, could fit a client's requirements exactly. Buffalo Forge printed the tables as a promotional piece, and editions were still in print when Carrier died almost forty years later.

In 1915, Carrier and six of his engineering associates spun off CACC as a separate entity, renamed the Carrier Engineering Company (CEC), and within days landed a contract to air condition a fuse-making facility for the American Munitions Corporation. After the war, CEC began making its own equipment in a factory in Newark, New Jersey, where the company's headquarters were moved and where they remained until 1937. During the 1920s, the company continued to provide systems for large-scale factory operations, but by then air conditioning was beginning to be seen as beneficial for the comfort of people as well as for manufacturing processes, and it gradually was being introduced in movie theatres and commercial establishments of various kinds. In 1922, Carrier air conditioned Grauman's Metropolitan Theater in Los Angeles, the first in the world to be so treated.[33] The bargain basement of J.L. Hudson's Detroit department store was air conditioned in 1924.[34] In 1928, the Milam Building in San Antonio, Texas, became the first skyscraper to have air conditioning — by Carrier, of course — included in its design and, in 1929, Carrier air conditioned both the Capitol and the White House. In 1930, the deepening Depression caused CEC to merge with two partial competitors, a refrigeration company and a heating and ventilation company, to become the Carrier Corporation (CC). In 1937, with cash and incentives provided by Syracuse, New York, the company's plants in New Jersey were sold and its operations were consolidated in one facility in that city.[35] Meanwhile, the company's continuing quest to develop ever-smaller and more efficient systems resulted in Carrier's ability to air condition railroad cars (1931),[36] hotel rooms (the Detroit Statler, 1934), and even submarines (1936).[37]

From Luxury to Necessity

Two Depression-era publicity events symbolized the coming democratization of air conditioning. In 1933, futurist and designer Buckminster Fuller's teardrop-shaped Dymaxion air-conditioned car became a major attraction at the Century of Progress Exhibition in Chicago.[38] Then, in 1936, Willis Carrier, interviewed on a New York radio show about the future of air conditioning, replied: "At any time of the year, the average businessman will rise, pleasantly refreshed, having slept in an air-conditioned room, he will travel in an air-conditioned train, and toil in an air-conditioned office, store, or factory — and dine in an air-conditioned restaurant."[39] It might have seemed slightly fantastic at the time, but Carrier's prediction was a harbinger of the coming migration of air conditioning from commercial, industrial, and governmental buildings to the wider consumer market. Carrier, indeed, had already tested a relatively small residential air conditioner in the homes of Irvine Lyle and another company employee in 1929.[40] But first, air conditioning would have to be made safe in the hands of ordinary consumers, and units would have to become smaller and more affordable.

One of the critical elements of any air-conditioning process is the refrigerant, a fluid which, when compressed, heated, and allowed to expand, becomes cool. Ammonia had been in common use in refrigeration since the 1870s, then later carbon dioxide, but both of these could be deadly if there was leakage. In 1928, Dr. Thomas Midgely, a researcher at General Motors, developed a non-toxic refrigerant gas he called Freon. It had a variety of uses, but Carrier noted that one type of Freon had excellent properties for compression in air conditioning. Frigidaire, a GM subsidiary, licensed the product to CC. Freon allowed CC to design smaller units, which could be used to air condition houses and even automobiles — introduced as an accessory by Packard as early as 1939. Unfortunately, by the 1970s Freon was discovered to have deleterious effects on the atmosphere's ozone layer, which shields the earth's surface from the sun's ultraviolet rays, and it was gradually phased out in favor of less noxious chemicals.

By the time of Carrier's death in 1950, air conditioning was beginning to become a relatively common addition to North American living, with window-sized air conditioners finding their way into offices and homes.[41] That year, Houston claimed to be the most air-conditioned city in America.

Workin' for the Yankee Dollar

One reason the South lost the Civil War is that it was primarily an agricultural economy, and much less industrialized than the North. By the end of the war, for instance, one New Jersey factory produced a railroad engine per day; the South never had the capacity to produce any at all. Fifty years later, as electricity began to free industry from the limitations imposed by water, steam, and coal, businessmen began to look to the still unindustrialized South for pools of cheap labor. First came the textile mills, built closer to cotton supplies — the first all-electric textile mill appeared in Charleston, South Carolina, in 1894. By the late 1940s, Southern textile mills were being built without windows, as air conditioning allowed both increased production efficiency and personal comfort.[42]

After the 1920s, with foreign immigration largely choked off by Congress, other industries began looking southward. Military expenditures on Southern bases during the Second World War also contributed to the migration of industry and people south, as did the growing winter tourism industry. By the 1970s, for the first time since the Civil War, more people moved into the region than left it, and the concept was advanced of the Sunbelt, a string of States through the near South out to the Pacific Ocean that by then had become the fastest-growing part of the country.[43] And it was made possible by the widespread adoption of air conditioning. As David Shi relates, "[i]n 1951 inexpensive window units were invented, and soon thousands of homes featured dripping, humming metal boxes hanging out bedroom windows. My siblings and I in Atlanta were so dazzled by our first window unit that we showed it to our friends as if it were an exotic attraction at a carnival….Mortality rates have dropped, and economic activity has soared. Working conditions have improved, along with productivity."[44] By 1955, about 10 percent of Southern homes had adopted air conditioning (as opposed to 5 percent of Northern homes).[45] By 1960, air conditioning could be found in 18 percent of homes in the South, and today virtually no new house is built without it.[46]

But air conditioning also changed the South in more subtle ways. As Shi also relates,

> [a]long with the nearly universal ownership of televisions, the spreading availability of air conditioning has reduced social interaction and made us a more private society. The tradition of families gathering on front porches or lawns on hot summer nights, sitting in rockers or swings, sipping lemonade or tea, listening to cicadas and tree frogs, telling stories and greeting neighbors, has faded from view...."When I was growing up," one of my neighbors remembers, "all of the families on the street would sit on their front porches, just like Aunt Bea and Sheriff Taylor in Mayberry. But now people stay inside and often folks don't even know their neighbors, much less speak to them."[47]

Indeed, Raymond Arsenault, a history professor at the University of South Florida, laments the impact of air conditioning on the South's traditional ways of life. "General Electric," he writes, "has proved a more devastating invader than General Sherman." Arsenault notes that few Southerners wanted to acknowledge the importance of this technology to their lives. In part, this may have been due to a reaction to environmental determinism, popular in the early decades of the Century in explaining the differences between the Yankee North and the Cavalier South.[48] Air conditioning changed the culture of the South, away from a rural and agricultural one towards an urban one that increasingly not only resembled the North, but also began to set the tone of life for all the country. By the twenty-first century, Texas and Florida had become sizable States in terms of population and most of the growth was in their cities.[49] They began to resemble Northern cities, and Northern cities began to resemble them.[50] As W.S. Kowinski has written, "can you imagine William Faulkner writing about the Yoknapatawpha Mall?"[51]

The change air conditioning brought to the American South also turned out to be exportable. In 1999, the *Wall Street Journal* asked a sample of world luminaries to name the greatest inventions of the millennium just ending. Singapore's former prime minister Lee Kwan Yew was emphatic about including air conditioning as one. Without it, he said, the tropical countries could never achieve the productivity and development that cooler countries enjoyed.[52] When the wartime Japanese occupation ended in 1945, Singapore was a tropical mudflat with a strategic port. By the end of the Vietnam War in 1975, it had established itself as a modern city-state, with personal incomes approaching North American standards and air-conditioned high-rise offices and apartment buildings and shopping malls to shield its residents from the city's equatorial heat and humidity. As one study notes,

> air conditioning has become so common that almost all Singaporeans now

enjoy its benefits in one way or another....In public areas, it is now common to enjoy this kind of cool comfort by taking advantage of the ambient environments found within shopping malls whilst, in the domestic setting, air conditioning is now almost as common as the video player amongst even the poorest of groups. Beyond the home, office and mall, conditioned air has also started to spread through bus interchanges, tennis courts and even crematoria. Arguably the Singaporean fascination with this form of thermal consumption scaled new heights in 2005,when the opening of the first sub zero degree bar presented excited tropical consumers with the hitherto unavailable opportunity to combine social drinks with the embodied delight of a freezing ambient environment.[53]

Like the American South, Singapore and other large cities in southeast Asia, Japan, and southern China could not have developed into commercial or industrial powerhouses without air conditioning.[54] Urban areas in the Middle East, too, have been transformed by air conditioning. The technology centers of Bangalore, India, and Tel Aviv, Israel, could not exist without controlled environments for both people and the silicon chips they manufacture. Indeed, air conditioning has changed peoples' lives the world over. Once seen as the privilege of the rich, now only the poor lack air conditioning in hot climates. Willis Carrier's invention not only is harmonizing the world's urban architecture and redistributing its economic activity; it has also spread its social effects across the globe and is eroding traditional cultures just as it did in the American South. Thanks to air conditioning, Asian countries are becoming more and more like North America and Europe in their attitudes toward youth, marriage, home hospitality, transportation. and even in the adoption of Western-style clothing. Now, except in sports or fitness environments, sweat is seen as crude or indicative of illness. And all the while, the hum of the ever-present air conditioner does its unobtrusive work in the background.

Unlike the past, people in air-conditioned societies spend most of, if not virtually all, their time inside climate-controlled structures. Encounters with unpredictable geographic and environmental effects are confined to vacations and traveling. Even in these pursuits, air conditioning has made tropical life more appealing by providing relief from constant heat. As well, the increase in travel has brought new food and drinks into the global diet.[55]

People have become more private and less tolerant of different temperature environments. Clothing, whether business suits or fashion garments, has become globalized, at the expense of more traditional garb. Environments with air conditioning only a few years ago, were seen as the privilege of the rich: now, those without air conditioning are seen as the result of poverty. Air conditioning is a populist symbol, even in China.

Air conditioning has even also found its way into cultural criticism.[56] Henry

Miller's *The Air-Conditioned Nightmare*, a 1945 literary rant about an automobile trip he took through part of the United States in 1940, *The Air-Conditioned Nightmare* was a complaint about the "soulless" technological culture he found there after a decade spent in Europe, rather than about air conditioning itself — for him, the term was a symbol of an American culture he disliked.[57] In 1978, Russell Baker, in a somewhat more humorous vein, wrote:

> If I were a conservative,...I would now forget the death penalty and the crusade against homosexuality for a while and attack one of the taproots of waste and big government. I refer to air conditioning in Washington. Air conditioning has contributed far more to the decline of the Republic than unexecuted murderers and unorthodox sex. Until it became universal in Washington after World War II, Congress habitually closed shop around the end of June and did not reopen until the following January. Six months of every year, the nation enjoyed a respite from the promulgation of more laws...Once air conditioning arrived, Congress had twice as much time to exercise its skill at regulating and plucking the population.[58]

21

Dunkirk, New York: The Yankees and the Railroad

Well, it beats the very devil. I never saw a critter go so fast with such short legs.
— Exclamation of an inmate of the Massachusetts State Lunatic Hospital, temporarily released to see the first train entering Worcester, July 4, 1835[1]

There will be icicles in hell when Erie common [shares] pays a dividend.
— Jay Gould, railroad financier and stock manipulator, who controlled the New York and Erie Railroad between 1868 and 1872[2]

Damn old Vanderbilt's time! We want God's time! The Vanderbilts cannot run me [even] if they run the rest of the country, by Jehosophat!
— Comment by an elderly man in 1883, upon hearing of the railroads' decision to set standard time zones for the country[3]

A Rail Road to the West

The small town of Dunkirk, New York, lies between US 20 and the Lake Erie shore, southwest of Buffalo. On the northeast side of town is the site where the tracks of the New York and Erie Railroad (hereafter called the Erie) once crossed those of the New York Central Railroad. Even today, the visitor can see where the two sets of present-day tracks dating from the famous old railroads come together, the crossing to the Dunkirk waterfront having been altered (and largely forgotten) years ago. It is a good place to start, given that the Lake Erie shoreline is responsible for the Yankee ventures into railroading.[4]

In the minds of most people who thought about transportation in the 1820s, the challenge, as always, was to find a way to move goods with the least amount of resistance and at the greatest speed. For thousands of years, water was the preferred mode, though one had to go when and where the currents, tides, and winds permitted, while weather had a great impact on reliability. Before the days of the steamboat, if one had access to a river, it made sense to go downstream and sell everything, boat and all, at the destination and walk home rather than fight the current. Artificial waterways, or canals, had also been in use for centuries. With their returnable boats and reliable water levels, they were handy where large rivers did not flow. But they were expensive to build and maintain, their routes depended on the availability of water sources above that could be tapped to fill locks and channels on a consistent basis, and, in more northerly climates, they froze over for months. Road surfaces in the 1820s were poor, though a Scotsman named John Loudon McAdam was experimenting with a system of building roads using crushed rock and good drainage that would later be called "macadamized." Speed was pretty much the same for roads, rivers, and canals: about 15 mph at most, depending on the pulling power of the horse or speed of the current.

Another method was to use rails, with loads pulled by horses, which avoided the surface problems of regular roads, but the composition of the tracks had to be durable enough to withstand the weight of carriages, and durability equaled high capital investment. Rails were in common use in the eighteenth century to haul ships onto shore for refit. In the Boston area in the 1780s, at least three were in use.[5] Rails were also used to haul earth and rock short distances. As early as 1800, some Boston developers used a downhill tramway to move earth from the crest of Beacon Hill to allow them to build in the area.[6] When a quarry in Quincy, Massachusetts, managed by Thomas H. Perkins, one of the textile investors in the Boston Associates,[7] won the contract to provide the granite for the Bunker Hill Memorial in 1826, it used a three-mile, horse-pulled railway to bring stones from the quarry to the Neponset River, there to be shipped by sea to the monument, a distance of seven miles, thus saving 80 percent of the project's estimated transportation costs. It was the first use of incorporation for a railroad in America.[8] Two years later, a Pennsylvania coalmine developed a nine-mile track to move coal to a nearby river. The rail line used gravity to go downhill and mules to haul the wagons back up to the mine. Indeed, it was in the mining industry that the hauling of heavy loads some distances by rail on a regular basis first developed.[9] The rough floors of mineshafts meant that pulling carts was a slow and expensive way to get ore to the surface, but rails could skip over irregularities and lower the cost of extraction.[10] When the steam engine was invented, its first use in mines, as early as 1712, was to pump water out of the shafts, but it was soon adapted to move ore carts, and the railroad was born.[11]

The Steam Railroad

The impulse for railroad building in America largely came with the successful completion of the Erie Canal in 1825. The canal's success shocked the merchants of Boston, Baltimore, and Philadelphia.[12] At a blow, New York, their competitor, had captured the trade from the West. Traveling the canal was slow and seasonal, but it made hauling freight cheap — a ton of freight could now be shipped from Buffalo to New York City for as little as $5, compared with $100 in 1810.[13] The canal put an end to Boston's influence westward, and severely damaged the ability of Baltimore and Philadelphia to capture the Western trade. New Orleans, too, served by a large fleet of river steamboats, watched uneasily as plans were made to connect the northern reaches of the Mississippi watershed, primarily the Ohio River, with the Erie Canal via other canals and Great Lakes steamers.[14] Even north of the border in Canada, Montreal began to see its commerce diverted from the St. Lawrence River into upstate New York, and entrepreneurs there began to plan another canal, the Welland, to bypass Niagara Falls and connect Lake Erie with Lake Ontario.[15]

In the late 1820s, Americans looked across the Atlantic and saw that, in Britain, steam-powered engines were beginning to be used to haul coal from mines to market.[16] As well, manufacturing advances were allowing rails to be made that sustained the weight of the engines.[17] The first modern railroad, the Stockton and Darlington, had opened in 1825,[18] and word of a steam railroad being constructed in 1828 across the thirty miles between Manchester and Liverpool stimulated more interest in the technology.[19] When George Stephenson's "Rocket" won a contest sponsored by the Manchester and Liverpool Railroad in 1829 by going as fast as 30 mph, steam power trumped horsepower on the rails conclusively.[20] Indeed, the effect was electric. As late nineteenth-century American railroader Charles Francis Adams Jr. noted,

> [n]ow the great peculiarity of the locomotive engine and its sequence, the railroad, as compared with other and far more important inventions, was that it burst rather than stole or crept upon the world. Its advent was of

the highest degree dramatic....It was this element of spontaneity, therefore — the instantaneous and dramatic recognition of success — which gave a peculiar interest to everything connected with the Manchester and Liverpool railroad. The whole world was looking at it, with a full realizing sense that something great and momentous was happening.[21]

In 1827, Erastus Corning and his associates in Albany gained a charter to develop a horse-pulled, short-haul railroad around the Erie Canal locks near the mouth of the Mohawk River, but it took him until mid-1831 to complete it, and his horses were soon replaced by a locomotive.[22] Then, in 1828, the Baltimore and Ohio Railroad was chartered, again first using horses and mules, but switched to steam in 1830.[23] Entrepreneurs in Charleston, South Carolina, opted to use steam before opening their railroad in 1831.

The Western Railroad

Massachusetts was slower to move, in part due to the time spent debating the merits of a proposed canal across the Berkshire Hills to Albany. In 1830, Boston investors received a charter for a rail line to Lowell, promoted by the owners of the mills there. The Middlesex canal was useless for freight through the cold months and too slow for passengers in any case. The Boston and Lowell railroad charter was the first to be given for an exclusively steam railroad.[24] Its first locomotive, and its engineer, were imported in 1832 from Stephenson's company in Britain. Later engines were built by the Lowell machine shop and a shop in Boston. The line was completed in June 1835, by Major George W. Whistler, the father of the painter James McNeill Whistler.[25]

Having started to charter railroads, the Commonwealth aggressively decided to divert some Erie Canal trade to its own port. In 1831, after a debate that took state ownership of a railway to Albany out of the picture, a private westward venture was chartered,[26] though only to Worcester, which was reached in 1835.[27] Three weeks later, another line linked Boston with Providence, Rhode Island, coordinating with a fast steamship to New York.[28]

Promoters then began to connect New England with Albany, New York, via the Western line.[29] After financing delays caused by the Panic of 1837, which led to lobbying for significant state and municipal financing,[30] Springfield, where new textile mills had opened in 1835, finally was reached in 1839 and Albany in December 1841.[31] Travelers and freight could go from Boston to Albany in fifteen hours, summer and winter.[32] An editorial in a Utica, New York, paper commented: "New England, the birthplace of many of us, and the home to many of our Fathers, which we regard as far removed from us, has, by this magical operation [the railroad] approximated to our border....Yankee travel and notions will pass through our city for western New York and for westward of New York....Here is a western world now, open to Yankees, which all your ingenuity and industry cannot fill."[33]

The Western, at 155 miles, was the longest and most expensive railroad in North America at that time.[34] It was the first to adopt some modern methods of management, technology, and financing. Though based on the experience of the New England textile mills, the challenge of coordinating distant divisions and branches changed known management techniques, the different complexities of steam engines altered the expertise of waterpower technicians, and the sheer scale of capital needed for construction and maintenance expanded the pool of owners and bondholders, and engineering and ironworking companies. Yankee talent used its textile background as a base on which to begin building railroads, a field they largely dominated at all levels for nearly half a century after the 1840s. Boston capital dominated the investment market from the end of the Second Bank of the United States in 1836 until almost 1850, when New York City became the country's financial hub.[35] During this period Yankee capital was free to follow Yankee labor to the West, and Western goods flowed easily into Boston.[36]

In 1842, New York City reacted to this Yankee head start by endorsing a New York to Albany railway:

> Heretofore, we have been accustomed to consider the Erie Canal as the property of our own State, and we have looked upon it as a proud monument of our enterprise, and as one of the richest sources of our wealth; but while we have been sleeping in our supposed security...we have been surprised by the stealthy march of our rivals, who have extended their Western Railway...to the depot of our supplies and thereby turned the channel of our trade into a new direction...Boston and not New-York is now reaping the great advantages of our internal improvements.[37]

The State, however, wedded to the Hudson River steamboats, was slow to react, and it took another decade before rails connected New York City with Lake Erie via Albany.[38]

In New England, meanwhile, competition for capital from the promoters of textile mills and other manufacturing concerns hampered the railroads' ability to finance themselves. The inability of Boston businesspeople to develop coherent investment and trade strategies meant that the city never achieved its possibilities as a real rival to New York City, and the railroad from Boston gradually worked to carry New England manufactures to Western markets, rather than Western products eastward.[39]

Helen of Ramapo: Building the Erie

The story of the New York and Erie Railroad begins like this, so we are told.[40] In January 1831, Henry L. Pierson and his bride, Helen, were on their honeymoon in Charleston, South Carolina. Helen learned that a steam engine, the "Best Friend of Charleston," was going to make its first commercial run pulling passenger carriages since being put on the tracks of the South Carolina Railroad a couple of months earlier. The railroad was one of the first in America designed for steam locomotives as well as for horse-drawn carriages, and had opened only four months after Stephenson's Manchester and Liverpool line.[41] Helen was keen to take the ride, and Henry indulged her. They were pulled by the "Friend" for six miles and returned without incident.[42] Helen was thrilled, and when she returned to New York City, the experience was all she could talk about.[43]

Helen's husband had caught some of her excitement as well. Henry Pierson was the son of a prominent New York businessman, Jeremiah Pierson, who had left Massachusetts for greener pastures in 1790, and a considerable part of his activities after 1795 focused on the development of the waterpower available in the town of Ramapo, northwest of New York City on the west side of the Hudson River.[44] For a decade and more, Jeremiah, his son Henry and his Yankee son-in-law Eleazar Lord, had been involved in building a turnpike from Ramapo southeast to the Hudson at a place called, appropriately, "Piermont" to improve transportation of their manufactured products, such as nails, to the river and then by boat into the city.[45]

Helen's excitement came at a propitious time, as another railroad line was about to open from above Albany to the Erie Canal town of Schenectady. The success of this project released a torrent of enthusiasm for this new mode of "road" transport all across New York State, resulting in the presentation of forty-two bills in the state legislature in 1832 alone to incorporate railroads.[46] One bill derived from a petition that Eleazar Lord circulated to create what was proposed to be the longest trunk line in the world. It would stretch five hundred miles through the ridges and rolling country of New York's "Southern Tier," from the Hudson River at Pierson's waterfront property, Piermont, to Dunkirk, just southwest of Buffalo. Throughout 1831, a series of town meetings had been held across the Southern Tier to pledge political and financial support for Lord's project.[47] On receiving its charter in 1832, however, the railroad was encumbered with obstacles, and the project was hampered by delays, as a result of opposition by the "Canal Ring."

The New York and Erie Railway Company was organized in 1833, with subscriptions for $1 million of the $3 million authorized.[48] The company asked for further subscriptions for stock and donations of land along the route, but, as the line had not been surveyed, little was forthcoming.[49] Generally, support rested not on the hope of a financial return, but on the benefits that might accrue to those located along the actual route.[50] Nonetheless, in 1834, using what little resources it had acquired — by then, the cost estimate had risen to $6 million, with only $1.5 million raised — the company began surveying west from Piermont.[51] Having been chartered by the New York state government, it was obliged to run its tracks entirely inside the State. This meant that its traffic had to be ferried up or down the Hudson River some twenty-four miles from New York City, as it was technologically impossible at the time to bridge the river much south of Albany, and the railroad could not use the direct ferry route across the Hudson to New Jersey.[52] Also, to direct upstate New York traffic into New York City and not elsewhere, the Erie's charter forbade it to have connections with any other railroad; it also had to be built using a wide gauge.[53] The antipathy felt by people in New York's Southern Tier for the Erie Canal lake port of Buffalo was such that Dunkirk, an alternative port and less favorably located on Lake Erie, was designated as the western terminus of the railway.[54] In between was a collection of small towns and farms that, alone, would be hard put to make the line pay.

Surveys and construction were slow and impeded by financial struggles and management turnover. The first wheelbarrow-load of soil was moved at a point on the Delaware River in November 1835, but then a fire in New York City a month later ruined many of the stock's subscribers and the political fight in Washington over the National Bank and the subsequent Panic of 1837 hurt the rest.[55] The railroad went to the state legislature for help, with suggestions ranging from a loan (which turned into a grant) to outright takeover of the project by the State. Ultimately, the State made a $3 million loan, and construction was restarted in 1839.[56] Lord, who had been removed as president in 1835, returned with a plan to build the line incrementally, starting with the piece near the Hudson of interest to his family's concerns.[57] In 1841, just four days before the Western Railroad was opened from Boston to Albany, an Erie train chugged from Piermont to Ramapo. Three months later, the railhead reached Goshen, 46 miles from Piermont. In 1842, the line was assured of completion by the creation of a construction syndicate that undertook to finance the next 76 miles for $4 million in second mortgage bonds, and the last 169 miles, from Corning to Dunkirk, was also financed with bonds.[58] It took seven more years, however, including three in receivership, before the railhead passed out of the southern part of the Catskill Mountains and reached Binghamton, 216 miles from Piermont.[59]

At last, on May 14, 1851, President Millard Fillmore and other guests were greeted at Piermont for the inaugural trip, and Helen Pierson and her husband were among the first passengers to travel the entire length of the line.[60] Soon thereafter, the Erie recognized reality and built a spur line to Buffalo. As well, in 1852, it managed to convince the State to allow it to buy a local railroad leading from Ramapo southeast to a new southern terminus in Jersey City, directly across from New York.[61]

Financially, however, the railroad was a mess. It had fallen into receivership a couple of times, and had issued every kind of security it could, got a lot of State aid, and depended on construction consortia for revenue. The railroad's capital needs were such that the New York Stock Exchange could not place all its bonds domestically, and some were sent overseas to the exchange in London. Six months after the Erie was completed in 1851, the directors declared a dividend of 4 percent on its shares, but the debt load proved too great, and the company went into receivership in 1859.[62]

When it was opened along its full length, twenty years after it had been conceived, the Erie was the longest continuous line in the world.[63] It was also the most expensive project financed up to that time — at $23.5 million, it cost three times as much as the Erie Canal and about equaled the federal budget for a year in the 1840s.[64] Although most of the lines built in America to that time, except the Western, were local in character, the Erie was a trunk line, carrying not only local products but also those from States farther west.[65] But competition was not slow to emerge.

A Yankee Builds the New York Central

At the end of his book, *Steelways of New England*, Alvin Harlow names about three dozen New Englanders who were involved in building railroads all over the country from the 1840s to the 1890s.[66] Yet he fails to mention one who had been born in the same Connecticut town as Eleazar Lord and made his fortune in New York: Erastus Corning.[67]

Corning's habitually unsuccessful father moved the family to the Troy, New York, area when Erastus was about eleven. The boy, who had been a cripple from an early age, was given a job in a relative's hardware store. He proved successful in the business and, after the War of 1812, set up shop himself with a partner in nearby Albany. He supported the development of the Erie Canal and prospered by providing ironwork for its facilities. In 1826, he bought an ironworks near Troy,[68] and joined a group of investors in backing a new and unproven technology: the railroad. That year, the group, which had been put together by an immigrant Englishman, George Featherstonehaugh,[69] secured a charter from the State to build the horse-drawn Mohawk and Hudson Railroad, which, the group hoped, would lure travelers willing to pay a premium to bypass the Erie Canal's slow locks at the mouth of the Mohawk River at its eastern terminus.[70]

It took over three years to raise the financing and another year to build the seventeen-mile line. By then, it had been decided to replace horses with a steam engine called, with political intent, the "DeWitt Clinton," after the politician who had promoted the Erie Canal.[71] When the railroad opened at the end of July 1831, it was an immediate sensation.[72] Among the large number of bills for charters of incorporation of railroads that were put before the State Legislature in the next few years, was one for the Utica and Schenectady, which would connect with the Mohawk and Hudson. The new line was chartered in 1833 and opened in 1836.[73] Corning became its major investor and was its president for twenty years.

Gradually, a series of other connecting lines began to appear along the route of the Erie Canal, reaching the Finger Lakes just as the Western Railroad connected Boston with Albany. The patchwork continued westward, and when a short connection was finally made between the two railroads that terminated in Rochester in 1844, the physical, though not organizational, connection from Boston to Albany and on to Buffalo was complete.[74] By that time, the Erie railhead was in the Corning, New York, area and the Western had been in operation for three years.

There were no through tickets in 1844, and the trip from Albany to Buffalo took thirty-six hours and required six changes of trains.[75] And all the while, the so-called 'Canal Ring' fought to compete, both on price and in lobbying. Passengers who could pay for speed were given up to the railroads early, but immigrants for the most part still went by the Erie Canal. At first, the State government prohibited the railroads from carrying freight, then they were allowed to do so during the winter months, when the canal was frozen over.[76] Later, they were allowed to carry local freight, but it was not until the end of 1851 that the railroads finally were allowed to compete with water along the whole range of services.[77]

Erastus Corning, meanwhile, began to assemble all the parts chartered originally as local lines. Using his Utica and Schenectady, he devised a method of leasing the infrastructure of connecting rail lines, and then he began buying most of the rolling stock of the various local lines and coordinating their schedules. The result, in February 1848, was the first "through" train from Albany to Buffalo.[78] In 1851, Corning convened a meeting of the directors of ten of the railroads that existed along the route from Albany to Buffalo and got them to agree to a merger framework. Since the proposal would affect all ten charters, it was sent to the State Legislature for approval, which was duly received on May 17, 1853. With what might be America's first great corporate merger, the New York Central Railroad came into being.[79] It extended from Troy, New York, to Erie, Pennsylvania, and, a year later, into Canada via the Niagara Falls Suspension Bridge. (The line to New York City was owned by a separate company until Cornelius Vanderbilt merged the two after the Civil War.[80])

Corning held the majority of shares in the New York Central, and was its president for twelve years, during which time it became the largest incorporated company in the country.[81] Under his guidance, the New York Central connected, through allied lines, with Cleveland and Chicago. He also bought shares in the Michigan Central and the Chicago, Burlington, and Quincy, which allowed the New York Central access to the upper Mississippi valley.[82] He also controlled the company that built and operated the Sault Ste. Marie ship canal and locks between Lake Superior and Lake Huron, which gave access to the iron mines in northern

Michigan and Minnesota.[83] Corning, and after him the Vanderbilts,[84] developed the most heavily used and most profitable line into Chicago, while the Erie, hampered by its problems accessing New York City, the sparse population along its line, and continual financial depredations, was not able to enter Chicago until long afterward, and when it did so, its route went down into central Ohio and up to Chicago from the southeast, all of which was also lightly populated country. The Erie found itself squeezed between the New York Central and the equally aggressive Pennsylvania Railroad.

The Railroads' Effects

POLITICAL UNION

It has been said that the railroad, as a technology, unified the nation. According to the *Omaha Daily Republican* in 1883, "they have made the people of the country homogenous, breaking through the peculiarities and provincialisms which marked separate and unmingling sections."[85]

It took a good twenty years after 1831 for the railroads from the East to get through the Appalachian barrier. By the outbreak of the Civil War in 1861, six railway systems connected the Atlantic with the Great Lakes and the Mississippi valley: one in Canada, one in the South (from Richmond, Virginia, and/or Charleston, South Carolina, to Memphis, Tennessee) and four (the New York Central, the Erie, the Pennsylvania, and the Baltimore and Ohio) in the North.[86] By 1871, connecting lines could take a passenger from Boston to San Francisco, through the barriers of deserts and mountain ranges in the West. When Helen Pierson had her first train ride in 1831, little was known of the West beyond the edge of the forest, Texas was still a part of Mexico and California was sparsely settled by Mexicans, no settled people lived around the Great Salt Lake, and the Mormon religion was but four years old and centered in western New York. By the time the rails reached northwestern Pennsylvania and the Ohio River and linked with those built in Ohio, Michigan, and Illinois, the United States had expanded to the Pacific, settlers were moving along wagon trails to Oregon, Utah, and California, the Gold Rush was a couple of years old, and tensions between North and South over slavery were beginning to rise.

A secondary debate about the route that a transcontinental railroad ought to take became caught up in the slavery issue, as such a line clearly would need federal support.[87] This form of connection between regions was noted by all sides when it came to the West Coast. In the 1840s and 1850s, Southern politicians saw how railroads might exacerbate the standoff between slave and non-slave parts of the country, since they would help bind Western regions to their Eastern geographic counterparts. Thus, the contest for approval of a transcontinental railway began not for economic purposes, but for political ones.

A northern route to Oregon, through the territories just to the south of British North America, had a certain appeal, but was impractical until the 1870s. The creation of Yellowstone National Park in 1874 was due in part to the desire by promoters of the northern route for a tourist attraction to justify a railroad through the long stretch of virtually uninhabited land between Minneapolis and the Pacific Coast.

A central route had many of the attractions the northern route lacked, not the least being the small but growing Mormon population and market halfway along. As well, the Oregon Trail had shown that a relatively level route existed across the Rocky Mountains, with only the Sierra Nevada blocking the way to the coast.

A southern route had been made possible by America's success in the Mexican War. The Santa Fe Trail indicated the feasibility of a railroad to the Rio Grande, and the overland expedition to California in the Mexican War had shown it was possible to cross the high Colorado Plateau and the Mojave Desert to the mountainous outskirts of Los Angeles. When an easier route was discovered across from El Paso, Texas, on the Mexican border, to California, a South Carolina railroad promoter was dispatched from Washington, DC, to Mexico City to revisit the location of the border. This effort resulted in the Gadsden Purchase of 1854, which moved the border from the mountains around the Gila River south into the flat deserts between El Paso and Tucson, Arizona.

The Civil War ended the arguments about the initial route. It definitely would not be going through the South, while the northern route led only to the Oregon farmland of the Willamette Valley, not to the goldfields of California and the silver mines of Nevada. In 1862, Congress authorized the terms of construction along the central route, from Omaha, Nebraska, to Oakland, across the bay from San Francisco, and by 1869, California was attached to the Union by rail. Later, as Western markets grew and transcontinental shipments became a viable alternative to the long sea voyage around Cape Horn, the other routes were also built.[88]

Another way railroads brought the country together was in the expansion of a continental web or network. The original concept of the US transportation system was to unite the East Coast with the waterways of the interior. Once it became clear that the cost per ton to transport goods by rail was similar to that of canal boats and steamboats, but at a considerably greater speed, the railroads quickly expanded past river and lake ports and took traffic away from boats.[89]

In 1854, the Rock Island Railroad began a bridge across the Mississippi River in west-central Illinois. The riverboat companies opposed project, and a court case was fought on the notion that the bridge would be an obstruction to navigation — indeed, once constructed, the bridge was the object of an apparent sabotage attempt by a riverboat, the *Effie Afton*. An up-and-coming Illinois railroad lawyer named Abraham Lincoln won the case,[90] and prairie products that previously had moved downriver to St. Louis and New Orleans began to go to Chicago and New

York. This cutting of economic ties between the old Northwest and the South meant that, when the Civil War broke out, two parts of the country, the Northwest and the Northeast, that otherwise might have been separated in their loyalties and interests by distance united to preserve the Union. The railroads also did their part in prosecuting the war by allowing the rapid movement of armies and supplies on a large scale, which particularly benefited the North, with its greater industrial and technological sophistication and much more dense rail network.

Corporate Organization

The railroads required organization of a complexity unknown in the world until then, mainly because they owned all parts of their systems, from the rolling stock to the roadways to the maintenance of both traffic control and freight handling; no other form of transportation had such singular responsibilities, and it led to their becoming America's first modern corporations, with all the good and bad points connected to this concept.[91]

The lengths of the trunk railroads also forced some organizational innovation. Such a line could not be run centrally, but had to be split into operating divisions.[92] At the same time, each section of the line had to be well coordinated with all the others to avoid accidents. The need for technical competence in operations also required professional management. In the Lowell textile mills, for example, there was generally only a technical division where the machine shop was concerned; the railroads, in contrast, needed divisional managers whose tasks replicated much of those performed in head offices, but on a narrower geographic scale.[93] The Western, for instance, was managed by three separate divisions. With three trains a day each way running on a single track, there were numerous opportunities daily for trains to collide — indeed, in October 1841, even before the line was finished, a collision killed two people and injured a number of others, leading to a State investigation of the accident and a series of organizational reforms, including the employment of full-time salaried officers to oversee the three divisions and an internal reporting system.[94] Because there was no method in the pre-telegraph days for communication faster than the trains themselves, the trains were put on rigidly fixed schedules between sidings where they were supposed to meet. If there was a delay, a train might have to wait until the one coming the other way passed before it could proceed. The setup caused delays, but saved lives.

Until the 1850s, only a couple of railroads — the Erie and the Pennsylvania — were long enough and had sufficient freight and passengers to require more complex management structures. Such long lines had to develop complex management structures to decide the number of cars to attach to a train, lay out logistical systems for moving people and freight efficiently, and provide interlocking schedules to attract passengers and freight and get them safely to their destinations.[95] It was one thing to have to bring some empty cars back from a destination

40 miles away; it was another to have to do it from 450 miles away. Mistakes could be costly.

In 1853, the Erie married the telegraph to a system of hourly, daily, and monthly reports, so that recurring delays could be eliminated and data used to evaluate divisional efficiency.[96] The Pennsylvania improved upon this system before the Civil War. The demands of the war itself resulted in the spread of these organizational forms to all the Northern railroads. It was no coincidence that many former military men, such as George Washington Whistler, Grenville Dodge, and George McClellan, ran railroads during their careers. Not only did they have a West Point or other engineering education; they had managed large and dispersed military organizations.[97]

In solving the problem of processing the data they needed to operate, the railroads developed statistical methods and the accurate measurement of receipts and expenditures in the form of cost accounting.[98] By the time large organizations appeared in other industries in the 1870s and 1880s, the railroads had many years' experience with modern systems of rules, reporting, and measurement, which non-rail owners and executives copied, adapted, and improved upon.[99] By the 1880s, consultants such as Arthur D. Little and Frederick Taylor were assisting these industries to enlarge their facilities to take advantage of scale economies.[100]

FINANCE

The basic financial challenge for the railroads, as for the canals, was their high fixed costs, most of which had to be "sunk" before operations and revenue generation could even begin. This problem was exacerbated for early railroads in that the technology was only in its infancy and the markets for their services were mostly untested.[101] The scale was simply unlike anything tried before. One of the largest industrial enterprises of the 1850s, consisting of three integrated textile mills in Maine, had annual operating expenses of just over $300,000; in contrast, in 1855, the Erie needed $2.86 million and the Pennsylvania $2.15 million.[102] The asset value alone of the New York Central in 1865 was equal to one-quarter of all the manufacturing investment in the country.[103]

Early railroad financing depended on selling shares to the public.[104] Because so much money was needed for construction, the earliest promoters, after having secured incorporation from State legislatures, would issue a prospectus that estimated the cost of the railroad and the number and price of shares — for instance, shares in the Boston and Worcester were set at 10,000 at $100 each. These shares were placed among local investors and in towns through which the railroad might pass. Investors would acquire the shares by paying a small "assessment" at the outset, then pay the balance, in the form of further assessments, as money was needed.

Unlike most corporations of the time, railroads had to have limited-liability status, meaning that the shareholders were not individually and severally liable for the company's debts and other obligations. The reason for the change was obvious: the sheer size of the operation meant that, if no individual shareholder was able to meet the company's debts, no one could afford the risk and the project could not be financed. Only where the distance between populated markets was short, as in Britain, could unlimited liability be contemplated. Often, towns would have public meetings at which potential investors heard tales of what the addition of a railroad might do for the local economy. People then might pledge to buy shares in the enthusiasm of the day, and town councilors might pledge funds from the town treasury.[105] Normally, these public funds came from bonds issued with the approval of the voters. The problem with these pledges was that there was always a difficulty in collecting the payments, which were scheduled installments or based on the railroad's calling for another installment when it was needed to pay for part of the construction.

State governments were also approached for assistance. Over the first three decades of the nineteenth century, the government in Washington swung between whether or not it should sponsor and finance internal improvements. In the 1820s, it supported roads and canals, but by the end of the Andrew Jackson administration in early 1839, the federal government had ceased to pursue that policy, at least temporarily.[106] This principle never deterred the States. Whether it was the Erie Canal in the 1820s or railroads in the 1830s and 1840s, the States generally had no compunction about lending a hand, either by agreeing to provide loans or subscribing shares.[107] The money would be made available in direct payments or through the issuance of scrip — in the form of either State-minted paper money or the more formal issuance of bonds. However the money was provided, it was a borrowing on the part of the State, with the proceeds going to the railroad. The railroad might try to "pay" foreign suppliers with scrip, appropriately discounted, for good quality rails, engines, or other goods and services. The scrip was not provided in full, but sold on the market as each "assessment" was required by the railroad company.[108]

The most common private sector construction-financing method turned out to be the syndicate, where contractors would take railroad bonds, either at an attractive discount or based on an inflated construction cost estimate, and complete the work.[109] The bonds could be held by the syndicate until the line was profitable, or rediscounted and sold to other investors, most often on the New York capital market, which was already collectively known as "Wall Street" and which had taken the leadership as the country's largest capital market from Philadelphia in the early 1800s.[110] Railroad bonds and shares soon became the largest single part of the market's activity, acquired by financiers for resale and trading purposes or exported to Europe.[111] Many of the first non-New York railroad companies were wary of this market, however, fearing that they might be strangled for cash by a

Wall Street that was looking out for the interests of New York State.[112] Only gradually, and with the help of the telegraph, did the notion of a rather more neutral, national exchange emerge.

If a State was lightly settled and capital short, railroad promoters would encourage its government to put its creditworthiness behind their projects in the form of bonds on which the State would guarantee payment. This was always a risky business, as State revenues were as equally likely to fall as were those of the project when an economic crisis occurred. The bonds were most often sold to European buyers, who naturally assumed that the State, or perhaps the federal government, would honor them. Numerous State defaults in the Panic of 1837 soon began to educate these buyers, and bond sales fell off. Eastern railroads looking to expand westward then swooped down on these bankrupt lines and added them to their collections at bargain prices.[113]

The early railroads also got involved in some financing schemes whereby the State would not invest or lend directly, but instead would provide title to public land that might or might not be contiguous to the railroad route. The precedent for this seems to have been the grant of land by a private company, the Holland Land Company of western New York, in 1813 to the Erie Canal Commission, on the condition that the Erie Canal was built through its holdings.[114] This method was more popular later in the West, where there was more public land available, than in the East.[115] The railroads then might mortgage or sell these land assets for construction cash. In Illinois, for instance, land given to the Illinois Central that the State could not sell for $1.25 an acre beforehand became saleable by the railroad company for $11.00 an acre simply because it was near the line.[116]

From 1830 to the start of the Civil War in 1861, government financial involvement in railroad construction, including grants and loans and the donation of acreage that the railroads could then sell to immigrants, amounted to approximately $300 million by States, $125 million by municipalities, and $7 million in cash by Washington plus 22.5 million acres of public land worth perhaps $3 an acre on average, for a total of about $500 million.[117]

DISTRIBUTION

At least until after the Civil War, the economics of railroads focused on the transport of relatively high paying passengers and high-value-added or perishable goods; other passengers and goods were left to steamboats and canal boats.[118] Only gradually — and particularly during the war, when men and supplies were moved to theatres of combat in huge numbers — did railroad executives come to realize that the success of their business would depend on moving low-unit-value commodities and masses of immigrant settlers across long distances. Improved technology and management, plus a denser network, allowed freight costs to continue to fall. From 1848, when no railroad yet connected the East Coast to the land west of the Appalachian Mountains, to 1870, costs dropped from 8 cents per ton per mile to 2 cents.[119] The initial costs of transported goods themselves were also dropping due to increasing economies of scale, again made possible by the movement of freight in and out of factories by rail.

The railroads also made it possible to deliver standardized finished and consumer goods to all parts of the country.[120] For instance, a system of clothing sizes, first adopted during the Civil War, spread to the postwar civilian market and allowed not only large-scale production but also the development of chain stores. Where stores often had to buy stock once or twice a year and guess at the coming fashions or preferences, the arrival of the railroad and the telegraph allowed clothing wholesalers to operate with relatively small inventories and get orders filled quickly. This meant, in turn, that manufacturers could dispense with agents and not have capital tied up waiting for payment at the end of a season.[121] These changes also largely brought an end to the traditional Yankee peddler.

These innovations affected the distribution of other products, too. In pre-railroad settlements, iron products were handmade by a blacksmith, with varying quantity and quality. Now, the railroads provided standardized substitutes from factories located near urban areas, leaving the blacksmiths to handle repair business and custom designs.[122] The distribution of agricultural products was affected at both the production and consumption ends. The ability of the railroads to provide supplies and take away large quantities of product meant that farms could be developed on a large scale and effectively "industrialized." At the other end, a product in demand, such as wheat, which once had to be produced close to its point of consumption, could now be sent to all corners of the world. As Arthur T. Hadley wrote in his 1886 book, *Railroad Transportation*, "[t]wo generations ago, the expense of cartage was such that wheat had to be consumed within two hundred miles of where it had grown. Today, the wheat of Dakota, the wheat of Russia, and the wheat of India come into direct competition. The supply at Odessa is an element in determining the price in Chicago."[123]

The nature of the railway business itself was also affected by scale economies. Once the cost of operating a line was covered, all extra business was very profitable. As well, whether the line had a lot of business or a little, the investment was the same, and costs differed little. These two considerations meant that competing lines had a motive to cooperate in pricing, as they all needed consistent throughput more than short-term advantages.[124] The logic led to "pools," or market sharing, where each railroad was allocated sufficient market to allow it to cover its fixed costs and its operating costs and to make a profit. It could do this by being the price setter, with all other competitors setting the same or a higher price in the designated region.[125] This, of course, violated the traditional notion of a free market and irritated shippers who had to deal with fixed transportation costs while they themselves were at the mercy of different market forces. The end result was political action and the creation of regulatory commissions.[126]

TECHNOLOGY

The demands that resulted in the more professional management of railroads were also transmitted to the technical staff. The trend away from artisanal learning to a more scientific and experimental basis led to an increasing number of technical staff having formal engineering education. West Point was the first to provide engineering and scientific talent through its civil and mechanical engineering programs (1824),[127] followed by the University of the City of New York (1834), Norwich University in Vermont (1834), Rensselaer Polytechnic in Troy, New York (1835), the University of Alabama (1837), the Virginia Military Institute (1838), Union College in Schenectady, New York (1845), Harvard (1846), and Yale (1846).[128]

At least one technological development can be traced to the construction of the Western Railroad: the first use of a steam shovel, devised by a local mechanic, William Otis, cousin of Elisha Otis, the inventor of the elevator. His machine was patented in 1839 and went into use that year. Unfortunately, William also apparently died that year, at twenty-six.[129] Another innovation, at least for American railroads, was the gradual use of coal to fire the locomotives, due to the increasing difficulty of acquiring local wood for the fireboxes (wood-short Britain had been using coal to generate steam for decades). Only when the western Pennsylvania coalfields were linked with the overall railroad system in the 1850s, however, did coal become popular, and by the time the railroads emerged from the Eastern forests into the wood-deficient prairies, coal became a necessity.[130]

In the 1870s, standardization, particularly in the haulage of freight, became the norm as railroads settled into a more mature world of operations rather than expansion.[131] As a result of friction among railroads during the Civil War, in 1866 Congress required that all freight in interstate commerce be freely exchanged.[132] Efficiencies derived from using standardized machinery and practices more proficiently than the competition became the basis for profitability. Differences in track gauges — the Erie's were six feet wide, for example, while Southern railroads used a five-foot width — meant that when a standard width of 4 feet 8 inches was adopted, lines with incompatible gauges suffered serious financial strain as they were obliged to rebuild their systems to conform.[133]

Even time had to become standardized, if only so that passengers knew when to arrive at the station. Instead of determining the local time by when the sun rose and set, in 1883 the railroads agreed to standardize time in broad zones so that all lines in the same zone would operate on the same time.[134] The *Indianapolis Daily Sentinel* noted: "The sun is no longer to boss the job. People — 55,000,000 people — must now eat, sleep and work, as well as travel by railroad time."[135] Indeed, an earlier comment by another writer foretold the standardization of time: "The locomotive is an accomplished educator. It teaches everybody that virtue of princes we call punctuality. It waits for nobody. It demonstrates what a useful creature a minute is in the economy of things."[136]

Standardization also came to railroad equipment, car sizes, couplings, and even seating in passenger cars when, in 1867, the Master Car-Builders' Association was formed from companies that built rail cars for the different railroads, although so great were the discrepancies that a *Car-Builder's Dictionary* was not published until 1879.[137] The association's project was matched by efforts of ticket agents, baggage handlers, and even accountants, all of whom wanted the adoption of a standard terminology so that they could understand each other.

Finally, the railroads' demands created new businesses. For instance, in 1846, the Erie Railroad took the step of contracting with two New Jersey brothers, the Scrantons, who had developed an iron works in northeastern Pennsylvania, to provide 12,000 tons of rails. The Scrantons' rails were of competitive quality to those imported from Britain but just half the price, since they were manufactured nearby rather than having to be shipped across the Atlantic. The contract laid the foundation for the Scranton area iron and steel complex a quarter of a century before Andrew Carnegie moved into rail manufacture.[138] By 1850, the railroads were absorbing 10 percent of the iron and steel produced in the United States, and the proportion grew to 50 percent by 1875.[139]

The Railroads in Maturity and Decline

By the early 1900s, the effects of the railroads dominated the national economy.[140] Even in the cities, passenger transportation was based on trains, whether elevated, below ground, commuter rail, or interurban electric — or all together. But the railroad system was approaching maturity, and the maximum number of railroad companies, 1,564, was reached in 1907, although most of these were local short lines or controlled by larger companies. By that time, there were more than 350,000 miles of track, although the maximum extent, 429,883 miles, would not be reached until 1930.[141]

However, the increasingly heavy regulation of the railroads[142] and the gradual inroads made by road-based traffic caused the freight and passenger loads hauled by rail to began to slide, except for a temporary increase during the First and Second World Wars. Railroads now had to find specialized niches in the broad sweep of truck traffic, so different from their dominance of a century earlier.

By the 1970s, passenger service had retreated to a transcontinental service kept alive mainly for vacationers by federal subsidies and to some regional commuter lines, although even the latter did not exist in the fast-growing cities of the Sunbelt. By then, too, freight moved by rail consisted mainly of large quantities of raw materials such as coal. The rise of intermodal services in the following decades, where truck trailers are put on flatcars, saw some other freight begin to return to the railroads and, as the twenty-first century began, containers of goods from Asia were increasingly being hauled by rail from West Coast ports to markets in the East. The explosion of shale oil production in the northern plains has led to increasing volumes of crude oil being hauled by rail, since pipeline construction was encountering the same regulatory obstacles that had once hampered the burgeoning rail system.

22

Chautauqua, New York: The Democratization of Education

It being one chief object of that old deluder, Satan, to keep men from the knowledge of the scriptures... it is therefore ordered that every township... after the lord hath increased them to the number of fifty householders... shall... appoint one within their own town to teach all children as shall resort to him to read and write.
— Old Deluder Satan Act, Massachusetts, 1647[1]

Education then, beyond all other devices of human origin, is the great equalizer of the conditions of men — the balance wheel of the social machinery.
— Horace Mann, 1848[2]

Soon the Methodists will be shaking out their tents and packing their lunch-baskets for their camp meeting grounds....Rev. Dr. J. H. Vincent, the silver-tongued trumpet of Sabbath Schoolism, is

marshalling a meeting for the banks of Chautauqua Lake which will probably be the grandest religious picnic ever held since the five thousand sat down on the grass and had a surplus of provisions to take home to those who were too stupid to go.
— T. DeWitt Talmage, 1874[3]

Democratizing Education

Lawrence Cremin notes that humans receive knowledge about their environment, the world about them, in five main ways. First is through the home, as babies turn into children and then into youths. Until recently, this was probably the sole method of education for most humans. They learned informally from relatives, friends, and others nearby. A second way is through a church, which teaches people how to relate to the parts of their environment that are unknown and in the hands of a superior being. A third way is through a formal school, where information is provided in a concentrated and organized format. A fourth way is through self-learning, by accessing content providers such as books, newspapers, and, more recently, electronic media. A fifth is the workplace, which teaches skills and other learning relevant to the tasks at hand.[4]

Formal education was a basic part of the Puritan community.[5] As early as 1647, the Massachusetts colony passed an act requiring each town — later called "townships" in other parts of the country — of a minimum population to designate someone to provide for the literacy of children.[6] Rather than have some kind of overall supervision of education, the New England colonies left the responsibility for administration to the towns.[7] In truth, because of the entanglement of Church and State, the responsibility for "quality control" was left to the local minister, arguably the best-educated person in the community. Effective separation did not come until 1789.[8]

There were no qualifications for teachers besides religious ones, and the obvious corollary was that they had to be literate. Because of the practical need for land surveyors, some mathematics was desirable as well. Education was generally done in someone's home, and children of all ages were mixed together, often with the older ones helping to instruct the younger ones. Often, though, the older children went to school only in winter, when they were not needed as much at the farm, while the younger ones went in summer, to get them out of the way during the busy time.[9] The schools were supported by taxation and tuition, as the method of voluntary donations had not worked well among the cash-strapped colonists. It was assumed that parents should teach their children basic literacy before they went to school, but, in 1818, Boston absorbed this function into its primary schools.[10]

The needs of the society at the time required most children to learn trades useful for a farming community.[11] Education beyond elementary school was more restricted, with most young people apprenticed to learn work skills or simply following their parents into trades.[12] Young women received less education and had few options to pursue besides the domestic arts.[13] In 1821, Boston opened an "English" high school for boys who did not intend to go on to college, and in 1826 opened a female high school.[14] There were also some opportunities for higher education — Harvard, for instance, was created in 1636, shortly after the founding of the Massachusetts Bay Colony, as a school for ministerial training,[15] though its graduates gradually filled other positions as the society grew more complex. Yale was founded in 1701 to serve the Connecticut towns, and Dartmouth College, in New Hampshire, was founded in 1769.[16]

Key to education was the availability of printed works. Printing in New England began around 1636 in Massachusetts Bay with a psalm book.[17] In the late 1680s, writer and printer Benjamin Harris produced *The New England Primer*, based on a Protestant catechism; it became the most popular textbook of its time, and was still being printed and sold in many American colonies as late as 1766.[18] The first successful newspaper was the *Boston News Letter*, which appeared in 1704, two years before Benjamin Franklin, America's most famous printer, was born there.[19] At first, printed materials were generally purchased from Britain and shipped over — by 1720, there were only eight printing shops in the American colonies: five in Boston, one in New London, Connecticut, one in New York City, and one in Philadelphia. After the 1720s, however, there was a rapid growth in printing, including newspapers, and more presses were put into use. Much of what was printed, outside of government notices, were republished books from British sources, primers and chapbooks, and even a cookbook.[20]

By the time of the Revolution, there were three types of schools in New England: "petty," held in homes or small schoolhouses and limited to one or two years of learning to read and write; "grammar," held in more formal buildings such as day schools or boarding schools and consisting of six or seven years of training in languages such as Latin and Greek;[21] and "colleges," such as Harvard and Yale, where the education gradually broadened from theology into a variety of subjects, such as science and law — although, for the most part, law was still learned in a semi-apprenticeship fashion, by "reading" in the office of a practicing lawyer; medicine had even fewer formal requirements.

Enter Horace Mann

Horace Mann was born in Franklin, Massachusetts, on the border with Rhode Island, in 1796.[22] He went to Brown University in Providence, Rhode Island, graduating in 1819 and tutoring there for two years. He then went to the Law School at Litchfield, Connecticut, which produced the most prominent of New England lawyers at the time. He was a successful Massachusetts legislator and lawyer who combined an enthusiasm for railroads and water technology with that for social causes such as mental asylums and education. He also nearly ruined himself supporting a brother in a failed textile mill.

At the time, towns such as Lowell and Boston were developing school systems. Lowell, by 1840, was judged as having "built one of the most complete systems of public schooling in any community its size,"[23] and by 1850, it had forty-six primary schools, thirteen grammar schools, and a high school. Meanwhile, in rural areas, the support of schools, legislated or not, was poor and declining, especially as the impact on agricultural prices of the Erie Canal and then the railroads added to the pressures of overcrowding on poor land. This was of concern to manufacturers in the mill towns, who needed literate and numerate employees, especially women, who generally flocked in from rural areas, but the turnover was high, as they would leave as soon as they had saved enough to bring a dowry to a marriage. The mills and the railroads also needed engineers and surveyors, men who could not be trained in these pursuits without a basic education.

With this concern in mind, in 1837, manufacturer Edmund Dwight approached Horace Mann, then a State legislator, to act as secretary for the new Massachusetts Board of Education.[24] Dwight owned the water rights around the Falls of the Chicopee River where it flowed into the Connecticut River, and was successfully developing textile mills there along the lines of those established in Lowell a decade and more earlier. Mann's own goals included the interests of mill owners in an educated workforce, but were also loftier.[25] The Puritans believed that general literacy would render the population more likely to read the Bible and be able to understand God's plans for them. Mann was convinced that general literacy, regardless of faith, would create good citizens. To that end, he was a proponent of "common schools," meaning a school that was common to all the people, as opposed to a school for the common people,[26] which would leave the elite free to pay for better schools for their children, if they preferred.

Mann's focus was on the primary school, after which the role of the State should stop. This notion was quickly outdated, as John D. Pierce, a New Hampshire preacher, in discussions with Isaac Crary, a Connecticut-born lawyer, both living in Marshall, Michigan, devised a whole public system for the proposed new State. Crary became part of the constitutional process that preceded Michigan's application for statehood. He was also instrumental in getting a clause inserted into the new constitution that was unique in the constitutions of the States emerging from the Northwest Territory in that it provided for a comprehensive State educational system. Pierce was named Michigan's first superintendent of public education in July 1836.[27]

Mann felt that schooling should be compulsory and that schools should admit students of both sexes. Their teachers needed to be trained and schoolhouses should be well designed and well supplied with materials, and there should be free public libraries in each town.[28] He spent twelve years meeting with citizens all over the Commonwealth, and produced annual reports that formed the intellectual underpinnings that led to the acceptance of a relatively standardized set of school systems in most of the country.[29] Mann succeeded in reforming the Massachusetts system and, despite political controversies and Michigan's claims to the contrary, he is considered the "Father of the American Public School System,"[30] although the attribution of "American" had to wait for the imposition of some Yankee values on the South after the Civil War. As his wife, Mary Peabody Mann, related, "Mr. Mann had frequent correspondence with Southern gentlemen upon the subject [of universal education]; but it always ended in the conviction there could be no common schools established in a region where equality before the law was not even desired for all classes of white men."[31]

One of Mann's major reforms was the creation of three "normal schools" in Massachusetts over the period from 1829 to 1842.[32] They were colleges of a practical sort, providing a one-year curriculum that trained people to teach. Until then, educated young people, including Mann himself, could get a first job by becoming schoolteachers in a town school, but it was always seen as a temporary position, and women used it much as did their sisters in mill jobs — to put aside some money for marriage. In Mann's case, he taught because he had needed money to go to Brown University. Mann felt that teaching in common schools should be regarded as a profession with a body of knowledge and technique. Today, most "normal" (referring to norms rather than a common condition) schools have become State colleges or universities, and instruct students in a range of disciplines besides teaching. Mann succeeded in sparking a common schools movement that had three aims: to provide a free elementary education to all (white) children; to create a profession of teaching; and to have some sort of State control over education, thus standardizing its practices.[33]

Mann then moved on, succeeding John Quincy Adams as a congressman upon the latter's death in 1849.[34] In Washington, Mann advocated for the creation of a National Office of Public Instruction and became involved in antislavery politics.[35] Three years later, he was nominated for governor of Massachusetts by the Free Soil Party, a somewhat radical breakaway faction of the Whigs, but lost. Then, in 1853, he accepted the founding presidency of Antioch College in Ohio, which he hoped to make the western Harvard, but he struggled unsuccessfully to keep it out of bankruptcy.[36] Mann died at Antioch six years later, apparently from milk from a cow that had eaten some poisonous snakeroot weed.[37]

Two things strike me as odd about Mann's moving his young family to southwestern Ohio. Why should he have chosen choose to be involved in a start-up of this nature, a thousand miles to the west of his lifelong home, especially without checking the financial strength of the pledges made for its development?[38] In the end, the college's financial strain was a continual drain on his attention, and he tended to receive less salary than promised or none at all, forcing him to go on the Lyceum circuit to raise money to maintain his family.[39] Second, why did he choose a school close to the slave States in this tense decade, given his known association with antislavery politicians, such as Charles Sumner, and his recent candidacy for governor on the Free Soil ticket?[40]

By 1875, the general structure of American education had been set in the mold that it maintains today. A decade or so after the Civil War, a more-or-less standard set of grade schools, high schools, and colleges and universities was established, ranging from the practical to the academic in focus. Mann's sister-in-law, Elizabeth Peabody, introduced the German innovation of kindergarten to Massachusetts, and it began to spread throughout the country.[41] The result of these reforms was a decline in illiteracy from around 20 percent at the time of the Civil War to 6 percent by the First World War.[42]

Once all these parts of the broader educational system had been standardized, the debate shifted from whether the various types of educational institutions should be supported out of the public purse to how they might become more effective at serving their respective missions, and how these missions could contribute to the society. This latter took on more importance as the economy became more complex, blurring the clear messages about "good citizenship" or the integration of immigrants into American society. The complexity of the economy especially affected the higher levels of this structure, requiring investment in research, laboratories, technology, and extended training.

Finally, by this time, the country was full of newspapers, magazines, journals, and books. The European indentured apprentice system had never worked well in America, primarily because Americans could move around so easily — apprentices could just move on and never be located, as did the young Benjamin Franklin. Instead, employers simply acquired machinery that reduced individual job tasks to as few simple functions as possible, which also allowed labor, whether immigrants or native-born, to be trained quickly for tedious jobs regardless of their education beyond minimal literacy.

The Lyceum Movement....

Like quicksilver, one type of education seems to slip through this categorization: a form of adult lifelong self-education or mutual education. Sometimes it is connected to religion, sometimes to politics, sometimes to science and letters. It was not a precondition for democracy or citizenship, but more a result of these — something based on a people that is already literate.

In early New England, town meetings acted as a kind of "school." Attendance at the meetings was open, but speaking was restricted to males over age twenty. Town meetings required, or led, their participants to learn about issues arising about a broad range of topics: school issues and property details, personal behavior, the use of alcohol, the condition of roads, and so on. These meetings were a form of direct democracy, where every adult male had a voice. As town populations grew, the town meeting fell away, to be replaced by the election of "selectmen," and expertise moved from the group to a smaller number of people who made this learning a part of their lives.

The interest of New Englanders in "practical knowledge" led to the Lyceum movement. In October 1826, Josiah Holbrook, a well-to-do farmer-turned-amateur-scientist, published an anonymous letter in the *American Journal of Education* advocating a method for mutual instruction among adults: "It seems to me that if associations for mutual instruction in the sciences and other branches of useful knowledge could once be started in our villages and upon a general plan, they would increase with great rapidity and do more for the general diffusion of knowledge, and for raising the moral and intellectual taste of our countrymen, than any other expedient that can possibly be devised."[43] Holbrook, a Yale graduate, spent his life trying out new methods of instruction, new types of schools, and even a utopian village. When his "Agricultural Seminary" failed in 1825, he began lecturing on scientific subjects in towns throughout New England.[44] The experience led him to advocate what then became known as a Lyceum, meant to be a lecture-demonstration program that contained elements of mutual instruction as well as some relevant outside scientific instruction.[45] Holbrook was a close personal friend of Horace Mann's, and Mann became a popular lecturer for Holbrook's Lyceum movement.

The reaction to Holbrook's letter was immediate, given the standards of the time. By the end of 1826, Lyceums had been organized by textile workers in Worcester County, Massachusetts, and then in parts of Connecticut.[46] By 1828, more than fifty branches had been organized, by 1831 as many as a thousand, and in 1839 there were perhaps five thousand, spread from New England to Missouri to Florida, though they never really flourished in the South.[47] One Lyceum, in Concord, Massachusetts, was organized in 1829 by, among others, Ralph Waldo Emerson, Henry David Thoreau, and Amos Bronson Alcott, the father of author Louisa May Alcott.[48] In 1838, a lawyer in Springfield, Illinois, was asked to speak at a local Lyceum against mob violence in politics. Abraham Lincoln prophetically titled his lecture "The Perpetuation of Our Political Institutions."[49]

Angela Ray considers that the Lyceum movement evolved through a process of expansion, diffusion, and eventual commercialization.[50] The Yankee culture incorporated a Puritan tradition whereby people came to listen to especially effective preachers. As the society lost its overwhelmingly religious overtones, this tradition changed to an interest in hearing people speak on all manner of topics, and foreign writers, scientists, and philosophers were asked to address crowds all over New England, not unlike today's TED Talks.

Similarly, as the Lyceum movement developed, the idea of mutual instruction gave way in part to addresses by traveling celebrities, which gradually took on the trappings of a commercial production. Speakers such as Emerson and Mann used the Lyceum circuit to supplement their incomes, especially as fees got larger.

This trend took off with the expansion of the telegraph and the railroad in the 1850s. The telegraph led to news organizations, such as the Associated Press, which could instantly supply member newspapers all across the country with stories about the sayings and doings of celebrities, while the railroads cut speakers' travel times to days instead of weeks. Where Emerson saw a speaking trip from Concord, Massachusetts, to New York City as a chore in the 1830s, in the 1850s Chicago was a relatively easy and comfortable trip. Mann could cover lecture engagements in Troy, Buffalo, Cleveland, Toledo, and Chicago all in about two weeks. Generally, the more celebrated personalities — such as Samuel Clemens (Mark Twain), Alexander Graham Bell, and explorer Henry Stanley — commanded fees so high that they were booked only in cities where the paid attendance by large crowds would cover the costs. These Lyceums were mostly held in comfortable auditoriums and became social events.

By then, the original intent of supporting mutual instruction by mechanics and other workers had largely disappeared. One commentator noted in 1868 that, "with the name 'Lyceum' is also passing away the [traditional] 'Lyceum lecture'. The scholar recedes from sight, and the impassioned orator takes his place. There is no time for Longfellow to analyze 'Dante' or for Lowell to explain *Hamlet*, while [Charles] Sumner thunders the terrors of the Lord against a delinquent President, or [suffragette] Anna Dickinson pleads for the enfranchisement of one half of the human race."[51] Though they had languished after the recession of 1857 until the postwar period, for more than a decade after the Civil War the Lyceum tours were a relatively big business.

One of the more successful Lyceum agents was a former antislavery journalist, James Redpath, who started the Boston Lyceum Bureau in 1868. Redpath's family had emigrated from England to Kalamazoo, Michigan, where he took up journalism and became heavily involved in the antislavery movement and related social causes. He came to know and like John Brown, the antislavery activist, in 'Bloody Kansas' in the 1850s and, after Brown's execution in Virginia in 1859, wrote the first biography of the Northern 'martyr'.

Redpath was one of the "carpetbaggers" who found work in the South during Reconstruction, but soon found himself out of a job for being too extreme in his sympathies for the freed slaves. Oddly enough, he later helped ghost-write Jefferson Davis's history of the Confederacy.

Redpath supposedly created his speakers' bureau after he found out that Charles Dickens had had a difficult time organizing an American tour,[52] and the Redpath Lyceum Bureau became the premier agent for the Lyceum circuit. In

1875, with his assumption of the editorship of the prestigious *North American Review*, Redpath sold the company to his employees and moved on to the cause of Irish independence. He died in New York City in 1891 after being run over by a horse-trolley. We will encounter the Redpath agency again in the story of Chautauqua.

....*and Sunday Schools*

The original rationale for Sunday schools is offensive to modern sensibilities. They were started in England in 1782 when Robert Raikes, a Gloucester newspaper publisher, began to support the teaching of child workers who were unable to attend regular school because they worked all week and, in any case, could not afford the tuition.[53] The teachers were volunteers and laypeople and the text was the Bible. Within four years, an estimated 250,000 children were attending English Sunday schools.

Child labor was common in the early textile mills in both Britain and America. Children as young as six were put to work around the spindles because they were small enough to get under the machinery. Besides, since whole families worked in the mills and had been used to sharing chores on their farms, the idea of children working did not seem wrong. Reaction set in against child labor only after a couple of generations of city dwellers began to see things in a different light and, at least in America, both the technology changed and educational demands took younger children out of factories. But even in the early 1900s, it was still acceptable for young boys to be sent down into mineshafts that were too narrow for full-grown men to work. My grandfather, a Pennsylvania coalminer, went into the mines in 1903 at age thirteen.

Raikes's concept crossed the Atlantic, first to Virginia, where the Methodist preacher and Bishop, Rev. Francis Asbury, founded a Sunday school in 1786. Another early effort was made in Rhode Island, where Samuel Slater had begun the American textile industry in the 1790s. He instituted a like program for his child laborers. A Sunday school was organized by the First Day Society, a group of merchants and clerics, in Philadelphia in 1791, and others gradually were organized in larger cities and towns. Over the next forty years, the idea of Sunday school as a way of making city children literate spread, especially in the central and southern parts of the country.[54] It was followed by increasing pressure to provide free public education and to restrict children's hours of work.

By the 1830s, however, this form of Sunday school began to die out, to be replaced by those of various church denominations, almost entirely Protestant, as a way to instruct children in matters of faith and morals.[55] In particular, the change was prompted by a feeling that, with the inflow of immigrants, such methods would help keep the country Protestant. In 1824, the First Day Society expanded into the American Sunday School Union. Its mission was to promote an interdenominational form of schools, materials, and training and to add this to the missionary work going on among the Americans moving west to the frontier. Over time, various denominations decided to develop their own programs within this framework; Methodists, for example, formed the Methodist Episcopal Sunday School Union in 1827.

After the Civil War, the Sunday school movement became a part of the reaction to the immorality that seemed to permeate the age, whether from war profiteering, public vice, or leftover violence in a society that had become inured to it on a grand scale. Sunday school seemed a way to redeem the children, at least. In 1872, a national Sunday School Teachers' convention agreed to continue the publication of common materials, in cooperation with the British Sunday School Union, but to encourage each denomination to develop its own interpretation of them. In response to this decision, the Reverend John Heyl Vincent was given the responsibility of training Methodist Sunday school teachers and, like educators before him, he thought of adapting to the task the "normal school" concept used for training teachers. His "national normal school" — the Chautauqua Institution — would be augmented by regional ones elsewhere.[56] Nonetheless, Sunday schools offered no certification and were not taught by certified teachers. The teachers were volunteers, mostly women, though a number of successful men also volunteered their time — John D. Rockefeller was an early example, and in our day author John Grisham, comedian Steven Colbert, and former president Jimmy Carter have taught Sunday school.

To Chautauqua

As one drives southwest from Buffalo along US 20, the plain along the shore of Lake Erie becomes narrower and the ridge behind, the edge of the Alleghenies, more defined, and it remains that way until the outskirts of Cleveland. At Westfield, some seventy miles from Buffalo and fourteen from the Pennsylvania border, if one turns inland on New York Route 394 and begins to climb the ridge, half a dozen miles farther on one clears the top and descends to Mayville, a small town at the northern end of Lake Chautauqua.

The lake is geologically the same as its Finger Lake counterparts, a hundred miles to the northeast, but it and a few other similar, small lakes do not share this appellation — there would be too many Fingers, then, I guess.[57] Lake Chautauqua curves in an arc now bisected by an expressway; its eastern end is a source of the Allegheny River, which flows south until it meets the Monongahela in Pittsburgh to form the Ohio River. It is an odd geography: the lake is but a few miles from Lake Erie, whose waters eventually join the Atlantic far to the northeast, but Chautauqua's waters flow south into the Mississippi and eventually to the Gulf of Mexico.

A few miles past Mayville on Route 394, along the west side of the lake, is the site of the Chautauqua Institution.[58] It is a gated community in a unique sense: a summer residence, by and large, but open to the paying public like a kind of educational Disney World. And though there are entertaining features such as golf courses, the beachfront, and boat rides, this is a village where people live for a day, a week, or a season, and some all year round.

The casual visitor enters the 250-or-so-acre village site by purchasing a ticket for the half-day or day one expects to be there. There is little motorized traffic, but bicycles are common. The streets are narrow and the buildings tend to be Victorian in style, if not in age. The village has a wooden hotel, the Athenaeum, which dates back to around 1880, and many of the houses or rooms in houses are available for rent. In the summer, as many as 7,500 people live on the grounds, in cottages, or hotels.[59] A large relief map of Palestine sits at the shore, a remnant of Chautauqua's origins as a religious camp meeting site — indeed, the place maintains a discreet religious demeanor in that a number of church groups are represented, and the tone is an inclusive one. The whole atmosphere of the summer season lends itself to thought and discussion, and the general atmosphere is one of calm and peace. The presentations are many and varied, with the "weeks" of the summer having individual themes. Seating is a democratic first-come, first-served, and the place has a tone of orderliness, both qualities that Reverend Vincent felt were needed in more modern times at a camp meeting. Other places, short of those inside sublime scenery, might generate similar emotions, but my wife and I have not seen them.

The Chautauqua Institution dates back to 1874, and was perhaps the single most influential purveyor of learning and culture to Americans of all ages before television.[60] At one point, three-quarters of a million people were involved in its programs and, in the circuit Chautauquas, perhaps 8 million would attend in a season. Today, thousands still flock to the Chautauqua Institution to get their culture and learning in person, much as people still go to sporting events they could more easily watch on television.

In 1872, Vincent began to look for a venue for his training project, and met with a potential backer, Lewis Miller, an Akron, Ohio, businessman and Sunday school teacher.[61] Miller dissuaded Vincent from his initial idea of city-based training programs in favor of a country leisure setting.[62] Their partnership turned out to be a good match, with Vincent providing the vision and oratory and Miller the organizing, practical execution. Miller felt that volunteers would be more likely to attend if sessions could be combined with summer vacations.[63] The combination of leisure and learning would also overcome a public morality problem. Puritan ethics about "idleness being the devil's workshop" still ran deep;[64] as well, antipathy toward the antebellum stereotype of a leisured Southern aristocracy living off the backs of its slaves added to the sensitivities of postwar workers and businesspeople. Given that sloth was sinful, a Sunday school assembly in a country setting could combine this new leisure activity with meritorious labor through learning.[65]

The idea of fashioning productive uses for leisure time did not, of course, begin with Chautauqua — Vincent and Miller were drawing on an American tradition that stretched back to the Puritans, found expression in the maxims of Benjamin Franklin, and manifested considerable strength throughout the early nineteenth century. Americans had been using their leisure in the pursuit of self-improvement since long before most would have contemplated a "vacation."[66]

Vincent had been contacted by a Reverend James G. Townsend about a site at Lake Chautauqua, where, coincidentally, Miller already had some involvement, and, in the summer of 1873, Miller and Vincent decided to investigate.[67] They were taken with the location and arranged to rent the campsite from the Fair Point Camp Meeting Association for two weeks the next summer.[68] From that initial idea and its promotion through the *Sunday School Journal*, the Chautauqua Institution took off. At the first meeting, in 1874, 2,000 people showed up, taught by 142 teachers.[69] Vincent and Miller decided to take the site for the whole summer the next year.[70] As an attraction for potential campers, Vincent prevailed upon an old friend from his Illinois days, President Ulysses S. Grant, to visit Chautauqua in 1875.[71] Grant started a trend that continues to this day: Bill Clinton came in 1995 to prepare for his second campaign; in total, nine presidents have spoken at Chautauqua over the years, as have countless other famous people.

Before the 1876 season, the Fair Point Camp Meeting Association, in financial difficulties, agreed to transfer its ownership (and presumably debts) to the Sunday School Assembly for a dollar. The village name was also changed to Chautauqua,[72] and its general layout and ambience were based on ideas from the Oneida Colony and other utopian communities of the time.[73] The 1876 season included a "scientific congress" to take advantage of the popularity of that year's Philadelphia Centennial Exhibition.[74] In general, talks on Bible history, geography, and Sunday school pedagogy outnumbered sermons three to one from the start.[75]

In 1878, Vincent announced the formation of the Chautauqua Literary and Scientific Society (CLSC), perceived as a book-of-the-month-type organization coupled with local discussion groups and a four-year cycle of reading leading to "graduation." He hoped that ten attendees might sign up; two hundred did in the next three days, and eight thousand eventually registered for the first "class." By 1882, 180,000 had registered, and by the 1920s, 300,000.[76] Eventually, 12 percent of all registrants completed the program of instruction. The success of the CLSC resulted in the decline of the original Sunday school impulse, since the same instruction could be given teachers through the materials provided throughout the year by this means.[77] By the late 1890s, ten thousand reading circles had been formed, a quarter of which were in small towns and villages of fewer than five thousand people, a testament to Vincent's observation that "there is a hunger of mind abroad in the land."[78] Iowa alone had a hundred CLSC circles in 1885. Members met either in private homes or in the local schoolhouse. They served another purpose as well, as shown in a note a young Iowa woman received: "Will you go to Literary with me a-Friday night? If so, please let me know by a-Tuesday night."[79] Henry Ford, once Miller's sales representative in Michigan, met his wife at a CLSC.

Both these readings and the summer program featured a theme often called the Social Gospel.[80] They presented material on labor problems, the emancipation of women, the peace movement, and the arts and crafts movement, among many other subjects. Referring to the range of subjects available at the summer session, Edward Eggleston, editor of the *Sunday School Teacher*, once half-jokingly observed: "You can learn Greek, Latin, Hebrew, and for aught I know Choctaw here. You can learn 'Americanized Delsarte' and Penmanship and Pedagogy and Exegetics and Homiletics and History and Rowing and Piano Music and fancy bicycling and singing and athletics and how to read a hymn in public and the art of writing family and Business letters [and] everything else except dancing and whist."[81] Nor drinking — no alcohol was allowed on the grounds --- and it is said that the Women's Christian Temperance Union got its start here.

The appeal of the CLSC and the various other manifestations of the Chautauqua Assembly to women, in particular, meant that the women's rights movement and progressive politics came together under this umbrella.[82] The only female senior member of the Chautauqua organization, however, was CLSC director Kate Kimball, who devoted her life from age eighteen to the promotion and continuing success of the CLSC.[83] Kimball's influence with the CLSC women gave her, and them, a large say in an otherwise male-dominated organization. For instance, although Vincent personally objected to female suffrage, he still found himself addressing a welcome to a sea of Chautauqua women sporting the yellow ribbons of suffrage at one summer Women's Day.[84] Another female personality employed at the Chautauqua establishment was the editor and writer Ada Tarbell, later a famous 'muckraker'.

Other innovations followed as the Chautauqua Institution kept reinventing itself.[85] In 1879, language training in Latin, Greek, and Hebrew was added to the program, which also marked the beginning of a summer school curriculum. Correspondence courses were introduced in 1881, and a charter giving Chautauqua university status was secured from the New York Legislature in 1883. A young Hebrew instructor, Baptist minister William Rainey Harper, whom John D. Rockefeller would later install as the first president of the University of Chicago, was made director of instruction.[86]

Soon, as many as a hundred thousand people — not necessarily all Sunday school volunteers — were spending part or all of their summer hearing sermons by Vincent and others and speeches by famous people and exchanging ideas among themselves.[87] By the mid-1880s, the initial reason for the creation of the campsite on Lake Chautauqua had given way to a broader mission: that of adult education mixed with the Social Gospel. As Vincent wrote in his *The Chautauqua Movement*, "[t]he theory of Chautauqua is that life is one, and that religion belongs everywhere. Our people, young and old, should consider educational advantages as so many religious opportunities. Every day should be sacred. The schoolhouse should be God's house. There should be no break between Sabbaths. The cable of divine motive should stretch through seven days, touching with its sanctifying power every hour of every day."[88] James Garfield, in his 1880 presidential campaign, said of the Chautauqua Institution: "It has been the struggle of the world to get more leisure, but it has been left to Chautauqua to show what to do with it."[89]

Chautauqua's golden age lasted from its inception in 1874 until about 1925.[90] It was a major force in adult education, and was even presented in novels of middle-class aspiration such as Sinclair Lewis's *Babbitt*. It was the national podium for politicians (President Theodore Roosevelt spoke there in 1905), progressive reformers,[91] and scientists. Although its assemblies were supposed to be interdenominational Protestant,[92] classless, and nonracial, this aspiration was never entirely attained, as its setting was most relevant to middle-class, increasingly suburban white families. In this sense, Chautauqua was caught between Coney Island and the White City of the 1893 Columbian Exposition, or a bit like today's Las Vegas and Orlando. Further, real estate speculation inside the Chautauqua grounds began pricing accommodation out of reach of the lower middle class.[93] Eventually, Chautauqua became a victim of its own success as it bred a host of specialized competitors. The registration for courses and for the CLSC fell off, many of the reading circles were absorbed into the Women's Clubs that were springing up in the cities, and correspondence courses were hurting from the growth of private sector adult education offerings. Meanwhile, the focus of the Chautauqua summer program shifted toward recreation and the more sophisticated arts. By the early 1900s, people were "choosing baseball over Whittier, or croquet over classes."[94] In great part, the change was due to the automobile, movies, and radio. The isolation of smaller communities had lessened with improved transportation and communication, while many rural people had moved to the city.[95]

Today, Chautauqua has reacquired a certain degree of its past status, and its role as a national podium. Each year, 150,000 people visit its public spaces, shops, hotels, and programs. The CLSC remains an outlet for readers, and the religious mission has embraced many more Protestant denominations and other faiths than before. According to David McCullough, "Chautauqua is part of the American imagination. It belongs with Concord Massachusetts, or Hannibal Missouri, or Springfield Illinois, as one of those places that help define who we are and what we believe in. It has its own mythic force."[96]

"Morally, We Roll Along" . . . .[97]

The Chautauqua experience motivated groups in other places to develop their own versions. Some were early — the South Framington, Massachusetts, camp meeting was reorganized in 1880 as a chautauqua, and the Hedding, New Hampshire. camp meeting was so reorganized in 1886.[98] In all, twenty-two camp meeting organizations became chautauquas between 1874 and 1899. By 1900, a hundred other places were sponsoring temporary chautauquas at camp meetings or other wooded sites, events that might last only a week or two.

For instance, the Gladstone, Oregon, chautauqua began with the formation of the Willamette Valley Chautauqua Association in 1894. Eva Emery Dye, a former student at Oberlin College and a member of the association, approached Judge Harvey Cross, an Oregon pioneer and fellow member, to lease the association a wooded tract by a lake for a dollar a year, and the first of thirty-four successive annual chautauquas, ranging in length from three days to two weeks, was held there. One of the first talks at the inaugural meeting was titled "Remove the Cause of Strikes," by equal rights advocate Abigail Scott Duniway.[99] Others were copies of the Chautauqua village, with meeting venues, permanent cottages, a hotel, and stores. My wife and I have visited a number of these sites, including sites at Lakeside, Ohio, on the shores of Lake Erie;[100] Bay View, Michigan, on Lake Michigan; Boulder, Colorado; and Pacific Grove,[101] near Monterey, California. All are patterned as late-Victorian villages, but are now in different states of existence. Lakeside is functional, with a summer gate, a hotel, and a village that is active in the summer. Bay View still has a few chautauqua activities, but the village seems more like a quaint suburb of the nearby town of Petoskey, and is bothered by a main highway east and north that cuts off the community from the shore.[102] Pacific Grove still has a Chautauqua Hall as the town centerpiece, as well as some programs, but has otherwise been made over into a small suburban seaside town. And Boulder's Chautauqua village and park on the south side of the city remains an attraction that features both a spring and a summer program.

Another spinoff of the movement was what became known as circuit chautauquas. At first, although towns developed their own chautauquas, which might include local and regional ministers as well as notables attracted by fees, the quality and length of these efforts varied according to organizers' abilities. Gradually, they were superseded by more professionally organized programs, though not without some resistance from locals who wanted to keep their own unique programs. The circuit chautauquas could offer a cost-competitive, quality program for less than a local community group could provide.[103] The concept was devised by Iowan Keith Vawter, who had purchased a one-third interest in the Redpath Lyceum Bureau in 1901 and took over the Redpath office in Chicago. Vawter saw the Lyceum as focused by then on larger towns and cities with comfortable auditoriums and the ability to charge high fees for celebrities, while the smaller towns struggled to put on chautauquas with low prices in camp-like settings.[104] He saw how to reconcile these differences by providing these towns a complete package: Redpath-Vawter would supply a large tent, a team of workers and managers, and a stable of speakers and other performers, all of whom would be committed for the summer season and sent out to different towns on a set schedule. In 1904, he lined up fifteen towns in Iowa as an experiment, and although he lost money, he was convinced there was a way to make it work. By 1907, he was running a circuit of thirty-seven towns, and the circuit chautauquas were returning a profit.

In 1909, Vawter offered the communities a seven-day chautauqua, akin to one of the "weeks" that the mother Chautauqua back in New York had devised to break up its long summer season. The train would arrive on Monday and the tent would be set up, with the first event taking place that afternoon. After a final Sunday morning event, the tent would be taken down and moved to its next location overnight. As time went on, even this procedure became more sophisticated, with performers tightly scheduled so that they could do a number of appearances a week in different communities, allowing Vawter to have a number of his packages going on at the same time.

Vawter realized, as well, that the chautauqua had to involve the community to succeed. In a letter to someone who had complained about his requirement of "contracts" with the community, when traveling circuses did not ask for one, he noted: "Over twenty years ago I told some friends that a Chautauqua would be a success just as long as it was of sufficient importance in the community that a group of men and women would do a lot of hard work for the good of the Community. Chautauqua never has [been], and never can be, self-supporting from a purely business standpoint."[105] It would proceed to offer a number of events spread over the next few days or perhaps a week before moving on to another site. Famous preachers and orators would be presented, there would be an orchestra, book discussions, artistic lessons, and possibly other forms of entertainment, all things that normally might be unavailable in the town and particularly not on the surrounding farms. They would contain perhaps two-thirds music and dramatics and one-third lectures and sermons.[106] Because actors were held in low esteem by good churchgoers, Vawter had some of them perform singly as "readers" or "elocutionists," thus providing an audience with some good dramatics, but not by actors in a play.[107] The enthusiasm would then build in a community. The local guarantors would work at selling tickets to the different events, and people would plan their lives around their Chautauqua week. One woman told journalist Ida Tarbell her thoughts:

> It is a great thing for us, particularly for us younger women with growing children. Here are none of us in this town very rich. Most of us have to do all our work. We have little amusement, and almost never get away from home. The Chautauqua brings us an entire change. We plan for weeks before it. There is hardly a woman I know in town who has not her work so arranged, her pantry so full of food, that she can get to the meetings at

half past two in the afternoon, and easily stay until five. She gets her work done up for Chautauqua week.[108]

Circuit chautauquas went throughout the West and the Middle West and into different parts of Canada. Most companies were either affiliated with the Redpath Bureau or simply copied the innovation. The First World War interrupted the growth of circuit chautauquas to a degree, but their real heyday was in the early 1920s. In 1920, twenty-one companies were operating in North America on ninety-three different circuits visiting more than eight thousand towns, and with total attendance estimated at more than 35 million. By the end of that decade, however, the concept of the circuit chautauqua, like the "mother" Chautauqua itself, was in decline for many reasons, including a vast increase in and oversupply of the number of companies, decreasing quality, and the shifting balance from education to entertainment. As one performer put it, "[t]he old pleasant idea of Chautauqua being a big community enterprise, a big picnic was gone. It was [now] regarded with a thoroughly professional attitude. We were a company of players who came in to give a show, and we were judged, condemned or acquitted entirely on the merit of our performances. The old friendly, personal note was gone."[109]

Indeed, the changes that America was experiencing in the 1920s were more deep-seated. As a new generation grew up with no direct connection to the family farm, the icons of the nineteenth century — the idyllic small town with its little red schoolhouse, evangelistic Christianity, and the platform oratory of a William Jennings Bryan — were giving way to an expanding rail net, rural free delivery, electrification, the telephone, radio, and talking pictures. Most important, the changing status of women affected the circuits, which, for many rural women, seemed to function in the same manner as women's colleges had for urban, upper-class women. This opportunity gap decreased after the First World War. Vawter, I think perceptively, once wrote to his fellow managers: "I still insist that the radio did not materially affect [the] lyceum and [the] Chautauquas, but rather [it was] the advent of Country Clubs and Dancing Mothers." The Great Depression brought a final end to the circuits, the last one folding in 1932. In effect, the chautauquas arose out of a passing need, and the need had passed.

Since the 1980s, however, local chautauquas have begun once again to be staged in some communities in the West and Midwest, perhaps out of nostalgia, perhaps out of boredom with "canned" entertainment on television, or perhaps because of the appeal of an outdoors, environmentally interesting, alternative vacation.

The Resurrection (Perhaps) of the Town Hall

Town Hall meetings descended from New England town meetings. This description of eighteenth-century Boston shows the unique cultural characteristics of effective town meetings:

> In Boston, 2500 men out of a total population of some 16,000 were entitled to vote at the Town Meeting, whose sessions at Faneuil Hall were rowdily eloquent. Pitt [a British Prime minister of the time] and his generation like to imagine themselves so many Ciceros. But Boston really was the size of the optimum face-to-face neo-Athenian democracy imagined by Rousseau in the 1760s; and its orators — James Otis, Sam Adams, Josiah Quincy, and John Hancock — believed they were so many Demostheneses....
>
> Boston, especially, was a book-crazy, news-hungry, astonishingly literate (70 per cent for men, 45 per cent for women), habitually litigious place, with a strong religious disposition to see the world in terms of the embattled forces of good and evil....The city prided itself on its civic culture: its grammar schools and colleges, gazettes and libraries. And its educated leaders made conscious efforts to reach a popular audience.[110]

While these meetings did have decision-making capability, as the population grew the practice of selecting representatives, or "selectmen," began to replace this form of direct democracy. As well, those who were permitted to speak gradually expanded to include anyone who wanted to attend, and Town Hall meetings became forums for the discussion of public issues, perhaps with an opening statement from a person of importance, but with the focus on people being able to present their side of questions of the day and to learn from each other. William Keith describes the lineage of Town Hall meetings as coming from New England to the Lyceums and through the Chautauqua.[111]

The modern version of the Town Hall forum began at the Cooper Union in New York City in 1894. Another venue was the Ford Hall in Boston, which today bills itself as "the nation's oldest continuing free public lecture series," with each lecture followed by an open debate. In 1935, NBC radio introduced "America's Town Meeting of the Air," and in 1948 ABC brought Town Hall meetings to television. The TV format consisted of prominent politicians, writers, and experts debating current, frequently controversial, events. The program was unique for its time in that it also allowed for questions from both the live audience and the viewing audience. It went off the air after less than four years. Since then, irregular Town Hall meeting formats have been used, most commonly by cable television and campaigning politicians. The Internet, too, has spawned an e-democracy movement that includes forms of participation.

Experiments of this kind have appeared on social media as well. In 2009, the White House media office devised an Online Town Hall featuring President Barack Obama answering the most common of a hundred thousand questions submitted online in a YouTube event. Indeed, the extraordinary number of questions posed points to the reason most Town Hall attempts in all media are not successful. They are hampered by the large number of citizens who want their opinions heard and the relatively short time that people are prepared to listen. Whether it is a relatively small New England town or a country of more than 300 million, the numbers make direct democracy, and even direct discussion, impossible.

Conclusion

Universal, free public schooling grew out of the combined effects of the Puritan religion and the creation of town political units around different congregations. Horace Mann transformed the aim of education from religion to citizenship, and his idea of common schools was enlarged upon in the Yankee frontier of Michigan, such that, by 1875, the structure of the present system was in place.

Interest in continuing, or adult, education ran alongside that in the schooling of children, but it was tied either to the politics of active citizens or to their educational needs while on the job. The Chautauqua movement represented a combination of all of these: the training of religious instructors, the expansion of democracy by improving the position of women, and the provision of new ideas and improved job skills to political activists. Whether Lyceums or chautauquas, the success of these activities led inexorably to their decline, as the providers of education found themselves taking a back seat to entertainment. Other bodies then picked up where these early educational movements left off, creating a welter of private and public adult educational opportunities to meet everyone's felt needs.

Epilogue

At Chautauqua, it is time to take a break and review what we have seen in our trip along the Yankee Road so far.

In the 1600s, a relatively large number of English people, along with a few Scots, Welsh, and northern Irish, bound together by a set of religious ideas they held to be the Truth, emigrated across the Atlantic Ocean to found new colonies. Because of their insistence on this Truth, emigrants from other lands avoided joining these colonists, regarding them in some ways as akin to the native North American tribes in their exclusiveness.

As they grew in number, they encountered others, who called them by the derogatory term "Yankees." One hundred and seventy years after the first of them arrived, many Yankees began to move westward across what is now upstate New York and northern Pennsylvania to take up better farmland and improve it, and then often to move on farther west to start again, a practice that continues to this day as people buy houses, fix them up for resale at a profit, then move on.

Along the way, the Yankees had some ideas about almost every part of

American life. Eventually, these notions became a significant part of the cultural underpinnings of society, at least in Northern States. The Civil War was fought largely, over whether everyone else in America should accept this complex of ideas. Appomattox settled that question. America would be organized in one way, not two. It would be urban, capitalist, and highly organized, and everything — religion, culture, literature — would be democratized. In short, America would become much like the country it is today.

Then, as mass immigration took hold from the 1860s until the end of the First World War, the few million Yankees were overwhelmed by Irish, Germans, Italians, Jews, and many others. For all practical purposes, the biological Yankees became subsumed by the newcomers, but not before they had trained them in "American" culture and practices. One way to see how the Yankees created what might be called the "American" cultural normal is to try to imagine an America in which these millions of immigrants after the Civil War had actually gone in equal numbers to both North and South, which were then still very different regions. But they did not, yet later historians led us to believe the America of the latter part of the nineteenth century was a whole unit. George M. Cohan expressed it best in 1917 when he said musically that "the Yanks are coming," a refrain picked up by people in all the Allied countries. To them, all of America was Yankee...but the reality — then, and to some extent now — is that it is not so.

In these volumes, I follow this "tribe" of Yankees along a road — which began as a series of native American trails and early pioneer tracks and was designated as US 20 in the great highway numbering of 1926 — that goes through the heart of their migrations. Today's Yankee Road is unique among America's highways in still reaching from the Atlantic to the Pacific, from tidewater in Boston to Newport, Oregon. Within a band of about sixty miles north and south of US 20, all the way across the country, are places that connect to important Yankee innovations of all sorts.

In this first of three volumes, most of the chapters have dealt with the Yankee ideas that emerged in the decades before the Civil War — the period of Yankee expansion. Three of the chapters — on Frederick Taylor and scientific management, the creation of the Route 128 technology core, and air-conditioning pioneer Willis Carrier — tell of ideas that came later, but each shows how Yankee ideas spread to the rest of America and then on to the rest of the world. The other chapters show a Yankee character dominated by practical concerns that grew out of the often hostile geography of New England: how to support many people on rocky and infertile soil by moving into industry, both manufacturing and cultural. These Yankees learned how to produce interchangeable parts, which made Taylorite management methods useful, if not absolutely necessary. They overcame the legal impediments to mass production by adapting corporate structures to their business needs. They democratized the economy with a capitalism that provided clocks, clothes, shoes, muskets, and a myriad other products that everyone wanted. They spread their ideas on learning — the education needed by a productive workforce and educational opportunities for rural people, especially women — across the country.

Yankee settlement patterns reflected their Puritan spirit of organization. From the surveying of the Northwest Territory in advance of most settlement, to the finding and purchasing of land by so many men before bringing out their families, to the moving west of whole colonies and even towns, using the Erie Canal and the Great Lakes, the Yankee experience is at odds with the myth of the lone Conestoga wagon bumping through the wilderness in search of a place to call home. The Yankee frontier was more like an extension of the Puritan frontier: organized and communal. The Mormon experience is the most extreme version of this pattern of settlement.

Yankees generally were not simply subsistence farmers, but were integrated into the global economy early on. Whether it was the sale of pot-ash from tree-fellings or the impulse to grow wheat, a grain that was more resistant to rot and other damage, Yankees had one eye on the world demand for what they could produce. Some who did not move west voyaged to markets in East Asia or to far-off oceans to look for whales whose blubber could be made into lighting oil. The capital for the Merrimack River textile mills came from profits made in this seagoing trade. Yankee insurance companies and banks formed the backbone of the American financial system of the time.

Yet, despite the broad coverage of Yankee innovations in these chapters, I have had to pass over many areas of Yankee endeavor, including most of the arts. Except for Samuel F.B. Morse, whose paintings are well regarded but whose fame rests with the telegraph, the conservationist-philosopher, Thoreau, and, to some extent, the Romantic non-Yankee, James Fenimore Cooper, I have neglected the writers, such as Ralph Waldo Emerson, Nathaniel Hawthorne, and Margaret Fuller. I gave scientists more attention: "Goody" Pincus, the Sillimans, father and son, and the researchers who created the field of information technology. I have also paid little attention in this volume to the plight of the slaves, but that part of the story is planned for later. Finally, I have devoted a relatively greater space to women's rights and the improvement of their status in Yankee culture, especially up to the Civil War. To me, the central innovation here was the erosion of the legal code of *coverture*, not the long struggle to achieve suffrage.

I also passed over some other important issues. Yankees fought a couple of nasty wars with their Indian neighbors in the 1600s, but with the possible exception of militia activity at the beginning of the Revolution, they have never dominated the American military. They largely sat out the War of 1812; others fought the Indians in Indiana and Michigan and later in Illinois. They were not prominent in the Mexican War, nor were Yankees particularly distinguished in the Civil War, exceptions like the defense of Little Round Top by the Mainers at Gettysburg notwithstanding. It is possible to trace partial Yankee ancestry in Grant and Sherman, but for the most part their success lay in the invention of technological total war, not in battlefield tactics.

Where Yankees did shine militarily was in the area of technology. The decades-long labor to produce genuine interchangeable parts for military use created the American industrial boom of the late 1800s. This was as big a research and development exercise as was the space program a century later. New England intellectual and institutional activity helped the military create the first usable computer in the Second World War, along with the nuclear industry.

So far, we have traced the Yankees to the Pennsylvania border along Lake Erie. In the next volume, we will continue along the general route of US 20 towards Chicago; in the third, we will follow those who went west to the prairies and the Oregon country. Gradually, the picture of the classic Yankee fades a bit, as other people who adopted Yankee culture with enthusiasm come to the fore. But there's still room for Perry, Drake, Tarbell, Rockefeller, John Brown and a host of others as we travel the Yankee Road.

Notes

Introduction

1. Diogenes Laërtius, who lived in the third century AD, about 600 years after Xenophon, included a short sketch of him in his *Lives of Eminent Philosophers*, Book 2, trans. R.D. Hicks (Cambridge MA: Harvard University Press, 1925), chap. 6, n59.

2. Douglas Brinkley, *The Majic Bus: An American Odyssey* (San Diego: Harcourt Brace and Company, 1993), 12

3. Matthew Algeo, *Harry Truman's Excellent Adventure: The True Story of a Great American Road Trip* (Chicago: Chicago Review Press, 2009), 1.

4. Jack Kerouac, *On The Road* (1957; New York: Penguin, 1991), part 2, chap 3.

5. Words by Robert Hunter, first performed August 18, 1970.

6. Many books and articles have been written about this campaign. A useful, recent account is Robin Waterfield, *Xenophon's Retreat* (London: Faber and Faber, 2006).

7. Another Lowell diner, which held out until 1991, was Arthur's "Paradise," which Kerouac often visited and from which, it appears, he took his *On the Road* pseudonym "Sal Paradise"; John J. Dorfner, *Kerouac: Visions of Lowell* (Raleigh, NC: Cooper Street Publications, 1993), 52. Nearby Worcester is said to be the home of the first diner.

8. The title of the book was a takeoff on Neal Cassady's phrase for being high: "going on the road."

9. Steve Turner, *Angelheaded Hipster: A Life of Jack Kerouac* (New York: Viking, 1996), 17.

10. Ibid., 209. I felt this myself there.

11. Ibid., 23.

12. Brinkley, *Majic Bus*, 26.

13. Turner, *Angelheaded Hipster*, 101.

14. When Kerouac died, most of his books were out of print and his assets totaled $91; Turner, *Angelheaded Hipster*, 211.

15. Kerouac, *On The Road*.

16. In 1948, Kerouac described his project to a friend as "American picaresque"; John

Leland, *Why Kerouac Matters: The Lessons of On the Road (They're not What You Think)* (New York: Viking, 2007), 3.

17. Turner, *Angelheaded Hipster*, 17
18. Leland, *Why Kerouac Matters*, 5.
19. Turner, *Angelheaded Hipster*, 20–1.
20. Leland, *Why Kerouac Matters*, 8.
21. Kerouac saw hoboes as lost pioneers who represented the American drive to the frontier. But, by the 1930s, there was no frontier; Turner, *Angelheaded Hipster*, 214.
22. Ibid., 18.
23. Robert M. Pirsig, *Zen and the Art of Motorcycle Maintenance* (New York: Bantam, 1974).
24. Walden went along on the trip. See W. Barksdale Maynard, *Walden Pond: A History* (New York: Oxford University Press USA, 2004,2005) 286.
25. The mystique of motorcycling is described in Paul McHugh, 'The Pure Jones of It', in James O'Reilly, Sean O'Reilly, Tim O'Reilly, eds. *The Road Within: True Stories of Transformation and the Soul (Travelers' Tales Guides)* (San Francisco: Travelers Tales, 2002), 94-6.
26. See, for instance, Mark Richardson, *Zen and Now: On the Trail of Robert Pirsig and the Art of Motorcycle Maintenance* (New York: Alfred A. Knopf, 2008).
27. William Least Heat-Moon, *Blue Highways: A Journey into America* (Boston: Little, Brown, 1982), introduction.
28. Kris Lackey, *Road Frames: The American Highway Narrative* (Lincoln: University of Nebraska Press, 1997), 53.
29. Phil Patton, *Open Road: A Celebration of the American Highway* (New York: Simon and Schuster, 1986), 241–2.
30. Lackey, *Road Frames*, 87.
31. Brinkley, *Majic Bus*, 19.
32. Ibid., 15.
33. The first was published in Connecticut in 1935; the federal government liked the idea and spread it nationally — see Nick Taylor, *American-Made: The Enduring Legacy of the WPA: When FDR Put the Nation to Work* (New York: Bantam, 2009) p.292
34. See John Gunther, *Inside U.S.A.* (New York: Harper & Brothers, 1947); Jamie Jensen, *Road Trip USA* (Berkeley, CA: Avalon Travel Publishing, annual).
35. John Steinbeck, *Travels With Charley: in Search of America* (New York: Viking: 1962), 121–2.
36. Tang, developed as an artificial orange juice crystal for the US manned space program, led to a variety of artificial foods. The Internet grew out of an attempt by the US federal Defense Advanced Research Projects Agency to improve communication among researchers.
37. Science and science-fiction writer Isaac Asimov, who had some negative involvement with a presidential Strategic Defense Initiative advisory committee, commented, perhaps accurately, "Star Wars? It's just a device to make the Russians go broke"; David G. Hartwell and Kathryn Cramer, *The Hard SF Renaissance* (New York: Tor, 2002), 919.
38. An exception is Edmund Morgan, *The Puritan Family: Religion and Domestic*

Relations in Seventeenth Century New England (New York: Harper Torchbook, 1944, rev. ed. 1966), chap. 6, entitled "Puritan Tribalism." I suspect that the desire to see the melting pot everywhere and everywhen might have obscured the influence of Native American customs, especially on early colonial life.

39. See Joseph A. Conforti, *Imagining New England: Explanations of Regional Identity from the Pilgrims to the Mid-Twentieth Century* (Chapel Hill: University of North Carolina Press, 2001), introduction.

40. The term "Yankee" seems to have entered popular American parlance in the 1820s, with the publication of Washington Irving's *Rip Van Winkle* and *The Legend of Sleepy Hollow* and the novels and stories of Harriet Beecher Stowe. As well, Lydia Maria Child's *The American Frugal Housewife*, which went through thirteen editions between 1829 and 1842, contributed to the Yankee mystique. See ibid., 150–60.

41. Historians often use "the United States" as shorthand to refer to the time before 1865 when there was greater personal identification with one's state and most policy areas were dealt with at the state level; after the war, the federal government held undisputed primacy over the states in many areas of policy.

42. Susan E. Gray, *The Yankee West: Community Life on the Michigan Frontier* (Chapel Hill: University of North Carolina Press, 1996), 1.

43. Garry Wills, *Head and Heart: A History of Christianity in America* (New York: Penguin Books, 2007), 75.

44. Jim Webb, *Born Fighting: How the Scots-Irish Shaped America* (New York: Broadway Books, 2004).

Chapter 1

1. There are a number of somewhat or largely conflicting stories about the song. "Doodle" is apparently eighteenth-century German slang for a "fool" or "bumpkin," while "macaroni" was British slang referring to London dandies who imitated the latest Italian fashions. See also, Charles Panati, *The Extraordinary Origins Of Everyday Things* (New York: Harper Collins, 1987), 286-7

2. T.J. Stiles, *The First Tycoon: The Epic Life of Cornelius Vanderbilt* (New York: Alfred A. Knopf, 2009), 166.

3. Cait Murphy, *Crazy '08: How a Cast of Cranks, Rogues, Boneheads, and Magnates Created the Greatest Year in Baseball History* (New York: Smithsonian/Collins, 2007), 4.

4. From "Cape Cod," quoted in Stephen Mulloney, *Traces of Thoreau: A Cape Cod Journey* (Boston: Northeastern University Press, 1998), 123.

5. Gore Vidal notes that, "most comically, he [President Martin Van Buren] said, not 'Yankees,' but 'Jankes,' the Dutch word for a barking dog"; *Burr* (New York: Bantam Books, 1974), 546.

6. An alternative origin, only slightly less derogatory, is from the nickname "Janke," or "Little John":

> "Johnnies" the Dutch called them — "Johnnies" or Jankins." All the lads and gentry of the Connecticut settlement carried the Johnnie label. They might be fishermen testing their lines on the Shrewsbury banks in Yorker

waters, or trappers stalking the game runs east of Hartford; they might be coastal peddlers tying up at the jetties of New Amsterdam with deck cargoes of pelts and potashes, tobacco and turnips. Everywhere British paths crossed those of the Dutch, the same derisive sobriquet followed.

But the English "J" was difficult for Germanic tongues, and after a while the "Jankin" degenerated into "Yankee," and the neighbors in Manhattan had fixed for all time a title on the sons of Connecticut.

W. Storrs Lee, *The Yankees of Connecticut* (New York: Henry Holt, 1957), x,1.

7. Apparently, the first official use of the term was by General James Wolfe during the British expedition against Quebec, in mentioning the Yankee soldiers who accompanied it. A few years later, the British soldiers sent out to seize American arms thought to be located in Concord and Lexington, Massachusetts, sang the tune "Yankee Doodle Dandy" to mock the country bumpkins they found there. The song had been in derogatory use by British troops at least since the 1745 assault on the French Fort Louisbourg in Cape Breton, now part of Nova Scotia. See Connecticut State Library, "Yankee Doodle, the State Song of the State of Connecticut" (Hartford, 2004), available online at http://www.cslib.org/yankeedoodle.htm. See also Linda S. Watts, *The Encyclopedia of American Folklore* (New York: Facts on File, 2007), 425–6.

8. By 1920, 62 percent of New Englanders were either immigrants or had at least one parent who was, compared with 38 percent nationally; see Robert F. Dalzell, *Enterprising Elite: The Boston Associates and the World They Made* (Cambridge, MA: Harvard University Press, 1987), 173.

9. The Puritans were influenced by Protestant Dutch refugees from Spanish persecution who settled in the textile manufacturing towns of East Anglia. See D. Plooij, *The Pilgrim Fathers from a Dutch Point of View* (New York: New York University Press, 1932), 13–14.

10. Russell Shorto, *The Island at the Center of the World: The Epic Story of Dutch Manhattan and the Forgotten Colony That Shaped America* (New York: Doubleday, 2004),156

11. A Royal charter appears to have been issued as early as 1606, dividing colonization into two companies: the first, the London Company, with lands below the Potomac River; and the second, the Plymouth Company, with lands above Long Island. An attempt at a colony by the Plymouth Company failed in 1608, and the charter might have lain dormant thereafter. Colin Woodard, *The Lobster Coast: Rebels, Rusticators, and the Struggle for a Forgotten Frontier* (New York: Viking, 2004), 78. For an in interesting and readable account of the goings-on that led to the creation of the Plymouth colony and its commercial side, see Godfrey Hodgson, *A Great and Godly Adventure* (New York: Public Affairs, 2006).

12. The landing at Plymouth Rock has become a kind of regional creation myth. For 150 years, the Plymouth colony was largely downplayed. Then, in 1769, a group of wealthy Plymouth residents began to celebrate "Forefathers' Day." Eventually, having an ancestor who had sailed on the *Mayflower* became a status symbol. See Joseph A Conforti, *Imagining New England: Explanations of Regional Identity from the Pilgrims to the Mid-Twentieth Century* (Chapel Hill: University of North Carolina Press, 2001), 203–5; and idem, *Saints*

and Strangers: New England in British North America (Baltimore: Johns Hopkins University Press, 2006), 38–40.

13. They had been invited to settle in Dutch North America before sailing, but decided they wanted to remain English rather than being assimilated; see Plooij, *Pilgrim Fathers*, 29, 53.

14. They sent for supplies from a couple of fishing outposts, on islands off the coast of what is now Maine, that had been established as long ago as 1607. By 1620 there were nine such ouposts, while illegal fishing settlements on Newfoundland, to the northeast, dated from as early as 1615. See Woodard, *Lobster Coast*,84–6, 92.

15. Larzer Ziff, *Puritanism in America: New Culture in a New World* (New York:Viking, 1973), 37–40.

16. As early as 1614, Captain John Smith, while working for the London Company, followed the coast north to Maine and suggested that fishermen and farmers might create a "new" England there (Woodard, *Lobster Coast*, 83); he subsequently published *A Description of New England in 1616*; see Conforti, *Imagining New England*, 14–16.

17. R.A. Billington, *Westward Expansion: A History of the American Frontier*, 4th ed. (New York: Macmillan, 1974), 70–1.

18. There was enough work in the colonies that the better-off emigrants were prepared to accept indentured servants who would be free to pursue their own fortunes after a period of years working for a master, in exchange for passage and sustenance; Ziff, *Puritanism in America*, 43.

19. Ibid., 49–50.

20. Apparently, a number of English "puritans" who left for New England found the local Puritans too puritanical and went on to New Amsterdam. The Dutch, who were apprehensive of a New England population that was 10 times their own, and expansive to boot, welcomed them, but with mixed feelings. Shorto, *Island* pp.158-9.

21. Conforti, *Saints and Strangers*, 82–6.

22. John Winthrop noted on the eve of his departure for Massachusetts Bay in 1629 that England could provide employment for but half its people. The economy remained weak from declining textile sales and poor harvests, a good impetus for this emigration. See Margaret Ellen Newall, "The Birth of New England in the Atlantic Economy: From Its Beginning to 1770," in *Engines of Enterprise: An Economic History of New England*, ed. Peter Temin (Cambridge, MA: Harvard University Press, 2000), 19.

23. The decline of emigration to New England led to an economic recession (Conforti, *Imagining New England*, 32).

24. This Biblical reference was made in a speech by the Puritan leader John Winthrop before departing England for Massachusetts in 1630. It has resonated through American political thought ever since. Winthrop's speech, in part, is quoted in Stephen Innes, *Creating the Commonwealth: The Economic Culture of Puritan New England* (New York: Norton, 1995), 14. See also Conforti, *Imagining New England*, 27–8.

25. For an analysis of this process in Connecticut, see Richard L. Bushman, *From Puritan to Yankee: Character and the Social Order in Connecticut, 1690–1765* (New York: W.W. Norton, 1967).

26. Conforti, *Imagining New England*, 18.

27. R. Nash, *Wilderness and the American Mind* (New Haven, CT: Yale University Press, 1967), 35, 37.

28. "Outlivers" would claim or purchase lands on the boundaries of towns, begin to develop them, and then ask to be part of the town's decision-making bodies. Their need for roads to be extended to them and for a second meetinghouse a few miles from the original added to overall costs and tax rates and prompted a clash between the outlivers and townspeople. See Bushman, *From Puritan to Yankee*, chap. 4.

29. Conforti (*Saints and Strangers*, 50) maps the dispersion of settlements over time.

30. By 1700, Connecticut towns were using land sales to generate funds to provide local sales. Nearly twice as many towns were created in the colony in the thirty years after 1690 as in the thirty years before. This was roughly comparable to population growth, but leaves out the town-infilling process. A lot of new land came on the market before the old land was completely sold. See Bushman, *From Puritan to Yankee*, 77–9, 82.

31. See Richard Slotkin, *Regeneration through Violence: The Mythology of the American Frontier, 1600–1860* (New York: Harper Perennial, 1996), 189–90.

32. For a history of the colonial economies of New England and their evolution into the Yankee culture, see Innes, *Creating the Commonwealth*.

33. Quoted in ibid., 18.

34. Conforti, *Saints and Strangers*, 15.

35. Ibid., 73–9.

36. Paul Johnson, *A History of the American People* (London: Phoenix, 1997), 56.

37. Bushman, *From Puritan to Yankee*, chap. 7, 135.

38. Conforti, *Imagining New England*, 58, 68.

39. The colonies began to hire agents in London to represent their interests *vis-à-vis* their competitors in England, but realized they had no elected voice, as English merchants did; see ibid., 67.

40. Ibid., 58–61.

41. Bushman, *From Puritan to Yankee*, 20, 44–6.

42. See Lewis Leary, "Introduction," in Benjamin Franklin, *The Autobiography of Benjamin Franklin* (1817; New York: Simon and Schuster/Touchstone, 1962), v. Both Jedediah Morse, the father of Samuel Morse, and Timothy Dwight, president of Yale University, recorded the dimensions of a New England cultural identity around the end of the eighteenth century, but did not use the term "Yankee" in their writings; see Conforti, *Imagining New England*, 131.

43. Franklin, *Autobiography of Benjamin Franklin*, 16–32.

44. John Mack Faragher, *Rereading Frederick Jackson Turner: "The Significance of the Frontier in American History" and Other Essays* (New York: Henry Holt, 1994), 71.

45. Garry Wills, *Henry Adams and the Making of America* (New York: Houghton Mifflin Harcourt, 2005), 30–31.

46. James Fenimore Cooper, *The Pioneers* (1823), quoted in Hugh C. MacDougall, "James Fenimore Cooper: Pioneer of the Environmental Movement" (James Fenimore Cooper Society, 1999), available online at www.oneonta.edu/external/cooper/articles/informal/hugh-environment.html. See also Alexis de Tocqueville's reaction to settlers and

trees in George Wilson Pierson, *Tocqueville in America* (Baltimore: John Hopkins University Press, 1996), 193.

47. When the Acadians were expelled from Nova Scotia in 1755, because they would not swear allegiance to the British Crown, their lands were sold off to New England farmers, called Planters locally. Haliburton's grandfather came from Massachusetts with the Planter emigration.

48. See Richard A. Davies, *Inventing Sam Slick: A Biography of Thomas Chandler Haliburton* (Toronto: University of Toronto Press, 2005), 53–4.

49. Ibid., 55.

50. Ibid.

51. Quoted in Lee, *Yankees of Connecticut*, 143.

52. Billington, *Westward Expansion*, 536.

53. Mark Twain, *A Connecticut Yankee in King Arthur's Court* (1889; New York: Washington Square Press, 1964), 4–5.

54. Pierson, *Tocqueville in America*, 497.

55. See Conforti, *Imagining New England*, 150–96. See also David Hackett Fischer, *Albion's Seed: Four British Folkways in America* (New York: Oxford, 1989). Closer to my own interpretation is a skeptical look at Fischer's thesis in Virginia DeJohn Anderson, "The Origins of New England Culture," *William and Mary Quarterly* 48 (2, 1991): 231–7.

56. Quoted in Gavin Weightman, *The Industrial Revolutionaries: The Making of the Modern World, 1776–1914* (New York: Grove Press, 2007), 229.

57. Sallie A. Marston, "Contested Territory: An Ethnic Parade as Symbolic Resistance", in Robert Weible ed. *The Continuing Revolution: A History of Lowell, Massachusetts* (Lowell: Lowell Historical Society), 1991, 220.

58. A common example is noted in Johnson, *History of the American People*, 182.

Chapter 2

1. Quoted in Hugh Brogan. *Alexis de Tocqueville: A Life* (New Haven, CT: Yale University Press, 2006), 180.

2. Andrew Roberts, *A History of the English-Speaking Peoples since 1900* (London: Weidenfeld & Nicolson, 2006), 447.

3. Quoted in Richard Lyle Power, *Planting Cornbelt Culture: The Impress of the Upland Southerner and the Yankee in the Old Northwest* (Indianapolis: Indiana Historical Society, 1953), 45.

4. See the 2000 US census ancestry map, online at http://upload.wikimedia.org/wikipedia/commons/thumb/a/a7/Census-2000-Data-Top-US-Ancestries-by-County.svg/1000px-Census-2000-Data-Top-US-Ancestries-by-County.svg.png. The ethnic ancestry designations are based on the plurality, however small, of the largest group. Along US 20 from the east, there are English only in northern New England, but Irish along the highway to New York. Then, except for a few counties dominated by African Americans, the rest, until the Mormon- (English-) dominated Idaho and eastern Oregon, is German, as is the rest of Oregon. It is interesting that counties in the South are dominated either by African Americans or by Scots-Irish.

5. Burr himself was of Yankee descent: his mother was the daughter of the famous Puritan preacher Jonathan Edwards and his father came from Fairfield, Connecticut. The third vice president of the United States, he killed Alexander Hamilton in a duel and was implicated, though not convicted, of trying to detach Mississippi Valley territory from the country.

6. Gore Vidal, *Empire* (New York: Random House, 1987), 403.

7. See, for example, Jack Whyte, *The Skystone* (Toronto: Viking Canada, 1992), the first of five novels in the series, *A Dream of Eagles*, which imagines how Roman Britons attempted to maintain their culture in the face of the disappearance of the Empire and an influx of Saxons between AD 367 and 448.

8. One geographer who has traced New England place names is John Leighly, "Town Names of Colonial New England in the West," *Annals of the Association of American Geographers* 68 (2, 1978): 233–48.

9. Before the Civil War, Michigan was considered the most Yankee state in the Union.

10. My approach is roughly similar to that pioneered by Daniel J. Elazar in his *Cities of the Prairie: The Metropolitan Frontier and American Politics* (New York: Basic Books, 1970), chap. 4.

11. Wilbur Zelinsky, *The Cultural Geography of the United States* (Englewood Cliffs, NJ: Prentice-Hall, 1973), esp. chap. 1; see also Jared M. Diamond, *Guns, Germs, and Steel: The Fates of Human Societies* (New York: W.W. Norton, 1997).

12. Only a small minority (5 percent) of emigrants chose New England as a destination before 1700, but the region constituted 40 percent of the English population of North America by then; see Conforti, *Saints and Strangers*, 49.

13. See Bernard Bailyn, *Voyagers to the West: A Passage in the Peopling of America on the Eve of the Revolution* (New York: Vintage Books, 1986).

14. See William Labov, *Principles of Linguistic Change*, vol. 3 (New York: Wiley-Blackwell, 1994; reprinted 2010), chap. 10.

15. It was institutionalized by the federal practice of surveying the land of the then Northwest Territory *before settlement* into squares generally called townships.

16. To Washington, the economic development of the Potomac River valley and its rise to be the main route into the interior was often said to be "the favorite object of his heart"; see Merritt Roe Smith, *Harpers Ferry Armory and the New Technology: The Challenge of Change* (Ithaca, NY: Cornell University Press, 1977), 27.

17. See Webb, *Born Fighting*.

18. Morgan notes that the early Puritan settlers made God so much their own that God "took on the character of a tribal deity" (*Puritan Family*, 168), and they struggled with the realization that, even among themselves, too many people were not faithful enough. Immigrants who lacked even the rudiments of this attachment to God were hardly welcome.

19. Margaret Ellen Newell, "Birth of New England in the Atlantic Economy", 26.

20. Winifred Barr Rothenberg, "The Invention of American Capitalism: The Economy of New England in the Federal Period," in *Engines of Enterprise: An Economic History of New England*, ed. Peter Temin (Cambridge, MA: Harvard University Press, 2000), 75.

21. Peter Temin, "Introduction," in *Engines of Enterprise: An Economic History of New England*, ed. Peter Temin (Cambridge, MA: Harvard University Press, 2000), 8.

22. Alan Taylor, *William Cooper's Town: Power and Persuasion on the Frontier of the Early American Republic* (New York: Vintage Books, 1995), 90.

23. Paul Horgan, *Great River: The Rio Grande in North American History* (New York: Wesleyan University Press, 1984), 624.

24. Bailyn, *Voyagers to the West*, 486.

25. Marietta was begun by Rufus Putnam, a land surveyor, veteran of what Americans call the French and Indian War (part of the wider conflict known as the Seven Years' War) and the American Revolution, and a founder of the Ohio Company. Putnam was instrumental in getting Congress to ban slavery in the Northwest Territory and from 1786 to 1803 served as America's first surveyor-general. See Billington, *Westward Expansion*, 213.

26. Yankees moved into what is now Canada as well. When the Acadians were expelled from their Nova Scotian farms in 1755, their places were occupied by Yankee "planters." Loyalist Yankees moved north from Vermont and New Hampshire after the Revolution and their influence contributed to rebellions in Upper and Lower Canada in 1837. Still others moved west across the Niagara frontier after the War of 1812

27. Bill Bryson, *A Walk in the Woods* (New York: Broadway Books, 1998), 210–12.

28. George F. Kennan, *Sketches from a Life* (New York: Pantheon Books, 1989), 207.

29. For a succinct description of the Auburn system, see Milton M. Klein, *The Empire State: A History of New York* (Ithaca, NY: Cornell University Press, 2001), 344. Its evolution is best described in Pierson, *Tocqueville in America*, 94–8. William Brittin built the prison at Auburn and was its first warden. Given the construction contract for the Genesee River bridge at Rochester in 1820, he proposed to bring 150 convicts to help build it. The locals in what was then a small town of 1,500 were not happy, and in the end he employed only thirty. The convicts were told they would be pardoned by Governor DeWitt Clinton if they worked for canal wages for the duration of their sentences, but that they would be punished if they tried to escape — a quarter of them nevertheless got away. See Gerard Koeppel. *Bond of Union: Building the Erie Canal and the American Empire* (Cambridge, MA: Da Capo Press, 2009), 327–8. Ice flows in the spring breakup on the river and Brittin's death in 1822 led to a new and different contract. See Peter L. Berstein, *Wedding of the Waters: The Erie Canal and the Making of a Great Nation* (New York: W.W. Norton, 2005), 271.

30. At the beginning of the twentieth century, a New Yorker, Thomas Mott Osborne, warden at Auburn and Sing Sing prisons, led a new impulse for prison reform, arguing for indeterminate sentences, a degree of prisoner self-government, and a greater focus on reintegration of prisoners into society; see Klein, *Empire State*, 524–5.

31. See Leo Damrosch, *Tocqueville's Discovery of America* (New York: Farrar, Straus & Giroux, 2010), 11.

32. There seems to be little in de Tocqueville's background to indicate a prior interest in America, other than his exposure to a distant cousin's novels about the country. The cousin, François-René de Chateaubriand, had written his latest story about America four years before de Tocqueville's visit. A relatively famous book on penal systems in Europe

and the United States had been published in Paris in 1827 might have acted as a basis for their research. See Brogan, *Alexis de Tocqueville*, 136–40, 142–3.

33. Ibid., 142. De Tocqueville and de Beaumont wasted no time. Gaining approval from the government in February 1831, they were on their way by the end of March.

34. Ibid., 145.

35. He and de Beaumont read everything they could lay their hands on about penal reform and visited some of the French facilities before presenting a well-organized brief to the minister; see ibid., 142–3.

36. Ibid., 141.

37. Ibid., 159.

38. One of his frequent hosts, Nathaniel Prime, a Yankee from Massachusetts, was reputed to be the richest man in New York City; see Damrosch, *Tocqueville's Discovery of America*, 19.

39. Pierson, *Tocqueville in America*, 100–3.

40. De Tocqueville noted: "We hoped to obtain valuable information on whatever central government there is in this country. The offices and registers were all open to us, but as far as *government* goes, we're still looking for it. Really, it doesn't exist at all"; Damrosch, *Tocqueville's Discovery of America*, 47.

41. Brogan, *Alexis de Tocqueville*, 162–3.

42. Later in the 1830s, another French traveler, Michel Chevalier, did pay attention to economic and cultural matters, but his subsequent publications did not receive the same attention as de Tocqueville's; see Robert Weible, *The World of the Industrial Revolution: Comparative and International Aspects of Industrialization* (Lanham, MD: Rowman & Littlefield, 1987), 21. See also Thomas Southcliffe Ashton and Pat Hudson, *The Industrial Revolution, 1760–1830* (Oxford: Oxford University Press, 1997), 169–70; and Pierson, *Tocqueville in America*, 174–5.

43. One commentator has noted that the flour barrel began to replace the beaver pelt as the symbol of New York in the 1820s; see Berstein, *Wedding of the Waters*, 369. The cost of transporting a ton of flour from Buffalo to New York City dropped by 95 percent as a result of the completion of the canal; see John Steele Gordon, *The Great Game: The Emergence of Wall Street as a World Power, 1653–2000* (New York: Scribner, 1999), 56–7.

44. Three weeks after the two Frenchmen left Albany, the locomotive "The DeWitt Clinton" hauled its first trainload of politicians from Albany to Schenectady.

45. De Beaumont was startled that the governor could receive them in the boardinghouse parlor without ceremony and easily discuss the affairs of his brother the grocer and his cousin the salesman; Damrosch, *Tocqueville's Discovery of America*, 58.

46. Ibid., 60–1.

47. Brogan, *Alexis de Tocqueville*, 170–2.

48. Having taken a few steamboats in America, he commented to his mother: "these same Americans, who have never discovered a single law of mechanics, have given to navigation a new machine that is changing the face of the earth"; Damrosch, *Tocqueville's Discovery of America*, 45.

49. Brogan, *Alexis de Tocqueville*, 138.

50. De Tocqueville later mentioned to a friend that Boston is where he would go were he to live in America. He felt its upper classes to be most like the cultured gentry of Europe; ibid., 180.

51. Ibid., 200.

52. Witold Rybczynski, *City Life* (New York: Harper Perennial, 1996), 108–9.

53. Brogan, *Alexis de Tocqueville*, 164.

54. Ibid., 86, 89, 351–2.

55. Attitudes toward free blacks actually had deteriorated since the Revolution; see Damrosch, *Tocqueville's Discovery of America*, 28–9.

56. Washington Irving characterized the attitude when he coined the term "the almighty dollar, that great object of devotion throughout the land"; ibid., 25.

57. Historian Merritt Roe Smith documents the problems that a Maine Yankee, John Hall, had in attempting to mechanize the process of making muskets at the federal arsenal in Harpers Ferry, then part of Virginia. The rural and relatively uneducated workers resisted attempts to introduce new technology from the time Hall was sent there until the arsenal was destroyed in 1861 at the outset of the Civil War. Although it was not his intent, Smith highlights the contrast between what the locals called "Yankeeism" and their way of life. See Smith, *Harpers Ferry Armory*.

58. As late as the 1970s, my father-in-law, a city manager in Michigan, wore "prison shoes," and we all knew that the state's auto license plates were made by prisoners.

59. Rybcynski, *City Life*, 112.

60. "Navvies," short for "navigators," was the name mockingly given to laborers in England who dug the canal system. In America, the term was applied to Irish laborers. See, for instance, Gavin Weightman, *The Industrial Revolutionaries: The Making of the Modern World, 1776–1914* (New York: Grove Press, 2007), 162–3.

61. For a similar discussion of New England in the mid-1900s, see Conforti, *Imagining New England*, 206–14

62. Ibid., chap. 3, explores this identity of New England morés with American ones. For a more detailed expression of this process, see Power, *Planting Cornbelt Culture*.

63. Sherwood Anderson, *Winesburg, Ohio: A Group of Tales of Small-town Ohio Life* (New York: B.W. Huebsch, 1919), 62–3.

64. Zelinsky, *Cultural Geography of the United States*, 120.

Chapter 3

1. Peter Aleshire, *Arizona Highways* (April 2005), 3.

2. Rybcynski, *City Life*, 34.

3. Quoted in John W. Oliver, *A History of American Technology* (New York: Ronald Press, 1956), 484.

4. *Collier's Encyclopedia*, vol. 12 (New York: P.F. Collier, 1993), 113–4.

5. For a succinct explanation of the problems encountered when the federal government got involved in early nineteenth-century road financing, see Algeo, *Harry Truman's Excellent Adventure*, 86. See also Carter Goodrich, *Government Promotion of American Canals and Railroads, 1800–1890* (New York: Columbia University Press, 1960), 19–38.

6. See United States, Department of Transportation, Federal Highway Administra-

tion, *America's Highways: 1776–1976* (Washington, DC: US Government Printing Office, 1977), 17.

7. Jack Beatty, *Colossus: How the Corporation Changed America* (New York: Broadway Books, 2001), 60; Klein, *Empire State*, 311.

8. By comparison, the Romans, 1,500 years earlier, could boast of a paved road system 70,000 miles in length; see Christopher Finch, *Highways to Heaven: The Autobiography of America* (New York: HarperCollins, 1992), 20.

9. Eric Hobsbawn, *The Age of Revolution, 1789–1848* (New York: Barnes & Noble, 1962), 171.

10. See, for example, the fall and rise of roads in Connecticut (Lee, *Yankees of Connecticut*, 100–2).

11. United States, *America's Highways*, 14–15.

12. Ray Spangenburg and Diane Moser, *The Story of America's Roads* (New York: Facts on File, 1992), 41.

13. In 1900, Iowa's roads were said to have more steep grades than those in Switzerland; ibid., 40.

14. Laura Byrne Paquet, *Wanderlust: A Social History of Travel* (Fredericton, NB: Goose Lane Editions, 2007), 134–5.

15. David Harlick, *No Place Distant* (Washington, DC: Island Press, 2002), 15–16.

16. Dan McNichol, *The Roads that Built America: The Incredible Story of the U.S. Interstate System* (New York: Sterling, 2006), 36.

17. Dixon Ryan Fox and Robert V. Remini, *The Decline of Aristocracy in the Politics of New York: 1801–1840* (New York: Harper & Row, 1965), 6; Thomas J. Schlereth, *Victorian America: Transformations in Everyday Life, 1876–1915* (New York: HarperCollins, 1991), 221; Spangenburg and Moser, *Story of America's Roads*, 41. See also Harlick, *No Place Distant*, 15–16.

18. Spangenburg and Moser, *Story of America's Roads*, 42–4, 50.

19. Merritt Ierley, *Traveling the National Road: Across the Centuries on America's First Highway* (Woodstock, NY: Overlook Press, 1990), 180. The Office of Road Inquiry's first "Agent" was a League of American Wheelmen activist, General Roy Stone; see McNichol, *Roads that Built America*, 39.

20. Patton, *Open Road*, 145–8. The first use of asphalt as a road surface was on Fifth Avenue in New York City in 1872; McNichol, *Roads that Built America*, 39

21. Patton, *Open Road*, 55–6.

22. United States, *America's Highways*, 80. See also William Leach, *Land of Desire: Merchants, Power, and the Rise of a New American Culture* (New York: Vintage, 1993), 182–5.

23. Robert Hendrickson, *The Grand Emporiums: The Illustrated History of America's Great Department Stores* (New York: Stein and Day, 1980), 49.

24. Harvey Wish, *Society and Thought in Modern America: A Social and Intellectual History of the American People from 1865*, 2nd ed. (New York: David McKay, 1962), 106.

25. Ibid., 107.

26. See Pete Davies, *American Road: The Story of an Epic Transcontinental Journey at the Dawn of the Motor Age* (New York: Henry Holt and Company, 2002), 14. By 1925,

Jackson's sixty-three-day trip could be duplicated in about five days. See also Tom Lewis, *Divided Highways: Building the Interstate Highways Transforming American Life* (New York: Penguin Group, 1997), 50; Spangenburg and Moser, *Story of America's Roads*, 50. A year later, in 1904, a caravan of seventy cars left the East Coast for the St. Louis World's Fair; fifty-eight made it, producing a publicity sensation when they arrived; see Algeo, *Truman's Excellent Adventure*, 87–8. In 1909, Alice Huyler Ramsey became the first woman to drive across the country, taking three female companions on a forty-one-day trip from New York to San Francisco.

27. Spangenburg and Moser, *Story of America's Roads*, 47. By 1917, all the states had some form of road program

28. As late as 1908, Kansas did not have a single road that joined two of its borders; ibid., 44.

29. In September 1899, a New York City real estate agent was struck by an electric taxi as he alighted from a streetcar, becoming the first person killed by an automobile in America; Algeo, *Harry Truman's Excellent Adventure*, 36. In a reverse of that accident, my grandfather was killed in 1926 when he was hit by a car and thrown into the path of a streetcar in Saginaw, Michigan.

30. Spangenburg and Moser, *Story of America's Roads*, 109.

31. E.W. James, "Making and Unmaking a National System of Marked Roads," *American Highways* 12 (4, 1933): 16–18.

32. Fox and Remini, *Decline of Aristocracy*, 51; Spangenburg and Moser, *Story of America's Roads*, 80–1.

33. United States, *America's Highways*, 109. Apparently, there was a proliferation of "Dixie" and "Yellowstone" Highways; see Patton, *Open Road*, 44–5. Bruce Springsteen mentions the Dixie highway in his song, "Darlington County."

34. Schlereth, *Victorian America*, 222.

35. David Haward Bain, *The Old Iron Road: An Epic of Rails, Roads, and the Urge to Go West* (New York: Penguin, 2005), 56; Oliver, *History of American Technology*, 485.

36. Fox and Remini, *Decline of Aristocracy*, 50. The corruption that came with the building of the transcontinental railroads had not been forgotten; see Patton, *Open Road*, 145.

37. Davies, *American Road*, 222; Schlereth, *Victorian America*, 222. See also Paquet, *Wanderlust*, 140.

38. Quoted in Paquet, *Wanderlust*, 139.

39. Fox and Remini, *Decline of Aristocracy*, 34.

40. McNichol, *Roads that Built America*, 61.

41. Fox and Remini, *Decline of Aristocracy*, 11; Patton, *Open Road*, 191.

42. Oliver, *History of American Technology*, 566.

43. Fox and Remini, *Decline of Aristocracy*, 89.

44. McNichol, *Roads that Built America*, 61.

45. Davies, *American Road*, 216.

46. Ierley, *Traveling the National Road*, 180. It was a 1/3–2/3 shared-cost program of which only thirteen states and twenty-eight counties availed themselves.

47. An alternative story is that President Woodrow Wilson was about to veto the

act, but changed his mind after a report of a sighting of a German submarine in Baltimore harbor brought the road issue into the national defense picture (Oliver, *History of American Technology*, 486).

48. Fox and Remini, *Decline of Aristocracy*, 13; Patton, *Open Road*, 45.

49. Willis F. Dunbar and George S. May, *Michigan: A History of the Wolverine State*, 2nd ed. (Grand Rapids, MI: William B. Eerdmans, 1965), 493.

50. Another source refers to the National League for Good roads, formed in 1893, and the National Good Roads Association, organized around the turn of the century; see Johnson, *History of the American People*, 428, 485.

51. Dunbar and May, *Michigan*, 492–3. The federal funds were matching ones and focused on roads where mail was delivered. A similar story can be seen in Pennsylvania; see Davies, *American Road*, 45–7.

52. Spangeburg and Moser, *Story of America's Roads*, 51–2. Part of that industry was the traffic light, invented by a Cleveland tinkerer, Garrett Morgan, the son of former slaves. He devised a light with red and green signals and a buzzer, which was installed in Cleveland in 1914; see Stephen Van Dulken, *Inventing the 20th Century: 100 Inventions that Shaped the World from the Airplane to the Zipper* (New York: Barnes & Noble, 2007), 78.

53. Some ideas related to the core industry of trucking can be found in Finch, *Highways to Heaven*, 331–3.

54. Davies, *American Road*, 216.

55. Seventy-five years earlier, in 1847, when railroads were still new, the editor of the Toledo Blade, J.W. Scott, called for a similar grid to link the sections of the country — two railroads north to south and four east to west — before anyone had completed a rail line five hundred miles long. See Archie Hobson, *Remembering America: A Sampler of the WPA American Guide Series* (New York: Collier Macmillan, 1987), 2–3.

56. Ibid., 110. See also Lewis, *Divided Highways*, 12–21; and Patton, *Open Road*, 45.

57. Fox and Remini, *Decline of Aristocracy*, 20.

58. "The United States Numbered System," *American Highways* 11 (3, 1932): 16.

59. James, "Making and Unmaking a System of Marked Roads," 17.

60. Paquet, *Wanderlust*, 141.

61. Finch, *Highways to Heaven*, 77. Olmstead and Vaux provided prototypes of parkways in their designs for Central Park and later Prospect Park, Brooklyn. They added parkways to their designs for other major cities before the turn of the twentieth century — see Lewis, *Divided Highways*, 28–9; see also Patton, *Open Road*, 67. They seem to have derived their inspiration in part from Baron Haussmann's Parisian boulevards and from some divided American streets, such as Commonwealth Avenue in Boston, later a part of US 20 for a number of years.

62. Patton, *Open Road*, 70.

63. Fox and Remini, *Decline of Aristocracy*, 68.

64. Patton, *Open Road*, 71–2, 141 The first stretch of California freeway was opened in 1940.

65. Finch, *Highways to Heaven*, 158.

66. Spangenburg and Moser, *Story of America's Roads*, 57.

67. Algeo, *Harry Truman's Excellent Adventure*, 170–1.

68. Ierley, *Traveling the National Road*, 203. By this time, many states had copied Pennsylvania's turnpike approach, so that by the year after the passage of the *Defense Highway Act*, one could travel from the East Coast to Chicago without meeting a traffic light. Meeting a toll booth was another matter. See Algeo, *Harry Truman's Excellent Adventure*, 171.

69. Fox and Remini, *Decline of Aristocracy*, 108.

70. Finch, *Highways to Heaven*, 226.

71. Rybcynski, *City Life*, 160–1.

72. The federal system never stopped growing despite the Interstates, and by the late 1980s totaled 301,000 miles. The Interstate system also grew a bit by then, to 44,000 miles. See Ierley, *Traveling the National Road*, 203.

73. Patton, *Open Road*, 267.

74. President Dwight Eisenhower saw the system's potential for economic development, though not this implication. It was, however, an extension of the interest in internal improvements that had driven the Whigs and early Republicans; see ibid., 87, 115.

75. There is an earlier book that follows US 20 as it crosses the continent. See Mac Nelson, *Twenty West: The Great Road across America* (Albany: State University of New York Press, 2008), which looks at themes, such as literature and political power, as well as pointing out roadside attractions and containing many photographs. Its approach is to follow these themes across the continent.

76. For a description of the experience of being a passenger on the post road, see Stewart H. Holbrook, *The Old Post Road: The Story of the Boston Post Road* (New York: McGraw-Hill, 1962), 13–16.

77. Spangenburg and Moser, *Story of America's Roads*, 6–7. The early roads were often no more than paths, sometimes widened to accommodate a wagon. Coaches — from a Hungarian named Kocz, a renowned maker of enclosed carriages — were not used much because of the state of the roads. See John H. Lienhard, *Inventing Modern: Growing Up with x-Rays, Skyscrapers, and Tailfins* (New York: Oxford University Press, 2003), 117; and Thomas Crump, *A Brief History of the Age of Steam: The Power that Drove the Industrial Revolution* (New York: Carroll & Graf, 2007), 3–5.

78. Originally, three post roads ran from Boston to New York. US 20 parallels or replaces some of the northerly, or "old" post road. Apparently parts of the original route were later sold off as railroad right-of-way; Holbrook, *Old Post Road*, 11.

79. Ibid., 10.

80. Klein, *Empire State*, 160.

81. Holbrook, *Old Post Road*, 8. Because postage was assessed by distance at the time, Franklin established milestones on the post roads, eliminating controversy (7).

82. A Corridor Commission was also established to provide technical services and financial assistance to communities and organizations interested in forming partnerships to preserve the Valley's heritage.

83. Worcester has many claims to industrial fame, including the company that produced the first barbed wire, which had a lot to do with the "taming" of the Western plains (Holbrook, *Old Post Road*, 116–8).

84. Taylor, *William Cooper's Town*, 31–2; see also Peter J. Hugill, *Upstate Arcadia:*

Landscape, Aesthetics, and the Triumph of Social Differentiation in America (Lanham, MD: Rowman & Littlefield, 1995), 62–3. Elsewhere (see, for example, Klein, *Empire State*, 267), the Great Western Pike is identified as the road paralleling the Erie Canal. Yet the third Great Western Pike definitely ran along the higher ground to the south.

85. There is some confusion about the original designation of US 20. Lewis (*Divided Highways*, 290) suggests it was first located along the Erie Canal route and not uphill to the south.

86. A 1932 promotional piece for the turnpike noted that it had been completed as far as Syracuse in 1811, partly as a make-work project during the depression caused by the 1807 US trade embargo with the warring European powers leading up to the War of 1812; during the war, it played an important role in getting supplies to the Niagara frontier; see Cherry Valley Turnpike Association, "The Cherry Valley Turnpike" (Waterville, NY: Cherry Valley Turnpike Association, 1932), 2–3.

87. Hugill, *Upstate Arcadia*, 176, 192.

88. Carol Poh Miller and Robert A. Wheeler, *Cleveland: A Concise History, 1796–1996*, 2nd ed. (Bloomington: Indiana University Press, 1997), 19.

89. Richard F. Weingroff, "From Names to Numbers: The Origins of the U.S. Numbered Highway System" (Washington, DC: Department of Transportation, Federal Highways Administration, n.d.); available online at http://www.fhwa.dot.gov/infrastructure/numbers.cfm. Before this the highway that connected Portland, Oregon, and the Pacific to the east was US 26; this later portion of US 20, in defiance of the national grid, crosses US 26 in Idaho and reaches the ocean to its south.

90. See Bill London, *Country Roads of Idaho* (Castine, ME: Country Roads Press, 1995).

91. Ibid., 67–8.

92. Least Heat-Moon, *Blue Highways*, 222.

Chapter 4

1. Quoted in Holbrook, *Old Post Road*, 138. Longfellow composed this after he and his bride visited the Springfield Armory (not Watertown) in 1843. She had likened the stacked musket barrels to a set of organ pipes.

2. Quoted in James C. Scoville, "The Taylorization of Vladimir Ilich Lenin," *Industrial Relations* 40 (4, 2002): 620–1.

3. Kirsten Downey, *The Woman Behind the New Deal: The Life and legacy of Frances Perkins* (New York: Anchor Books, 2010), 32.

4. The site of Springfield's armory was said to have been personally selected by George Washington in 1776 and was the site of a major clash in Shay's Rebellion in 1787. When Washington toured New England as president in 1789, he inspected the armory (Holbrook, *Old Post Road*, 138–9).

5. Oliver, *History of American Technology*, 93.

6. I am not sure of the difference, if any, between an "arsenal" and an "armory." I suspect an arsenal made weapons and an armory could both make and store weapons. At any rate, the words seem to be used pretty interchangeably in the literature.

7. Smith, *Harpers Ferry Armory*, 28. Washington was determined that the main arsenal be located at Harpers Ferry, upstream from the new capital.

8. Friends of Watertown Free Publications and Watertown Historical Society, *Watertown* (Charleston, SC: Arcadia Publications, 2002), 56. In 1911, the Watertown Arsenal employed about four hundred people and its facilities and equipment were mostly of Civil War vintage and obsolete; see Robert Kanigel, *The One Best Way: Frederick Winslow Taylor and the Enigma of Efficiency* (New York: Viking Penguin, 1997), 450–1.

9. Smith (*Harpers Ferry Armory*, 272–5) notes another short walkout at the Arsenal in 1842, the "clock strike," when workers protested the introduction of a time clock as an intrusion on their traditional freedom to come and go as they pleased.

10. Kanigel, *One Best Way*, 500.

11. Ibid., 74.

12. Max Weber, *The Protestant Ethic and the Spirit of Capitalism* (1904; London: Routledge Classics, 2001).

13. "The American boy of 1854," Henry Adams wrote, "stood nearer the Year 1 than the Year 1900"; quoted in Beatty, *Colossus*, 97.

14. Robert B. Gordon and Patrick M. Malone, *The Texture of Industry: An Archaeological View of the Industrialization of North America* (New York: Oxford University Press, 1994), 386–8. By 1828, a couple of arms manufacturers could make interchangeable parts for muskets, but these depended on artisans filing the pieces down until they fit standardized gauges. The gauge was the innovation; precision-cutting machine tools were a long way off.

15. Andro Linklater, *Measuring America: How the United States Was Shaped by the Greatest Land Sale in History* (New York: PLUME Penguin Group, 2003), 237.

16. This interest was shared by a young sawyer, John Muir, whose study of his employers' sawmill produced observations much like those of Taylor. Muir went on to fame as a conservationist and founder of the Sierra Club. See Frederick Turner, *Rediscovering America: John Muir in His Time and Ours* (San Francisco: Sierra Club Books, 1985), 123–5.

17. Kanigel, *One Best Way*, 351, 379.

18. Shop culture, especially when employees have been rewarded for their performance under one set of rules, is hard to alter, except around the edges; Schlereth, *Victorian America*, 56, 62.

19. Ibid., 56.

20. In this, Taylor was preceded by Charles Babbage, the Englishman who first conceived the computer, and who is credited with producing the first text on operational analysis in 1832, well before its emergence in America. Babbage's ideas, however, did not have much effect on contemporary British production practices; see Wade Rowland, *Spirit of the Web: The Age of Information from Telegraph to Internet* (Toronto: Somerville House, 1997), 219.

21. Frank Gilbreth, the model of the father in the film, *Cheaper by the Dozen*, measured individual movements in a task, reducing standardized partial movements to "therbligs," which, by changes in the task and placement of tools, could be reduced, thus speeding up the task without adding to the worker's task; see Darren Wershler-Henry, *The Iron Whim: A Fragmented History of Typewriting* (Toronto: McClelland & Stewart, 2005), 144.

22. Henry Ford later could say that, of the 7,882 actions needed to build a Model T, only 949 required whole bodies; the rest could be done by disabled people. Of course,

this meant that workers' ethnicity and race also could be ignored — and eventually were (Wershler-Henry, *Iron Whim*, 143).

23. Gail Cooper, *Air-Conditioning America: Engineers and the Controlled Environment, 1900–1960* (Baltimore: Johns Hopkins University Press, 1998), 47.

24. Kanigel, *One Best Way*, 262.

25. Schlereth, *Victorian America*, 65–6.

26. Taylor's paper has been republished numerous times; the edition used in this writing is in Frederick Winslow Taylor, *The Principles of Scientific Management* (New York: W.W. Norton, 1967).

27. Hugh G.J. Aitken, *Taylorism at Watertown Arsenal: Scientific Management in Action, 1908–1915* (Cambridge, MA: Harvard University Press, 1960), 172–3.

28. Ibid., 40–1.

29. Ibid., 49–57.

30. Ibid., 56–7, 70.

31. Barth later worked at two other arsenals as well; ibid., 81–2.

32. The text of the petition is in ibid., 150.

33. Frank and Lillian Gilbreth had published their *Motion Study* in 1911; see Wershler-Henry, *Iron Whim*, 146.

34. Kanigel, *One Best Way*, 472.

35. Ibid., 472–4.

36. Taylor, *Principles of Scientific Management*, 6–8.

37. Kanigel, *One Best Way*, 482–4.

38. Anthony Sampson, *Company Man: The Rise and Fall of Corporate Life* (New York: Random House, 1995), 42.

39. Ibid., 506.

40. Kanigel, *One Best Way*, 12.

41. Ibid., 501.

42. Kanigel, *One Best Way*, 489–90.

43. Cooke went on to head the rural electrification program of the New Deal in the 1930s. He once stated, "[w]e shall never fully realize…the dreams of democracy until the principles of scientific management have permeated every nook and cranny of the working world"; quoted in *Business Week*, 18 April 18, 1964, 132.

44. Kanigel, *One Best Way*, 493.

45. Ibid., 486.

46. Ibid., 488.

47. Ibid., 489.

48. Ibid., 501–2.

49. Charles Petersen, "Google and Money," *New York Review of Books* 57 (19, 2010): 60, 62.

50. Downey, *Woman Behind the New Deal*, 33–4.

51. The fire was the worst workplace disaster in New York City for ninety years, until 9/11. At the time, workplace accidents killed an average of a hundred people a day across the country; see David Von Drehle, *Triangle: The Fire that Changed America* (New York: Grove Press, 2003), 3.

52. Ibid., 3–4.

53. Downey, *Woman Behind the New Deal*, 6–8.

54. This was consistent with the times, as the use of social science information in dealing with social problems had gained serious credibility only in the early 1900s (Von Drehle, *Triangle*, 196–8)

55. Perkins was part of a broader progressive movement that attracted thousands of bright, young people to work to alleviate social problems (ibid., 15–16). Her approach in Philadelphia in 1907 was one of researching an issue about women, which her Addams House friend, Grace Abbott's sister Edith, had pioneered with her PhD work at the University of Chicago in 1905. See Ellen Fitzpatrick, *History's Memory: Writing America's Past, 1880–1980* (Cambridge, MA: Harvard University Press, 2004), 58. By 1910, there were over four hundred such facilities in America (Von Drehle, *Triangle*, 197; see also Klein, *Empire State*, 504–8).

56. Von Drehle, *Triangle*, 195.

57. About this time, the value of scientific, or at least rigorous, well-presented facts began to affect court cases as well as public opinion (ibid., 197).

58. In a tactic that presaged the fair trade coffee movement in our day, the Consumers' League provided a white tag for clothes that were made in factories with progressive work practices and encouraged retailers and consumers to look for it when buying clothes (Downey, *Woman Behind the New Deal*, 29–30).

59. Many of New York's feminine elite joined with the strikers (Klein, *Empire State*, 498).

60. Edwin G. Burrows and Mike Wallace, *Gotham: A History of New York City to 1898* (New York: Oxford University Press, 1999), 309–10.

61. Von Drehle, *Triangle*, 15; Klein, *Empire State*, 497.

62. The use of steam power and, by the end of the nineteenth century, electricity, plus the invention of the sewing machine freed the textile industry from the confines of water power and the New England rivers. Making garments moved to where the market was — in large cities such as New York (ibid., 8, 15, 38–9). The shirtwaist was a type of blouse made for working women that became very fashionable in the 1890s and beyond when popularized by the magazine artist Charles Dana Gibson as the blouse of the "Gibson girl" (44–7).

63. Tammany had lost the 1909 city elections badly to progressives because of inattention to its voting base (ibid., 53, 189–90).

64. Ibid., 173–5. He lost the Triangle case, but got as far as becoming governor of the State.

65. Klein, *Empire State*, 499.

66. The League had been founded in 1890 and both employers and employees were ineligible to join, which meant that middle- and upper-class women dominated it. It was not a mass organization, but focused on research and publicity in order to change legislation (ibid., 497).

67. Von Drehle, *Triangle*, 201–5.

68. Klein, *Empire State*, 499.

69. Downey, *Woman Behind the New Deal*, 45. She hated the nickname, especially its variation, "Ma" Perkins.

70. Von Drehle, *Triangle*, 212.

71. Ibid., 210–12.

72. Downey, *Woman Behind the New Deal*, 41–3.

73. Klein, *Empire State*, 500.

74. In an exchange between Tammany boss Charles Murphy and Perkins, Murphy got her to admit that she was behind the bill limiting work hours for women, which Murphy unsuccessfully opposed. At the end he said, "[i]t is my observation that the bill made us many votes. I will tell the boys to give you all the help they can with this new bill. Good bye" (Von Drehle, *Triangle*, 218).

75. Klein, *Empire State*, 521.

76. Downey, *Woman Behind the New Deal*, 49–52.

77. Klein, *Empire State*, 545.

78. Ibid., 564.

79. Ibid., 577.

80. Ibid, p.579

81. My late father-in-law worked in the Civilian Conservation Corps as a young man and, from the number of times he recounted his adventures, it clearly was a defining moment in his life.

82. In 1961, she helped dedicate a fiftieth anniversary plaque on the building where the Triangle fire occurred; the building is now part of New York University (Von Drehle, *Triangle*, 263).

83. Beatty, *Colossus*, 271; it would be forgotten in the wake of the 2007–08 Crash.

84. Weber, *Protestant Ethic*, xxxi–xxxix.

85. David Brooks, "The responsibility deficit," *New York Times*, September 23, 2010.

Chapter 5

1. Temin, "Introduction," 1.

2. Quoted in Bernard DeVoto, *The Year of Decision, 1846* (1942; New York: St. Martins Press, 2000), 210.

3. The term "industrial revolution" seems to have been invented by the French economist Louis Auguste Blanqui in 1827.

4. See, for instance, Burrows and Wallace, *Gotham*, 170. For an interesting history of many of the British, and some French and Americans, whose efforts led to this revolution, see Weightman, *Industrial Revolutionaries*.

5. See, for instance, Peter Padfield, *Maritime Power: The Struggle for Freedom* (Woodstock, NY: Overlook Press, 2003).

6. See David S. Landes, *The Wealth and Poverty of Nations: Why Some Are So Rich and Some So Poor* (New York: W.W. Norton, 1998), 186.

7. These were not incorporated, as British policy discouraged the use of this instrument. The only one of note to have been incorporated was that of Robert Owen at New Lanark, in Scotland; see Weible and Lowell Historical Society, *Continuing Revolution*, 67n5.

8. Weible, *World of the Industrial Revolution*, 16. As well, the product being processed had to be right, at least for the first essay into mechanized production. Cotton proved to be most useful, with wool a close second (Landes, *Wealth and Poverty of Nations*, 191).

One commentator states that the cotton textile industry was, in itself, the prime motivator of the industrial revolution; see Weible, *World of the Industrial Revolution*, 16.

9. Alfred D. Chandler Jr., *The Visible Hand: The Managerial Revolution in American Business* (Cambridge, MA: Belknap Press of Harvard University Press, 1977), 54–5.

10. John Steele Gordon, *The Business of America* (New York: Walker Publishing, 2001), 19.

11. Apparently, the Chinese had water mills for spinning hemp in the 1100s; Landes, *Wealth and Poverty of Nations*, 55.

12. See, for instance, Johnson, *History of the American People*, 314.

13. "Putting out" had a mixed history in Europe, where weaving guilds in Italian and Dutch towns attacked village weavers and broke their looms to prevent them from undercutting their business. This apparently did not happen, or succeed, in England, where, in the fifteenth century, half the woollen cloth was "put out" to rural people. This so lowered costs that it reduced exports of raw wool and moved England into exporting the finished product, eventually stimulating the eighteenth-century revolution in textile manufacturing that grew out of cotton processing. See Landes, *Wealth and Poverty of Nations*, 43.

14. Eric Hobsbawm, *The Age of Capital, 1848–1875* (London: Cardinal, 1991), 48–9; and idem, *Age of Revolution*, 34–5.

15. Weible and Lowell Historical Society, *Continuing Revolution*, 6.

16. Burrows and Wallace, *Gotham*, 170.

17. Landes, *Wealth and Poverty of Nations*, 296.

18. A British visitor to New England, Lord Adam Gordon, complained in 1765 that "the levelling principle here, everywhere, operates strongly and takes the lead. Everybody has property, and everybody knows it" (ibid.).

19. Johnson, *History of the American People*, 219.

20. Jefferson did write to a Bostonian in 1816, however, that, "[e]xperience has taught me that manufactures are now as necessary to our independence as to our comfort"; see Dalzell, *Enterprising Elite*, 42; see also Arthur M. Schlesinger Jr., *The Age of Jackson*, 2nd ed. (Old Saybrook, CT: Konecky & Konecky, 1971), 18.

21. Jonathan Prude, "Capitalism, Industrialization, and the Factory in Post-revolutionary America," in *Wages of Independence: Capitalism in the Early American Republic*, ed. Paul A. Gilje (Lanham, MD: Rowman & Littlefield, 1997), 90–2.

22. Johnson, *History of the American People*, 218–9.

23. Ibid., 232.

24. Gordon and Malone, *Texture of Industry*, 95–7.

25. The works were operated sporadically after 1652 and closed in 1676; see Gordon and Malone, *Texture of Industry*, 68.

26. Newell, "Birth of New England in the Atlantic Economy," 42.

27. Government support for economic development can be traced back to Medieval times, helping to develop contractual law, provide labor, and enforce commercial codes to foster business activity and generate tax revenue (Landes, *Wealth and Poverty of Nations*, 44).

28. Newell, "Birth of New England in the Atlantic Economy," 47.

29. Temin, "Introduction," 7.

30. Holbrook, *Old Post Road*, 191.

31. A Welshman, John Adams Dagyr, arrived in Massachusetts in 1750 and started a small shoemaking business that eventually grew into a large industry that employed a quarter of the Massachusetts labor force in the mid-1800s. It derived some technology from the textile mills that came later; Oliver, *History of American Technology*, 48.

32. Newell, "Birth of New England in the Atlantic Economy," 62.

33. Temin, "Introduction," 7.

34. The basic technology was provided by Thomas Somers, who had been sent by some Baltimore merchants to Britain to learn about the mechanized cotton-carding process. When he returned, their interests had shifted, so he took his knowledge to Boston, where he impressed the Cabots. Incorporated in 1789, the mill contained roller cards and spinning jennies. The mill closed in 1807. See David J. Jeremy, *Transatlantic Industrial Revolution: The Diffusion of Textile Technologies between Britain and America, 1790–1830* (Oxford: Blackwell, 1981), 17, 83; and Dirk J. Struik, *Yankee Science in the Making* (New York: Collier Books, 1962), 191.

35. Burrows and Wallace, *Gotham*, 232.

36. These had been struggling, as Napoleon's closing the European continent to British goods had led to their being dumped in America (Jeremy, *Transatlantic Industrial Revolution*, 83).

37. Rothenberg, "Invention of American Capitalism," 96.

38. Ibid., 93.

39. Chandler Jr., *Visible Hand*, 56.

40. New York, in contrast, did not adopt factory methods in any serious way until the Civil War. Its textile industry was virtually driven out of business by the big New England mills, with production declining from an estimated 16.5 million yards in 1825 to less than 1 million in 1855 (Klein, *Empire State*, 317).

41. Rothenberg, "The Invention of American Capitalism," 93.

42. Ibid., 100.

43. Most of the following story about Slater is adapted from Paul E. Rivard, *Samuel Slater: Father of American Manufactures* (Pawtucket, RI: Slater Mill Historic Site, 1974; repr. 1988).

44. He apparently read that a Philadelphia mechanic had been given a sizable bounty for devising a carding machine (Jeremy, *Transatlantic Industrial Revolution*, 79).

45. Meanwhile, Britain was forcing the emigration of convicts to Australia as a way of relieving the pressure of population driven off the land by enclosures. Before 1776, the American colonies had been used for this purpose.

46. Burrows and Wallace, *Gotham*, 306.

47. Gordon, *Business of America*, 15–16; Harlick, *No Place Distant*, 48–51.

48. Burrows and Wallace, *Gotham*, 307–10.

49. Various entrepreneurs who had profited from the shipping trade after the Revolution reinvested much of their money in enterprises that served the local and domestic markets; for one perspective related to Boston, see Lisa B. Lubow, "From Carpenter to Capitalist: The Business of Building in Postrevolutionary Boston," in *Entrepreneurs: The*

Boston Business Community, 1700–1850, ed. C.E. Wright and K.P. Viens (Boston: Massachusetts Historical Society, 1997), 197.

50. Some details of Slater's start-up are provided in Jeremy, *Transatlantic Industrial Revolution*, 79–91.

51. Gordon and Malone, *Texture of Industry*, 348.

52. The mill was small by later standards, being 47 feet by 29 feet and 2½ storeys high.

53. In 1833, President Andrew Jackson visited Slater at Pawtucket and formally honored him as the "father of American manufactures (Gordon, *Business of America*, 17).

54. Beatty, *Colossus*, 65.

55. The invention of the cotton gin in 1793 and its quick, wide diffusion sparked a huge growth in Southern cotton production, unfortunately reviving the use of slavery. See Thomas Dublin and United States National Park Service, *Lowell: The Story of an Industrial City — A Guide to Lowell National Historical Park and Lowell Heritage State Park, Lowell, Massachusetts* (Washington, DC: Department of the Interior, 1992), 20; and Chandler Jr., *Visible Hand*, 57.

56. Even in 1880, the Blackstone was considered the most intensively used river in the region (Gordon and Malone, *Texture of Industry*, 62–3).

57. Johnson, *History of the American People*, 396.

58. Chandler Jr., *Visible Hand*, 57–8.

59. Holbrook, *Old Post Road*, 202.

60. Lowell was appalled by the dirty and unhealthy conditions of English mill towns. More attractive were the Scottish planned villages, of which there were a relatively large number by 1810, though mostly concerned with agriculture and fishing; see Dalzell, *Enterprising Elite*, 12, 20.

61. Benjamin W. Labaree, "The Making of an Empire: Boston and Essex County, 1790–1850," in *Entrepreneurs: The Boston Business Community, 1700–1850*, ed. C.E. Wright and K.P. Viens (Boston: Massachusetts Historical Society, 1997), 352. Lowell and Appleton were involved in a variety of ventures, including the export of potash, land speculation in Maine, and Boston land development. See Rothenberg, "The Invention of American Capitalism," 97; see also Jeremy, *Transatlantic Industrial Revolution*, 93–6. In 1858, Appleton wrote a memoir entitled *Introduction of the Power Loom and Origin of Lowell*; see Weible and Lowell Historical Society, *Continuing Revolution*, 1, who note that others in New England were also developing power looms at the same time, some of which incorporated better technology.

62. Susan Rosegrant and David Lampe, *Route 128: Lessons from Boston's High-Tech Community* (New York: Basic Books, 1992), 41.

63. Dalzell, *Enterprising Elite*, 27.

64. A power loom had been invented as early as 1785, but it took until the early 1800s before a truly workable machine was in operation. Its inventor, Edmund Cartwright, had received a £10,000 bounty from the British Parliament in 1809, which Lowell could not have missed when he visited the next year (ibid., 5). Moody began his career as a weaver and was later trained in machinery by Jacob Perkins, the most noted mechanic of his time.

65. Beatty, *Colossus*, 69.

66. The concept of limited liability was not a feature of incorporation in the companies owned by these men until 1830 (Dalzell, *Enterprising Elite*, 49). Instead, the aim of the incorporation was more to provide for the long-term existence of the operation and the integrity of the capital investment. The incorporation petition also spoke to a larger public purpose: to "secure the establishment of manufactures upon a more permanent foundation than has hitherto been found practicable in this commonwealth," this last perhaps referring to the unhappy textile experience of the Cabots in Beverly a generation earlier (28). The name of the company and its proposed product are reminiscent of the New York City company that employed Samuel Slater briefly in 1790.

67. Weible and Lowell Historical Society, *Continuing Revolution*, 42.

68. Ibid., 48

69. Wills, *Henry Adams and the Making of America*, 223–44, 350–6.

70. Weible and Lowell Historical Society, *Continuing Revolution*, 40.

71. Some of the basic technicalities of Lowell's and Moody's designs are described in Jeremy, *Transatlantic Industrial Revolution*, 99–101.

72. Dublin and United States National Park Service, *Lowell*, 40. See also David Freeman Hawke, *Nuts and Bolts of the Past: A History of American Technology, 1776–1860* (New York: Harper & Row, 1988), 91–3.

73. Dublin and United States National Park Service, *Lowell*, 33.

74. While American producers considered only the domestic market, by 1814 British producers exported four yards of cloth for every three consumed at home (Hobsbawm, *Age of Revolution*, 35).

75. One estimate of the capitalization of Rhode Island spinning mills was but a tenth of this; see Weible and Lowell Historical Society, *Continuing Revolution*, 42. The willingness of Boston merchants to invest in manufacturing on this scale seems to have been unique in America at this time (54).

76. Dalzell, *Enterprising Elite*, 27. The cash was collected from the investors in installments, as the money was needed.

77. Ibid.

78. "The curve of this change [toward support for industry] could be traced in the permutations of Daniel Webster's views on the tariff, as his thunderous briefs for the merchant and free trade began to hesitate and quaver, till they gave way to equally thunderous briefs for the mill owner and protection" (Schlesinger Jr., *Age of Jackson*, 144; see also Weible and Lowell Historical Society, *Continuing Revolution*, 2; and Chandler Jr, *Visible Hand*, 58). Tariffs on imported goods traditionally had been the US government's means of financing itself, but with the nascent industrialization prompted by the embargo and the War of 1812, the protectionist aspect of the tariff began to gain in importance.

79. Some of the Associates must have had severe conflicts of interest. As shippers and importers, they would have opposed tariffs, but, as manufacturers, they would have favored them. The commercial leaders not involved in manufacturing expressed their unhappiness in a 1820 *Report of the Committee of Merchants and Others, of the Tariff*, which must have been produced for the debate over the proposed new tariff; see Dalzell, *Enterprising Elite*, 41. As late as 1833, one commentator claimed that, had the tariff been eliminated, three-quarters of the production in New England would have been curtailed

and half the industrial sector would have been bankrupted (101–2). This is probably an exaggeration.

80. Ibid. There is disagreement over which country's workforce was the most productive in the years before the Civil War. Peter Temin ("The Industrialization of New England, 1830–1880," in *Engines of Enterprise: An Economic History of New England*, ed. Peter Temin [Cambridge, MA: Harvard University Press, 2000], 126) argues that Americans were twice as productive as British workers. This would justify the continuing wage disparities, but implies that the tariff, as long as it protected low-cost American production, was really aimed at low-wage Indian cloth producers.

81. Dalzell, *Enterprising Elite*, 113.

82. Chandler Jr, *Visible Hand*, 58. Women were not used intensively on New England farms, so they could be spared to go to the mills and add to the family income — or at least not put pressure on it (Dalzell, *Enterprising Elite*, 115).

83. Beatty, *Colossus*, 69. The boardinghouses were leased to matrons who, in turn, charged the women a fixed price for room and board (Dalzell, *Enterprising Elite*, 33).

84. In 1850, there were ten thousand workers in the Lowell mills: in 1845 the average size of Boston manufacturing firms was eight workers; see Ronald J. Zboray and Mary Saracino Zboray, "The Boston Book Trades, 1789–1850: A Statistical and Geographical Analysis," in *Entrepreneurs: The Boston Business Community, 1700–1850*, ed. C.E. Wright and K.P. Viens (Boston: Massachusetts Historical Society, 1997), 218n11. The sheer scale attracted many visitors, including Andrew Jackson and Charles Dickens.

85. The machine shop was later relocated to Lowell; Moody went with it as its head (Weible and Lowell Historical Society, *Continuing Revolution*, 141).

86. Their wages tended to be higher than prevailing rates to make sure they attracted the "best sort" of workers. Waltham soon had an employment waiting list (Dalzell, *Enterprising Elite*, 32–3). Wages were also as much as 84 percent higher than those for equivalent British workers (101–2).

87. Weible and Lowell Historical Society, *Continuing Revolution*, 17.

88. The economics of the trade-off between power costs and labor costs of the time are noted in Isaac Cohen, *American Management and British Labor: A Comparative Study of the Cotton Spinning Industry* (New York: Greenwood Press, 1990), 48–9.

89. Dalzell, *Enterprising Elite*, 30.

90. Sales grew from $3,000 in 1815 to $34,000 in 1817 and $345,000 in 1822; see Richard D. Brown, *Massachusetts: A Bicentennial History* (New York: Norton, 1978), 137.

91. The canal went through property that had been deeded to the Permacook Indians in 1655, when the town of Chelmsford was created by the Massachusetts colony. The Indians sold it to the town in 1685, following King Philip's War. It later became the town of East Chelmsford and, a century later, Lowell. See Weible and Lowell Historical Society, *Continuing Revolution*, 4.

92. The company had been incorporated in 1792 (ibid., 11). Apparently, the first canal built in New England was the South Hadley and Montague Canal just north of Springfield, Massachusetts, on the Connecticut River, opened in 1792; see Charles Frederick Carter, *When Railroads Were New* (New York: Simmons-Boardman, 1926), 5.

93. Wright and Viens, *Entrepreneurs*, 352.

94. Dublin and United States National Park Service, *Lowell*, 18.

95. Funded privately, it was supported by many who would be political leaders of Massachusetts in the next decades. In 1808 Secretary of the Treasury Albert Gallatin said it was "the greatest work of its kind which has ever been completed in the United States." Although its profitability was affected by the embargo and the War of 1812, it did pay dividends from 1819 until 1843. It closed in 1852, superseded by the railroads. See Weible and Lowell Historical Society, *Continuing Revolution*, 12–13; see also Struik, *Yankee Science in the Making*, 159–62.

96. It seemed never to have paid for itself until activity picked up at the falls in 1818. The promoters of the Erie Canal were often reminded of this fact; see Berstein, *Wedding of the Waters*, 131. Even after the construction of the mills in Lowell, the canal was used mainly to haul raw materials to them, while finished goods were sent back by wagon; see Stephen Salsbury, *The State, the Investor, and the Railroad: The Boston and Albany, 1825–1867* (Cambridge, MA: Harvard University Press, 1967), 41; see also Dalzell, *Enterprising Elite*, 84.

97. The farms in the area were not particularly fertile; through the eighteenth and nineteenth centuries, families left them for better land north and west, a typical scene in what was becoming overcrowded New England (Weible and Lowell Historical Society, *Continuing Revolution*, 5).

98. Ibid., 11.

99. Beatty, *Colossus*, 70–1. Half of the fourteen early shareholders were not on the list of the Boston Manufacturing Company; two were other Appletons and one was Daniel Webster. Later, all the shareholders in the Boston Manufacturing Company were allowed to purchase shares, and twenty-seven exercised this option (Weible and Lowell Historical Society, *Continuing Revolution*, 43).

100. Although the mills were incorporated separately and shareholders in each mill differed somewhat, they were largely the same and the mills functioned as a unit. The Merrimack mill (1823) was followed by the Hamilton (1826), the Appleton (1828), the Lowell (1829), the Middlesex (1831), the Suffolk and Tremont (1832), the Lawrence (1833), the Boott (1836), and the Massachusetts (1840); see Dublin and United States National Park Service, *Lowell*, 39.

101. Weible and Lowell Historical Society, *Continuing Revolution*, 138.

102. He was connected to the Boott Scottish trading family resident in Boston (Brown, *Massachusetts*, 138).

103. The standard mill was four storeys high, 150 feet long by 40 feet wide, and contained about 6,000 spindles; see Temin, "Industrialization of New England, 134.

104. Dublin and United States National Park Service, *Lowell*, 39.

105. Dalzell, *Enterprising Elite*, 47.

106. Weible and Lowell Historical Society, *Continuing Revolution*, 43–4; Dalzell, *Enterprising Elite*, 49.

107. Weible and Lowell Historical Society, *Continuing Revolution*, 143.

108. Dalzell, *Enterprising Elite*, 57. The Boston merchants, Amos and Abbott Lawrence, named the site after their family.

109. Struik, *Yankee Science in the Making*, 321–2; Hawke, *Nuts and Bolts of the Past*, 93, 196–8.

110. Most of the machinery in American mills was built of wood, partly because of its cheapness as a material and because it could be formed easily. Metal was used for fittings because of its strength and resistance to wear. The use of metal turbines in the 1840s was the beginning of the shift to metal machines. Metal allowed for steam power as well; see Brooke Hindle, *America's Wooden Age: Aspects of Its Early Technology* (Tarrytown, NY: Sleepy Hollow Restorations, 1975), 38–9, 184–6.

111. Dublin and United States National Park Service, *Lowell*, 36, 81. The Lowell Machine Shop was incorporated in 1845.

112. Weible and Lowell Historical Society, *Continuing Revolution*, 175. Manchester, New Hampshire, was created later, in the 1850s. The company that had mills at Lawrence successfully approached the Massachusetts legislature to amend its charter to allow it to own property in New Hampshire; see Theodore Steinberg, *Nature Incorporated: Industrialization of the Waters of New England* (New York: Cambridge University Press, 1991), 108.

113. Rothenberg, "The Invention of American Capitalism," 99–100. By 1889, the water control system had been taken apart as the downstream companies abandoned waterpower and the dams were sold to electricity producers (Steinberg, *Nature Incorporated*, 269).

114. The interlocking nature of the Boston Associates can be seen in a study of a similar structure that ran the mills at Fall River, Massachusetts (Cohen, *American Management and British Labor*, 122–5).

115. Weible and Lowell Historical Society, *Continuing Revolution*, 44; Dalzell, *Enterprising Elite*, 49.

116. Hobson, *Remembering America*, 123, table 4.7. Technology improvements led to price declines between 1820 and 1860 to one-quarter that of early production (Jeremy, *Transatlantic Industrial Revolution*, 105).

117. The shareholders at Waltham, the earliest and least productive mill, got their investment back in dividends in seven years (Chandler Jr, *Visible Hand*, 59).

118. Nathan Appleton gave his children a number of shares in the companies over his lifetime and even gave four shares to his son-in-law, the poet Henry Wadsworth Longfellow (Weible and Lowell Historical Society, *Continuing Revolution*, 52).

119. Chandler Jr, *Visible Hand*, 60.

120. Dalzell, *Enterprising Elite*, 47.

121. Chandler Jr, *Visible Hand*, 65.

122. For a very good picture of Lowell society in the 1830s, see Lawrence A. Cremin, *American Education: The National Experience, 1783–1876* (New York: Harper and Row, 1980), 415–25.

123. Johnson, *History of the American People*, 396.

124. Weible and Lowell Historical Society, *Continuing Revolution*, 143. It reminds me of my home town of Flint, Michigan, in the 1950s, when it was said that 50,000 of its 150,000 people worked for General Motors.

125. Ibid., 149.

126. Ibid., 237.

127. Fall River passed Lowell by 1890 as the largest center for textile manufactur-

ing, with 19,000 employees; see Dublin and United States National Park Service, *Lowell*, 65. By 1909, more people were employed in making clothes, not cloth, in Manhattan alone than were engaged in the mills and plants of Massachusetts (Von Drehle, *Triangle*, 15–16). For an interesting study of the evolution of the town itself up to the late 1930s, see Margaret Terrell Parker, *Lowell: A Study of Industrial Development* (Port Washington, NY: Kennikat Press, 1940).

128. Landes, *Wealth and Poverty of Nations*, 301.

129. Albro Martin, *Railroads Triumphant: The Growth, Rejection, and Rebirth of a Vital American Force* (New York: Oxford University Press, 1992), 306.

130. One of these, the *Patrick*, ran on the line when it opened in 1835. By 1838, the machine shop at Lowell had built and sold thirty-two locomotives, eight of them for the Western Railroad, pushing through the Berkshires to Albany, New York (Weible and Lowell Historical Society, *Continuing Revolution*, 149; see also Salsbury, *State, the Investor, and the Railroad*, 175).

131. Another view of this process sees the investment in the mills as a way to preserve family fortunes and status through "sure things" such as cloth production, not as budding capitalism (Weible and Lowell Historical Society, *Continuing Revolution*, 63). This may have been true after 1830, when the industry was established, but does not explain either the start of the textile manufacturing business or the subsequent rise of Boston as a venture capital center for railroads and other manufacturing in the 1840s and after.

132. Rothenberg, "The Invention of American Capitalism," 98.

133. Weible, *World of the Industrial Revolution*, 51.

134. One student of the period notes that the completion of the Erie Canal allowed factories to follow the frontier, and a number of relatively small (40-200 employees) textile plants did appear in the late 1820s along the falls of water from streams that came down from the plateau south of the canal (Berstein, *Wedding of the Waters*, 358). In 1831, the average mill near the Erie Canal employed forty-nine people (Klein, *Empire State*, 317). Yankee migration into the area probably brought technical expertise and financial connections. Most of these small mills were driven out of business by the big New England mills.

135. Chandler Jr, *Visible Hand*, 60. One New England advantage that was negated by the railroads was that of cheap sea transport. As long as population was concentrated along the coast, the major markets were easily accessible. The railroads not only opened up the West for settlement; they also equalized locational advantages with respect to this population shift (Weible and Lowell Historical Society, *Continuing Revolution*, 8–9).

136. Dublin and United States National Park Service, *Lowell*, 65.

137. Weible and Lowell Historical Society, *Continuing Revolution*, 50; and Dalzell, *Enterprising Elite*, 55. Even so, a Lowell mill became the first in the world to figure out how to produce carpets on a power loom in 1842. As well, in 1846, the first turbines were installed in place of breastwheels to improve the power efficiency gained from falling water (Weible and Lowell Historical Society, *Continuing Revolution*, 142).

138. In 1860, New York City produced 40 percent of the country's clothing, and 35 percent of its manufacturing employment was in the clothing trades (Burrows and Wallace, *Gotham*, 310).

139. The use of electricity rather than waterpower or steam, when combined with air conditioning, was a centerpiece of Southern attractiveness. The new sites could be designed from the ground up, as opposed to expensive reconstruction in the North (Gordon and Malone, *Texture of Industry*, 318).

140. Frederick Turner, *Spirit of Place: The Making of an American Literary Landscape* (San Francisco: Sierra Club Books, 1989), 64–5.

141. David Lampe, *The Massachusetts Miracle: High Technology and Economic Revitalization* (Cambridge, MA: MIT Press, 1988), 277.

142. Weible and Lowell Historical Society, *Continuing Revolution*, 283–316.

143. In Britain, the ratio was about even, not counting child labor (Jeremy, *Transatlantic Industrial Revolution*, 105).

144. At one point in the growth of Lowell, the Lowell Savings Bank reported that 978 of the women employed had deposited $100,000 of their savings (Dalzell, *Enterprising Elite*, 46).

145. Dublin and United States National Park Service, *Lowell*, 56.

146. Lee, *Yankees of Connecticut*, 237.

147. Weible and Lowell Historical Society, *Continuing Revolution*, 236–7. The second generation of French-Canadians was not so enthusiastic about mill work (Ashton and Hudson, *Industrial Revolution*, 136–7, 142).

148. Dublin and United States National Park Service, *Lowell*, 67. Jack Kerouac's French-Canadian parents were latecomers to Lowell, arriving in the early 1900s. His father was a printer and businessman, not a millhand.

149. Mill towns such as Lowell converted "prefactory" immigrants into city people with industrial discipline and attitudes from the 1840s through to the Depression, nearly a century later; see R. Douglas Hurt, *The Ohio Frontier: Crucible of the Old Northwest, 1720–1830* (Bloomington: Indiana University Press, 1996), 55.

Chapter 6

1. Struik, *Yankee Science in the Making*, 25.

2. Quoted in Paul Freiberger and Michael Swaine, *Fire in the Valley: The Making of the Personal Computer*, 2nd ed. (New York: McGraw-Hill, 2000), 3

3. Quoted in Smithsonian Institution, National Museum of American History, "Lighting a Revolution — Lamp Inventors 1880–1940: G.E.M. Lamps" (Washington, DC, 2013); available online at http://americanhistory.si.edu/lighting/bios/whitney.htm.

4. Quoted in James S. Huggins, "James S. Huggins' Refrigerator Door"; available online at http://www.jameshuggins.com/h/quo1/quotations_general.htm#hopper.

5. See Eastern Roads, "Yankee Division Highway: Historic Overview"; available online at http://www.bostonroads.com/roads/MA-128/.

6. Yanni Kosta Tsipis and David Kruh, *Building Route 128* (Charleston, SC: Arcadia, 2003), 7. The choice of the actual number was haphazard.

7. There was already a controversy at the national level over the creation of a southwestward-trending highway from Chicago to Los Angeles, eventually numbered US 66. It broke the grid and the grid numbering system.

8. Paul Sutter, "'A Retreat from Profit': Colonization, the Appalachian Trail, and the

Social Roots of Benton MacKaye's Wilderness Advocacy," *Environmental History* 4 (4, 1999): 553–7; see also Patton, *Open Road*, 216.

9. Susan Rosegrant and David Lampe, *Route 128: Lessons from Boston's High-Tech Community* (New York: Basic Books, 1992), 107.

10. Tsipis and Kruh, *Building Route 128*, 8. In 1949, he noted: "This new highway will cause the relocation of business establishments and open new residential sections."

11. Rosegrant and Lampe, *Route 128*, 107.

12. Tsipis and Kruh, *Building Route 128*, 8.

13. The politicians cut the ribbon to open it in 1951. When the completed highway was opened (again), some parts already were seeing traffic numbers three times higher than the road had been designed for (ibid., 9).

14. See Mark Sardella, "The History of Route 128"; available online at http://marksardella.wordpress.com/2008/05/09/route-128/.

15. The basic story of the making of Route 128 can be found in a number of places; see, for example, Eastern Roads, "Yankee Division Highway: Historic Overview." at http://www.bostonroads.com/roads/MA-128

16. Tsipis and Kruh, *Building Route 128*, 116

17. By 1973, 186 new companies had established in Waltham (Rosegrant and Lampe, *Route 128*, 130).

18. Ibid., 108.

19. Michael T. Peddle, "Planned Industrial and Commercial Developments in the United States: A Review of the History, Literature, and Empirical Evidence Regarding Industrial Parks and Research Parks," *Economic Development Quarterly* 7 (1, 1993): 107–24.

20. Tsipis and Kruh, *Building Route 128*, 116.

21. Alan R. Earls, *Route 128 and the Birth of the Age of High Tech* (Charleston, SC: Arcadia, 2002), 7

22. Beatty, *Colossus*, 6. The concentration of nineteenth-century arms manufacturers in the Connecticut River valley was an earlier example of this phenomenon.

23. Eastern Roads, "Yankee Division Highway: Historic Overview."

24. Rosegrant and Lampe, *Route 128*, 130.

25. Sardella, "History of Route 128."

26. Rosegrant and Lampe, *Route 128*, 130.

27. Ibid., 109–10.

28. Patton, *Open Road*, 212–13.

29. United Nations Educational, Scientific and Cultural Organization, "Science Policy and Capacity-Building: Science Parks around the World" (Paris: UNESCO, n.d.); available online at http://www.unesco.org/new/en/natural-sciences/science-technology/university-industry-partnerships/science-parks-around-the-world/.

30. As early as 1635, some Puritan settlers were experimenting with local alternatives to needed chemicals, such as gunpowder, which otherwise had to be ordered from Europe (Oliver, *History of American Technology*, 64–6).

31. Rosegrant and Lampe, *Route 128*, 37.

32. Weible and Lowell Historical Society, *Continuing Revolution*, 160.

33. Ibid., 185.

34. Ibid., 162.

35. Nathan Rosenberg and L.E. Birdzell, *How the West Grew Rich: The Economic Transformation of the Industrial World* (New York: Basic Books, 1986), 252–3.

36. Weible and Lowell Historical Society, *Continuing Revolution*, 147.

37. Andrew Carnegie has been credited as the first to introduce science to production by employing chemists in his steel plants (see, for example, Johnson, *History of the American People*, 563), but this was decades after Francis and Roswell in Lowell and Springfield.

38. Massachusetts tried early on to include some mechanical instruction in its education system, which attracted the interest of New York educators. Then New York was given an endowment for the creation of the Rensselaer Institute in Troy (the Yankee town in the Albany area) in 1825, its mission being to instruct in "the application of science to the common purposes in life" (Fox and Remini, *Decline of Aristocracy*, 323; see also Bruce Mazlish, *The Railroad and the Space Program; An Exploration in Historical Analogy* [Cambridge, MA: MIT Press, 1965], 64). Renssalaer was the only school besides West Point at the time to offer engineering training. Harvard began to offer engineering in 1847, with an endowment from Amos Lawrence. Railroad construction apparently spurred this interest in early engineering education (Mazlish, *Railroad and the Space Program*, 169–70). Other donors followed suit at Yale, Dartmouth, and in Brooklyn, New York (Oliver, *History of American Technology*, 244).

39. Struik, *Yankee Science in the Making*, 442–3.

40. In 1847, another Yankee, Professor Jonathan B. Turner, one of the founders of the University of Illinois, had proposed that a system of such colleges be set up on the basis of an equal grant for each state. Morrill, in contrast, favored allocating public lands to each state based on the number of its congressional representatives, a method that favored eastern states at the time. For the aims of this act, see Rosegrant and Lampe, *Route 128*, 32–3; and for more on Turner and Morrill, see 35–6.

41. Ibid., 32–3.

42. John D. Pulliam and James J. Van Patten, *A History of Education in America* (Upper Saddle River, NJ: Merrill Prentice Hall, 2003), 139.

43. Rosegrant and Lampe, *Route 128*, 36.

44. The term "scientist" was first used by a Cambridge don in 1833 to distinguish those interested in the material world from those more concerned with things spiritual and artistic; see Doron Swade and Charles Babbage, *The Difference Engine: Charles Babbage and the Quest to Build the First Computer* (New York: Viking, 2001), 24.

45. Palmer C. Ricketts described Rennselaer as "the first school of science and school of civil engineering, which has had a continuous existence, to be established in any English-speaking country"; see the Rennselaer website at http://www.rpi.edu/about/history.html.

46. Rosegrant and Lampe, *Route 128*, 50.

47. Ibid., 50–1. Whitney is regarded as one of the fathers of industrial research.

48. Even automobile manufacturers became involved; see, for instance, Alfred P. Sloan, John McDonald, and Catharine Stevens, *My Years with General Motors* (New York: Doubleday/Currency, 1990), 248–51, 259, 262.

49. Rosegrant and Lampe, *Route 128*, 51.
50. Landes, *Wealth and Poverty of Nations*, 201, 283–4.
51. Rosenbloom, "Challenges of Economic Maturity," 169–71.
52. Ibid., 171.
53. Chandler Jr, *Visible Hand*, 40.
54. Rosegrant and Lampe, *Route 128*, 38.
55. Hobsbawn, *Age of Capital*, 57–8.
56. Support was not all driven by military needs. For example, the National Institutes of Health, the Bureau of Mines, and the Bureau of Standards grew out of other concerns, as did the Smithsonian Institute; see Rosenberg and Birdzell, *How the West Grew Rich*, 253; and Rosegrant and Lampe, *Route 128*, 54.
57. Rosegrant and Lampe, *Route 128*, 75–85.
58. He had developed an electric calculating machine to solve differential equations.
59. Lynn Elaine Brown and Steven Sass, "The Transition from a Mill-Based to a Knowledge-Based Economy: New England, 1940–2000," in *Engines of Enterprise: An Economic History of New England*, ed. Peter Temin (Cambridge, MA: Harvard University Press, 2000), 210.
60. We would know it today as a four-function calculator; see Swade and Babbage, *Difference Engine*, 98.
61. A good survey of Babbage and his contemporaries is Swade and Babbage, *Difference Engine*. In his time, a computer was not a machine but a person who worked out numerical tables and calculations, much like we would refer today to a programmer, meaning someone who writes programs to operate computers.
62. Ibid., 11–17, 27–9.
63. Ibid., 67. Although Swade dismisses the question of precision in terms of parts technology in the 1830s, he notes that the cost of producing a large quantity of identical parts added to the time needed for construction and therefore the cost.
64. Ibid., chap. 10. When his father attempted to promote orders for the machine, he found no takers. In 1859, a Scheutz machine was built in Britain and used to help calculate actuarial tables, but only did 28 of 600 pages of numbers.
65. See ibid., part 3, for an account of the construction of the modern machine at the Science Museum in London. The 1942 Mark I was the first such analytical engine, but it came seventy years after Babbage died and used different technology than was available to him (Rowland, *Spirit of the Web*, 237).
66. A lot of mythology seems to be attached to Ada, if only because of her parentage. One of the early computer programming languages was named Ada. The relationship between Babbage and Ada Lovelace is described in Swade and Babbage, *Difference Engine*, chap. 8. Dr. Grace Hopper, who was the oldest serving member of the US Navy when she retired as a rear admiral in 1986, just shy of her eightieth birthday, was Ada Lovelace's logical successor, having been the chief developer of the first robust computer language, COBOL.
67. Babbage used punched cards in his engines, copied from the Jacquard textile loom.
68. When this design was taken over by IBM, the machine, called the Automatic

Sequence calculator, used punched tape rather than cards; see Rowland, *Spirit of the Web*, 237.

69. The movement from analog, where the machine uses continuous physical phenomena (like a steering wheel on a car) to solve a problem, to digital, where everything is reduced to numbers and solved that way, was a momentous step. It led to the standard adoption of the base 2 numerical system as the bedrock of the computing age instead of the base 10 used by most human minds. Today, if it is digital, it works off a numerical system unknown to most people.

70. ENIAC, built in 1945, cost $800,000, contained 17,500 tubes, required a dedicated power line, occupied 1,800 square feet of space, weighed 30 tons, and its front panel was 100 feet long; it could perform all of 37 calculations in a second (Rowland, *Spirit of the Web*, 237). A controversy has developed in which one side claims that John Atanasoff, a professor at the University of Iowa, invented the first digital computer. Like the automobile, there were many inventors, using a variety of technologies, one of which eventually becomes dominant (Lienhard, *Inventing Modern*, 255; Van Dulken, *Inventing the 20th Century*, 110).

71. Rowland, *Spirit of the Web*, 245. By 1953, the SAGE weighed 250 tons, but could track as many as 48 targets at a time; the system remained in service until 1984.

72. Lampe, *Massachusetts Miracle*, 240–3. Employment was small at first; DEC, for instance, employed about 5,000 people in 1970 and 10,000 in 1974. But by 1984 it employed 25,000 in Massachusetts alone. See Brown and Sass, "Transition," 213.

73. Once the marvel of the age, the last Polariod instant film was produced at the end of 2008, a victim of digital photography.

74. Raytheon developed the microwave oven in 1945 as a spinoff of its military-sponsored radiation lab; see Earls, *Route 128 and the Birth of the Age of High Tech*, 38.

75. Apparently Compton had had an idea like this before the war put it on hold (Rosegrant and Lampe, *Route 128*, 111.

76. Beatty, *Colossus*, 6.

77. Paul Gompers and Joshua Lerner, eds., *The Venture Capital Cycle*, 2nd ed. (Cambridge, MA: MIT Press, 2004), 4. It had trouble raising its first $3 million tranche in 1946, as well (Rosegrant and Lampe, *Route 128*, 111).

78. Rosegrant and Lampe, *Route 128*, 119. By the 1960s, the bank had accounts with 85 percent of the Route 128 tech companies (120).

79. It folded in the mid-1960s, and it would be a while before others sprang up (ibid., 114); by the 1980s, Silicon Valley was attracting the lion's share of venture capital money (126–7).

80. Ibid., 121–2.

81. In all, ADRC had 1,200 investors over its autonomous life (ibid., 111).

82. Earls, *Route 128 and the Birth of the Age of High Tech*, 7–8.

83. Ibid., 87.

84. See Ross DeVol and Anita Charuwon, *California's Position in Technology and Science: A Comparative Benchmarking and Assessment* (Santa Monica, CA: Milken Institute, 2008); Lampe, *Massachusetts Miracle*, 243; and Rosegrant and Lampe, *Route 128*, 13.

85. Brown and Sass, "Transition," 232.

86. Herman Hollerith had the idea of adapting punch cards to a machine-readable format that could be programmed to analyze data, such as that on census forms. Punch cards had been in use by textile mills to "program" their production for calculations and arithmetical functions for decades. Babbage and Lovelace also had the same idea. Hollerith received his PhD for a dissertation on the subject in 1889 and then succeeded in using his cards and machines to sort data from the 1890 census returns. As late as the mid-1960s, as a graduate student, I used a "counter-sorter" with punch cards for the data in my master's thesis.

87. Flint began with a New York shipper, became a US consul in Chile, got into banking and selling warships to developing countries, and then became a "disinterested intermediary" in putting together trusts; see Sean Dennis Cashman, *America in the Gilded Age: From the Death of Lincoln to the Rise of Theodore Roosevelt* (New York: New York University Press, 1984), 57–8.

88. Grace Hopper's team on the early Mark I found a moth that had short-circuited part of the computer, the legendary first "bug," a picture of which is in Earls, *Route 128*, 43.

89. Rowland, *Spirit of the Web*, 246, 278.

90. When Bell applied for a patent, its lawyers found that a scientist living in Canada had patented a similar device in 1930, but nothing had come of it. The Bell device was different enough that the company was able to get around the patent.

91. They all got the Nobel Prize for their effort. It is important to recognize, however, that "invention" is never a lone process, but normally consists of the useful model of something that has been thought about by others in the past. The word "utility" is crucial to what we see as "invention"; see Diamond, *Guns, Germs, and Steel*, 244–5.

92. Raytheon claims to be the first commercial producer of transistors, applying them to hearing aids as early as 1947 (Earls, *Route 128 and the Birth of the Age of High Tech*, 8, 40–1).

93. Rowland, *Spirit of the Web*, 266.

94. See Lampe, *Massachusetts Miracle*, 265; and Rowland, *Spirit of the Web*, 266.

95. In 1971, Ted Hoff at Intel developed the microprocessor, a programmable chip that allows the variety of uses that a computer user might want to have performed without redesigning the circuitry. Five years later, the microprocessor was at the heart of the fledgling personal computer.

96. In 1968, Intel's chips held 1,024 bits of data, and in 1991, 1,024,000 bits, while the chips were sold at roughly the same price, unadjusted for inflation; see Robert X. Cringley, *Accidental Empires: How the Boys of Silicon Valley Make Their Millions, Battle Foreign Competition, and Still Can't Get a Date* (New York: Harper, 1996), 41.

97. Moore's Law does not have any intrinsic driving force, but it has been an uncanny predictor of integrated circuit/ chip development, which, in turn, has led to increasingly sophisticated products and services, rendering earlier ones obsolete.

98. It had 4,100 integrated circuits, the most advanced instrumentation of its kind then (Earls, *Route 128 and the Birth of the Age of High Tech*, 65). Apollo 8 had a computer the size of a suitcase (Rowland, *Spirit of the Web*, 269).

99. An example is the daily newspaper industry, which has seen its margins eroded by mobile online readership, online want ads, alternative news sources, and constant

"breaking" news. None of this was possible with the technology of a decade ago — the bottleneck being computing power.

100. Lampe, *Massachusetts Miracle*, 246.

101. Olsen was one of Jay Forrester's MIT Whirlwind group that developed the magnetic core memory. After he got his master's degree in 1952, he went to work on SAGE, and left in 1957 to found DEC (Rosegrant and Lampe, *Route 128*, 91).

102. Ibid., 5–6.

103. Lampe, *Massachusetts Miracle*, 247.

104. An Wang earned his PhD from Harvard in 1948 and went to work on computer development. He came up with a magnetic core memory about the same time as Jay Forrester at MIT. He set up Wang Laboratories in 1951 with $600 and a disputed patent on the magnetic core memory (Rosegrant and Lampe, *Route 128*, 92, 115–7).

105. Earls, *Route 128 and the Birth of the Age of High Tech*, 123.

106. The National Park was the first created east of the Mississippi in the twentieth century (Lampe, *Massachusetts Miracle*, 275).

107. Ibid., 275–9.

108. Earls, *Route 128 and the Birth of the Age of High Tech*, 105.

109. Rosegrant and Lampe, *Route 128*, 6.

110. Freiberger and Swaine, *Fire in the Valley*, 24–5, 27. DEC was not prepared to sell or provide customer support to individuals.

111. Ibid., 379–80.

112. Rosegrant and Lampe, *Route 128*, 116.

113. Ibid., 136–7.

114. Ibid., 171–84.

115. Rosenberg and Birdzell, *How the West Grew Rich*, 128.

116. Rowland, *Spirit of the Web*, 81–6.

117. Technology, including the types of products chosen to offer the market, is not value neutral. Neither is the segment of the market chosen. In traditional societies, television, McDonald's, and Wal-Mart are all subversive of leadership structures (ibid., 277–8).

118. The British followed a roughly similar path, though not to the extreme that the United States has done; see Landes, *Wealth and Poverty of Nations*, 222.

Chapter 7

1. Simon Schama, *Landscape and Memory* (Toronto: Vintage Books, 1996), 572.

2. Edmund Morris, *Theodore Rex* (New York: Modern Library, 2002), 487.

3. Aldo Leopold. *A Sand County Almanac* (1949; repr. New York: Ballantine Books, 1966), 157. The "forty freedoms" is a reference to President Harry Truman's postwar statement about "Four Freedoms" that the world's people ought to enjoy, which was issued about the time of Leopold's book was originally published.

4. Henry David Thoreau, *Walden and Civil Disobedience* (1854; repr. New York: Penguin, 1980), 9. Walden is one of 1,100 lakes and ponds in Massachusetts. There is even another Walden Pond elsewhere in the State; see Maynard, *Walden Pond*, 4–5.

5. The area is probably more heavily wooded now than it has been for the past three hundred years.

6. Thoreau was not really a back-to-the-land type, but shared Ralph Waldo Emerson's feeling that the railroad meant progress and utility. He found farming at Walden Pond to be distracting because it required so much time; see George H. Douglas, *All Aboard: The Railroad in American Life* (New York: Smithmark, 1996), 90. The railroad reached its terminus in Fitchburg in March 1845, just before Thoreau went to Walden Pond; see Alvin F. Harlow, *Steelways of New England* (New York: Creative Age Press, 1946), 238.

7. Maynard, *Walden Pond*, 160, 195.

8. Thoreau lived at Walden Pond from July 4,1845, to September 6, 1847. Friends helped him raise the house, and he apparently left because Emerson needed someone to look after his own house while he went on a European tour.

9. Maynard, *Walden Pond*, 192.

10. Ibid., 198. By the end of the century, wilderness was being ascribed to the city, not the country, in books such as Upton Sinclair's *The Jungle*, about Chicago; see Nash, *Wilderness and the American Mind*, 143.

11. Thoreau, *Walden and Civil Disobedience*, 91. A bill was introduced in the State legislature that year to make the donation into a reservation, with the Middlesex County Council as trustees (Maynard, *Walden Pond*, 229).

12. Maynard, *Walden Pond*, 248.

13. Ibid., 256–60. Also, in 1958, Concord residents voted to establish a dump near the Pond.

14. Ibid., 307–16. Henley's fund-raising for the Walden Woods Project secured sponsorship from a number of large corporations, leading critics to feel the site was being "saved" by the kind of institutions that it should stand against.

15. Thoreau, *Walden and Civil Disobedience*, 94–5.

16. Wallace Stegner and Page Stegner, *Marking the Sparrow's Fall: The Making of the American West* (New York: Henry Holt, 1998), 148.

17. Thoreau, *Walden and Civil Disobedience*, 8.

18. Emerson felt he could "pace sixteen rods more accurately than another man can measure it by tape"; see Carlos Baker, *Emerson among the Eccentrics: A Group Portrait* (New York: Viking, 1996), 328.

19. Ibid., 327–8.

20. His aunt Maria wished he would "find something better to do than walking off every now and then" (Turner, *Spirit of Place*, 23).

21. Emerson saw him as "how near to the old monks in their ascetic religion!" (Baker, *Emerson among the Eccentrics*, 437).

22. Fred Setterberg, *The Roads Taken: Travels through America's Literary Landscapes* (Athens: University of Georgia Press, 1993), 133–4. Thoreau was noted for his curt and blunt words, and liked "no" more than "yes" (Baker, *Emerson among the Eccentrics*, 100–4).

23. Johnson (*History of the American People*, 424–7) places Thoreau in the context of writers of his time.

24. Baker, *Emerson among the Eccentrics*, 263–4.

25. Maynard, *Walden Pond*, 25.

26. Ibid., 203.

27. Baker, *Emerson among the Eccentrics*, 265–6; Turner, *Spirit of Place*, 45.

28. Baker, *Emerson among the Eccentrics*, 270–1.

29. Maynard, *Walden Pond*, 8.

30. Thoreau had just read Charles Darwin's *On the Origin of Species*, and it moved him from writing travel-oriented material toward explaining and measuring natural phenomena; see Barbara Kingsolver, *High Tide in Tucson: Essays from Now or Never* (New York: HarperCollins, 1995), 238.

31. Turner, *Spirit of Place*, 39.

32. Emerson seemed to see the Walden Pond area as having some parallels to the English Lake district, home to the Romantic poets William Wordsworth and Samuel Taylor Coleridge (Maynard, *Walden Pond*, 29).

33. Ibid., 220.

34. Ibid., 67. He might also be considered the founder of the civil disobedience movement: while at Walden Pond, he spent a night in jail for refusing to pay his taxes because he opposed the was with Mexico. Abbey points out, however, that this movement goes back at least to the Boston Tea Party.

35. Ibid., 121.

36. Ibid., 254.

37. Ibid., 140; see also Nash, *Wilderness and the American Mind*, 101–6. Thoreau was not the first to suggest this prospect. Painter George Catlin, after traveling in the Dakotas in 1832, wrote that there should be a national park there to preserve the western wilderness: "[W]hat a beautiful and thrilling specimen for America to preserve and hold up to the view of her refined citizens and the world, in future ages! A *nation's park*" (101). In the 1980s, there was a proposal to reserve a large area of the Dakota grasslands as a "buffalo commons," reminiscent of Catlin's dream.

38. Schama, *Landscape and Memory*, 578.

39. Nash, *Wilderness and the American Mind*, 89.

40. Stegner and Stegner, *Marking the Sparrow's Fall*, 125. He did go north, however, taking a railroad and boat excursion to Montreal and Quebec City in 1850.

41. Setterberg, *Roads Taken*, 137–8.

42. Nash, *Wilderness and the American Mind*, 84.

43. Bernard-Henri Levy (*American Vertigo: Traveling in the Footsteps of Tocqueville* [New York: Random House, 2006], 173) notes that this is one of the most enigmatic aspects of Americans.

44. Stegner and Stegner, *Marking the Sparrow's Fall*, 122–5; Nash, *Wilderness and the American Mind*, xi–xiv, 24.

45. One can see this in Thoreau's imitation of the Romanticist Wordsworth in his long walks, his regard for nature, poetry writing, and solitude.

46. Roderick Nash relates European Romanticism to American ideas in *Wilderness and the American Mind*, 47–55. The Indians themselves were baffled by the idea of wilderness. Sioux chief Standing Bear commented: "We did not think of the great open plains, the beautiful rolling hills, and the winding streams as 'wild'." Only to the white man was nature a "wilderness" (xiv).

47. Slotkin, *Regeneration through Violence*, 518–9.

48. There is an alternate possibility, that Cooper modeled Natty after a Connecti-

cut squatter, David Shipman, whose cabin stood by a lake on a property that Cooper's father William came to look over; indeed, Cooper retrospectively acknowledged Shipman (Taylor. *William Cooper's Town*, 63).

49. Some of Cooper's political views are reflected in his novels. See Schlesinger Jr., *Age of Jackson*, 375–80; and Holbrook, *Old Post Road*, 228.

50. Klein, *Empire State*, 353–4.

51. Mark Twain, writing seventy years after the Leatherstocking Tales, was scathing in his criticism of Cooper's style. Even Cooper's later defenders acknowledged his stylistic deficiencies, but he did give us the stock characters of the frontier myth; see John Foreman, *Frommer's Comprehensive Travel Guide to New York State* (New York: Prentice-Hall Travel, 1994), 272.

52. Quoted in Edward Abbey, *Desert Solitaire: A Season in the Wilderness* (New York: McGraw-Hill, 1968), 167–8.

53. Cooper's romanticism played an important role in American myth development; see Johnson, *History of the American People*, 412–3; Nash, *Wilderness and the American Mind*, 75–7; Richard Slotkin, *Gunfighter Nation: The Myth of the Frontier in Twentieth-Century America* (Norman: University of Oklahoma Press, 1998), 15–16; idem, *Regeneration through Violence*, 230, 286, 294, 310, 467–8, 512; Stegner and Stegner, *Marking the Sparrow's Fall*, 123, 199–202; and Taylor. *William Cooper's Town*, 412.

54. Maynard, *Walden Pond*, 29.

55. Nash, *Wilderness and the American Mind*, 86–7.

56. To the thinker, once the process of exploitation and reduction had pushed the frontier, especially the forested frontier, back from the coast, wilderness became something to be protected. The prospect of an America without wilderness was, in its way, un-American. For me, living in a Canada that is mostly still wilderness puts a rather different "spin" on this idea.

57. Most of the biographical details that follow are from Witold Rybczynski, *A Clearing in the Distance: Frederick Law Olmstead and America in the Nineteenth Century* (New York: Scribner, 1999); and Frederick Law Olmstead, *Civilizing American Cities: Writings on City Landscapes*, ed. S.B. Sutton (New York: Da Capo Press, 1997).

58. Rybczynski, *City Life*, 125.

59. Olmstead, *Civilizing American Cities*, 4–5. For his writings on the South, see idem, *The Cotton Kingdom: A Traveler's Observations on Cotton and Slavery in the American Slave States*, ed. Arthur M. Schlesinger (New York: Knopf, 1953).

60. Olmstead, *Civilizing American Cities*, 6–7, 9.

61. Rybczynski, *Clearing in the Distance*, 285–6.

62. Anne Whiston Spirn, "Constructing Nature: The Legacy of Frederick Law Olmsted," in *Uncommon Ground: Rethinking the Human Place in Nature*, ed. William Cronon (New York: W.W. Norton, 1997), 93.

63. Olmstead, *Civilizing American Cities*, 12.

64. Schlereth, *Victorian America*, 235.

65. Olmstead, *Civilizing American Cities*, 101–5. Olmstead had done some work for the Pinchot family. See also Spirn, "Constructing Nature," 99–101.

66. Spirn, "Constructing Nature," 91–3, 110–13.

67. Turner, *Rediscovering America*, 37–9.
68. Much of the biographical details on Muir are found in ibid., 29–40.
69. Ibid., 284–5.
70. Nash, *Wilderness and the American Mind*, 106.
71. He had previously visited Wordsworth in England; see Rebecca Solnit, *Wanderlust: A History of Walking* (New York: Penguin Books, 2001), 153.
72. Turner, *Rediscovering America*, 214.
73. Schama, *Landscape and Memory*, 573. Muir relates that he told Emerson, "you are a Sequoia yourself. Stop and get acquainted with your big brothers."
74. Nash, *Wilderness and the American Mind*, 160.
75. Turner, *Rediscovering America*, 292–3.
76. Nash, *Wilderness and the American Mind*, 132–3.
77. Turner, *Rediscovering America*, 330–1.
78. Char Miller, *Pioneers of Conservation: Gifford Pinchot and the Making of Modern Environmentalism* (Washington, DC: Island Press, 2001), 121.
79. Turner, *Rediscovering America*, 230–1.
80. Congress had passed a law reserving forest reserves in 1891 (Nash, *Wilderness and the American Mind*, 133).
81. Ibid., 122.
82. A ubiquitous sign on federal lands in the West today is "Land of Many Uses."
83. Miller, *Pioneers of Conservation*, 137–8.
84. Turner, *Rediscovering America*, 284–5, 293.
85. Nash, *Wilderness and the American Mind*, 129.
86. One author claims that Muir saw himself as a John the Baptist, leading Americans to immerse themselves in God's Nature (ibid.).
87. Miller, *Pioneers of Conservation*, 7.
88. Maynard, *Walden Pond*, 203.
89. Nash, *Wilderness and the American Mind*, 133.
90. Solnit, *Wanderlust*, 153.
91. Dave Foreman, *Confessions of an Eco-Warrior* (New York: Crown Trade Paperbacks, 1991), 200.
92. Stegner and Stegner, *Marking the Sparrow's Fall*, 125–6.
93. Miller, *Pioneers of Conservation*, 55.
94. Ibid., 57.
95. Ibid., 95. Prominent New Yorkers were concerned in the 1880s that deforestation in the Adirondacks and drought in the western part of the State would hamper trade with the West, and were pushing the idea of a State reserve north of the Hudson River. This led to the creation of the Adirondack Park in 1892 (Nash, *Wilderness and the American Mind*, 118–21).
96. Miller, *Pioneers of Conservation*, 103–4.
97. Marybeth Lorbiecki, *Aldo Leopold: A Fierce Green Fire* (Guilford, CT: Falcon, 2005), 24.
98. Miller, *Pioneers of Conservation*, 130–8.
99. Miller, *Pioneers of Conservation*, 169–70.

100. Ibid., 155.

101. One of the prime movers of this spate of organizing came as a result of a major Governors' Conference on the Conservation of Natural Resources, organized by Pinchot in 1908 and followed by a presidential conference in late 1909. See United States, Library of Congress, "American Memory: The Evolution of the Conservation Movement, 1850–1920" (Washington, DC); available online at http://memory.loc.gov/ammem/amrvhtml/cnchron5.html. See also Faragher, *Rereading Frederick Jackson Turner*, 211.

102. Miller, *Pioneers of Conservation*, 333.

103. Ibid., 374–6.

104. Firestation, "A Brief History of Leopold Desk Company" (Burlington, IA, n.d.); available online at http://burlingtonfirestation.com/leopold.htm.

105. Growing up in Michigan in the 1950s, we called a "davenport" what is known elsewhere as a "sofa" or a "chesterfield," reflecting the importance of furniture from another Mississippi River town, Davenport, Iowa.

106. Lorbiecki, *Aldo Leopold*, 65–6.

107. Ibid., 68.

108. Ibid., 90.

109. Ibid., 112.

110. Marshall was a plant physiologist working in the Wind River area of Wyoming, who in 1930 issued a ringing manifesto for the conservation of wilderness areas wherever they existed, not just on the Gila River (Stegner and Stegner, *Marking the Sparrow's Fall*, 132).

111. Lorbiecki, *Aldo Leopold*, 132.

112. Foreman, *Confessions of an Eco-Warrior*, 178.

113. Stegner and Stegner, *Marking the Sparrow's Fall*, 132.

114. Lorbiecki, *Aldo Leopold*, 182.

115. Nash, *Wilderness and the American Mind*, 272.

116. Ibid., 248. An interesting discussion of a "natural" as opposed to a "made" environment can be found in William Cronon, *Uncommon Ground: Rethinking the Human Place in Nature* (New York: W.W. Norton, 1996), 34–5, 76, 79–80, 284–7.

117. Leopold was thinking both more broadly and contrary to the land ethic that had driven the settlement of the frontier: "There is no ethic dealing with man's relation to land and the animals and plants which grow upon it. Land, like Odysseus' slave girls, is still property. The land relation is still strictly economic, entailing privileges but no obligations"; quoted in Linklater, *Measuring America*, 234–5.

118. See, for instance, Aldo Leopold et al., *Aldo Leopold's Southwest* (Albuquerque: University of New Mexico Press, 1995), vii.

119. It was published in 1950, and went out of print a few years later, only to be revived by the environmental movement in the 1960s; by the late 1980s, over a million copies had been sold. Thoreau's *Walden* went through the same process of death and resurrection a century earlier (ibid., 1).

120. "Ecology is a science that attempts the feat of thinking in a plane perpendicular to Darwin… The outstanding discovery of the twentieth century is not television, or radio, but rather the complexity of the land organism" (Leopold, *Sand County Almanac*, 189–90).

121. Nash, *Wilderness and the American Mind*, 193–6.

122. Federal Bureau of Investigation, "Edward Abbey" (Washington, DC, n.d.); available online at http://www.fbi.gov/fbi-search#output=xml_no_dtd&client=google-csbe&cx=004748461833896749646%3Ae41lgwqry7w&cof=FORID%3A10%3BNB%3A1&ie=UTF-8&siteurl=www.fbi.gov%2F&q=Edward+Abbey.

123. Edward Abbey, *The Brave Cowboy: An Old Tale in a New Time* (New York: Dodd, Mead, 1956).

124. Edward Abbey, *The Monkey Wrench Gang* (New York: Lippincott, Williams & Wilkins, 1975).

125. See Edward Abbey, "In Defense of the Redneck," in Edward Abbey, *Abbey's Road* (New York: Dutton, 1979).

126. James Bishop Jr., *Epitaph for a Desert Anarchist: The Life and Legacy of Edward Abbey* (New York: Touchstone, Simon and Schuster, 1995), 209.

127. Maynard, *Walden Pond*, 263; and Abbey, *Desert Solitaire*, 129–31.

128. Bishop Jr., *Epitaph for a Desert Anarchist*, 201–2; and Foreman, *Confessions of an Eco-Warrior*, 174–5.

129. In 1980, he descended the Green River through Utah accompanied by a "water-soaked, beer-stained, grease-spotted, cheap paperback copy of Walden" (Maynard, *Walden Pond*, 288).

130. Bishop Jr., *Epitaph for a Desert Anarchist*, 185.

131. Ibid., 132.

132. Ibid., 122.

133. Nash, *Wilderness and the American Mind*, 269–70; Bishop Jr., *Epitaph for a Desert Anarchist*, 228; and Abbey, *Desert Solitaire*, 184.

134. Abbey, *Desert Solitaire*, 169.

135. See T. Coraghessan Boyle, "A voice griping in the wilderness," *New York Times*, February 10, 2002.

136. Bishop Jr., *Epitaph for a Desert Anarchist*, 14. Dave Foreman notes that monkeywrenching has a hallowed past in America, including the Boston Tea Party and the Underground Railway (*Confessions of an Eco-Warrior*, 119).

137. See the town's website at http://www.discovermoab.com/biking.htm.

138. Foreman, *Confessions of an Eco-Warrior*, 70. Roads on public lands total thirteen times the mileage of the Interstate system (Harlick, *No Place Distant*, 2).

139. Harlick, *No Place Distant*, x.

140. Nash, *Wilderness and the American Mind*, 261.

141. Abbey, *Desert Solitaire*, xiv.

142. Stegner and Stegner, *Marking the Sparrow's Fall*, 133.

Chapter 8

1. Lyrics available online at http://allspirit.co.uk/mercy.html.

2. Lawrence Lader and Milton Meltzer, *Margaret Sanger: Pioneer of Birth Control* (New York: Thomas Y. Crowell, 1969), 137–8.

3. Loretta McLaughlin, *The Pill, John Rock, and the Catholic Church: The Biography of a Revolution* (Boston: Little, Brown, 1982), 92.

4. John A. McCracken, "Editorial: Reflections on the 50th Anniversary of the Birth Control Pill," *Biology of Reproduction* 83 (4, 2010): 685.

5. It survived beyond the lives of its creators, ceasing operations only in 1997.

6. The oldest guide to conception is the *Petrie Papyrus*, an 1850 BC recipe book for contraceptive devices and suppositories, including one made from crocodile dung. See James Reed, *The Birth Control Movement and American Society: From Private Vice to Public Virtue* (Princeton, NJ: Princeton University Press, 1978), ix; and Andrea Tone, *Devices and Desires: A History of Contraceptives in America* (New York: Hill and Wang, 2001), 13–14.

7. An example is the development of thin rubber by Connecticut Yankee Charles Goodyear, which others used instead of pig's intestines to produce condoms. Goodyear even mentioned them in his 1853 book, *Gum-Elastic and Its Varieties*. The term "rubbers" has been slang for condoms for 150 years; see Reed, *Birth Control Movement*, 13,15–16, although Tone (*Devices and Desires*, 14) contradicts him; see also idem, 53–5.

8. The ovarian cycle was identified only in 1840. The passive role of women was put forward in 1677 by the Dutch inventor of the first microscope, who found "little animals" in sperm; see Bernard Asbell, *The Pill: A Biography of the Drug that Changed the World* (New York: Random House, 1995), 13–14.

9. "By the turn of the nineteenth century, the fertility of American women had begun a long, secular trend downward" (Reed, *Birth Control Movement*, x).

10. Ibid., 4, 202.

11. Susan E. Klepp, *Revolutionary Conceptions: Women, Fertility, and Family Limitation in America, 1760–1820* (Chapel Hill: University of North Carolina Press, 2009), 28–9, 168–9.

12. Tone, *Devices and Desires*, 15.

13. Elizabeth Frost and Kathryn Cullen-DuPont, *Women's Suffrage in America: An Eyewitness History* (New York: Facts on File, 1992), 2, 18.

14. Reed, *Birth Control Movement*, 5.

15. Elizabeth Siegel Watkins, *On the Pill: A Social History of Oral Contraceptives, 1950–1970* (Baltimore: Johns Hopkins University Press, 1988), 9.

16. Klepp, *Revolutionary Conceptions*, especially chap. 4.

17. John M. Riddle, *Eve's Herbs: A History of Contraception and Abortion in the West* (Cambridge, MA: Harvard University Press, 1997).

18. Lara V. Marks, *Sexual Chemistry: A History of the Contraceptive Pill* (New Haven, CT: Yale University Press, 2001), 41.

19. Klepp, *Revolutionary Conceptions*, 193.

20. The first State law banning abortion came in 1821; two decades later, such laws had been passed in ten States (Tone, *Devices and Desires*, 16).

21. Klepp, *Revolutionary Conceptions*, chap. 4.

22. Ibid., 280.

23. Reed, *Birth Control Movement*, x.

24. The decline in fertility began long before the urbanization of the American population (ibid., 4).

25. Ibid.

26. Watkins, *On the Pill*, 10.

27. A good description of the times is Tone, *Devices and Desires*, chaps. 1–4.

28. Reed, *Birth Control Movement*, 34–9.

29. Since members of Congress know already who will go and who will stay after an election, they tend to become very active in the period between the election and the swearing-in of the new Congress. Such was the case in early 1873 (Tone, *Devices and Desires*, 3–5). One member of that Congress protested the inclusion of a clause prohibiting birth control, but his amendment was voted down in the rush.

30. Ibid., 5.

31. For a description of a number of the more popular birth control books, which sold into many editions before the Comstock repression, see Reed, *Birth Control Movement*, 6–13. In 1839, a Massachusetts physician and graduate of Dartmouth wrote one of the most popular books, *Fruits of Philosophy*, which was still in print in the late 1870s.

32. In 1880, American fertility rivaled that of France as the lowest in modern countries (Reed, *Birth Control Movement*, 17).

33. In 1936, the Supreme Court modified the Comstock laws by removing contraceptive information from the obscenity list (Watkins, *On the Pill*, 14).

34. Over the course of the twentieth century, the complications and hypocrisy introduced into American life by the political pressures of this period were gradually worked out (see Reed, *Birth Control Movement*, 55). As late as 1960, thirty States had laws prohibiting or restricting the sale of contraceptives (Watkins, *On the Pill*, 12). Outside the United States, similar laws were nonexistent until Canada banned the advertisement or sale of contraceptives in 1892 (Tone, *Devices and Desires*, 23).

35. Tone, *Devices and Desires*, 26.

36. Ibid., 106–7. This gave Sanger a legal way to open the country's first birth control clinic in 1923.

37. Watkins, *On the Pill*, 16.

38. Reed, *Birth Control Movement*, 202–5.

39. Elaine Tyler May, *America and the Pill: A History of Promise, Peril, and Liberation* (New York: Basic Books, 2010), 16.

40. Ellen Chesler, *Woman of Valor: Margaret Sanger and the Birth Control Movement in America* (New York: Anchor Books, 1993), 300–1.

41. Reed, *Birth Control Movement*, 239. Birth control was a $250 million business in 1937, including condoms and a catch-all, "feminine hygiene." The Consumers Union provided a quality report on contraceptives to members who wrote they were married and that their physicians advised them to use contraceptives.

42. Ibid., 245. When the Comstock laws were first introduced, the medical profession was trying to get itself legitimized, and was reluctant as a whole to do anything that might seem to endanger this mission (Tone, *Devices and Desires*, 18, 81, 138).

43. Tone, *Devices and Desires*, chaps. 7–8.

44. Chesler, *Woman of Valor*, 391.

45. Reed, *Birth Control Movement*, 231.

46. Much of the following biography is taken from Dwight J. Ingle, "Gregory Goodwin Pincus, 1903–1967: A Biographical Memoir," in *Biographical Memoirs*, vol. 64 (Washington

DC: National Academy of Sciences, 1971), 229–270; available online at http://books.nap.edu/html/biomems/gpincus.pdf.

47. Reed, *Birth Control Movement*, 317.

48. In high school he was the editor of a literary journal and then, at Cornell, founded and edited *The Cornell Literary Review* (ibid.).

49. He also found time to marry a social worker; the couple had two children.

50. At one point, Harvard cited his work as "one of the greatest scientific achievements in history," but soon soured on this line of research (May, *America and the Pill*, 23).

51. His work was compared to the scientific dystopian novel Brave New World, by Aldous Huxley, which came out in 1932 (Tone, *Devices and Desires*, 209).

52. These were titular professorships, which allowed him to supervise graduate students at the WFEB.

53. At first, money was tight. Hoagland cut the lawn to save money, while Pincus was the janitor (Tone, *Devices and Desires*, 209). In 1945, when M.C. Chang came to the WFEB, he was asked to serve as the night watchman; see M.C. Chang, "Recollections of 40 Years at the Worcester Foundation for Experimental Biology," *Physiologist* 28 (5): 400.

54. Gregory G. Pincus, *The Control of Fertility* (New York: Academic Press, 1965). The book is dedicated to "sister of mercy" Katharine Dexter McCormick.

55. Ingle, "Gregory Goodwin Pincus," 239.

56. Ibid.

57. Leon Speroff, *A Good Man: Gregory Goodwin Pincus* (Portland OR: Arnica Publishing, 2009).

58. The town had been created by a land speculation company whose members included Erastus Corning, the creator of the New York Central Railroad. It was felt by the directors that using his prestigious name for the town would help attract purchasers of the lots there. A similar approach by others was made much later in the mountain West, when Cody, Wyoming, named after Buffalo Bill Cody, was laid out at the eastern entrance to Yellowstone Park.

59. May, *America and the Pill*, 18. Aldous Huxley, in *Brave New World Revisited* (1958), referred to a birth control "Pill," and has been credited with the term's injection into common parlance (Van Dulken, *Inventing the 20th Century*, 138).

60. Asbell, *Pill*, 10–11.

61. Much of the following is derived from Armond Fields, *Katharine Dexter McCormick: Pioneer for Women's Rights* (Westport, CT: Praeger, 2003).

62. Tone, *Devices and Desires*, 204.

63. She traveled often to Europe and smuggled supplies of diaphragms back to Sanger in her extensive luggage (Chesler, *Woman of Valor*, 431).

64. Marks, *Sexual Chemistry*, 56.

65. Fields, *Katharine Dexter McCormick*, chaps. 28–31. She said one winter that she was "freezing in Boston for the pill" (Marks, *Sexual Chemistry*, 57).

66. Tone, *Devices and Desires*, 214–15. Pincus said about her later, "[l]ittle old woman, she was not. She was a grenadier (215–16).

67. Chesler, *Woman of Valor*, 434.

68. Watkins, *On the Pill*, 6.

69. Tone, *Devices and Desires*, 229–30.

70. Much of this section is derived from two biographies; Lader and Meltzer, *Margaret Sanger*, and Chesler, *Woman of Valor*.

71. She coined the term "birth control" about this time (May, *America and the Pill*, 17).

72. Chesler, *Woman of Valor*, 145–52.

73. May, *America and the Pill*, 17.

74. Sanger lost interest in the variety of devices and jellies that were available and focused her efforts on finding a medical means of birth control that would be under medical supervision (Watkins, *On the Pill*, 14; May, *America and the Pill*, 19).

75. May, *America and the Pill*, 22.

76. The most useful layperson's account of the science behind the Pill is probably Marks, *Sexual Chemistry*. See also Reed, *Birth Control Movement*, 112, 313–16.

77. Asbell, *Pill*, 14–17.

78. A good sketch of Marker's career can be found in ibid., 82–107.

79. In the 1980s, before his death, he was given an honorary doctorate by the university; see American Chemical Society, "National Historic Chemical Landmarks: The Life of Russell Marker" (Washington, DC, 2004); available online at http://acswebcontent.acs.org/landmarks/marker/life.html.

80. In 1980, we visited my wife's brother in what is now Silicon Valley. He worked for Syntex at the time and took me through parts of its "campus," which seemed the most alien factory facility that someone from an automobile manufacturing town could conceive.

81. Watkins, *On the Pill*, 24.

82. They published some preliminary results as early as 1953 (ibid., 28).

83. Tone, *Devices and Desires*, 217.

84. This is now known as the "Rock rebound effect" (Watkins, *On the Pill*, 28).

85. Djerassi claimed that Pincus opted for the Searle steroid over the one he had synthesized for Syntex because Searle was paying him a retainer. Pincus denied this, and other evidence suggests that it was John Rock who recommended it during the trials (Marks, *Sexual Chemistry*, 72–4).

86. The story of the testing is summed up in Tone, *Devices and Desires*, 216–25.

87. Katharine McCormick kept pressing for large-scale testing (May, *America and the Pill*, 27).

88. Watkins, *On the Pill*, 36.

89. May, *America and the Pill*, 31.

90. Quoted in Tone, *Devices and Desires*, 226.

91. Competing versions of the Pill were approved in 1960, 1962, and 1964 (Watkins, *On the Pill*, 38).

92. Reed, *Birth Control Movement*, 321; Watkins, *On the Pill*, 8. There had been potions prescribed to prevent conception since the dawn of history, but no one knew how they actually worked, if at all.

93. In 1970, Searle's then medical director, Irwin Winter, recalled: "The attitudes prevailing in 1959…were vastly different from those which exist today….No major phar-

maceutical manufacturer had ever dared to put its name on a 'contraceptive'....Open association with 'contraception'... was an activity which our government shunned like the plague" (quoted in Tone, *Devices and Desires*, 227).

94. Watkins, *On the Pill*, 53–5.

95. In 1955, a survey indicated that 70 percent of married women had used some form of contraception; the proportion rose to 80 percent in 1960, the year the Pill was approved (Marks, *Sexual Chemistry*, xv). By 1970, 40 percent were using the Pill, and by 1990, 80 percent of women born since 1945 had used the Pill. The Pill thus appears to mark a shift in method rather than a "revolution. See Watkins, *On the Pill*, 61–2, 133; see also May, *America and the Pill*, 167–8.

96. Watkins, *On the Pill*, 71.

97. Ibid., 5, chap. 5.

98. The connection between the Pill and physicians was reinforced by State laws that its sale required a prescription (ibid., 12).

99. Tone, *Devices and Desires*, 248–9.

100. Watkins, *On the Pill*, 1–4.

101. May, *America and the Pill*, 38–44. Hugh Moore wrote a pamphlet in 1956 called *The Population Bomb*, which predated Paul Erlich's 1968 book of the same title (Chesler, *Woman of Valor*, 438).

102. "Combined Oral Contraceptive Pill," *Wikipedia: The Free Encyclopedia*; available online at http://en.wikipedia.org/wiki/Combined_oral_contraceptive_pill, accessed January 7, 2014.

103. Reed, *Birth Control Movement*, xiii, xvii.

104. Mazlish, *Railroad and the Space Program*, 215–6; the internal quote is from Lowell Tozer, "A Century of Progress, 1833–1933: Technology's Triumph over Man," *American Quarterly* 4 (1, 1952): 78–81.

Chapter 9

1. Adapted from a quotation in Hawke, *Nuts and Bolts of the Past*, 112.
2. Swade and Babbage, *Difference Engine*, 132.
3. Quoted in Struik, *Yankee Science in the Making*, 175.
4. It was built at 44 Taylor St.; the building no longer exists. See Richard P. Scharchburg. *Carriages without Horses: J. Frank Duryea and the Birth of the American Automobile Industry* (Warrendale, PA: Society of Automotive Engineers, 1993), 54.
5. Charles was born in 1862 and Frank in 1869 (ibid., 8).
6. These were high-wheelers, or what in Britain were called penny-farthings, with a large (36-inch to 46-inch) wheel in front and a smaller one in the rear.
7. Scharchburg, *Carriages without Horses*, 8, 51.
8. This venture was in the tradition of Moses Brown's financing Samuel Slater's machinery and mill, those who backed Fulton's steamboat, and various government backers of railroad enterprises (Hawke, *Nuts and Bolts of the Past*, 40).
9. By this time, buggy construction was such that an engine conceivably could be mounted on one, provided the engine was light enough (Finch, *Highways to Heaven*, 26, 37).

10. The first trial was held on September 21 (Scharchburg, *Carriages without Horses*, 77).

11. There were a lot of automobile developers before this, but they tended to concentrate on electric and steam vehicles (ibid., 2).

12. Ibid., 94.

13. Finch, *Highways to Heaven*, 39.

14. The Duryeas participated in two other US races in 1896, one in New York and the other in Providence, Rhode Island, the first in the country to be done on a track (Scharchburg, *Carriages without Horses*, 75, 108, 121, 126).

15. Their argument was probably over focusing on selling licenses for production in Europe and Canada versus concentrating on design, quality, and production. The Duryea was assembled by students from a local industrial school, and quality became an issue (ibid., 143, 151).

16. The company was sold in 1898 to the National Carriage Co. of New York City, which did nothing with it (ibid., 150).

17. Scharchburg (ibid., 184) says, however, that the company folded in 1915.

18. Diamond, *Guns, Germs, and Steel*, 243.

19. Scharchburg, *Carriages without Horses*, 7–8).

20. Lenoir had also used coal gas to power an internal combustion engine in 1863 (Finch, *Highways to Heaven*, 27).

21. See, for instance, David Traxel, *1898: The Birth of the American Century* (New York: Vintage Books, 1999), 57.

22. Karl Benz drove his single-cylinder Motorwagen in the streets of Mannheim in 1886 (Finch, *Highways to Heaven*, 29). In 1896, he developed the first truck (Diamond. *Guns, Germs, and Steel*, 243).

23. Daimler adapted a carriage with an engine and a tiller steering apparatus and drove it in 1886 (Finch, *Highways to Heaven*, 29).

24. Ibid., 37. The article appeared in 1889 and inspired, at least, the Duryea brothers to develop their car.

25. Davies, *American Road*, 233–5.

26. Finch, *Highways to Heaven*, 56. They used Daimler engines.

27. Wish, *Society and Thought in America*, 275.

28. Oliver, History of American Technology, 487.

29. Roper died at age seventy-three in 1896; he had a heart attack while driving his motorcycle at 40 mph near Harvard University (Lienhard, *Inventing Modern*, 128).

30. Mazlish, *Railroad and the Space Program*, 166

31. Oliver, *History of American Technology*, 479.

32. Similar choices for gasoline engines were made in motorizing the bicycle and the airplane about the end of the 1800s as well (Lienhard, *Inventing Modern*, 117).

33. Traxel, *1898*, 57.

34. Ron Chernow, *Titan: The Life of John D. Rockefeller, Sr.* (New York: Random House, 1998), 101.

35. This is not too far off the choice of computer technology that was available in the 1980s. During the course of that decade, I moved from a Commodore 64 to an Atari to an IBM, skipping the technology of Apple, the first popularizer of the personal computer.

The principles were the same in each computer, but the operating systems were quite different, and eventually society, led by business desires for standardization, settled on one basic system. Almost thirty years later, the majority of computers still use the Microsoft operating system.

36. Hawke, *Nuts and Bolts of the Past*, 32. Historians suggest that a mill worker's family could save as much as a sixth of its income and soon afford to move West and buy land and the necessities for a frontier farm (128–9).

37. Chandler Jr., *Visible Hand*, 54–5.

38. The Lowell shops were involved in the early railroad industry and developed a proficiency in metalworking as well as woodworking (Struik, *Yankee Science in the Making*, 319). These shops gradually lost their primacy in machinery as other centers in New England and elsewhere began to develop (Dalzell, *Enterprising Elite*, 124).

39. See, for instance, Christopher Clark, *The Roots of Rural Capitalism: Western Massachusetts, 1780–1860* (Ithaca, NY: Cornell University Press 1990); see also Rothenberg, "Invention of American Capitalism," 93–4.

40. These were the "Wyoming War" or the "Pennamite Wars"; see, for instance, Paul B. Moyer, *Wild Yankees: The Struggle for Independence along Pennsylvania's Revolutionary Frontier* (Ithaca, NY: Cornell University Press, 2007).

41. "And through them all was a strain of respect for expediency: whether their affairs were of the pulpit or of the purse, the conviction persisted that attainment of a glorious end could justify the employment of an opportune means. Such were the enduring qualities of the Yankees of Connecticut" (Lee, *Yankees of Connecticut*, 6).

42. David S. Landes, *Revolution in Time: Clocks and the Making of the Modern World* (New York: Barnes & Noble, 1993), 313.

43. He was using a waterpowered cutting machine by 1803 (ibid., 310). In general, wood-cutting machines tended to be wasteful of material, but that was cheap; what was expensive was labor (idem, *Wealth and Poverty of Nations*, 302).

44. Hawke, *Nuts and Bolts of the Past*, 70–3. In 1838, Chauncey Jerome began to manufacture brass clocks that were big sellers for less than 50 cents — probably something like $325 today (Landes, *Revolution in Time*, 311; see also Temin, "Industrialization of New England," 117).

45. Landes, *Wealth and Poverty of Nations*, 191.

46. Even a generation before the Revolution, Connecticut had more metal shops (foundries, smithies, and so on) than the rest of the colonies combined (Newell, "Birth of New England," 60).

47. Dalzell, *Enterprising Elite*, 166–71.

48. Depending on the size and the risk, outside capital, usually from Boston, might be brought in to help with financing; see Brown, *Massachusetts*, 140.

49. See, for example, Chandler Jr., *Visible Hand*, 270; and Lee, *Yankees of Connecticut*, 118–19. In terms of creativity and diversity of employment, Connecticut residents consistently exceeded the ratio of patents to population, sometimes by enormous margins. By the 1860 census, Massachusetts residents were employed in three hundred of the six hundred industrial, commercial, and agricultural occupations used by government statisticians. (Brown, *Massachusetts*, 142).

50. Edward Deming Andrews, *The People Called Shakers* (1953; New York: Dover, 1963), 124–5.

51. Lee, *Yankees of Connecticut*, 118–19.

52. James M. McPherson, *Battle Cry of Freedom: The Civil War Era* (New York: Oxford University Press, 1988), 19.

53. Hawke, *Nuts and Bolts of the Past*, 32–3, 40.

54. Transportation difficulties both protected local craftspeople and restricted their markets; for a description, see Richard Stott, "Artisans and Capitalist Development," in *Wages of Independence: Capitalism in the Early American Republic*, ed. Paul A. Gilje Lanham, MD: Rowman & Littlefield, 1997), 103.

55. A hundred years later, Frederick Taylor's assignment at the Watertown Arsenal was "to provide stock, tools and materials for keeping [his 250 employees] employed, to preserve order, subordination, regularity of exertion, [and] to retain every branch of business in a relative state of progression with the rest" (Aitkin, *Taylorism at Watertown Arsenal*, 57).

56. Except for Oliver Hazard Perry's naval victory on Lake Erie and the belated repulse of British soldiers at New Orleans, the war had proven to be a series of American mishaps largely traceable to unpreparedness both in leadership and materiel. At the end of the war, the United States had had to re-equip all aspects of its military, and in 1815 the new Ordnance Department, along with Whitney and the heads of the Springfield and Harpers Ferry arsenals agreed on a standardized musket to replace the various (and largely broken) muskets used in the war (Smith, *Harpers Ferry Armory*, 107).

57. The legislation required the new department "to draw up a system of regulations…for the uniformity of manufactures of small arms" (Hawke, *Nuts and Bolts of the Past*, 104). This was much easier said than done. For instance, it took until 1823 before Harpers Ferry agreed to use gauges to measure parts for even relative uniformity, something that had been a practice at Springfield for years (Smith, *Harpers Ferry Armory*, 108–11).

58. The story of the twenty-five-year quest to produce muskets in quantity with interchangeable parts is told in detail in Merritt Roe Smith, "Military Entrepreneurship," in *Yankee Enterprise: The Rise of the American System of Manufactures*, ed. Otto Mayr and Robert C. Post (Washington, DC: Smithsonian Institution Press, 1981, rev. 1995).

59. Beatty, *Colossus*, 91; see also Gordon and Malone, *Texture of Industry*, 87.

60. Hawke, *Nuts and Bolts of the Past*, 97–103. Whitney's ruse was confirmed by researchers only in the 1960s when they discovered discrepancies between some of his parts still extant and the documentation surrounding them.

61. Lee, *Yankees of Connecticut*, 113; and McPherson, *Battle Cry of Freedom*, 16.

62. Harold Livesay, quoted in Hawke, *Nuts and Bolts of the Past*, 102.

63. Struik, *Yankee Science in the Making*, 187.

64. Gordon and Malone, *Texture of Industry*, 87.

65. He insisted, almost from the beginning, that his smaller armory could outproduce the larger establishment at Harpers Ferry (Smith, *Harpers Ferry Armory*, 63).

66. In any case, the continual movement of machine-shop workers meant that new processes would not be secret for long. An example from the latter part of the nineteenth century is the career of Henry Leland:

Henry M. Leland grew up on a New England farm, began as a mechanician in a loom factory, moved on to the armory at Springfield, to the Colt factory in Hartford, to a wrench company in Worcester, then to Providence and Brown & Sharpe, where he headed the sewing-machine department. At every stop along the way, "My vision of the possibilities of manufacturing broadened," he said in old age. "I realized that manufacturing was an art and I resolved to devote my best endeavors and my utmost ability to the Art of Manufacturing." He ended up in Detroit where he designed and built the first engines for the Cadillac automobile. (Hawke, *Nuts and Bolts of the Past*, 33, 133)

Lee was not averse to buying rights from patent holders and synthesizing their machines into ones that were more effective; see, for instance, Smith, *Harpers Ferry Armory*, 117–24.

67. Gordon and Malone, *Texture of Industry*, 354–5. The use of the trip-hammer to make iron shapes in Massachusetts dates back to 1808 (Smith, *Harpers Ferry Armory*, 114).

68. Chandler Jr., *Visible Hand*, 73.

69. It was not until the 1850s and 1860s that calipers and micrometers were introduced into manufacturing in any large way. Of course, by then, machinists were able to produce parts accurate to 1/10000 of an inch, so they were necessary. Their widespread use reflected Yankee educational standards (Gordon and Malone, *Texture of Industry*, 377).

70. A technical discussion of the meaning and implications of interchangeability and the use of gauges can be found in Paul Uselding, "Measuring Techniques and Manufacturing Practice," in *Yankee Enterprise: The Rise of the American System of Manufactures*, ed. Otto Mayr and Robert C. Post (Washington DC: Smithsonian Institution Press, 1981, rev. 1995).

71. Springfield quickly adapted the water-driven trip-hammer to make barrels more efficiently, but Harpers Ferry resisted, and continued doing them by hand. The failure rate between 1823 and 1829, before Harpers Ferry finally adopted the trip-hammer, was 26 percent; with the machine it was 8–12 percent (Smith, *Harpers Ferry Armory*, 95, 114–5).

72. A good rendition of Blanchard's career is found in ibid., 124–38; see also Struik, *Yankee Science in the Making*, 190.

73. The process used to guide the lathe-cutting tool is described in Hawke, *Nuts and Bolts of the Past*, 131–2; see also Hindle, *America's Wooden Age*, 151–3.

74. Hawke, *Nuts and Bolts of the Past*, 105–6; Temin, "Industrialization of New England," 117. In England, the engineer Isambard Kingdom Brunel had sequenced twenty-one machines prior to this to make pulleys for warships. The importance of sequencing single-purpose machines is crucial as they led eventually to the twentieth-century assembly line.

75. The importance of wood as an industrial material in the United States can hardly be overstated. American Iron making, by contrast, lagged behind British innovation because the only real demand for the product in the United States was as bars that local blacksmiths made into domestic products (Johnson, *History of the American People*, 371). Typical was the career of Ephilalet Remington in Ilion Gorge, New York, whose father

was a blacksmith with a sideline of forging rifle barrels for local farmers, something that over the next generation he grew to be E. Remington and Sons, manufacturers of a famous line of rifles.

76. Eli Terry was using waterpowered cutting tools to shape wooden parts as early as 1803. In 1807, he produced three thousand wooden clocks (Landes, *Revolution in Time*, 310). By 1820, other producers were in the business, making as many as fifteen thousand clocks a year (Chandler Jr., *Visible Hand*, 56). It took until 1838 for Chauncey Jerome, a pupil of Terry's, to produce brass parts in quantity that replaced wooden clock mechanisms (Landes, *Revolution in Time*, 311). The British and other Europeans did not respond to American mass production techniques, preferring to concentrate on the higher-margin upscale market (313).

77. They went as far west as Buffalo and as far south as Richmond, Virginia, in the late 1820s (Chandler Jr., *Visible Hand*, 56; see also Hawke, *Nuts and Bolts of the Past*, 69–73).

78. Hawke, *Nuts and Bolts of the Past*, 69.

79. Brown, *Massachusetts*, 143.

80. Smith, *Harpers Ferry Armory*, 156.

81. Ibid., 222–41.

82. North had acquired his first arms contract in 1813, in the midst of the War of 1812, to provide pistols for the war effort (Lee, *Yankees of Connecticut*, 113). He was also the inventor of the milling machine in 1816 (Hawke, *Nuts and Bolts of the Past*, 131). North became involved as a contractor as there was a demand by State militias for Hall's arms, but federal armories were prohibited from selling arms to anyone other than the federal government (Smith, *Harpers Ferry Armory*, 210–12).

83. Gordon and Malone, *Texture of Industry*, 297, 387. At the 1851 Great Exhibition in the Crystal Palace, London, the talk was about Sir Joseph Whitworth's machines, which could cut screw threads to 1/10000 of an inch. It was possible for artisans at the extreme edge of technology to match this feat in the mid-1820s, but not in any quantity (Linklater, *Measuring America*, 207).

84. Hall spent what was then the fabulous sum of $2 million on a set of machines, tools, and gauges that could be used by relatively unskilled labour to produce muskets (Hawke, *Nuts and Bolts of the Past*, 154).

85. McPherson, *Battle Cry of Freedom*, 16. The British imported American machinery for their Enfield Armoury in time to use its products in the Crimean War.

86. For Colt's story, see, for instance, Hawke, *Nuts and Bolts of the Past*, 145–7.

87. DeVoto, *Year of Decision*, 219–21.

88. Temin, "Industrialization of New England," 119.

89. Colt found, when he established a plant in Britain, that none of the equipment manufacturers could produce machinery to the precision he needed, so he imported it from America, to the chagrin of the British government (DeVoto, *Year of Decision*, 219–20).

90. In 1828, one of the Lowell mills became the first to install leather belts in place of wood shafts to run machines in a plant. Belts could be repaired or replaced faster than wooden gears and shafts, and could be rearranged more easily. The British resisted this innovation, as they had developed better iron gears for their shafts (Gordon and Malone, *Texture of Industry*, 312–13).

91. Quoted in Hawke, *Nuts and Bolts of the Past*, 155.

92. Hawke, *Nuts and Bolts of the Past*, 131–2.

93. When the first buildings were constructed using this method, someone said they looked less sturdy than a balloon, and the epithet stuck. The method has been superseded by others since the Second World War. See "Framing (Construction): Platform Framing," *Wikipedia: The Free Encyclopedia*; available online at http://en.wikipedia.org/wiki/Platform_framing#Platform_framing, accessed January 7, 2014. See also McPherson, *Battle Cry of Freedom*, 17. Tolerances were forgiving in houses; one simply used a generous amount of putty to fill in the cracks between windows and studs (Landes, *Wealth and Poverty of Nations*, 302).

94. Hawke, *Nuts and Bolts of the Past*, 204; McPherson, *Battle Cry of Freedom*, 17.

95. The Ames brothers had been employees of the Lowell shops before branching out on their own (Struik, *Yankee Science in the Making*, 319).

96. Chandler Jr., *Visible Hand*, 77; Temin, "Industrialization of New England," 119. Ames Brothers was one of the first machine-tool companies to sell a standard line of machine tools to the general public. Besides being major suppliers to the textile and arms manufacturers, they made mining equipment, and various forms of lathes and other cutting tools for businesses all across the country (Smith, *Harpers Ferry Armory*, 288). Pratt and Whitney is a part of today's manufacturing conglomerate, United Technologies. See also Hawke, *Nuts and Bolts of the Past*, 178–81.

97. The cost of the government armory program has been estimated at $2 million, a very large sum in the first half of the nineteenth century. Its extension into the private sector was both partial and gradual in the second half, increasing as the average cost of interchangeability declined. See Hawke, *Nuts and Bolts of the Past*, 154–6; and Landes, *Wealth and Poverty of Nations*, 191, 304.

98. A journeyman machinist in Boston, Elias Howe, had built the first such machine (McPherson, *Battle Cry of Freedom*, 19).

99. A British fact-finding commission, sent primarily to look at the production methods in Colt's Hartford factory, also noted the wide application of similar tools in dissimilar production enterprises (Beatty, *Colossus*, 91).

100. Gordon and Malone, *Texture of Industry*, 297, 299.

101. Landes, *Wealth and Poverty of Nations*, 303.

102. A British visitor in 1841 noted about the machine process: "The skill of the eye and the hand, acquired by practice alone, is no longer indispensable; and if every operative were at once discharged from the Springfield armoury, their places could be supplied with competent hands within a week" (quoted in ibid.).

103. Johnson, *History of the American People*, 316; Landes, *Wealth and Poverty of Nations*, 320–1. Even so, mass production also increased demand for skilled labour as well; see Stott, "Artisans and Capitalist Development," 106.

104. DeVoto, *Year of Decision*, 218–19.

Chapter 10

1. Sampson, *Company Man*, 17.

2. Kinderhook, New York, just south of Albany, was the home town of "Old Kin-

derhook," who was running for re-election; see Schlesinger Jr., *Age of Jackson*, 298; and George P. Olin, *The Story of Telecommunications* (Macon, GA.: Mercer University Press, 1992), 211. Allan Metcalf, in *OK: The Improbable History of America's Greatest Word* (New York: Oxford University Press, 2011), claims that "OK" was invented by a Boston editor in 1839 as a joke involving a trivial spat between Boston and Providence, Rhode Island, playing off their local pronunciation of "all correct." Old Kinderhook's campaign crew ignored the joke when OK became part of his advertising. Then, as the telegraph system grew in the 1840s, OK became common shorthand for telegraphers — and on it went.

3. Beatty, *Colossus*, 1.

4. Paul A. Gilje, "The Rise of Capitalism in the Early Republic," in *Wages of Independence: Capitalism in the Early American Republic*, ed. Paul A. Gilje (Lanham, MD: Rowman & Littlefield, 1997), 4.

5. Klein, *Empire State*, 665.

6. Albany had at least two other names before 1686. It was called Beverly, "beaver district," at its founding as an inland stockade, then Willemstad as it grew into a town site andgiven a charter of incorporation by the British in 1686; see William Kennedy, *O Albany!* (New York: Penguin, 1983), 155. In the early nineteenth century, it was described as "a third- or fourth rate town…indeed Dutch, in all its moods and tenses; thoroughly and inveterately Dutch. The buildings were Dutch — Dutch in style, in position, attitude, and aspect. The people were Dutch, the horses were Dutch, and even the dogs were Dutch" (79).

7. Klein, *Empire State*, xix, 252, 287.

8. It had doubled in size in the preceding decade from 12,630 to more than 25,000 (ibid., 309-10). Beaumont compared it to Amiens and the non-estuarial Hudson to the Somme River; see Pierson, *Tocqueville in America*, 176.

9. Ibid., 348–9.

10. Frederick Marryat, *A Diary in America: With Remarks on Its Institutions* (Philadelphia: Carey & Hart, 1839), 64–5.

11. Begun in 1823 to anticipate the canal's traffic, it was a pier that paralleled the river, creating a basin that could hold a thousand canal boats (Kennedy, *O Albany!*, 59).

12. Fox and Remini, *Decline of Aristocracy*, 304.

13. Klein, *Empire State*, 480.

14. Kennedy, *O Albany!*, 60.

15. Ibid., 63.

16. Johnson, *History of the American People*, 72.

17. Fernand Braudel, *The Wheels of Commerce, Civilization and Capitalism 15th–18th Century*, vol. 2 (New York: Harper and Row, 1979, 1982), chaps. 3–4.

18. Max Weber, *The Protestant Ethic and the Spirit of Capitalism* (London: Allen and Unwin, 1930; repr. Routledge Classics, 2001), xxxi–xxxii.

19. The attempt to build a communal society at Plymouth failed early on with the near-demise of the colony; until the Mormon experiment two hundred years later, this was not considered a large-scale option.

20. For an example of the value of partnerships in early America, see Naomi Lamoreaux, "The Partnership Form of Organization: Its Popularity in Early Nineteenth Century

Boston," in *Entrepreneurs: The Boston Business Community, 1700–1850*, ed. C.E. Wright and K.P. Viens (Boston: Massachusetts Historical Society, 1997).

21. Rosenberg and Birdzell (*How the West Grew Rich*, 190, 197) trace their use back to Roman law, where any unlicensed association of citizens was considered a conspiracy against the state. However, their use for economic purposes is much more recent.

22. John Micklethwait and Adrian Wooldridge, *The Company: A Short History of a Revolutionary Idea* (New York: Modern Library Chronicles Book, 2003), 4–5.

23. Ibid., 12. Early municipal corporations were established to allow towns to create and autonomously manage market places and other commonly used facilities; see H. Hartog, *Public Property and Private Power: The Corporation of the City of New York in American Law, 1730–1870* (Chapel Hill: University of North Carolina Press, 1983).

24. In 1819 Chief Justice John Marshall defined it as "an artificial being, invisible, intangible and existing only in contemplation of law" (quoted in Johnson, *History of the American People*, 571). It was a legal convenience for many to act as one.

25. The term "company" comes from the medieval Italian, when merchants would agree to share trade risks while they broke bread, a *compagnie*, or, from the Latin *cum panis* (Sampson, *Company Man*, 16). The term "bank" can be traced to the Italian *banchi*, referring to the benches where financiers sat while doing their business (Micklethwait and Wooldridge, *Company*, 8).

26. Lamoreaux, "Partnership Form of Organization," 273.

27. Apparently, the oldest business corporation still in existence is a Swedish company called Stora Enso, whose ancestor was a mining company that began in 1288 and received a royal charter in 1347 (Micklethwait and Wooldridge, *Company*, 12). Stora operated a paper plant in eastern Nova Scotia for a number of years. Some of the concerns about the depersonalization of economic life and the divorce of its morals from those in private life are expressed in Schlesinger Jr., *Age of Jackson*, 334–5. This is also the theme that runs through Charles Sellers, *The Market Revolution: Jacksonian America, 1815–1946* (New York: Oxford University Press, 1995).

28. Micklethwait and Wooldridge, *Company*, 19–20.

29. See Gordon, *Great Game*, 22–3; and Newell, "Birth of New England," 17.

30. Gordon, *Great Game*, 22–3.

31. The British had some prior experience. The joint-stock Muscovy Company had been chartered in 1555 to do business with Russia and to explore for a northeast passage to China (Micklethwait and Wooldridge, *Company*, 18).

32. Ibid., 21–8.

33. The seven hundred "adventurers" or investors in the Virginia company did not get their money out of their attempts, so encouraging other companies to take up part of their territory seemed like a good idea (Beatty, *Colossus*, 1).

34. Many of the features of English trading company charters were perpetuated in later State constitutions and in the US constitution; see John P. Davis, *Corporations* (1905; New York: Capricorn Books, 1961), 201.

35. Sir Walter Raleigh's attempts to establish colonies in Virginia led him and others to believe that personal resources could not be adequate to meet colonial capital and operating needs, and that a corporation was necessary; see Billington, *Westward Expansion*, 49.

36. The English government assumed that the directors of these New England companies would remain in England, but this proved not to be the case, as most of their shareholders emigrated to the new colonies. Those who put up the money, the "adventurers," did remain there, but most of the directors did not. The adventurers were the historical antecedents of today"s "venture capitalists"; see Gordon, *Business of America*, 4. As the Puritan colonists in these "corporate" colonies became the controlling shareholders, the corporate "annual meeting" was turned into a kind of legislature. In effect, these colonies became self-governing corporations, a conception that caused trouble later on.

37. The so-called Bubble Act was passed in 1720 (see Landes, *Wealth and Poverty of Nations*, 257; see also Micklethwait and Wooldridge, *Company*, 31–3). A Scotsman, John Law, managed to reduce the economy of France to depression with his Mississippi Company, which created a similar bubble in 1718–20 (Micklethwait and Wooldridge, *Company*, 28–31). For a succinct discussion of this law's effects in Britain, see Crump, *Brief History of the Age of Steam*, 180.

38. Sarah Gordon, *Passage to Union: How the Railroads Transformed American Life, 1829–1929* (Chicago: Ivan R. Dee, 1996), 16–22.

39. As early as the 1650s, leading Massachusetts merchants were incorporated to speculate in lands in Rhode Island and eastern Connecticut (Billington, *Westward Expansion*, 86).

40. In English law, a corporation could be formed without a charter, but then it was considered a form of partnership, with unlimited liability accruing to the shareholders as if they were partners (Rosenberg and Birdzell, *How the West Grew Rich*, 195–7).

41. A broader rationalization and explanation for the freeing of enterprises from hierarchical and elite control is found in ibid., 22–4, 113–15.

42. Dalzell (*Enterprising Elite*, 28) discusses this logic in terms of the formation of the Lowell mill at Waltham and its successors in Lowell.

43. These are sometimes described using the archaic term of "franchised" corporations, and are best seen today as utilities, where there is often a need for a legislated monopoly (Rosenberg and Birdzell, *How the West Grew Rich*, 194–5).

44. It should be kept in mind that the overwhelming number of enterprises, then as now, were small firms with a single location and proprietorship. What changed with incorporation reform was the ability to amass capital for large-scale activities. Where the locations were diverse, as in trade, partnerships were more common; see Chandler Jr., *Visible Hand*, 14, 16.

45. Peter F. Drucker, *The Concept of the Corporation* (New York: John Day, 1946), 209–12. Corporations still derive their legal existence from government's belief that they serve a broader public purpose, something forgotten at times in our society.

46. For the general story, see Gordon, *Great Game*, 49–50; and for an evaluation of the importance of the act, see W.C. Kessler, "A Statistical Study of the New York General Incorporation Act of 1811," *Journal of Political Economy* 48 (6, 1940): 877–82. The idea of general incorporation had preceded the act, as it had been used in the 1700s to permit easy formation of a church facility; see Ted Nace, *Gangs of America: The Rise of Corporate Power and the Disabling of Democracy* (San Francisco: Berrett-Koehler, 2005), 72. Massachusetts passed a similar act in 1818; see William J. Fowler Jr., "Marine Insurance

in Boston: The Early Years of the Boston Marine Insurance Company, 1799–1807," in *Entrepreneurs: The Boston Business Community, 1700–1850*, ed. C.E. Wright and K.P. Viens (Boston: Massachusetts Historical Society, 1997), 165.

47. See Richard Hofstadter, "William Leggett, Spokesman of Jacksonian Democracy," *Political Science Quarterly* 58 (4, 1943): 587; and Davis, *Corporations*, viii. There were also some strange benefits: startup loans to eligible entrepreneurs were authorized, as was exemption from militia or jury duty for owners of textile companies; see Berstein, *Wedding of the Waters*, 152.

48. Beatty, *Colossus*, 45–6. Even so, Massachusetts had been early off the mark in granting around a hundred charters for incorporation in the decade drom 1780 to 1790, though only a few were for for-profit companies. The first insurance company in Massachusetts to gain a charter did so in 1795. Insurance companies were required to place most of their capital in bank stock; see Fowler Jr., "Marine Insurance in Boston," 165–6.

49. Schlesinger Jr., *Age of Jackson*, 188–9.

50. Hofstadter, "William Leggett," 589. Leggett's philosophy would be seen as conservative today, but in an era of monopolies, imprisonment for debt and property qualifications in order to vote, this was heady stuff.

51. Ibid., 589. The "Locofocos" — the left wing of the Democratic Party in New York, was especially vociferous about the Democratic-dominated legislature's tendency to charter banks that benefited its members; see Fox and Remini, *Decline of Aristocracy*, 382–3. One democratic fallout of the reform movement was the low cost of incorporation: one Connecticut entrepreneur was said to have paid out 50 cents to acquire his incorporation papers (Beatty, *Colossus*, 85).

52. For Pennsylvania, see Susan Pace Hamill, "From Special Privilege to General Utility: A Continuation of Willard Hurst's Study of Corporation," *American University Law Review* 49 (1999): 101. For Connecticut, see Schlesinger Jr., *Age of Jackson*, 337; Micklethwait and Wooldridge, *Company*, 46; there were limits of $200,000 of capital (Davis, *Corporations*, ix). For New York, see Hofstadter, "William Leggett," 594; Kessler, "Statistical Study," 878; Gordon, *Passage to Union*, 21. New York dropped the categorization of purposes eligible for incorporation in favor of "any lawful purpose" in 1866, followed gradually by other States (Nace, *Gangs of America*, 77).

53. By 1875, over 90 percent of the States had general incorporation laws (Hamill, "From Special Privilege to General Utility," 87).

54. Ibid., 101–7; Micklethwait and Wooldridge, *Company*, 46. These continuing restrictions led more ambitious groups to pursue special charters (Davis, *Corporations*, xi).

55. Gordon, *Passage to Union*, 34. Even so, the use of special charters did not die out — between 1875 and 1996, the States issued more than twenty thousand special charters (Hamill, "From Special Privilege to General Utility," 86n17).

56. Nace, *Gangs of America*, 73.

57. In the 1400s, English courts had already defined the concept. "If something is owed to the group, it is not owed to the individuals, nor do the individuals owe what the group owes" (Beatty, *Colossus*, 6).

58. The practice dates at least back to the Middle Ages (Micklethwait and Wooldridge, *Company*, 8).

59. It is thought that one in five households in early nineteenth-century America faced insolvency at least once; see Jane Kamensky, *The Exchange Artist: A Tale of High-Flying Speculation and America's First Banking Collapse* (New York: Viking, 2008), 11.

60. Schlesinger Jr., *Age of Jackson*, 48; Fitzpatrick, *History's Memory*, 48.

61. Fox and Remini, *Decline of Aristocracy*, 353, 357.

62. The concentration on imprisoning people for a particular crime has not changed a lot. Today, about 20–30 percent of State prison inmates and as much as 60 percent of federal inmates are there for drug-related crimes; see Common Sense for Drug Policy, "Drug Offenders in the Correctional System"; available online at http://drugwarfacts.org/cms/node/63. As far as can be found, nobody today is in these prisons for not repaying debt.

63. Limited liability did not become entrenched in corporate law until the 1830s (Hamill, "From Special Privilege to General Utility," 92n42). In 1996, there were just under 4.5 million corporations — or 1.6 percent of all legal/physical persons — in the United States, of which slightly more than half had assets of $100,000 (83n1). See also Gilje, "Rise of Capitalism," 4. It could be a political football, however: the law on liability changed five times in Maine between 1823 and 1857, depending on whether the Democrats or Whigs were in power (Nace, *Gangs of America*, 78).

64. This was not an unmixed blessing, as many lenders wanted personal security and not collateral. They were also suspicious of this new use of the law. See Lamoreaux, "Partnership Form of Organization," 275.

65. Nace (*Gangs of America*, 74) cites a study by Royal Dutch/Shell indicating that a typical *Fortune* 500 corporation survives about fifty years before it disappears through bankruptcy or merger.

66. Often, incorporation allowed for interfamily alliances while preserving the separate family fortunes. This was the case with many Beacon Hill and other prominent Boston families; see Peter Dobkin Hall, "What the Merchants Did with Their Money," in *Entrepreneurs: The Boston Business Community, 1700–1850*, ed. C.E. Wright and K.P. Viens (Boston: Massachusetts Historical Society, 1997), 370.

67. Trading was largely centered on the transportation industry until 1890, when the rise of the trusts led to more Wall Street involvement in trading shares of industrial corporations (Rosenberg and Birdzell, *How the West Grew Rich*, 200). Mohawk and Hudson RR shares first traded there in 1830 (Burrows and Wallace, *Gotham*, 567). The Boston Stock Exchange, in fact, was the leading exchange for industrial stocks until 1890, because Wall Street resisted listing them until then (Rosenbloom, "Challenges of Economic Maturity," 188–9).

68. Nace, *Gangs of America*, 79.

69. Burrows and Wallace, *Gotham*, 1045–6. This shifted activity away from creating trusts, which were declared illegal in 1889, toward using interstate incorporation via New Jersey.

70. See "Making a Success of Failure," *Economist*, January 9, 2010, 68.

71. Chief Justice John Marshall is often given credit for helping to establish the outlines of the modern corporation (Johnson, *History of the American People*, 244).

72. Schlesinger Jr., *Age of Jackson*, 324; see also Nace, *Gangs of America*, 84–5.

73. Schlesinger Jr., *Age of Jackson*, 324–6.

74. Alexander Hamilton and Robert Morris attempted to create a "congressional bank" in 1781, during the Revolutionary War, but Congress did not feel it had the authority, so in 1782 they then secured common charters from a number of colonies for a Bank of North America. In 1791, after the Constitution was ratified, it was replaced by the federally chartered Bank of the United States (Hamill, "From Special Privilege to General Utility," 89–90n28,32,36.

75. Ibid., 108n102.

76. Martin, *Railroads Triumphant*, 325.

77. Rosenberg and Birdzell, *How the West Grew Rich*, 201, 212; Gordon, *Great Game*, 170.

78. The trust was not an invention of the 1880s. As early as 1765, Patrick Henry created a trust, the North American Land Company, to protect the speculative assets of Robert Morris; other trusts were devised after the Revolutionary War for similar purposes (Linklater, *Measuring America*, 149–50).

79. Roberts, *History of the English-Speaking Peoples since 1900*, 40.

80. Controversies between the colonies — those in New England, in particular — with Britain over local banking and the use of paper money were a contributing factor to the Revolution. See Margaret E. Newell, "A Revolution in Economic Thought: Currency and Economic Development in Eighteenth-Century Massachusetts," in *Entrepreneurs: The Boston Business Community, 1700–1850*, ed. C.E. Wright and K.P. Viens (Boston: Massachusetts Historical Society, 1997), 1.

81. The difference between specie and scrip is the difference between whether a user believes the money is worth what it says it is to the wider public. Specie was usually metal coins made by a government, and thus regarded as "hard" currency (Kamensky, *Exchange Artist*, 8). Scrip — say, paper currency printed by a frontier bank with few tangible assets — engenders a lot less trust than a federal government "greenback" and would be exchanged for a greenback on;y at a discount. A discount of 100 percent meant the local money was worthless, but most bills from banks with good reputations were discounted only slightly.

82. Each colony had the usual British-designated currency of pounds, shillings, and pence, but they were not exchanged equally with British currency (Berstein, *Wedding of the Waters*, 147). The "dollar," a corruption of a Spanish corruption of a Bohemian word for coins made from the metal from a particular European silver mine, did not become the official US currency until the 1790s.

83. Kamensky (*Exchange Artist*, 14–17) claims this was the first issue of paper money by a government in the Western world.

84. Massachusetts issued its first Revolutionary War bonds a couple of weeks after the events at Concord and Lexington. The Continental Congress issued $2 million in paper currency in May 1775, fourteen months before the Declaration of Independence. By the end of 1779, there was $226 million of them in circulation. Someone or something truly worthless was said to be "not worth a Continental" (ibid., 23). By then, the continental was discounted to about 3 cents relative to specie; it declined to about 150:1 by 1781 (24).

85. Rothenberg, "Invention of American Capitalism," 76.

86. Merchants and merchant bankers financed international trade, while the incorporated banks concentrated on domestic finance (Chandler Jr., *Visible Hand*, 29).

87. For a short explanation of the evolution of merchants' bills of exchange into a rudimentary banking institution, see Rosenberg and Birdzell, *How the West Grew Rich*, 117). Part of the 2007–8 problem with banks arose from overleveraging. Instead of a more normal 10:1 ratio, some of the more reputable banks let their ratios go as high as 30:1. The same thing happened in 1837.

88. After the demise of the First Bank of the United States in 1811, the number of State banks rose two and a half times in two years, and though their leverage was only supposed to be 3:1, no one checked into the actual capital in the banks or whether they were restraining themselves to the legal leverage limit (Johnson, *History of the American People*, 291). As the country slipped into war, the money supply grew by about 20 percent per year. New England banks refused to honor banknotes from the South and West, while their own notes commanded a premium in Philadelphia. By the end of the War of 1812, virtually all banks outside New England had to stop payment on their notes and the federal government was near financial collapse. Not surprisingly, the peak number of banks in the United States came in the 1920s, just before the Great Depression (Gordon, *Great Game*, 225).

89. Said to be located so far back in the woods that only a wildcat could find them (Hamill, "From Special Privilege to General Utility," 100n73).

90. Dunbar and May, *Michigan*, 96. One scam was to have chests of bullion to display as capital in which the coins were only a layer or two deep and the rest just sand or iron bars. Another was to have a load of valid capital moved around from bank to bank ahead of inspectors. Counterfeiting was a cottage industry — "all you needed to start a bank issuing paper money were plates, presses and paper" (Johnson, *History of the American People*, 283).

91. Most banks did not have branches and were small in size. As a result, there were many of them; in New England, there were 172 in 1830 and 505 in 1860 (Temin, "Industrialization of New England," 150). Most banks did not take deposits; instead, people bought share in them (151).

92. For an interesting study of an early New England case of financial shenanigans, including a fraudulent Michigan bank, lots of paper money issues, and the land development business in early nineteenth-century Boston, see Kamensky, *Exchange Artist*. A Rhode Island bank involved in the speculation issued $600,000 in notes based on a capital of only $86.48 (9).

93. Marryat, *Diary in America*, 29.
94. Kamensky, *Exchange Artist*, 17.
95. Ibid., 52–5.
96. Temin, "Industrialization of New England," 149–51.
97. Chandler Jr., *Visible Hand*, 30.
98. Fowler Jr., "Marine Insurance in Boston," 161–2; and Dalzell, *Enterprising Elite*, 93. Needless to say, the Commonwealth came under pressure to allow others to create "State" banks, and after 1792 a number were given charters. By 1837, there were 129.
99. Twenty percent of the shares were owned by the US government, the rest pri-

vately held. Notes from "The Great Game: The Story of Wall Street" (CNBC televsion, 2000).

100. Gilje, "Rise of Capitalism," 3.

101. A variation of this same debate went on in 2007–9 between conservatives, mostly from the South, who argued that banks should be allowed to fail as a result of their own greed and others who felt that this would lead to the collapse of the American and global financial systems.

102. Hamilton's Bank of the United States was federally chartered for twenty years, from 1791 to 1811; the charter was not renewed. A Second Bank was chartered for twenty years in 1816 after the mess of the War of 1812 demonstrated the need for it. Madison had opposed the chartering of the First Bank, and it was during his first term as president that the charter was not renewed. In his second term, he signed the bill creating the Second Bank. The lesson Madison learned was forgotten by the 1830s; see Cathy Matson, "Capitalizing Hope: Economic Thought in the Early National Economy," in *Wages of Independence: Capitalism in the Early American Republic*, ed. Paul A. Gilje (Lanham, MD: Rowman & Littlefield, 1997), 124–5; Schlesinger Jr., *Age of Jackson*, 10.

103. Jackson had said before his electionthat he opposed all banks in principle, but later political necessity did not allow him to go that far; see Hamill, "From Special Privilege to General Utility," 98n68.

104. Fox and Remini, *Decline of Aristocracy*, 382–3. The transfer of federal funds out of the United States Bank into State banks only improved their capitalization and made them more vital to entrepreneurs.

105. Hofstadter, "William Leggett," 586.

106. Ibid., 587. In the years since then, it is clear that overleveraging by New York banks remains a problem, imperiling the savings of, among others, retired workers.

107. The law said any twelve landowners in the State could form a banking association, as long as the capital stock subscribed was not less than $50,000 (Dunbar and May, *Michigan*, 223). None of the rules wase enforced, and most of the banks collapsed as the Panic tightened credit. By 1839, only three older banks and four "general charter" banks — out of the forty-nine created in 1837 — remained (237).

108. Schlesinger Jr., *Age of Jackson*, 286; Klein, *Empire State*, 321.

109. It is possible that the use of slaves kept the capital needs of plantations at a minimum, so that banking was needed mostly to facilitate trading and exports.

110. The estimated per capita amount of paper money in circulation rose from $6.69 in 1830 to $13.87 in 1837. The number of banks doubled in the same period (Dunbar and May, *Michigan*, 223; Gordon, *Great Game*, 65–6). One oddity was that the revenues from the purchase of public lands, for speculation or settlement, during this time allowed Jackson to pay off the national debt in 1835 for the only time in history. Another was that, in Michigan at least, the only money circulating in the 1840s was mint coinage in gold and silver (Dunbar and May, *Michigan*, 237).

111. Martin, *Railroads Triumphant*, 337–8.

112. See Gordon S. Wood, "The Enemy Is Us: Democratic Capitalism in the Early Republic," in *Wages of Independence: Capitalism in the Early American Republic*, ed. Paul A. Gilje (Lanham, MD: Rowman & Littlefield, 1997), 147.

113. Quoted in Beatty, *Colossus*, 6.

114. Ibid., 273. The Peruvian economist Hernando De Soto has stated for decades that one of the features of a poor society is the inability of its citizens easily to form companies and go into the production of goods and services.

Chapter 11

1. Lincoln did go to watch a game with his son during the Civil War; see Gerald Astor et al., *The Baseball Hall of Fame 50th Anniversary Book* (Englewood Cliffs, NJ: Prentice-Hall, 1988), 4. Another myth was that Lincoln had summoned Abner Doubleday to his deathbed and implored him to keep the game of baseball alive; see Zev Chafets, *Cooperstown Confidential* (New York: Bloomsbury, 2009), 28.

2. W.P. Kinsella, *Shoeless Joe* (1982; New York: Mariner Books, 1999), 252–3. The novel was the basis for the movie *Field of Dreams*.

3. Sinclair Lewis, *Babbitt* (1922; repr. Mineola, NY: Dover Publications, 2003), 73.

4. Apparently, the colonies accepted the validity of pre-Revolutionary Royal grants, or patents, but it was another thing for diehard Tories to hang on to them when their physical presence place them in jeopardy, at least during the first decade after the end of the war.

5. Taylor, *William Cooper's Town*, 4, 30–5, 44–5, 52.

6. Johnson, *History of the American People*, 412.

7. A short biography of Cooper may be found in Klein, *Empire State*, 353–4.

8. Schlesinger Jr., *Age of Jackson*, 378.

9. Cooper married into a prominent New York family in Westchester County.

10. Schlesinger Jr., *Age of Jackson*, 375–9.

11. Foreman, *Frommer's Comprehensive Travel Guide*, 272–3.

12. Steve Rashin, *Road Swing: One Man's Journey into the Soul of American Sports* (New York: Doubleday, 1998), 60–1, captures the Cooperstown experience. For less-focused tourists, there are a number of interesting attractions in and around the town.

13. Doubleday was a descendant of Connecticut Yankees who migrated to New York, first to New Lebanon and then to the Otsego Lake area. His father was born near Cooperstown and then moved to near Saratoga, where Abner was born. He moved on to Auburn and became a newspaper publisher and congressman. Abner was sent to live with an uncle and attend school in the Cooperstown area. In the Civil War, he was a fairly prominent Union general, who ordered the first Union fire from Fort Sumter.

14. He was at prep school there a couple of years earlier; see, especially, Peter Morris, *But Didn't We Have Fun? An Informal History of Baseball's Pioneer Era, 1843–1870* (Chicago: Ivan R. Dee, 2008), 226–7.

15. Astor et.al., *Baseball Hall of Fame 50th Anniversary Book*, 1–12.

16. Ibid., 35–7. The claim is still made; see Schlereth, *Victorian America*, 224. There are a number of other claims as to its antiquity, such as a baseball game played in Canada between a team from Beachville, Ontario, and the Zorras, a team from a neighboring township; a newspaper article about a baseball game in lower Manhattan; a 1791 Pittsfield, Massachusetts, ordinance banning baseball from being played near the town square; and various games. and religious rites going back through medieval Spain to classical Egypt.

17. Murphy, *Crazy '08*, 131.

18. Chafets, *Cooperstown Confidential*, 26.

19. From 1813 on, the town trustees, led by Cooper descendants, enacted rules against "rowdies" playing rounders and town ball on its streets and in its fields (Taylor, *William Cooper's Town*, 383–4).

20. "Baseball has a remarkable ability to create myths about itself that endure despite the fact that they wither under the slightest scrutiny (Murphy, *Crazy '08*, 49).

21. Much of the following is adapted from a number of sources, but they are all found in succinct fashion in Morris, *But Didn't We Have Fun?*.

22. Apparently, their favorite field was at Twenty-seventh Street and Madison Avenue (Burrows and Wallace, *Gotham*, 733).

23. A concise, but detailed account of early New York City baseball is in ibid.

24. See, for instance, John Thorn, "Doc Adams," SABR Baseball Biography Project (Phoenix: Society for American Baseball Research, n.d.); available online at http://bioproj.sabr.org/bioproj.cfm?a=v&v=l&bid=%20639&pid=16943.

25. Members came from fourteen clubs at first (Gordon, *Business of America*, 249). A National Association of Professional Baseball Players Association was set up in 1871 (Wish, *Society and Thought in America*, 277).

26. Most of the club's players came from New York City (Burrows and Wallace, *Gotham*, 975).

27. Schlereth, *Victorian America*, 225, ; Burrows and Wallace, *Gotham*, 974.

28. The only mass sport of totally American origin is basketball, created in Springfield, Massachusetts, in 1891 by Dr. James Naismith, a Canadian immigrant working with the Young Men's Christian Association (Wish, *Society and Thought in America*, 279) — well, maybe not totally. The National Basketball Hall of Fame is appropriately located in the town of the sport's birth.

29. This, despite the brother of the then patriarch of the Clark family's violent opposition to the New Deal programs.

30. The model for the naming was based on a 1900 Hall of Fame for Great Americans in New York City, which included a colonnade populated with busts of the inductees (Chafets, *Cooperstown Confidential*, 30).

31. As early as 1917, locals with an eye towards economic development had created a Doubleday Memorial Fund to build a ball field as well as a retirement home for players (ibid., 28).

32. Eldon E. Snyder, "Sociology of Nostalgia: Sport Halls of Fame and Museums in America," *Sociology of Sport Journal* 8 (3, 1991): 228–38.

33. This is just what Mills said about Cooperstown later. Chafets (*Cooperstown Confidential*, 8, 27) refers to the town as an "American Brigadoon."

34. Levi, *American Vertigo*, 27–8.

35. Chafets, *Cooperstown Confidential*, 4–5. Cooperstown attracts about 350,000 tourists each year.

36. Michael Patrick Allen and Nicholas L. Parsons, "The Institutionalization of Fame: Achievement, Recognition and Cultural Consecration in Baseball," *American Sociological Review* 71 (5, 2006): 808.

37. Seymour Martin Lipset, *American Exceptionalism: A Double-Edged Sword* (New York: W.W. Norton, 1996), 19. The others are liberty, egalitarianism, populism, and *laissez-faire*.

38. Lipset would disagree, seeing individualism as the basis for the American propensity to form voluntary organizations (ibid., 276–8).

39. Rashin, *Road Swing*, 187.

40. He was caught between a love for wilderness and a recognition that advancing civilization would overcome it (Stegner and Stegner, *Marking the Sparrow's Fall*, 123).

41. Slotkin, *Regeneration through Violence*, 467–8, 512; idem, *Gunfighter Nation*, 16; Stegner and Stegner, *Marking the Sparrow's Fall*, 123.

42. Stegner and Stegner, *Marking the Sparrow's Fall*, 200.

43. Lewis, *Babbitt*, 108.

44. Stegner and Stegner, *Marking the Sparrow's Fall*, 199–200.

45. Francis Parkman wrote of Hawkeye: "There is something admirably felicitous in the conception of this hybrid offspring of civilization and barbarism, in whom uprightness, kindliness, innate philosophy and truest moral perceptions are joined with the wandering instincts and hatred of restraint which stamp the Indian or Bedouin"; quoted in Slotkin, *Regeneration through Violence*, 512.

46. See Slotkin, *Regeneration through Violence*, 294.

47. David Brooks notes that turnover in plants in the early 1900s often topped 100 percent per year and that the average worker moved to a new job every three years or so. The rate might be lower now, but it is still high. Brooks attributes this not to the nature of the job, but to the aspirations of the worker: perhaps a different job would be more satisfying or present a new opportunity; see David Brooks, *On Paradise Drive: How We Love Now (And Always Have) in the Future Tense* (New York: Simon & Schuster, 2004), 232.

48. See, for instance, a later edition, *Smolensk under Soviet Rule* (Boston: Unwin Hyman, 1989).

49. Jon R. Katzenbach and Douglas K. Smith, "The Wisdom of Teams," *Small Business Reports* 18 (7, 1993): 68–71.

50. Ibid.

51. Even among those Hall of Fame players who played in major markets, with presumably major amounts of cash to buy the best talent, almost 15 percent never played in a single World Series game (Allen and Parsons, "Institutionalization of Fame," 817).

52. Michael Lewis, *Moneyball: The Art of Winning an Unfair Game* (New York: W.W. Norton, 2004), 37–8.

53. Ibid., 287–301.

54. Ibid., 34.

55. James's analysis of baseball statistics led him into rank heresy. As Lewis puts it, "[t]he statistics were not merely inadequate: they lied. And the lies they told led the people who ran major league baseball teams to misjudge their players and mismanage their games" (ibid., 67).

56. Apparently, the Boston Red Sox used a similar approach in building the team that broke "Babe Ruth's curse" and in 2004 brought a World Series Championship to Boston after eighty-six years.

57. It might be that the value of a star has more to do with drawing power — that is, the star's presence increases attendance receipts and television receipts. In 1927, Babe Ruth's quest for the home-run record is said to have added a million dollars to American League receipts (Wish, *Society and Thought in America*, 451). Ken Burns's film, *The Tenth Inning*, notes how the steroid-fueled home-run hitters of the late 1990s were responsible for raising revenue after the disastrous strike in 1994. Likewise, in basketball, see Ira Boudway, "A Problem Like LeBron," Bloomberg *Business Week*, June 21–27, 2010, 7–9.

Chapter 12

1. Wershler-Henry, *Iron Whim*, 243; and Setterberg, *Roads Taken*, 13. Capote apparently loved to repeat this wisecrack when the opportunity presented itself.

2. Freiberger and Swaine, *Fire in the Valley*, xv.

3. During the 1800s in upstate New York, there was a trend to name towns and cities after classical places, such as Troy, Rome, Utica, Syracuse, Palmyra, and Rochester, to name a few. Local citizens wanted to name the town after Remington, when the Post Office pressed for an official name in 1843, but Eliphalet II refused, and Ilium, the alternate name for the classic city of Troy, was chosen. The Post Office misspelled it and it became officially Ilion (Berstein, *Wedding of the Waters*, 262).

4. The nineteenth-century Remington story largely follows an account in George Hardin, ed., *A History of Herkimer County* (Syracuse, NY: D. Mason, 1893), 204–7; available online at http://www.herkimer.nygenweb.net/remington/remingtonfam5.html.

5. The first innovation acted to improve the range of guns by cutting spiral grooves in their barrels so that the shot had a spin as it left the barrel. The term "musket" was replaced by "rifle" as that practice caught on. Accuracy demanded that the rifle fire a shot that was propelled by a standard amount of powder and that had less air resistance as it went toward its target. "Cartridges" and "bullets" thus replaced "powder" and "balls" in the firearm lexicon.

6. William Least Heat-Moon describes the town and its industry as he passes through on his boat, "River-Horse"; see Least Heat-Moon, *River Horse: A Voyage Across America* (New York: Houghton Mifflin, 1999), 46–7.

7. Chandler Jr., *Visible Hand*, 308.

8. There is considerable literature on the sewing machine. A short piece that relates to interchangeable parts and the production of complex mechanisms is in Harlick, *No Place Distant*, 148–53.

9. Ibid., 308–9.

10. In the 1920s, an Underwood machine contained 3,200 parts, according to the obituary of a famous typewriter customizer and repairman; see "Martin Tytell," *Economist*, September 20, 2008, 106.

11. See How Products Are Made, "Christopher Latham Sholes Biography (1819–1890)"; available online at http://www.madehow.com/inventorbios/13/Christopher-Latham-Sholes.html.

12. Schlereth, *Victorian America*, 69.

13. Wershler-Henry, *Iron Whim*, 70–1; and Torbjorn Lundmark, *Quirky QWERTY: A Biography of the Typewriter & Its Many Characters* (New York: Penguin, 2002), 11.

14. It took ten years before this design feature was changed (Lundmark, *Quirky QWERTY*, 15; see also Wershler-Henry, *Iron Whim*, 71. The design that was to be used throughout the twentieth century was largely standardized by 1900 and featured a businesslike black metal frame and black keys with white lettering. Fashionable coloring and portability would creep in over the century. See Schlereth, *Victorian America*, 69.

15. Lundmark, *Quirky QWERTY*, 16.

16. Diamond, *Guns, Germs, and Steel*, 248; and Wershler-Henry, *Iron Whim*, 273.

17. The product's third version, the Remington II, was a success in 1881 (Chandler Jr., *Visible Hand*, 308).

18. Ibid.

19. See Lisa K. Slaski, "The Remington Family and Works of Ilion, NY" (2001); available online at http://www.herkimer.nygenweb.net/remington/remingtonfam5.html.

20. Eliphalet Remington III apparently made his fortune in the Canadian West. At the start of the 1900s, he lived on a "substantial and innovative" farm near the village of Cazenovia, near Syracuse. He retired there to build model boats and summer cottages around Cazenovia lake (Hugill, *Upstate Arcadia*, 142).

21. "How we cling to clickety-clack," *Guardian*, May 6, 2011, 31

22. Wershler-Henry, *Iron Whim*, 134.

23. Lundmark, *Quirky QWERTY*, 12.

24. Many of the early attempts were made to emboss letters so that the blind could read (ibid., 13). It is interesting how many communications innovations were made with the motivation of helping the disabled.

25. Wershler-Henry, *Iron Whim*, 35. The keyboard, so vital to typewriters and computers, was developed and used by those involved in telegraphy some twenty-five years earlier (Oliver, *History of American Technology*, 440).

26. Wershler-Henry, *Iron Whim*, 41, 65.

27. Lundmark, *Quirky QWERTY*, 12.

28. Wershler-Henry, *Iron Whim*, 66.

29. Sholes received a patent in 1868, a few months before another person patented a typing ball-based concept, a precursor to the IBM Selectric machine a century later. Oddly enough, he received his first patent for a second model using piano keys, before getting approval for his fist model, which used telegraph keys (ibid., 63-4, 68).

30. Ibid., 67.

31. How Products Are Made, "Christopher Latham Sholes Biography (1819–1890)"; see also Wershler-Henry, *Iron Whim*, 68.

32. Lundmark, Quirky QWERTY, 14; Wershler-Henry, *Iron Whim*, 68.

33. Apparently, the company wished to order a large number of machines, as it dealt in a large volume of words (370 million in 1869) to be transmitted at speed. Western Union did not want to make them, however, so Sholes and Densmore eventually contacted the Remington Company. The half-sold Western Union must have been an appealing incentive for Remington (Oslin, *Story of Telecommunications*, 202).

34. It apparently did contain the seeds of what would much later become the electric typewriter.

35. Wershler-Henry, *Iron Whim*, 70.

36. See, for instance, Kevin Laurence, "The *Exciting* History of Carbon Paper!" (1995); available online at http://www.kevinlaurence.net/essays/cc.php.

37. Wershler-Henry, *Iron Whim*, 135.

38. Sampson, *Company Man*, 56–8.

39. Wershler-Henry, *Iron Whim*, 136.

40. Ibid., 149. The connection between time-and-motion studies and typewriters and computers is reinforced by the fact that Charles Babbage, the developer of early computer models in the 1830s, is considered the father of operational analysis — jargon for time-and-motion studies — as well as having written the first text on the subject in 1832 (Rowland, *Spirit of the Web*, 219).

41. Quoted in Wershler-Henry, *Iron Whim*, 80.

42. Sampson, *Company Man*, 58.

43. Wershler-Henry, *Iron Whim*, 85–7; Schlereth, *Victorian America*, 69.

44. Wershler-Henry, *Iron Whim*, 262–3.

45. Ibid., 231–2.

46. The first typewriter sales outlet was apparently in New York City (Lundmark, *Quirky QWERTY*, 10).

47. Wershler-Henry, *Iron Whim*, 226–8.

48. Lundmark, *Quirky QWERTY*, 15; Least Heat-Moon, *River Horse*, 47.

49. The poet Allen Ginsberg claimed Kerouac could do 110 to 120 words per minute; see myTypewriter.com, "Jack Kerouac 1922–1969" (2006); available online at http://www.mytypewriter.com/authors/featured/kerouac.html.

50. Wershler-Henry, *Iron Whim*, 238.

51. Ibid., 240–1.

52. Ibid., 239.

53. Klein, *Empire State*, 559.

54. See "James Rand, Jr.," *Wikipedia: The Free Encyclopedia*; available online at http://en.wikipedia.org/wiki/James_Rand,_Jr., accessed January 8, 2014.

55. Wershler-Henry, *Iron Whim*, 253–6.

56. Freiberger and Swaine, *Fire in the Valley*, 9. The naming sequence here seems a bit garbled relative to other sources on this corporate history.

57. Patton, *Open Road*, 15; Rowland, *Spirit of the Web*, 271.

58. Engelbart and others around him recognized where Moore's Law would take them in terms of ever-cheaper computing power.

59. Rosegrant and Lampe, *Route 128*, 171–2.

60. Freiberger and Swaine, *Fire in the Valley*, 24–5, 77.

61. The individuals who could get their hands on minicomputers were university students who were assigned time on those purchased for training and research (Rowland, *Spirit of the Web*, 281).

62. Freiberger and Swaine, *Fire in the Valley*, 27.

63. Wershler-Henry, *Iron Whim*, 273.

64. Cringely, *Accidental Empires*, 80–92.

65. Sampson, *Company Man*, 190.

66. John Markoff, *What the Dormouse Said: How the Sixties Counterculture Shaped the Personal Computer Industry* (New York: Viking, 2005), 248.

67. Cringely, *Accidental Empires*, 91.

68. Jim Cullen, *The Art of Democracy: A Concise History of Popular Culture in the United States* (New York: Monthly Review Press, 1996), 136.

69. Earls, *Route 128*, 100.

70. Ibid., 8.

71. The most famous spinoff in the valley was that of the "traitorous eight," engineers who left Shockley to found the dominant chipmaker, Intel.

72. A version of the story in the preceding paragraphs can be found online at http://www.netvalley.com/.

73. Lampe, *Massachusetts Miracle*, 214, 243. There is now a lot of similar employment elsewhere, as the "computer industry" has come to include a variety of activities formerly in other industry groups, especially in the information sector. It is hard to distinguish Google's advertisement system from those in the traditional advertising industry, for instance; see Paul Krugman, "The Future of New England," in *Engines of Enterprise: An Economic History of New England*, ed. Peter Temin (Cambridge, MA: Harvard University Press, 2000), 273.

74. Lampe, *Massachusetts Miracle*, 243, 244.

75. Ibid., 244.

76. Ibid., 266.

77. Rosegrant and Lampe, *Route 128*, 127.

78. Data General, a 1968 spinoff from DEC, had an organizational culture that resembled that in California, but it was the exception; Tracy Kidder describes it in his *Soul of a New Machine* (New York: Avon Books, 1981). See also Sampson, *Company Man*, 187.

79. As an example, the response of the chip business in Silicon Valley to a Japanese "invasion" around 1980 was more dynamic and aggressive than that of Route 128 to the development of the PC (Sampson, *Company Man*, 198).

80. The parallel between Ford and Gates is made in Cringely, *Accidental Empires*, 98–104.

81. Ibid., 53. A French microcomputer, the "Micral," appeared about a year before the Altair 8800 (Freiberger and Swaine, *Fire in the Valley*, 196.

82. BASIC allowed for the translation of English-language commands into digital instructions that the computer could follow (Rowland, *Spirit of the Web*, 282–3).

83. Freiberger and Swaine, *Fire in the Valley*, 77.

84. It was not their first venture together. Apparently, they had made and sold telephone-hacking "blue boxes" in Berkeley before this (Markoff, *What the Dormouse Said*, 272).

85. Rowland, *Spirit of the Web*, 284.

86. Freiberger and Swaine, *Fire in the Valley*, 140.

87. Oddly enough, until it caught on with the wider public, the fully loaded version cost about $3,000 and its first programs — a spreadsheet and a database program — were aimed at the business market (Cringely, *Accidental Empires*, 62–4).

88. Rowland, *Spirit of the Web*, 286.
89. Freiberger and Swaine, *Fire in the Valley*, 277.
90. Ibid., 276–7, 314.
91. Rowland, *Spirit of the Web*, 287.
92. Freiberger and Swaine, *Fire in the Valley*, 314.
93. Cringely, *Accidental Empires*, 166–77.
94. See ibid. (119–38) for the story of how IBM and Microsoft found each other. IBM failed to appreciate the value of the integrated circuit and did not tie down Intel. The combination of Microsoft Windows software and Intel chips, "Wintel," went on to dominate the computer market.
95. Rowland, *Spirit of the Web*, 283.
96. Eventually Microsoft offered its proprietary versions of these and other business-related programs as a suite, or single package, making life impossible for single-application competitors.
97. See, for instance, Leander Kahney, *The Cult of Mac* (San Francisco: No Starch Press, 2004); Rowland, *Spirit of the Web*, 291.
98. Freiberger and Swaine, *Fire in the Valley*, xv.
99. Even the microcomputer industry on Route 128 had its share of the counterculture. Data General, founded by DEC employees in 1968, was led by an Amherst guitar-playing temporary dropout who became a computer engineer partly to escape the draft and Vietnam (Sampson, *Company Man*, 188).
100. Markoff, *What the Dormouse Said*, chap. 5.
101. Apparently, the first ecommerce transaction took place between Stanford and MIT students in 1971 or 1972. An undetermined amount of marijuana was exchanged between them (ibid., 109).
102. Ibid., 165.
103. Brand was one of the video cameramen who helped Engelbart put on his presentation in December 1968 (Markoff, *What the Dormouse Said*, 152). The Portola Institute also published Ted Nelson's *Computer Lib* (Freiberger and Swaine, *Fire in the Valley*, 77).
104. Freiberger and Swaine, *Fire in the Valley*, 77.
105. Ibid., 124.
106. Ibid., 77.
107. Rowland, *Spirit of the Web*, 282.
108. It was no coincidence that, in the film, the song HAL 9000, the spacecraft computer sang as he was shut down was "Daisy Bell." This was the first song ever played by a computer, done at the Bell Labs around 1960 (Freiberger and Swaine, *Fire in the Valley*, 160–1). The song was written in 1892 by Harry Dacre, an Englishman, who was challenged to write a song that included a reference to a bicycle.
109. The quote is attributed to Silicon Valley venture capitalist John Doerr in Markoff, *What the Dormouse Said*, 197.

Chapter 13

1. See "The Skinner and Webb Families" (2008); available online at http://www.skinnerwebb.com.

2. Quoted in Theodore Steinberg, *Slide Mountain, or the Folly of Owning Nature* (Berkeley: University of California Press, 1995), 157, 159.

3. Mazlish (*Railroad and the Space Program*, 76) gives it as 40 percent; Berstein (*Wedding of the Waters*, 350) estimates it at 33 percent in 1850.

4. Hawke, *Nuts and Bolts of the Past*, 128–9.

5. The phrase apparently began with a toast made at the annual dinners of the New England Society of the City of New York after its creation in 1805 (Berstein, *Wedding of the Waters*, 337).

6. Ibid., 350–1.

7. Though the National Road was extended across Ohio, Indiana, and Illinois, this segment was overshadowed by the speed and cost savings of the Ohio River; see James E. Davis, *Frontier Illinois* (Bloomington: Indiana University Press, 1998), 220.

8. Billington, *Westward Expansion*, 290–1.

9. Attempts to gain large tracts and "flip" pieces of them to others for a quick profit generally proved to be financial failures. Only subdivision, patient sales, and development worked, though even then the prices could not reflect potential demand because of the existence of cheap land in the territories farther west (Hugill, *Upstate Arcadia*, 18–19).

10. The Dutch were able to pay in cash and got better terms than the local speculators, who needed credit to buy (ibid., 33–5). Cazenove apparently had been advised by Robert Morris, Congress's Secretary of Finance during the Revolution and one of the largest landowners in New York, to buy the tract. Morris also advised the Pulteney Associates to buy their tract near the Finger Lakes and the Holland Land Company to make its purchase of virtually the whole of far-western New York (18–19).

11. Ibid., 38; there is some suspicion that the Dutch actually transferred ownership of some of the Cazenovia tract to Talleyrand for his services (85).

12. Billington, *Westward Expansion*, 247.

13. The canal drastically lowered shipping costs and thereby increased the value of farm land. A ton-mile cost from Buffalo to New York City declined from 19 cents to 2 cents between 1817 and the 1830s (Davis, *Frontier Illinois*, 224).

14. Hugill, *Upstate Arcadia*, 35–40.

15. One of Lincklaen's major complaints was the continual tendency of his buyers to move on from their half-cleared lands and sell to others — people he might have wanted as his own buyers.

16. Russell A. Grills, *Upland Idyll: Images of Cazenovia, New York, 1860–1900* (Albany: New York State, Office of Parks & Historic Preservation, 1993), 30. Much of the material in the following section is based on this source.

17. Hugill, *Upstate Arcadia*, 11.

18. Billington, *Westward Expansion*, 244. Some Yankees had moved into New York as early as the 1750s, though not in great numbers, especially into disputed lands along the western Massachusetts border (Klein, *Empire State*, 160).

19. Burrows and Wallace, *Gotham*, 20; Taylor (*William Cooper's Town*, 5) puts the 1820 population at 1.37 million, making it the most populous State.

20. Only around Geneseo, in western New York, was a family that purchased a large

tract in the 1790s able to attract renters and manage to hold on to its land through the twentieth century (Hugill, *Upstate Arcadia*, 19).

21. Hernando de Soto, *The Mystery of Capital: Why Capitalism Succeeds in the West, but Fails Everywhere Else* (New York: Basic Books, 2000).

22. Cooper noted in 1786, as he began to sell land, that "the land appears to be, take it altogether, Pretty good, but will not Sell for cash, tho at Ever So Low A Price, but have Plenty of Applications, but they have no money nor Provisions." Most of his early buyers were clusters of poor "hill Yankees" from the Berkshires or Vermont; see Taylor, *William Cooper's Town*, 71–2, 90.

23. Much economic activity depended on barter between the farmer and the storekeeper. Even in Puritan times, the store played a central part in the life of a frontier New England community; see Ziff, *Puritanism in America*, 292.

24. Wheat was the product of choice in the markets of America and Europe, so it was the crop first planted, and successive territories took their turns as the nation's premier supplier (Martin, *Railroads Triumphant*, 163).

25. Cooper also used his control over the farmers to press his political convictions on them as voters. In 1792, his tactics backfired when the legislature decided that his area had been intimidated into voting for one gubernatorial candidate and threw out their votes, giving the election to the other (Klein, *Empire State*, 254–5).

26. Taylor, *William Cooper's Town*, 105–6. In the early years of his activities, cattle receipts amounted to three-fifths of all "payments" for his land.

27. "Ashes were silver and gold to the young or poor farmer," according to one settler's son (ibid., 108–10).

28. For many years, Cooper could count on the political support of the Yankee newcomers, but not that of the more established communities to the east of the Otsego tract (ibid., 188–9).

29. He made sure the road bypassed the older villages around Cherry Valley (ibid., 162).

30. Ibid., 107.

31. Ibid., 260. US 20 was known traditionally as the Cherry Valley Pike as it passed through Cazenovia.

32. As well, Williamson founded the first newspaper in the region and the first theatre; Emerson Klees, *People of the Finger Lakes Region, the Heart of New York State* (New York: Friends of the Finger Lakes Publishing, 1995), xv.

33. Taylor, *William Cooper's Town*, 120. This was an idea that had some circulation, as the combination of hurricane devastation in the Caribbean and the uncertainties of sugar supplies during the wars of 1792–95 also interested the Dutch developers of Cazenovia in the same idea (Hugill, *Upstate Arcadia*, 34).

34. Davis, *Frontier Illinois*, 297.

35. "What did this movement westward mean for everyday life? For many, it entailed almost continual transience. Frontier historians calculate that nineteenth-century Westerners moved on average four to five times as adults. Richard Garland, father of novelist Hamlin Garland, aptly represents, in William L. Barney's estimate, an entire generation of Northern farmers. Born into a Maine farm family, Garland clerked for Amos Lawrence in

Boston, "went west with his parents on the Erie Canal in 1840, worked as a lumberjack in the forests of the upper Midwest, cleared a 160-acre farm in Wisconsin, marched off to fight for the Union, returned home, sold the farm, and moved to Minnesota, to Iowa, and finally to South Dakota" (Schlereth, *Victorian America*, 13). Not without reason did his son title one of his writings *Main Travelled Roads* (1891).

36. If there were no mill, the farmer and his family would have to grind their own grain by hand with a mortar and pestle. Normally, across the frontier, the miller was entitled to 10 percent of the flour as payment (Davis, *Frontier Illinois*, 109).

37. Ellicott employed two of his brother Andrew's sons as clerks. Both saw action in the War of 1812; see Catharine van Cortlandt Mathews, *Andrew Ellicott: His Life and Letters* (1908; repr. New York: Ralph Roberts, 2001), 183.

38. At the time, there were only 17,000 people in the whole area of western New York, half of whom were children (Berstein, *Wedding of the Waters*, 122).

39. Buffalo had only 2,400 people when the Erie Canal opened in 1825; by 1850, it was the world's largest grain port; see Koeppel, *Bond of Union*, 1, 394.

40. Even in 1800 they were little more than Indian trails (Berstein, *Wedding of the Waters*, 122).

41. Billington, *Westward Expansion*, 248–9.

42. Ellicott settled in Batavia in 1798 and coordinated the survey from there. His team laid out the town in long, narrow lots. In 1801, the first town lot was sold to an Abel Rowe; a school teacher settled soon after. Large-scale land sales started in 1802. See ibid., 249; and Barbara Ann Toal, *Batavia* (Charleston, SC: Arcadia Publishing, 2000), 7–8, 15–19.

43. Billington, *Westward Expansion*, 248–59.

44. Half of them, including Ellicott, fled the area when British forces attacked and burned Buffalo and Black Rock on New Year's Eve, 1813.

45. Ellicott's promotion of the canal involved one of the first land grants to finance a major public transportation work. He persuaded the Company's owners to offer a large tract of land, 100,000 acres, to the State if the canal were built through or near the Company's holdings rather than go into Lake Ontario, with a barge canal around Niagara Falls, as some were advocating; see Mathews, *Andrew Ellicott*, 16–18.

46. His relationship with the canal is documented in Koeppel, *Bond of Union*.

47. Ibid., 144.

48. Ellicott had a political machine in western New York called the "Big Family," which came apart in the face of settler protests over the lack of investment he and the Holland Land Company had made in their area. Ellicott, who had earlier resigned from the Erie Canal commission because of ill health, also was forced to resign as the manager (ibid., 346).

49. The first canal shipping company was opened on the finished portion of the canal in 1820 by two Yankees, and prospered from the start (ibid., 144).

50. In the fall of 1836, William Seward, the future governor of the State and secretary of state under Presidents Abraham Lincoln and Andrew Johnson, was hired by the Holland Land Company to be its manager in Chautauqua County, near the Pennsylvania border. He rented a house near the Buffalo-Erie road and worked at land development for about fifteen months, until the Panic of 1837 hit the area; see Doris Kearns Goodwin,

Team of Rivals: The Political Genius of Abraham Lincoln (New York: Simon & Schuster, 2005), 78–80.

Chapter 14

1. Quoted in Johnson, *History of the American People*, 119.
2. Quoted in Whitney R. Cross, *The Burned-over District; The Social and Intellectual History of Enthusiastic Religion in Western New York, 1800–1850* (New York: Harper and Row, 1950), 210.
3. J. Wilbur Chapman, *The Life and Work of Dwight Lyman Moody* (1900), chap. 11; available online at http://www.biblebelievers.com/moody/06.html.
4. The term is supposed to have been used first by Finney. An almost-poetic description of the area and its effects can be found in Klees, *People of the Finger Lakes Region*, xvi–xvii. Harold Bloom, *The American Religion: The Emergence of the Post-Christian Nation* (New York: Simon & Schuster, 1992), 191, suggests that California was the late-twentieth-century equivalent of the Burned-over District.
5. A key element in promoting the construction of the Erie Canal was nationalism, fanned by the experience of the War of 1812 and concern over the possible British seizure of the Western US trade through the St. Lawrence River to Europe. The canal was begun in 1817, only two years after Andrew Jackson's victory at New Orleans. The British completed the competing Welland Canal in 1829, four years after the opening of the Erie Canal. See Ronald R. Shaw, *Erie Water West: A History of the Erie Canal, 1792–1854* (Lexington: University of Kentucky Press, 1966), 371–415.
6. The Onondagas still have a strong presence in the area.
7. The canal led to a huge increase in population in the 1820s, albeit from a small base: Syracuse grew by 282 per cent and Utica by 183 per cent; farther west, Rochester grew by 512 per cent and Buffalo by 314 per cent (Klees, People of the Finger Lakes Region, xviii). For a short survey of the canal's development, see Robert H. Smith, *Clinton's Ditch: The Erie Canal, 1825* (Del Mar, CA: C Books, 2006).
8. Klein, *Empire State*, 316.
9. Cross (*Burned-over District*, 67) notes that Syracuse was the prototypical canal [or railroad] town: it did not exist until the canal came, and then mushroomed in size.
10. Cross (ibid., 72) describes the canal towns of Rochester and Utica of the time as manufacturing ones and Syracuse and Buffalo as devoted to commerce.
11. Ibid., 74–5.
12. C-SPAN and Ann Bentzel, *C-SPAN's Traveling Tocqueville's America* (Baltimore: Johns Hopkins University Press, 1988), 32; Pierson, *Tocqueville in America*, 191.
13. See, for instance, Mary Ryan, *Cradle of the Middle Class: The Family in Oneida County, New York, 1790–1865* (New York: Cambridge University Press, 1981). Utica had its first revival in 1813–14 and its last in the depression year of 1838.
14. The Puritans in America were acting out a variation on the English myth that they were the successor to the Jews as the Divinely appointed "Chosen People," popularized in Elizabethan times by John Foxe's *Book of Martyrs*. It continues to be a strain in Americans' thinking about their role in the world (Johnson, *History of the American People*, 20).
15. "Fused" might be too strong a word, as the Puritans were mindful of European

states where the church and state were run by the same people. Instead, they opted for different people in civil and church governance. It was expected that both would be drawn from among believers and both would support each other, but remain separate to a degree (Ziff, *Puritanism in America*, 52).

16. Brown, *Massachusetts*, 60–1.

17. But not so quickly that New England could not put its stamp on the myth of American origins. Edmund Morgan noted that, "[l]ong before 1860 New Englanders laid claim to the national consciousness and gave their own past as a legacy to the nation, whether it wanted it or not"; quoted in Wills, *Head and Heart*, 75.

18. The Halfway Covenant was a modification of the rule that only "converted" — or "born-again," as we would say today — people could be full members of the church. It allowed sons and daughters of the converted membership in anticipation of their own conversions. Massachusetts pastors resisted this reform in large part, while Connecticut ones adopted it. The Halfway Covenant had been prompted by a letter from King Charles II to Massachusetts informing the Puritans that they could no longer restrict the vote to full church members nor reserve the Lord's Supper to consciously saved members. See, for instance, Jon Butler, *Awash in a Sea of Faith: Christianizing the American People* (Cambridge, MA: Harvard University Press, 1990), 63; Conforti, *Saints and Strangers*, 102; and Wills, *Head and Heart*, 21.

19. Bailyn, *Voyagers to the West*, 166–89.

20. Butler, *Awash in a Sea of Faith*, 61, 63

21. Wills, *Head and Heart*, 61–4.

22. Butler, *Awash in a Sea of Faith*, 62.

23. See, for instance, Peter N. Carroll, *Puritanism and the Wilderness: The Intellectual Significance of the New England Frontier, 1629–1700* (New York: Columbia University Press, 1969).

24. Ibid., 125, 147, 165, 201–5.

25. Ibid., 215–16.

26. Ibid., 216, 217, 221.

27. Northampton was one of a number of places reluctantly founded by the Massachusetts government, which was caught between wanting a single, compact, holy city, and the personal desires of Puritans to expand into good farmland. In 1653, colonists from both Connecticut and Massachusetts already in the area petitioned that the site would afford "a Comfortable subsistence whereby people may Live and Attend upon god in his holy ordinances without distraction" (Carroll, *Puritanism and the Wilderness*, 122).

28. David Levin, "Edwards, Franklin, and Cotton Mather: A Meditation on Character and Reputation," in *Jonathan Edwards and the American Experience*, ed. Nathan O. Hatch and Harry S. Stout (New York: Oxford University Press, 1988), 43.

29. Conforti, *Saints and Strangers*, 181–2. Butler (*Awash in a Sea of Faith*, 174–7) notes a number of other years in the 1600s and 1700s when revivals happened in New England.

30. Much of the biographical information in this section comes from Alfred Owen Aldridge, *Jonathan Edwards* (New York: Washington Square Press, 1964), 1–66.

31. Johnson, *History of the American People*, 41.

32. The Great Awakening attracted more women than men and more youths than

their seniors. As well, New England society was on the edge of population growth that could not be accommodated within traditional structures, given the lack of good land. Revivals let these second-class people "step out of their place" see Amy Schrager Lang, "'A Flood of Errors': Chauncy and Edwards in the Great Awakening," in *Jonathan Edwards and the American Experience*, ed. Nathan O. Hatch and Harry S. Stout (New York: Oxford University Press, 1988), 170.

33. Roger Finke and Rodney Stark, *The Churching of America, 1776–1990* (New Brunswick, NJ: Rutgers University Press, 1992), 22–3.

34. The first Great Awakening was but the North American version of a transatlantic Awakening; see Ann Taves, *Fits, Trances, and Visions: Experiencing Religion and Exploring Experience from Wesley to James* (New Brunswick, NJ: Princeton University Press, 1999), 20.

35. Finke and Stark, *Churching of America*, 46–53. Whitefield made seven tours of British North America between 1738 and his death in Newburyport in 1770 (Butler, *Awash in a Sea of Faith*, 187).

36. Mark Noll refers to Daniel Pals's observation on this in his *The Scandal of the Evangelical Mind* (Grand Rapids, MI: William B. Eerdmans, 1994), 61–2. See also Butler, *Awash in a Sea of Faith*, 191.

37. A century later, a conservative minister wrote of Finney and his imitators that "[e]very theological vagabond and peddler may drive his bungling trade, without passport or license, and sell his false wares at pleasure"; see Jackson Lears, *Fables of Abundance: A Cultural History of Advertising in America* (New York: Basic Books, 1994), 57.

38. Finke and Stark, *Churching of America*, 51.

39. This term was never used by Edwards or his successors, but was invented a century later in 1842 as the title of a bestseller by Joseph Tracey, *The Great Awakening: A History of the Revival of Religion in the Times of Edwards and Whitefield*; see Johnson, *History of the American People*, 112.

40. There was almost no religious activity, churches, or preachers in the Carolinas before 1700. After this, the Anglican revival began to provide some (Butler, *Awash in a Sea of Faith*, 64, 98, 165–7).

41. Stephen Miller, *The Peculiar Life of Sundays* (Cambridge, MA: Harvard University Press, 2008), 178. See also Louis Menand, *The Metaphysical Club: A Story of Ideas in America* (New York: Farrar, Straus and Giroux, 2001), 62–3. Apparently, the feeling was not quite reciprocated, as Whitefield indicated he was more interested in Edwards's late and famous grandfather (Butler, *Awash in a Sea of Faith*, 179).

42. Whitfield wrote 1,500 sermons, nine volumes of "Miscellanies," and 10,000 separate entries in his "Blank Bible," along with other volumes. Nathan O. Hatch and Harry S. Stout, "Introduction," in *Jonathan Edwards and the American Experience*, ed. Nathan O. Hatch and Harry S. Stout (New York: Oxford University Press, 1988), 4.

43. Noll, *Scandal of the Evangelical Mind*, 77.

44. Wills, *Head and Heart*, 103.

45. See ibid. Ann Taves (*Fits, Trances, and Visions*, 16) notes that there was a division between "formalists," who saw religion as a set of forms that reflected the conclusions of experts and other authorities, and "enthusiasts," who claimed that God could be experienced directly in some manner by individuals.

46. After Edwards had identified himself with Whitefield's work, he later began to criticize it (Wills, *Head and Heart*, 103).

47. This was not a singular occurrence, as many ministers even before 1700 had been given only one-year contracts, which could be renewed or not on the decision of the board (Butler, Awash in a Sea of Faith, 171).

48. Paul Johnson (*History of the American People*, 118) claims it was "the proto-revolutionary event, the formative movement in American history, preceding the political drive for independence and making it possible." Hatch and Stout ("Introduction," 4) note that Timothy Dwight, an eighteenth-century president of Yale University, called Edwards, "that moral Newton, that second Paul."

49. The problem with revivals arose from the kind of social pressure that was exerted on laggards who might have other reasons than religious one for being saved. The nineteenth-century experience of Frederick Olmstead, the developer of Central Park, is useful in this regard; see Rybczynski, *Clearing in the Distance*, 64.

50. Johnson, *History of the American People*, 41.

51. Hatch and Stout ("Introduction," 10) feel that Edwards was America's first post-Millennial theologian.

52. Butler, *Awash in a Sea of Faith*, 195–205.

53. Donald Weber, "The Recovery of Jonathan Edwards," in *Jonathan Edwards and the American Experience*, ed. Nathan O. Hatch and Harry S. Stout (New York: Oxford University Press, 1988), 56–9.

54. Butler (*Awash in a Sea of Faith*, 206) estimates that three-quarters of the Anglican pastors lost their churches. They were also hampered by the fact that an Anglican bishop was never appointed to lead the church in America (Johnson, *History of the American People*, 41).

55. Wills, *Head and Heart*, 7.

56. Ibid., 229–30.

57. Finke and Stark, *Churching of America*, 61.

58. Much of this section on disestablishment in New York is taken from John Webb Pratt, *Religion, Politics, and Diversity: The Church-State Theme in New York History* (Ithaca, NY: Cornell University Press, 1967).

59. Breaking the tie to London cost the American church its British subsidies. The Wesleyans, embarrassed by their founder's support for London, likewise created the American Methodist church.

60. There is an interesting vignette about the failed attempt of some Sabbatarians to operate an Erie Canal transportation company on their religious principles in Paul Johnson, *A Shopkeeper's Millennium: Society and Revivals in Rochester New York, 1815–1837* (New York: Hill and Wang, 1978), 83–8.

61. By 1850, disestablishment tended to arise in a different way: the debate about the relationship between Christianity — in particular, Protestant Christianity — and government came to the fore as a general concept favoring no denomination, but disfavoring Catholic immigrants (Butler, *Awash in a Sea of Faith*, 258–68).

62. This prohibition irritated political and other elites, who sensed that enthusiasm threatened the social order. A Methodist example was John Fanning's 1814 book, *Methodist*

Error, or Friendly Christian Advice to those Methodists who indulge in extravagant emotions and bodily exercises. Fanning was a sober and studious Philadelphian who objected to the fits and trances of Methodist camp meetings and other gatherings' see Taves, *Fits, Trances, and Visions*, 34, 46, 76–7; and Nathan O. Hatch, *The Democratization of American Christianity* (New Haven, CT: Yale University Press, 1989), 64–6.

63. See Finke and Stark, *Churching of America*, 16, 26. This apparently had been the case during the colonial period as a whole (24).

64. Lears, *Fables of Abundance*, 43.

65. Richard Bassett, a Methodist from Delaware, said nothing while attending the sessions (Wills, *Head and Heart*, 225).

66. De Tocqueville noted the complex relationship between formal religion and spirituality in the West during his visit to Cincinnati on his way south (Pierson, *Tocqueville in America*, 564).

67. Hatch, *Democratization of American Christianity*, 59–60.

68. Finke and Stark, *Churching of America*, 63. The Plan was formally abandoned in 1852, but in the Burned-over District, it had been a dead letter by the 1830s (Cross, *Burned-over District*, 257).

69. Finke and Stark, *Churching of America*, 75.

70. Taves, *Fits, Trances, and Visions*, 71, 84. Hatch (*Democratization of American Christianity*, 81–93) gives a short, but sobering account of Asbury's work and that of the itinerants.

71. Butler, *Awash in a Sea of Faith*, 236–7.

72. Finke and Stark, *Churching of America*, 153. See also Wills, *Head and Heart*, 290–1.

73. Chernow, *Titan*, 19–20.

74. Mormon founder Joseph Smith responded to Free Will Baptist preacher Nancy Towle's criticism of his willingness to lead a new creed and preach with little formal education by saying, "[t]he gift has returned back again, as in former times, to illiterate fishermen" (quoted in Hatch, *Democratization of American Christianity*, 49).

75. Baptist and Methodist preachers might average $60–$100 a year in salary, while the Congregationalists and Presbyterians, better educated, could command between $400 and $3,000. The latter were either subsidized by the missionary societies or gravitated to larger settlements and toward the better-off people in the community (Finke and Stark, *Churching of America*, 82).

76. Ibid., 75–6.

77. Butler, *Awash in a Sea of Faith*, 280–1. One of the most prominent from 1815 to 1832 was Nancy Towle, a Free Will Baptist preacher of Yankee origin, who even preached before Congress. Most of the other women preachers of the time were also Yankees (345n44). Hatch (*Democratization of American Christianity*, 79) discusses Towle's influence; see also Cross, *Burned-over District*, 177.

78. There was a great revival as early as 1771 at Cub Creek, Virginia, just south of Appomattox. The date suggests that revivals spread with the migration of the Scots-Irish, especially given that the best record of this particular revival was by a Presbyterian, Caleb Wallace (Taves, *Fits, Trances, and Visions*, 86).

79. Even political campaign styles were affected by revival techniques (Wills, *Head*

and Heart, 293–4). Bloom (*American Religion*, 59–63) describes much of the emotional reactions at Cane Ridge and similar meetings. Much later, and in a secular way, the 1969 Woodstock rock festival took on many aspects of a nineteenth-century camp meeting. See also Andrews, *People Called Shakers*, 137–9.

80. Hatch (*Democratization of American Christianity*, 6) attributes the idea to Robert Wiebe.

81. The proportion who are adherents seems to have plateaued for the past sixty years at 59 to 62 percent (Wills, *Head and Heart*, 8).

82. Finke and Stark, *Churching of America*, 54–6.

83. Bloom, *American Religion*, 64.

84. Emerson Klees lists ten new religions, sects, and communes established in the Burned-over District; see *The Crucible of Ferment: New York's "Psychic Highway"* (New York: Friends of the Finger Lakes Publishing, 2001), 10–12.

85. Quoted in Hatch, *Democratization of American Christianity*, 32.

86. One can see this indirectly through sex ratio statistics. Massachusetts (without Maine), Rhode Island, and Connecticut all had ratios of 96 to 98 males for every 100 females; males, generally younger ones, were on the move west. See Finke and Stark, *Churching of America*, 33–5, who also note there was an almost perfect negative correlation between a high male-to-female ratio and rates of religious adherence — the frontier was unchurched and irreligious, at least until the "nice" women arrived (35).

87. The model for these societies seems, ironically, to have been the Society for the Propagation of the Gospel in Foreign Parts (1701), an Anglican organization that began to improve that church's fortunes in the South before the Revolution (Butler, *Awash in a Sea of Faith*, 234). In 1798, the General Assembly of the Congregationalists in Connecticut authorized the creation of a missionary society to work in the West. It was run by two ministers residing in the West along with one from Vermont. In 1826, the American Home Missionary Society was created to support the Christianization of Yankee lands in New York and beyond (Finke and Stark, *Churching of America*, 64–5). Its primary work was to send out missionaries and to subsidize settled ministers where the population was sparse. For an account of its activities in western New York, see Cross, *Burned-over District*, 188–90.

88. In 1831, de Tocqueville noted the nature of the itinerant Methodist preachers and the general religiosity of the people in Ohio (Pierson, *Tocqueville in America*, 564).

89. See the observations made about the Congregationalists in Finke and Stark, *Churching of America*, 40–1.

90. Ibid., 74.

91. Klein, *Empire State*, 222, 282.

92. Menand, *Metaphysical Club*, 80. Finney notes in his autobiography that he spent six weeks to two months in Buffalo after leaving Rochester in the spring of 1831, and then spent the fall of that year in Providence and Boston when de Tocqueville would have passed through there, so it is unlikely that the French commentator on American values was unaware of his work. An interesting account of the effect of "enthusiastic" revivals on Rochester is given in Johnson, *Shopkeeper's Millennium*.

93. Cross, *Burned-over District*, 67.

94. As Cross (ibid., 82) says, "[f]or whatever reason, the New York descendants of the Puritans were a more quarrelsome, argumentative, experimenting brood than their parents and stay at home cousins."

95. Ibid., 187–8.

96. Wills, *Head and Heart*, 57–61.

97. Millennialism is not an uncommon sentiment. St. Paul reassured early Christians that they would see the end of the world. The early believers in the Koran in the 700s were convinced by some of its passages that the world would end very soon; see Malise Ruthven, "The Birth of Islam: A Different View," *New York Review of Books*, April 7, 2011, 82. The Puritans as well felt that the "end time" was near (Wills, *Head and Heart*, 33–4. One commentator has noted, "[b]y 1815, Americans were 'drunk on the millennium'"; see Leonard I. Sweet, *The Evangelical Tradition in America* (Macon, GA: Mercer University Press, 1984), 114. The Mormons also have some Millennial flavor, as evidenced in their formal title of the Church of Jesus Christ of Latter-day Saints. Cross devotes a chapter to the Millerites and other Millennialists; see *Burned-over District*, chap. 17.

98. A different group in Britain, the Irvingites, had no better luck with their predictions of the end of the world in 1835, revised to 1838; see Hobsbawm, *Age of Revolution*, 228.

99. The main group migrated west to Battle Creek, Michigan, where John H. Kellogg created a cereal industry/health food empire (Johnson, *History of the American People*, 303–5.

100. Cross (*Burned-over District*, 79–80) argues that Millennialism, of the pre- or post- variety, is a reflection of American optimism.

101. Sweet, *Evangelical Tradition in America*, 37.

102. Bloom (*American Religion*, 65) notes: "Cane Ridge and the following Southern camp meetings were more spectacular manifestations than the individual revivalist events of the Burned-over District, but the Southerners were far surpassed in Gnostic intensity and Enthusiastic zeal for Millennial innovation by the transplanted children of the Puritan tradition." See also Dunbar and May, *Michigan*, 299. Cross (*Burned-over District*, 79–80) calls social perfectibility the "holy enterprise of minding other people's business," and notes it is a continuing feature of American society.

103. Cross, *Burned-over District*, 113–35.

104. The publication and subsequent popularity of Harriet Beecher Stowe's Uncle Tom's Cabin in 1852 resonated among the "saved" of the North. It helped that she was the daughter of a Yankee Congregational preacher who had moved to Cincinnati and the wife of another preacher; see McPherson, *Battle Cry of Freedom*, 88. Slotkin (*Regeneration through Violence*, 441–3) notes how the early Puritan "captivity" myth was adapted by abolitionist writers such as Stowe to add to the moral indignation about slavery.

105. Cross (*Burned-over District*, 217–26) gives an account of the rise of abolitionism.

106. Ibid., 168.

107. For an account of the rise of temperance movement in western New York, see ibid., 213–17. Temperance activists also felt that coffee, tea, and tobacco were wrong, but eventually fixed on alcohol as the single cause. The Mormons later picked up on part of this.

108. Sweet, *Evangelical Tradition in America*, 57–66.

109. Cross, *Burned-over District*, 85–8.

110. Cullen, *Art of Democracy*, 76.

111. Hurt, *Ohio Frontier*, 289.

112. See Camp Meeting (Dot) Org, "What Is a Camp Meeting?" (n.d); available online at http://www.campmeeting.org/whatis.htm.

113. Wish, *Society and Thought in America*, 87.

114. The reaction of the official church to this format resulted in the founding of the Primitive Methodist church.

115. Menand, *Metaphysical Club*, 188.

116. Ibid.

117. Hatch (*Democratization of American Christianity*, 20, 125, 128–32) relates a couple of stories about Dow's preaching. Dow gave at least one saying to the American lexicon, describing Calvinism as: "You can and you can't, You will and you won't, You'll be damned if you do, And you'll be damned if you don't."

118. Butler, *Awash in a Sea of Faith*, 241; see also Hatch, *Democratization of American Christianity*, 34–40.

119. She wrote a book entitled *Vicissitudes in the Wilderness* that was still in print after Dow died. Following him, the reality of her vicissitudes can only be imagined.

120. Much of this section depends on Charles G. Finney, *Charles G. Finney: An Autobiography* (Westwood, NJ: Fleming H. Revell, 1876, 1908). See also Klein, *Empire State*, 330–1; and Cross, *Burned-over District*, 152–175.

121. Finney, *Charles G. Finney*, 4.

122. Hatch (*Democratization of American Christianity*, 195) claims Finney transferred Methodist techniques to Presbyterianism.

123. Sweet, *Evangelical Tradition in America*, 139.

124. Cross, *Burned-over District*, 185.

125. Johnson, *Shopkeeper's Millennium*, 109. As a post-Millennialist, Finney noted that, "if they were united all over the world, the Millennium might be brought about in 3 months."

126. Sweet, *Evangelical Tradition in America*, vii. A Finney Festival was held in Rochester in October 1981, one hundred and fifty years after the revival.

127. Finney noted: "In New England, I have found a high degree of general education, but a timidity, a stiffness, a formality and a stereotyped way of doing things, that has rendered it impossible for the Holy Spirit to work with freedom and power." A New York preacher friend commented to him in Hartford, "Why, Brother Finney, your hands are tied [here], you are hedged in by their fears and by the stereotyped way they do everything. They have even put the Holy Spirit in a strait-jacket." See Finney, *Charles G. Finney*, 439.

128. Lears, *Fables of Abundance*, 56. By 1835, he was attacking Calvinist ideas in New York City (Hatch, *Democratization of American Christianity*, 197). He wanted religious life to be audience-centered, not clergy-centered.

129. Sweet, *Evangelical Tradition in America*, 33–9.

130. Much of the biographical information in this section comes Chapman, *Life and Work of Dwight Lyman Moody*, although it is a rather disorganized work. A capsule profes-

sional biography can be found in Wills, *Head and Heart*, 342–8. There is a considerable literature on Moody's life and career.

131. By the 1850s, about one-third of Americans had some kind of religious affiliation, but this number was lower in the cities (Butler, *Awash in a Sea of Faith*, 294).

132. The spread of full-time, settled Methodist clergy followed settlement in the West, growing from one-third of the Methodist preachers in 1843 to one-half in 1882. The average length of tenure in one church increased from two years in 1804 to three years in 1864 to five years in 1888 (Finke and Stark, *Churching of America*, 154). By the 1850s, Methodists were noting the institution of pew rentals and ministers with some formal education. Complaints were raised when the majority of Methodist clergy in the Burned-over District were subjected to the church's hierarchy and discipline, much as happened to the Congregationalists a century earlier, resulting in the formation of the Free Methodists in 1860 (150–2).

133. Wills, *Head and Heart*, 341–2.

134. If one wanted to consider Jonathan Edwards and Methodist bishop Francis Asbury in this line, he could be seen as part of a fifth generation, dating back to the 1730s.

135. Chapman, *Life and Work of Dwight Lyman Moody*, chap. 6.

136. Wills (*Head and Heart*, 346) quotes him as saying, "I look upon this world as a wrecked vessel. God has given me a lifeboat and said to me, 'Moody, save all you can'."

137. Quoted in Marshall W. Fishwick, *Great Awakenings: Popular Religion and Popular Culture* (New York: Haworth Press, 1995), 37.

138. An interesting modern, yet timeless, description of evangelist logic is Robert D. Kaplan, *An Empire Wilderness: Travels into America's Future* (New York: Random House, 1998), 278–80. See also Noll, *Scandal of the Evangelical Mind*, 61.

139. Hatch, *Democratization of American Christianity*, 213.

140. Ibid., 81.

141. Bloom, *American Religion*, 15.

142. Maynard, *Walden Pond*, 63.

143. Henry F. May, "Jonathan Edwards and America," in *Jonathan Edwards and the American Experience*, ed. Nathan O. Hatch and Harry S. Stout (New York: Oxford University Press, 1988), 30.

144. Weber, "Recovery of Jonathan Edwards," 64–5.

145. Bloom, *American Religion*, 25. Bellah et al. interviewed a woman identified as Shiela who espoused a religion she called "Shielaism" — the principle of which was "Try to love yourself and be gentle with yourself" — which the authors say was not uncommon in 1980s America and was are related to the Puritan Ann Hutcheson's ideas, which got her expelled to Rhode Island and then to Long Island in the mid-1600s. See Robert N. Bellah et al., *Habits of the Heart: Individualism and Commitment in American Life* (New York: Harper and Row, 1985), 221.

Chapter 15

1. John Locke, *Second Treatise of Civil Government*, 1690, chap. II.

2. William Blackstone, *Commentaries on the Laws of England*, 1765–69, Book I, chap. 15, section III. Blackstone's *Commentaries* constituted the basic document used to

train lawyers and guide judges throughout the first century and more of an independent America.

3. Neall was introducing the Hutchinson Family Singers to Lucretia Mott. The Singers were one of the most popular entertainment acts of the 1840s, featuring themes about abolition, temperance, labor concerns, and women's rights in their music; see "Hutchinson Family Singers," *Wikipedia: The Free Encyclopedia*; available online at http://en.wikipedia.org/wiki/Hutchinson_Family_Singers, accessed January 9, 2014. See also Otelia Cromwell, *Lucretia Mott* (Cambridge, MA: Harvard University Press, 1958), 97.

4. Deliberately patterned after Jefferson's Declaration of Independence, this is part of the Declaration of Sentiments adopted at the 1848 Seneca Falls Women's Rights Convention; see, for example, Fordham University, "Modern History Sourcebook: The Declaration of Sentiments, Seneca Falls Conference, 1848"; available online at http://www.fordham.edu/halsall/mod/Senecafalls.html.

5. The canal has been referred to as a "psychic highway" between New England and its "colonies" to the west; see E. Keith Melder, *Beginnings of Sisterhood: The American Woman's Rights Movement, 1800–1850* (New York: Schocken Books, 1977), 146.

6. Peggy A. Rabkin notes that this was the conclusion of Stanton, Anthony, and others in their multi-volume *History of Woman Suffrage*; see "The Silent Feminist Revolution: Women and the Law in New York State from Blackstone to the Beginnings of the American Women's Rights Movement" (PhD diss., State University of New York at Buffalo, 1975), 3.

7. A speech he gave on the subject in January 1848 essentially cost him his congressional seat. The speech is reproduced online at http://medicolegal.tripod.com/lincolnv-mexwar.htm.

8. See Constance Buel Burnett, *Five for Freedom* (New York: Greenwood Press, 1968), foreword. Basically, women, African Americans, and Native Americans were left out of this natural rights argument.

9. Norma Busch, *In the Eyes of the Law: Women, Marriage, and Property in Nineteenth-Century New York* (Ithaca, NY: Cornell University Press, 1982), 25. In 1789, some thirteen years after he was the major drafter of the American Declaration, Jefferson was involved in a repeat performance when, as American ambassador to France, he advised the French government about its proposed Declaration of the Rights of Man.

10. Sally G. McMillen, *Seneca Falls and the Origins of the Women's Rights Movement* (New York: Oxford University Press, 2008), 14. Even Adams's descendants were rebuked for their attitudes toward women; see Henry Adams, *The Education of Henry Adams* (New York: Random House, 1999), 442.

11. Mary Wollstonecraft, *A Vindication of the Rights of Woman*, 2nd. ed., ed. Carol H. Poston (New York: W.W. Norton, 1988). This was actually the second "vindication" Wollstonecraft wrote. The first, *A Vindication of the Rights of Man* (1790), was written in response to Edmund Burke's *Reflections on the Revolution in France* (1789), which attacked her circle of friends and her patron. It was forgotten in the controversy caused by one of these friends, Thomas Paine, with his *Rights of Man* (1791). *A Vindication of the Rights of Woman* was widely read in America around the middle of the nineteenth century; Lucretia Mott called it "my pet book" (McMillen, *Seneca Falls*, 32); see Carol Hymowitz and Michaele Weissman, *A History of Women in America* (New York: Bantam, 1978), 77.

12. Melder, *Beginnings of Sisterhood*, 1–2.

13. Plato wrote of having a dream of a society where all intelligent beings had equal rights, as noted in Eugene A. Hecker, *A Short History of Women's Rights* (Westport, CT: Greenwood Press, 1914), 236.

14. Busch, *In the Eyes of the Law*, 18; Hecker (*A Short History of Women's Rights*, 121), notes that the common law evolved toward more male dominance in the seventeenth and eighteenth centuries.

15. In special circumstances, such as being of age and unmarried, or widowed, a woman might be granted the status of *feme sole*, which allowed her to function in society; see ibid., 26–7.

16. Rabkin, "Silent Feminist Revolution," 31.

17. The idea that a sanctified marriage resulted in a man and woman becoming "one flesh" was translated into medieval law as a merger of the woman's rights and obligations into those of the man, since somebody in this union had to be responsible for its behavior.

18. Some colonies provided for joint deeds of land ownership and recognized women as business owners/operators in some areas of endeavor, such as taverns, inns, and shops. In 1787, Massachusetts allowed abandoned wives legally to become "sole traders" without requiring a special petition from the legislature, in part because of the hazards to men of the rising seaborne trade. Many States also shortened the time until a man could be presumed dead; see Busch, *In the Eyes of the Law*, 24–5.

19. As happened to Benjamin Franklin, leading him to escape from his native Boston to Philadelphia.

20. Busch, *In the Eyes of the Law*, 170.

21. Blackstone might even have distorted the common law in this regard out of personal feelings; such was the claim of Mary Beard in her *Woman as a Force in History: A Study in Transitions and Realities* (New York: Macmillan, 1946), according to Rabkin, "Silent Feminist Revolution."

22. Busch, *In the Eyes of the Law*, 40–9. Tappan Reeve founded the first law school in Litchfield, Connecticut, in 1784, and used a somewhat modified version of Blackstone in his teachings on marriage. In an 1816 publication, he upheld Blackstone's interpretation of marriage, though he questioned certain parts (57–8).

23. New York abolished property qualifications for suffrage in 1826 (ibid., 121).

24. Sara M. Evans, *Born for Liberty; A History of Women in America* (New York: Free Press, 1997), 72.

25. Frost and Cullen-DuPont, *Women's Suffrage in America*, 83.

26. Johnson, *History of the American People*, 671.

27. McMillen, *Seneca Falls*, 52.

28. Johnson, *History of the American People*, 402.

29. Melder, *Beginnings of Sisterhood*, 148.

30. Cullen, *Art of Democracy*, 85.

31. Temin, "Industrialization of New England," 115.

32. Melder, *Beginnings of Sisterhood*, 3.

33. W. Elliott Brownlee and Mary M. Brownlee, *Women in the American Economy* (New Haven, CT: Yale University Press, 1976), 3. A perceptive article that shows how *cov-*

erture and the specie shortage in the early nineteenth century worked to underrepresent both the numbers of women involved in the labor force and to misrepresent the value of their work is Margaret Coleman, "Homemaker as Worker in the United States," *Challenge* 41 (6, 1998): 75–87.

34. McPherson, *Battle Cry of Freedom*, 33, 36. Some three-fourths of Massachusetts teachers were women by that time as well.

35. Even though some men felt that women should be restricted to domestic life in any case, the needs of a growing economy, especially in the cities, led many others to sponsor charities to help working women find jobs and provide them some training.

36. Busch, *In the Eyes of the Law*, 141. As late as the 1850s, very few postsecondary institutions admitted women; Oberlin and Antioch colleges in Ohio were perhaps the best known.

37. Evans, *Born for Liberty*, 4.

38. Wish, *Society and Thought in America*, 123; Wershler-Henry, *Iron Whim*, 85–7. In the late 1800s, the typewriter was seen as a means of female liberation.

39. Frost and Cullen-DuPont, *Women's Suffrage in America*, 60.

40. A prominent feminist, Abby Kelley, noted that the "petition is the only mode of access which the women of this country have to Congress" (quoted in Melder, *Beginnings of Sisterhood*, 98).

41. Nancy A. Hewitt, *Women's Activism and Social Change: Rochester, New York, 1822–1872* (Ithaca, NY: Cornell University Press, 1984), 41.

42. These persisted over time; I recall my mother going off to her "sewing club" in the 1950s. Sewing was always connected to women's reading and discussion groups; see Debra Gold Hansen, *Strained Sisterhood: Gender and Class in the Boston Female Anti-Slavery Society* (Amherst: University of Massachusetts Press, 1993), 51. In Warren, Massachusetts, the local literary circle suggested in 1840 that "ladies ought to mingle in politics, go to Congress, etc.," which was not what the clergy or town leaders wanted to hear. These societies also sometimes invited female lecturers. See Nancy Isenberg, *Sex and Citizenship in Antebellum America* (Chapel Hill: University of North Carolina Press, 1998), 60.

43. Melder, *Beginnings of Sisterhood*, 70. It made $300 for the cause; see Hansen, *Strained Sisterhood*, 14.

44. Melder, *Beginnings of Sisterhood*, 119.

45. One scriptural piece used to justify equal rights is from Paul's Letter to the Galatians (Gal. 3:28): "There was neither bond nor free, neither male nor female, for all were one in Christ Jesus"; see Isenberg, *Sex and Citizenship in Antebellum America*, 10.

46. The converts in the First Great Awakening were primarily men, but the Second appealed mainly to women (Evans, *Born for Liberty*, 65). Hewitt (*Women's Activism and Social Change*, 17–22) notes the heavy Yankee involvement in the women's voluntary groups of the time.

47. Evans, *Born for Liberty*, 67; see also Isenberg, *Sex and Citizenship in Antebellum America*, 59.

48. Johnson, *History of the American People*, 672.

49. As early as 1766, "alienation of affection" began to appear as a reason for petitioning for a divorce (Evans, *Born for Liberty*, 42).

50. Quoted in Ralph F. Young, ed., *Dissent in America: Voices That Shaped a Nation* (New York: Pearson Longman, 2006), 205.

51. See the whole of Hansen, *Strained Sisterhood*, but esp. 7, 118–19.

52. McMillen, *Seneca Falls*, 138.

53. James Fenimore Cooper saw feminine subordination as a critical part of building a conservative, hierarchical society in America; he was destined to see his world evolve in a different way.

54. Rabkin, "Silent Feminist Revolution," 16, 32, 34, 40, 48.

55. Ibid., 158. The existing equity rules were complex and subject to the perceptions of individual judges. As well, they allowed land, but not financial instruments, such as bonds and cash, to be protected, so many of the wealthy who were not associated with the land found the equity rules deficient in any case.

56. Other parts of the *coverture* relationship did not change, but what did change was due to the rise of a commercial system and the needs of the middle-class people involved in it (Busch, *In the Eyes of the Law*, 38, 137). Rabkin ("Silent Feminist Revolution," 147) notes that the legal code that began to change the system as early as 1827 defeudalized business relationships but left the marriage hierarchy untouched for the next twenty years.

57. An act relating to this problem was passed in 1840 by the New York legislature, which had the effect of stimulating this new industry (Busch, *In the Eyes of the Law*, 137).

58. A savings bank law was passed in 1850, and an 1851 law allowed women who owned shares in a company to vote them on her own account (ibid., 160).

59. Rabkin ("Silent Feminist Revolution") presents a detailed legal history of these reforms. A more popular version of the forces leading up to the act can be found in Burrows and Wallace, *Gotham*, 817–20. Mississippi passed the first such statute in 1839, largely at the behest of plantation owners who worried about wastrel sons-in-law. Michigan exempted a wife's property from her husband's debts in 1844, and Massachusetts limted a wife's separate estate to the provisions of an antenuptial (or "prenuptial," in modern-day parlance) agreement (Busch, *In the Eyes of the Law*, 27).

60. One of its proposers was a yankee State assemblyman, Ansel Bascom, who represented the Seneca Falls area. He and his daughter attended the Seneca Falls Convention (Busch, *In the Eyes of the Law*, 168).

61. Rabkin, "Silent Feminist Revolution," 172, 202, 230; see also Busch, *In the Eyes of the Law*, 156. In later years, a number of amendments broadened its intent.

62. Busch, *In the Eyes of the Law*, 161.

63. Ibid., 200–3, 206. Legislatures also curtailed the reach of some acts, as the New York Legislature did in 1862 relative to the 1860 act. Many jurists felt it was necessary for society to preserve marital unity at all costs, even if this meant turning a blind eye to abuses and injustices; see Isenberg, *Sex and Citizenship in Antebellum America*, 162.

64. Rabkin, "Silent Feminist Revolution," 249. Property rights "resurrected" women after being "legally dead" in Blackstone"s interpretation.

65. This, more than other reforms, would imply that the wife had a separate identity from that of the husband.

66. Rabkin, "Silent Feminist Revolution," 231–2.

67. Busch, *In the Eyes of the Law*, 28, 165, 187, 194–5. By the end of the Civil War, twenty-nine States had passed some form of married women's property act.

68. Giving immigrant males the right to vote while denying it to native-born women was galling, but then to give the vote to ex-slaves, who were not seen as on a par even with immigrants, enraged feminists. Although most feminists were abolitionists, they, like the broader Northern society, did not necessarily support equal rights for blacks, as witnessed by the discriminatory laws against free African Americans even in Northern States.

69. Busch (*In the Eyes of the Law*, 171) notes that what was new after 1848 was not the analysis, but the organized political milieu of the women's rights campaign.

70. In 1921, the so-called Portrait Monument honoring these three women was unveiled in the Capitol Rotunda. Later, the monument was relegated to the Capitol Crypt and forgotten until relocated in 1997 to a visible and honorable place.

71. Quakers settled on Nantucket in the late 1600s. It was considered second only to Philadelphia as a Quaker center; see Lloyd C.M. Hare, *The Greatest American Woman: Lucretia Mott* (New York: American Historical Society, 1937), 25. Burnett (*Five for Freedom*, 13–48) provides a short biography, from which some of the details in this section are taken. See also Margaret F. Bacon, *Valiant Friend: A Life of Lucretia Mott* (Philadelphia: Quaker Books, 1999); and Anna Davis Hallowell, *James and Lucretia Mott: Life and Letters* (Boston: Houghton Mifflin, 1884).

72. "In the same sense in which the greatest man ever produced in this country was Benjamin Franklin [her distant cousin], the greatest woman ever produced in the country is Lucretia Mott"; quote attributed to nineteenth-century newspaper editor Theodore Tilton in Hare, *Greatest American Woman*, intro.

73. Her mother's uncle, Captain Timothy Folger was the first to chart the course of the Gulf Stream, at the request of his cousin, Benjamin Franklin, and her mother's brother, Captain Mayhew Folger, found the *Bounty* mutineers on Pitcairn Island (ibid., 33). Evans (*Born for Liberty*, 28) notes that these early economic activities by American women have remained largely undocumented and "invisible" to historians.

74. The Nine Partners referred to a group of speculators who had gained a British patent to land in Dutchess County, now almost a New York City suburb up the Hudson River. Mott noted in later years that her tuition was the same as that of the boys, but she was given less schooling, perhaps the first awakening of the unequal status of women in that time (Burnett, *Five for Freedom*, 23). The school had been built just over the border from Connecticut by Quakers from Nantucket and Cape Cod (Hare, *Greatest American Woman*, 35).

75. She kept a copy of Mary Wollstonecraft's *Vindication* "front and center" in her home for forty years and would lend it to anyone she could convince to read it (Cromwell, *Lucretia Mott*, 28–9).

76. Her husband conducted most of the family correspondence, except for her letters to her sister (Hare, *Greatest American Woman*, 96).

77. Ibid., 47.

78. In the late 1820s, the Motts joined the Free Produce Movement, which boycotted products produced by slave labor, including cotton and sugar. James Mott also

dropped slave-produced products from his wholesale business, even though this hurt it. See Cromwell, *Lucretia Mott*, 45-6, 47; and Hare, *Greatest American Woman*, 77.

79. Women's use of the petition appears to date to the mid-1830s; by the 1850s, they were using it in a sophisticated way (Busch, *In the Eyes of the Law*, 189–92).

80. Anti-feminist abolitionist James Birney called her "the Dangerous Mrs. Mott" (Hare, *Greatest American Woman*, 125).

81. Melder, *Beginnings of Sisterhood*, 113. On the three-week voyage to Europe, Lucretia spent much time down in steerage helping seasick mothers attend to their ill children (Cromwell, *Lucretia Mott*, 77).

82. Cromwell, *Lucretia Mott*, 74. There is the apocryphal story of another attendee, Wendell Phillips, who, as he left his boardinghouse in London, was told by his wife, "Wendell, don't shilly-shally." At the convention he delivered what is said to have been the first speech by a man in defense of women's rights; see Cleveland Amory, *The Proper Bostonians* (Orleans, MA: Parnassus Imprints, 1984), 104–5.

83. Hare, *Greatest American Woman*, 34.

84. Ibid., 163–4; and Melder, *Beginnings of Sisterhood*, 125.

85. Cromwell, *Lucretia Mott*, 116.

86. Hare, *Greatest American Woman*, 185. One of Mott's Philadelphia friends was Emily Winslow Taylor, the mother of Frederick Winslow Taylor, the "father" of scientific management.

87. Hewitt (*Women's Activism and Social Change*, 36), that fifteen to twenty Hicksite families had migrated into the Rochester area in the early 1840s, providing a core radical group

88. A number of such conventions held in places west of the Alleghenies at the time, but since Stanton's Declaration of Sentiments became the guiding light of the movement, pride of place as the first national convention went to Seneca Falls (McMillen, *Seneca Falls*, 71–2).

89. Hewitt, *Women's Activism and Social Change*, 131. She quotes the *Rochester Daily Democrat* as saying of this meeting: "[The] congregation of females…seem to be really in earnest in their aim at revolution…verily, this is a progressive era!" (134).

90. Burnett (*Five for Freedom*, 49—128) provides a short biography from which some of the details in this section are taken; see also Jean H. Baker, *Sisters: The Lives of America's Suffragists* (New York: Hill and Wang, 2005), 94–129.

91. Melder (*Beginnings of Sisterhood*, 15–17) has a short description of the school. It was an early example of government aid for women's education, as Governor DeWitt Clinton used public funds to attract Willard to Troy to set up her school (Hecker, *Short History of Women's Rights*, 169). McMillen (*Seneca Falls*, 45) says the legislature refused the funds, so the town of Troy made the offer. Between 1821 and 1872, the Troy Women's Seminary educated over twelve thousand women (Frost and Cullen-DuPont, *Women's Suffrage in America*, 4).

92. Hymowitz and Weissman, *History of Women in America*, 90. Prior to the founding of Cornell University in 1868, no New York college admitted women (Klein, *Empire State*, 280).

93. In 1852, Smith noted that "political rights are not conventional, but natural —

inhering in all persons, the black as well as the white, the female as well as the male" (Busch, *In the Eyes of the Law*, 171).

94. Rabkin, "Silent Feminist Revolution," 9. The first woman lawyer in America practiced in Baltimore in 1647, while the second, 222 years later, was Arabella Babb Mansfield, in Iowa in 1869, who gained a license but never practiced. By 1879, women could plead in front of the US Supreme Court (Hecker, *Short History of Women's Rights*, 179).

95. Stanton visited the Brook Farm commune just outside Boston to see a supposed example of gender equality (McMillen, *Seneca Falls*, 41).

96. Frost and Cullen-DuPont, *Women's Suffrage in America*, 86.

97. Stanton claimed in later years that her father said, "when you are grown up, you must go down to Albany and talk to the legislators….[I]f you can persuade them to pass new laws, the old ones will be a dead letter"; quoted in Rabkin, "Silent Feminist Revolution," 224.

98. Busch, *In the Eyes of the Law*, 137, 167. Women had been petitioning for such an act since 1839; see Rabkin, "Silent Feminist Revolution," 185, who also suggests (226) that Stanton had been involved in the campaign for the act since the early 1840s.

99. William Leach, *True Love and Perfect Union: The Feminist Reform of Sex and Society* (Middletown, CT: Wesleyan University Press, 1980), xiv.

100. Busch, *In the Eyes of the Law*, 141.

101. Burnett, *Five for Freedom*, 65.

102. Leach, *True Love and Perfect Union*, 143.

103. She had a precedent in that a petition by forty-four women from Wyoming and Genesee counties, adjacent to Seneca Falls, to the legislature four months earlier had made a similar reference to the Declaration of Independence (Busch, *In the Eyes of the Law*, 156).

104. Hecker (*Short History of Women's Rights*, 289) notes that Kentucky had given school suffrage to widows with school-age children in 1838; Upper Canada (now Ontario) did likewise in 1850.

105. Baker, *Sisters*, 6.

106. See Smithsoniam Institution, National Portrait Gallery, "The Seneca Falls Convention, July 19–20, 1848" (Washington, DC); available online at http://www.npg.si.edu/col/seneca/senfalls1.htm.

107. McMillen, *Seneca Falls*, 106, 113. Detractors would call the annual conventions "petticoat parliaments"; see Isenberg, *Sex and Citizenship in Antebellum America*, 32, for details of this convention.

108. Hymowitz and Weissman, *History of Women in America*, 117.

109. Leach, *True Love and Perfect Union*, 52.

110. See Burnett, *Five for Freedom*, 177–256, for a short biography, from which some of the details in this section are taken.

111. The dual standard was justified by claiming that women were working only until they got married, while male teachers had other opportunities (Melder, *Beginnings of Sisterhood*, 26).

112. Busch (*In the Eyes of the Law*, 174) notes that, in the 1850s, the temperance struggle seemed to be failing and many of its activists had switched to promoting women's rights.

113. Leach, *True Love and Perfect Union*, xvii, xix.

114. Rabkin, "Silent Feminist Revolution," 317. The Equal Rights Amendment was proposed in 1923, but not passed by Congress until 1972; by 1982 only thirty-five of the needed thirty-eight States had passed it, with five of these having rescinded their agreement, and the effort disappeared.

115. See Dee Brown, *The American West* (New York: Charles Scribner's Sons, 1994), 392–3; Dayton Duncan, *Miles from Nowhere: In Search of the American Frontier*, 2nd. ed. (New York: Penguin, 1994), 193–4; and Ted Morgan, *A Shovel of Stars: The Making of the American West, 1800 to the Present* (New York: Simon & Schuster, 1995), 355.

Chapter 16

1. Andrews, *People Called Shakers*, xiii.

2. Klees, *Crucible of Ferment*, 127.

3. See, for example, Marc Reisner, *Cadillac Desert: The American West and Its Disappearing Water* (1846; rev. ed. New York: Penguin, 1993), p.2.

4. Like other communal societies to come, the Plymouth colonists believed in the perfectability of mankind; see William M. Kephart, *Extraordinary Groups: The Sociology of Unconventional Life-Styles* (New York: St. Martin's Press, 1976), 285.

5. Ziff, *Puritanism in America*, 37–9. The failure of most of these communal experiments was first and foremost an economic one: if the colony could not turn a profit — that is, produce more than it consumed — it failed; see Kephart, *Extraordinary Groups*, 292, 294.

6. Ziff, *Puritanism in America*, 37–40. Equality also arose from the notion that each person was a priest unto himself and equal to all others. A leader was accepted because of his superior learning and wisdom, not because he was special in the eyes of God (279).

7. Robert V. Hine, *Community on the American Frontier: Separate but Not Alone* (Norman: University of Oklahoma Press, 1980), 184–5. Success first depended on the colonies' finding ways to earn money to sustain themselves, at which they sometimes became so successful that it threatened their future (Klees, *Crucible of Ferment*, 132).

8. Andrews, *People Called Shakers*, 5.

9. Bloom, *American Religion*, 67.

10. Ibid., 13.

11. A second party came in 1775; see John L. Scherer, ed., *A Shaker Legacy: The Shaker Collection at the New York State Museum*, rev. ed. (Albany: New York State Education Department, 2000), 1.

12. Andrews, *People Called Shakers*, 15; see also Klees, *Crucible of Ferment*, 143.

13. Andrews, *People Called Shakers*, 17.

14. Kephart, *Extraordinary Groups*, 161. Her name was Margaret Vedder (Scherer, *Shaker Legacy*, 1).

15. Nevertheless, females tended to outnumber males by two to one among Shaker members (Kephart, *Extraordinary Groups*, 165).

16. Andrews, *People Called Shakers*, 18–20; Klees, *Crucible of Ferment*, 140.

17. Andrews, *People Called Shakers*, 32–33.

18. Ibid., 35–8.

19. Most of the traditional Congregationalists in New England regarded Shakers as extreme heretics; see ibid., 44, 52.

20. New Lebanon is just over the Massachusetts line at the foot of the Berkshires on US 20. The large Shaker village there was abandoned in 1947. Parts were taken over by a school and the rest by the Shaker Museum and Library,

21. Klees, *Crucible of Ferment*, 152.

22. Meacham died in 1796, but Mother Lucy carried on into the 1820s.

23. Andrews, *People Called Shakers*, 58.

24. Ibid., 60, quoting from a Shaker document. James Fenimore Cooper stated he had not seen "in any country, villages as neat, and so perfectly beautiful, as to order and arrangement, without, however, being picturesque or ornamented, as those of the Shakers" (127).

25. Ibid., 120–1.

26. Expansion into Ohio came as a result of the rift between New Light Presbyterians and the more orthodox Presbyterians (Hurt, *Ohio Frontier*, 291–3, 298). The Shaker colony of North Union, Ohio, later became Shaker Heights, a suburb of Cleveland, and is only a couple of miles south of US 20. In Kentucky, the revivalist movement, with its emotional emphasis, was close to the "dance" of the Shakers (Andrews, *People Called Shakers*, 138–9).

27. Apparently most people rotated through jobs on a monthly basis, to improve their versatility (Burns and Burns, *Shakers*, 80–5).

28. Kephart, *Extraordinary Groups*, 168, 175.

29. Ann Lee emphasized labor and thrift. "If you are not faithful in the unrighteous mammon, how can you expect the true riches?"; quoted in Andrews, *People Called Shakers*, 24.

30. One result was that neighboring towns began to give the Shakers their share of "school money" for educating children (ibid., 66).

31. The text has been digitized and is available online at http://books.google.com/books?id=zPI2AAAAMAAJ&pg=PA1&source=gbs_toc_r&cad=4#v=onepage&q&f=false.

32. An 1847 reprint of this publication has been digitized and is available online at http://books.google.com/books?id=C1goAAAAYAAJ&printsec=frontcover&source=gbs_ge_summary_r&cad=0#v=onepage&q&f=false.

33. Andrews, *People Called Shakers*, 99–100.

34. The Shaker communities not only practiced celibacy, but planned the near-total segregation of the sexes in daily life (Kephart, *Extraordinary Groups*, 176).

35. In 1784, New York provided for the "free incorporation" of religious societies, which allowed them to apply for and receive money if they provided a school. New England provided similar rights for "dissenting" sects, which were colored by the continuing establishment of the Congregational church. See Curtis D. Johnson, *Islands of Holiness: Rural Religion in Upstate New York, 1790–1860* (Ithaca, NY: Cornell University Press, 1989), 16–17.

36. Andrews, *People Called Shakers*, 189.

37. Kephart, *Extraordinary Groups*, 172.

38. She said, "You must be kind to strangers, for that is the way that you can reward me" (Andrews, *People Called Shakers*, 24).

39. Pierson, *Tocqueville in America*, 178–9.

40. Apparently, there were few resignations, in part due to the careful process of selection of adherents (Kephart, *Extraordinary Groups*, 175).

41. Only one in ten children remained in the colonies after they grew up, which gives some indication of the problem (ibid., 165–6, 183).

42. Burns and Burns, *Shakers*, 119.

43. Andrews, *People Called Shakers*, 224. Another source puts their peak at nineteen colonies containing seventeen thousand adherents (Kephart, *Extraordinary Groups*, 164). See also "The Last Seven Shakers in the World," *Economist*, February 13, 1999, 11. The last male Shaker died in 1961 and the society decided to close its admissions shortly thereafter (Burns and Burns, *Shakers*, 120).

44. Klees, *Crucible of Ferment*, 77. Virtually all of the early Mormon leadership and membership were of Yankee origins; see Leonard J. Arrington, *Great Basin Kingdom: An Economic History of the Latter-day Saints, 1830–1900* (1958; repr. Cambridge, MA: Harvard University Press, 1993), 3.

45. See, for instance, Klees, *People of the Finger Lakes Region*, 191; and idem, *Crucible of Ferment*, 75–80. William Least Heat-Moon describes his reaction to the site in *Blue Highways: A Journey into America* (Boston: Little, Brown, 1999), 317.

46. Kephart, *Extraordinary Groups*, 196. There has been much controversy over its provenance, but, regardless of its inspiration, it was quite a feat for a basically unschooled young man to have dictated to his wife and some associates more than five hundred pages of a text done in a scriptural style. The book was published in an edition of five thousand copies in 1830; see Morgan, *Shovel of Stars*, 379–80.

47. Ralph Waldo Emerson, visiting Utah in 1871, saw them as an "afterclap of Puritanism" (Bloom, *American Religion*, 53).

48. A theologically contemptuous Bernard DeVoto (*Year of Decision*, 82–3) lists some of the parallels, which is a useful list if taken out of context.

49. Kephart, *Extraordinary Groups*, 216.

50. Ziff, *Puritanism in America*, 279.

51. The experiment with communistic organization was suspended early on, but a lower degree of communal activity and group settlement did make a success out of the settlements in Missouri, Illinois, and Utah.

52. There are suggestions that some Shaker influence was behind Smith's ideas to remove his people from physical integration with others (Andrews, *People Called Shakers*, 222–3).

53. A number of reasons have been advanced for this practice, including promoting a higher birth rate, an excess of females over males in the church, and so forth. None really stands up to examination, and the conclusion is that it arose from the belief that God wanted it. The rules governing polygamy were well defined; it was not simply a frivolous, personal decision by the husband (Kephart, *Extraordinary Groups*, 206–15). The proportion of polygamous families was always small, and each wife had an individual home with her children (Wish, *Society and Thought in America*, 82).

54. It gained a thousand members in its first year (Kephart, *Extraordinary Groups*, 201).

55. An interesting tale of one Mormon woman's life from the 1830s until teh creation of Deseret is found in Morgan, *Shovel of Stars*, 379–88.

56. Billington, *Westward Expansion*, 455–7.

57. Smith's announcement of a revelation that encouraged polygamy incensed the Mormons' neighbors, not just because it offended local religious tradition and custom, but also because of the more practical gender imbalance on the frontier (DeVoto, *Year of Decision*, 82).

58. Young started his working life in his teens as a painter and carpenter (Kephart, *Extraordinary Groups*, 202). When he was fifteen, he worked on William Seward's house in Auburn, New York, when he was 15 (C-Span and Bentzel, *C-SPAN's Traveling Tocqueville's America*, 40).

59. Ray Allen Billington, *The Far Western Frontier, 1830–1860* (New York: Harper, 1956), 195–7. Young had known in advance where the new colony would be located.

60. DeVoto, *Year of Decision*, 449.

61. By the fall of that year, there were 1,800 people in the valley (Billington, *Far Western Frontier*, 200).

62. See DeVoto, *Year of Decision*, 468. A compromise on creating the Utah Territory left the Mormon leadership largely in place (Billington, *Far Western Frontier*, 205). The original borders proposed by the Mormons took in nearly all of Utah, Nevada, and Arizona, plus portions of Oregon, Idaho, Wyoming, and Colorado (Duncan, *Miles from Nowhere*, 110). Congress continually cut the area down before 1896, when Utah was granted statehood.

63. Billington, *Far Western Frontier*, 202; see also Klees, *Crucible of Ferment*, 75, 88–90.

64. Town lots and farmland in the outskirts were allocated on the same basis as in Puritan New England in the 1600s and early 1700s (Hine, *Community on the American Frontier*, 211).

65. Billington, *Far Western Frontier*, 201–2.

66. Billington, *Westward Expansion*, 459–61.

67. Bloom, *American Religion*, 83.

68. Arrington, *Great Basin Kingdom*, 174.

69. Charles P. Roland, *Albert Sidney Johnston: Soldier of Three Republics* (Lexington: University Press of Kentucky, 2001), 204.

70. Arrington, *Great Basin Kingdom*, 83–4, 112–3, 133, 196.

71. Despite the arrival of so many non-Mormons, there was to be no relenting on the communal aspects of the society, an issue over which there was an internal Mormon struggle in 1869 (Arrington, *Great Basin Kingdom*, 233–4).

72. Klees, *Crucible of Ferment*, 88.

73. The 1862 anti-bigamy act was designed as well to prevent the church-business relationship from growing. As Arrington notes (*Great Basin Kingdom*, 257–8), "[t]he particular clause of the 1862 act which created difficulty was one which incorporated the Church of Jesus Christ of Latter-day Saints and limited the amount of real estate which it could hold to $50,000. Any property acquired by the church in excess of that value was

to be 'forfeited and escheat to the United States'. The object of this section, according to Senator Bayard of Delaware who sponsored it, was to prevent the accumulation, in Utah, of wealth and property in the hands of ecclesiastical corporations or "theocratic institutions inconsistent with our form of government."

74. Klees, *Crucible of Ferment*, 90.

75. Arrington, *Great Basin Kingdom*, 313, 321–2, 338.

76. Because the party in power in Washington was Republican, most Mormons opted to support that party; voting patterns suggest the Mormon-Gentile division remains today.

77. Seventy percent of Utah's population remains Mormon today. In 2002, all of its representatives in Congress, 90 percent of the members of the State legislature, all of its Supreme Court justices, and 80 percent of its state and federal judges were also Mormon; see "The Church of the West," *Economist*, February 9, 2002, 25. I doubt there has been much change to these proportions in the past decade.

78. Kephart, *Extraordinary Groups*, 235–6.

79. "Editorial notice in the Dial in 1840 by Elizabeth Palmer Peabody," in *The Brook Farm Book: A Collection of First-Hand Accounts of the Community*, ed. Joel Myerson (New York: Garland, 1987), 302–3.

80. Hine (*Community on the American Frontier*, 201–2) notes that, of ninety-two such experiments before the Civil War, only two were planted across the Mississippi; most were attempted on the frontier of the day, making Brook Farm something of an oddity.

81. Eventually the number grew to thirty-two. Thoreau, when approached, refused, claiming to be a "community of one" (Maynard, *Walden Pond*, 63).

82. Amory, *Proper Bostonians*, 103.

83. If the aim was to show a way of life superior to the industrial society of the time, then management had to be better than that "outside." It wasn't. See DeVoto, *Year of Decision*, 104.

84. Hawthorne was an original shareholder and participant in Brook Farm, and lasted over a year before leaving. At one point, he was effectively its chief financial officer. He sued to get his money back (ibid., 33).

85. Ibid.

86. A number of Fourier's "Phalansteries" were created in North America after his death in 1837, but almost none lasted long.

87. One, Isaac Hecker, became a Roman Catholic and in 1858 was the founder of the first American-based priestly order, the Paulists.

88. DeVoto, *Year of Decision*, 103.

89. Barbara L. Packer, *The Transcendentalists* (Athens GA: The University of Georgia Press, 2007, p.161.

90. The paper's editor, Horace Greeley, was a long-time advocate of Fourier's "associationism" ideas.

91. See Robert Twombly, ed., *Frederick Law Olmstead: Essential Texts* (New York: W.W. Norton, 2010), chap. 2.

92. Ibid., 33.

93. Noyes's mother was a relative of President Rutherford B. Hayes; his father was

a successful businessman and Vermont congressman (Kephart, *Extraordinary Groups*, 52, 252).

94. The term was applied to those ideas that claimed that faith essentially trumped human or church law. Perfectionism did not start with Noyes, but his approach to it was novel; see ibid., 53–4.

95. Lawrence Foster, *Religion and Sexuality* (New York: Oxford University Press, 1981), 77–9. He had received his license in 1833 (Kephart, *Extraordinary Groups*, 53).

96. Holbrook, *Old Post Road*, 131; see also Foster, *Religion and Sexuality*, 79–81.

97. Kephart, *Extraordinary Groups*, 54.

98. Ibid., 53–6.

99. He claimed he had done it to spare his followers from mob violence (ibid., 55).

100. Eventually, they acquired about six hundred acres (ibid., 56). Klein (*Empire State*, 339) believes it to be "one of the most successful communitarian experiment in American history."

101. Kephart, *Extraordinary Groups*, 67, 73.

102. Ibid., 74–5.

103. Klees, *Crucible of Ferment*, 132–3.

104. Traveler Charles Nordhoff noted that the colony also employed people from outside to help with its domestic tasks; see Charles Nordhoff, *The Communistic Societies of the United States* (1875; New York: Schocken, 1965), 286.

105. Noyes, as God's representative on Earth, played a dominant but not sole role in the larger managerial discussions. (Kephart, *Extraordinary Groups*, 61–3).

106. Nordhoff, *Communistic Societies*, 285. Nordhoff gives a good account of the daily life of the community.

107. Noyes became familiar with experiments in plant and animal breeding and Darwinian selection theory (Kephart, *Extraordinary Groups*, 89).

108. In 1873, they received over two hundred applications to join (ibid., 66).

109. Ibid., 59–60.

110. Ibid., 64.

111. Ibid., 58.

112. Ibid., 98.

113. The value of the colony and its factory was estimated to be $600,000 in 1881 (ibid., 77); see also Klees, *Crucible of Ferment*, 135.

114. Klees, *Crucible of Ferment*, 179–82. Since 2000, the company has struggled to survive globalization and other shocks; see Chana R. Schoenberger, "Tarnished," *Forbes*, March 15, 2004, 80–1.

115. Kephart, *Extraordinary Groups*, 292–5.

116. See Foster, *Religion and Sexuality*, 15–16.

117. Quoted in Andrews, *People Called Shakers*, xiii.

Chapter 17

1. Tom Standage, *The Victorian Internet: The Remarkable Story of the Telegraph and the Nineteenth Century's On-line Pioneers* (London: Phoenix, 1999), 96. Steam engines

were used to create air pressure variations in pneumatic tubes that ran from the telegraph office underground to large users downtown.

2. Oslin, *Story of Telecommunications*, 230.

3. Katie Hafner and Matthew Lyon, *Where Wizards Stay Up Late: The Origins of the Internet* (New York: Simon & Schuster Paperbacks, 2006), 192.

4. Numerous websites are devoted to the history of Rochester history; for a basic timeline, see http://www.cityofrochester.gov/app.aspx?id=8589937391.

5. Details of Reynolds's life can be found in M.B. Anderson and D.K. Bartlett, "Address and Sermon on the Death of William Abelard Reynolds with other memorial papers" (Rochester, 1872); available online at http://books.google.ca/books?id=03YPAAAAYAAJ&dq=Abelard+reynolds&printsec=frontcover&source=bl&ots=FcsGOyScBg&sig=XBS03U8UYqBIEspttmvt_h5XdhI&hl=en&ei=SDTnSpnJKcbDlAenyb3_Bw&sa=X&oi=book_result&ct=result&resnum=6&ved=0CBoQ6AEwBQ#v=onepage&q=&f=false.

6. Rochester was a classic boom town, growing from 1,500 in 1820 to 5,300 in 1825 and to 162,000 by the end of the century. In 1817, Rochester flour mills exported 26,000 barrels of wheat, mostly to Canada. In 1827, they sent 200,000 barrels of flour along the Erie Canal. By the late 1830s, twenty-four mills were shipping a half-million barrels east, and Rochester was the largest producer of wheat flour in the world at the time. See Koeppel, *Bond of Union*, 372.

7. For the Arcade's story, see Monroe County Library System, "The Reynolds Arcade" (Rochester, NY, n.d.); available online at http://www.rochester.lib.ny.us/rochimag/architecture/LostRochester/Reynolds/Reynolds.htm.

8. Elizabeth Brayer, *George Eastman: A Biography* (1996; Rochester, NY: University of Rochester Press, 2006), 19.

9. The precedent for this approach is Standage, *Victorian Internet*, introduction.

10. Ibid., 199–200.

11. A good biography of Morse is Kenneth Silverman, *Lightning Man: The Accursed Life of Samuel F.B. Morse* (New York: Alfred Knopf, 2003).

12. Robert L. Thompson, *Wiring a Continent: The History of the Telegraph Industry in the United States, 1832–1866* (New York: Arno Press, 1972), 5. His ancestors arrived in Newburyport in 1635.

13. Ibid., 6. Many years later, Silliman's son would be the Yale professor who verified the theories of some Connecticut businessmen that petroleum, or "rock oil" could be processed to make kerosene, a lighting substitute for increasingly hard-to-get "whale oil."

14. Ibid., 7.

15. As early as 1822, Morse recognized, in a letter to his wife, that the opening of the Erie Canal would bring wealth and a greater demand for portraits in New York City than anywhere else (Koeppel, *Bond of Union*, 309–10). In 1827, while living in New York, he befriended a Dartmouth College professor, James Freeman Dana, who, in 1825, had published a set of lectures entitled "Epitome of Chymical Philosophy." Dana was in New York lecturing on chemistry, including electricity, and he and Morse spent much time discussing the subject (Thompson, *Wiring a Continent*, 7).

16. At the time, a packet might take as long as a month to make the crossing, so there was time to discuss ideas and think about them; see K.G. Beauchamp and Institution of

Electrical Engineers, *History of Telegraphy* (London: Institution of Electrical Engineers, 2001), 51.

17. A detailed account of Morse's involvement in the development of the telegraph can be found in the text of a Memorial to S.F.B. Morse published by the Boston City council in 1972; available online at http://www.archive.org/stream/amemorialsamueloocoungoog/amemorialsamueloocoungoog_djvu.txt; seee also Oliver, History of American Technology, 218–19.

18. Oslin, Story of Telecommunications, 16.

19. Annteresa Lubrano, *The Telegraph: How Technology Innovation Caused Social Change* (New York: Garland Publishing, 1997), 6–7, notes that the "invention of the telegraph was the product of an accumulated store of knowledge." Closer to Morse personally was a Concord, Massachusetts, Yankee, Harrison Gray Dyar, who strung a line along the Lowell road out of Concord in 1826 as an experiment. Two years later, he strung a line several miles long from a New York racetrack that could carry signals marked on litmus paper. Finding himself in a controversy over his line, he left the country and made a career for himself as a chemist in France. Morse's brother-in-law, a lawyer, was Dyar's counsel at the time; see Oslin, *Story of Telecommunications*, 7–8; and Alfred Munroe, "Concord and the Telegraph" (paper presented to the Concord Antiquarian Society, Concord, MA, January 6, 1902), available online at http://www.archive.org/stream/preliminariesofc-ootolm/preliminariesofcootolm_djvu.txt.

20. Morse apparently knew little of others who had worked on parts of the telegraph system up to then, but Gale did and improved both his battery power source and his electromagnets. These allowed Morse to transmit messages for miles instead of feet (Thompson, *Wiring a Continent*, 9).

21. Ibid., 10–11.

22. Although claims after Vail's death were made that the code was his invention, both he and Morse had agreed that it was Morse's concept as early as his voyage home from Europe (Silverman, *Lightning Man*, 167, 434, 443). Brian Winston, in *Media Technology and Society: A History* (London: Routledge, 1998), 26, claims that Morse directed Vail to get the frequency distribution of letters from the printer and worked on the code revision himself. Beauchamp and Institution of Electrical Engineers (*History of Telegraphy*, 55) are more vague, noting that Morse originally began with his cipher system, but abandoned it after working with Vail. It would also seem that code was transmitted to a compass needle mechanism at first, as the buzzer or clicker announcing the dots and dashes was not invented until later (Crump, *Brief History of the Age of Steam*, 171). Beauchamp and Institution of Electrical Engineers (*History of Telegraphy*, 36) show a similar code developed in Germany by Carl August von Steinheil in 1836.

23. The government's interest was really in a proposal to build a semaphore system between Washington, DC, and New Orleans. Semaphore systems were in existence in the Boston and New York areas, and a national system had been proposed as a military aid in the years leading up to the War of 1812 but never implemented. Tensions with Mexico over Texas apparently were part of the appeal of this new proposal (Thompson, *Wiring a Continent*, 11–12, n18).

24. Hawke, *Nuts and Bolts of the Past*, 193; and Thompson, *Wiring a Continent*, 12–15.

Ever curious, Morse visited Louis Daguerre while in Paris and, upon his return, wrote the first American account of the new technology of photography.

25. The vote was 90 to 82, with 55 of the yeas coming from congressmen from "Yankee Road" States. The grant is a rare instance of the early federal government's supporting a technology demonstration (Hawke, *Nuts and Bolts of the Past*, 193).

26. The story has it that Cornell walked two and a half days from Boston to Portland, Maine, and, on entering the office of a small local newspaper, The Maine Farmer, he was greeted by its editor, "Fog" Smith, who was on his hands and knees trying to make a chalk outline of a plow to lay wire for the telegraph. He asked Cornell to design the plow, and Cornell went on to wealth and fame in the telegraph business. By 1860, Cornell had retired to become a gentleman-farmer near Ithaca, New York. He was then the largest shareholder in Western Union. He donated three hundred acres of land and $500,000 to endow the non-sectarian college that bears his name. See Walter P. Marshall, *Ezra Cornell, His Contributions to Western Union and to Cornell University* (New York: Newcomen Society in North America, 1951), 8–9. Morse, in fact, knew about the above-ground alternative, but feared that vandals would cut the lines (Silverman, *Lightning Man*, 222–3).

27. The phrase had been suggested to him by Annie Ellsworth, the daughter of his old friend, Commissioner of Patents H.L. Ellsworth (Oslin, *Story of Telecommunications*, 25, 32–3).

28. Ibid., 32–4.

29. Attempts to move information faster than animal power dated back thousands of years. Classical civilizations used fire signals and heliographs to warn of approaching ships and armies. Semaphore systems were in wide use by the military in Europe and America, with the French employing five hundred stations in the years before the telegraph; see Lubrano, *Telegraph*, 10; and Gordon, *Great Game*, 79. A British patent for an electric telegraph was issued in the year before Morse's in America (Lubrano, *Telegraph*, xiii, 6–9). A list of no fewer than forty-eight developers of telegraphic systems between 1753 and 1848 (including Morse) is found in Beauchamp and Institution of Electrical Engineers, *History of Telegraphy*, 27.

30. Thompson, *Wiring a Continent*, 42–7. Some of the first users of the line were the operators of a lottery in Philadelphia, who saw the telegraph as an efficient way to get their numbers sent to clients in New York City. Gambling was apparently the first "real-time" business to go on "the Victorian Internet." Stockbrokers were also early adopters.

31. Butterfield's father came from New Hampshire to the Albany area in the 1790s and married a local girl. John Butterfield went on to form the Butterfield Stage Lines in the West, later famous in the movies and, with Henry Wells and William Fargo, created the American Express Company to deliver packages on the railroads. His son, Civil War general Daniel Butterfield, is credited with composing the famous military salute, "Taps."

32. Standage, *Victorian Internet*, 55.

33. The postmaster general who had disparaged the invention now made an unsuccessful appeal to Congress to support the company for a subsidy for a line south to help improve military communications (Thompson, *Wiring a Continent*, 51–3).

34. For the story of this news syndicate and its battles with "Fog" Smith, see Lubrano, *Telegraph*, 41; and Thompson, *Wiring a Continent*, 217–39. Besides the news syndicate,

telegraph operators were encouraged to act as reporters along the lines, sending local news stories to regional papers (Beauchamp and Institution of Electrical Engineers, *History of Telegraphy*, 60).

35. The postmaster general also simply might have been politically opposed, as Clay had been a supporter of the telegraph, but had lost the 1844 presidential election to James K. Polk. As well, since Morse had already moved by then to create and finance new lines, there was a feeling in Congress that the competition would damage any public lines' profitability (Thompson, *Wiring a Continent*, 32–4). In 1868, Congress debated the possibility of either nationalizing Western Union or creating a government corporation to compete with it. After a great deal of controversy, the idea was shelved (Silverman, *Lightning Man*, 424–6).

36. Standage, *Victorian Internet*, 58, 60. It could be that the number of separate companies had more to do with the State regulations that governed incorporation than the desire of a multitude of entrepreneurs to get into the business. Both the railroads and Standard Oil ran into interstate activity and control problems from this legislative bias when dealing with national systems.

37. Thompson, *Wiring a Continent*, 240.

38. Ibid., 242. For another quote from this report, see Mazlish, *Railroad and the Space Program*, 95.

39. Mazlish, *Railroad and the Space Program*, 208.

40. These systems, including Morse's, employed keyboards that resembled the one Christopher Sholes developed for his typewriter twenty-five years later (Oliver, *History of American Technology*, 440).

41. One local businessman was Samuel Colt, who was busy stringing wire of his own manufacture around Long Island and into New Jersey until an order from the Army for a thousand of his revolvers for use in the Mexican War drew him into large-scale arms manufacturing (Thompson, *Wiring a Continent*, 90–2).

42. Ibid., 243.

43. Like many others in the post–Mexican War economy, they faced hard times, and O'Reilly had found alternative finance in Pennsylvania. He was then caught in a dispute between the two groups with claims to shares in his companies (ibid., 184–5).

44. Ibid., 264–5. Sibley worked in a textile mill as a youth, later setting up the mill's machine shop. In Rochester, he became rich in banking and real estate (Beauchamp and Institution of Electrical Engineers, *History of Telegraphy*, 66).

45. Oslin, *Story of Telecommunications*, 73.

46. Ibid., 75. Most of the following story comes from Thompson, *Wiring a Continent*, chaps. 16–18.

47. The House printing telegraph, which the Morse patentees fought from 1845 to 1848, was a formidable competitor (Thompson, *Wiring a Continent*, 54–5). This machine was faster than the human transcriber but also limited the range of the telegraph because of the electrical resistance caused by the machine.

48. Ibid., 270–5.

49. "Embraced" is perhaps too strong a word. The railroads gradually saw that the telegraph would allow them to use use their expensive capital stock more intensively

(Martin, *Railroads Triumphant*, 272), while telegraph operators found they could efficiently monitor their lines by using railroad crews to watch over them (Mazlish, *Railroad and the Space Program*, 95). The British were the first to realize that the two technologies were complementary, even though the first lines were built in both countries along railroad rights of way (Crump, *Brief History of the Age of Steam*, 171–2). American lines responded differently to the idea: as late as the 1870s, lines such as those emanating from Boston preferred to operate "by the book" — that is, with rigidly fixed schedules — rather than rely on telegraphers to coordinate things (Oliver, *History of American Technology*, 219–20; Martin, *Railroads Triumphant*, 271–4).

50. Thompson, *Wiring a Continent*, 275–7.

51. Cornell had worked for the Magnetic Telegraph Company, but left to go on his own as a result of disagreements with Vail. He then built the line across New York State for Faxton and Butterfield.

52. Cornell had built the line along the Erie Railroad across southern New York, which had gone bankrupt, and named it the New York and Western Union; presumably, he had an attachment to the name (Thompson, *Wiring a Continent*, 196).

53. For all practical purposes, Western Union emerged from the Civil War as a monopoly, its main competitor, the American Telegraph Company, having been hurt because its lines paralleled the Atlantic coast and had been subject to disruption by the war (Lubrano, *Telegraph*, 70; Beauchamp and Institution of Electrical Engineers, *History of Telegraphy*, 67). Even so, in 1886, there were 217 telegraph companies in America, nearly all with but a few miles of line (65).

54. Thompson, *Wiring a Continent*, viii. There are many examples of local services and industries that were "legislated" monopolies extending back to the national trading companies that led to the creation of many of the original colonies.

55. Both Thomas Edison and Andrew Carnegie had been telegraphers (Standage, *Victorian Internet*, 62).

56. Sibley's desire to develop a world-spanning network led him, during the Civil War, to explore the possibility of a line through Russian America (Alaska) and Siberia to Europe and China. This resulted in Western Union's seeking to buy or lease a right-of-way from the Russian-American Company. One thing led to another, and Secretary of State William Seward purchased the whole of the territory from the Russians in 1867. There were mixed motives: Seward and Sibley were friends, and Seward had a desire to see the United States acquire the whole of the continent north of Mexico.

57. Chandler Jr., *Visible Hand*, 189.

58. Lubrano, *Telegraph*, 15–16, 85, 133–7.

59. Beauchamp and Institution of Electrical Engineers, *History of Telegraphy*, 174–6.

60. Standage, *Victorian Internet*, 136–44; Lubrano, *Telegraph*, 72–3.

61. By 1860, New York was setting security prices for every major city in the United States, as the telegraph allowed for daily negotiations (Lubrano, *Telegraph*, 156). Forty years earlier, Andrew Dexter had carried fraudulent bonds from a then-remote Detroit bank to Boston to be used there as "capital" for his construction speculations. Such a way to commit that crime could not succeed in this new era — instead, new devices had to be invented for this purpose, as they still are; see Kamensky, *Exchange Artist*.

62. Lubrano, *Telegraph*, 156.

63. Standage, *Victorian Internet*, 125–6.

64. Oslin, *Story of Telecommunications*, 187–8. Western Union preserved Room 22 in the Arcade as it was, as a reminder of the company's origins.

65. Lubrano (*Telegraph*, xiv) notes that "it was the first step in the process of continuous improvement leading to the information superhighway." For a more detailed story of the evolution from the telegraph to the internet, with an emphasis on the role of AT&T and the telephone, see Alan Stone, *How America Got On-Line: Politics, Markets, and the Revolution in Telecommunications* (Armonk, NY: M.E. Sharpe, 1997).

66. Ironically, two patents were submitted by two different people on the same day (Lubrano, *Telegraph*, 7; Standage, *Victorian Internet*, 184–7). Alexander Graham Bell's had precedence, so he is the "inventor" of what others had developed as well, not unlike Morse's "invention" of the telegraph. Apparently, had Edison not been hard of hearing, he would have invented the telephone a month earlier than either of these (Oslin, *Story of Telecommunications*, chap. 14). A year later, Edison invented a "repeating telegraph," which became better known as the phonograph (Lubrano, *Telegraph*, 34n2). The second to submit a patent application for a telephone, Elisha Gray, had founded a telegraph equipment supplier in Cleveland in 1869 and moved it to Chicago, where it became Western Electric, eventually a large supplier to the telephone industry and a subsidiary of AT&T. Bell derived considerable assistance for his invention from contacts at MIT.

67. Western Union had bought out Elisha Gray's rights to the telephone. The 1879 agreement between Western Union and the Bell company was for a period of seventeen years (Lubrano, *Telegraph*, 16; see also Oslin, *Story of Telecommunications*, 227–31).

68. Lubrano, *Telegraph*, 17.

69. Standage, *Victorian Internet*, 191.

70. Ithiel de Sola Pool and Lloyd S. Etheredge, *Politics in Wired Nations: Selected Writings of Ithiel de Sola Pool* (New Brunswick, NJ: Transaction Publishers, 1998), 170. A line from Boston to New York followed in 1882.

71. Quoted from a 1914 magazine article in ibid., 168.

72. The first communications satellite, Telstar, had been launched in 1962 (Beauchamp and Institution of Electrical Engineers, *History of Telegraphy*, 403).

73. Lubrano, *Telegraph*, 90–2.

74. This was the date set by the International Maritime Organization for the end of Morse transmissions as distress calls by ships (Beauchamp and Institution of Electrical Engineers, *History of Telegraphy*, 389).

75. Twenty years before Marconi's transmissions, the principle of "inductive communication" between a wire laid alongside railroad tracks and a receiver system placed inside a moving railcar was known. It was a kind of very short-range radio, as an 1884 experiment in Wales between the island of Anglesey and a local lighthouse demonstrated. A longer-range radio transmission technique developed by the German Heinrich Hertz and eventually adapted by Marconi at the beginning of the twentieth century came to dominate (Beauchamp and Institution of Electrical Engineers, *History of Telegraphy*, 59–60, 184–92).

76. The air is irrelevant except for interferences such as lightning discharges. The Earth just gets in the way of the radio waves. Lee de Forest's 1907 invention of the audion

vacuum tube not only helped long-distance telegraphy and telephony, but also made long-distance radio and, therefore, broadcasting possible (Oliver, *History of American Technology*, 500).

77. Even so, AT&T and General Electric became engaged in a patent war that threatened to halt progress in both radio and telephony until a patent pool was formed between them and others in 1920.

78. See George Gilder, *Microcosm: The Quantum Revolution in Economics and Technology* (New York: Simon & Schuster, 1989), esp. chaps. 1–3.

79. Hafner and Lyon, *Where Wizards Stay Up Late*, 10.

80. Lubrano, *Telegraph*, 160.

81. Hafner and Lyon, *Where Wizards Stay Up Late*, 52, 57–63. The problem was not just switching speeds; analog phone technology distorts sounds, which the human ear cannot pick up, but which are a problem at the speed of digital communication.

82. Ibid., 38, 41.

83. Today, large information companies such as Microsoft and Google use numerous dedicated buildings full of heavy-duty servers located in "server farms" in the countryside.

84. Hafner and Lyon, *Where Wizards Stay Up Late*, 74–136.

85. Roberts, *History of the English-Speaking Peoples since 1900*, 556.

86. The first four servers were installed near the end of 1969 and another seven in 1970. It was not until 1971, however, that true communication was established between the computers (Hafner and Lyon, *Where Wizards Stay Up Late*, 151–4).

87. Ibid., 176.

88. Lubrano, *Telegraph*, 160.

89. Hafner and Lyon, *Where Wizards Stay Up Late*, 191–2.

90. It appears that this choice was made because the symbol was on the teletype keyboard, but not used much. Also, it had the advantage of being a commercial sign for "at," as in "tomatoes@20cents/lb," which could easily be seen as an Internet address as well.

91. Hafner and Lyon, *Where Wizards Stay Up Late*, 194.

92. Ibid., 191.

93. Ibid., 236–49.

94. Ibid., 248–56.

95. Beauchamp and Institution of Electrical Engineers, *History of Telegraphy*, 396–7; Standage, *Victorian Internet*, 193–4.

96. Lubrano, *Telegraph*, 160.

97. Roberts, *History of the English-Speaking Peoples since 1900*, 555. Not surprisingly, a number of others were working toward similar systems, but Berners-Lee had the advantage of being first and having a global, or at least an intercontinental, vision. Also, he was prepared to make his system available to everyone rather than exploit a patent for personal gain. This made it quicker and easier to implement. Few have had more effect on the behavior of the human race (555–6). He relates his experiences in Tim Berners-Lee, *Weaving the Web: Original Design and Ultimate Destiny of the World-Wide Web* (New York: HarperCollins, 1999, 2000).

98. Lubrano, *Telegraph*, 158.

99. Roberts, *History of the English-Speaking Peoples since 1900*, 555.

100. Metcalfe became interested in connecting computers after reading about the Alohanet experiments at the University of Hawaii. He went to work at PARC in Palo Alto, California, and, in 1973, tried to connect Xerox's computers there into a network. He got Xerox and others to make the system he devised, called Ethernet, into an open standard that everyone could use, then left to start 3Com to exploit the demand for related products; see "Technology Quarterly," *Economist*, December 12, 2009, 23–4.

Chapter 18

1. Thomas Pynchon, *Mason and Dixon* (New York: Henry Holt, 1997), 324.
2. Billington, *Westward Expansion*, 292. All three stanzas can be found in Dunbar and May, *Michigan*, 163–5.
3. Linklater, *Measuring America*, 171.
4. George Washington's order to Sullivan is similar to that which sent William T. Sherman off from Atlanta to the sea eighty years later: "it is proposed to carry the war into the heart of the country of the six nations, to cut off their settlements, destroy their next year's crops…so that the country may not only be overrun but destroyed." Sullivan reported at the end of his campaign that his troops had destroyed 40 villages, 160,000 bushels of corn, and acres of fruit trees and other crops (Klees, *People of the Finger Lakes Region*, xiii–xiv).
5. New York was given the sovereignty right, while Massachusetts gained the pre-emption right over the land (Klees, *People of the Finger Lakes Region*, xiv).
6. Billington, *Westward Expansion*, 245; and Klein, *Empire State*, 262.
7. Large parcels were bought by New York City speculators in an auction in 1791, when the State sold 5.5 million acres for a million dollars (Burrows and Wallace, *Gotham*, 335).
8. Billington, *Westward Expansion*, 245.
9. Phelps and Gorham, the speculators, got caught in a classic currency squeeze. They bought the land denominated in Massachusetts currency, which was virtually worthless at the time. Then the federal government announced the assumption of all the States' war debts, and the currency shot up in value, bankrupting them (Klees, *People of the Finger Lakes Region*, xv).
10. Hugill, *Upstate Arcadia*, 18–19.
11. Benjamin Franklin's grandson, William, was Morris's agent in London.
12. Morris's land flips are described in Billington, *Westward Expansion*, 246–7; see also Klein, *Empire State*, 262.
13. The following details of Ellicott's career also come from Patrick Weissend, *The Life and Times of Joseph Ellicott* (Batavia, NY: Holland Land Purchase Historical Society, 2002–3), 2–3.
14. He employed two of his brother Andrew's sons as clerks. They both saw action in the War of 1812. See Catharine van Cortlandt Mathews, *Andrew Ellicott: His Life and Letters* (1908; repr. New York: Ralph Roberts, 2001), 183.
15. At the time, there were only 17,000 people in the whole area of western New York, half of whom were children (Berstein, *Wedding of the Waters*, 122).

16. It had only 2,400 people when the Erie Canal opened in 1825, and by 1850 it was the world's largest grain port (Koeppel, *Bond of Union*, 1, 394).

17. Even in 1800, they were little more than Indian trails (Berstein, *Wedding of the Waters*, 122).

18. Billington, *Westward Expansion*, 248–9.

19. The British authorities in New York had simply made their peace with the *patroons* and eventually outdid them in gaining for themselves large tracts of land (Billington, *Westward Expansion*, 94–5).

20. Ellicott settled in Batavia in 1798 and coordinated the survey from there. His team laid out the town in long, narrow lots. In 1801, the first town lot was sold to an Abel Rowe, and a school teacher settled soon after. Large-scale land sales started in 1802 (ibid., 249). See also Barbara Ann Toal, *Batavia* (Charleston, SC: Arcadia Publishing, 2000), 7–8, 15–19.

21. Half of them, including Ellicott, fled the area when British forces attacked and burned Buffalo and Black Rock on New Year's Eve, 1813.

22. Billington, *Westward Expansion*, 248–59.

23. Ellicott's promotion of the canal involved one of the first land grants to finance a major public transportation work. He persuaded the Company's owners to offer a large tract of land, 100,000 acres, to the State if the canal were built through or near the Company's holdings rather than go into Lake Ontario, with a barge canal around Niagara Falls, as some were advocating. See Weissend, *Life and Times of Joseph Ellicott*, 16–18.

24. His relationship with the canal is documented in Koeppel, *Bond of Union*, especially 46–52, 59, 122, 138–9, 151, 212, 228–9, 341, 346.

25. Ibid., 144.

26. Ellicott had a political machine in western New York called the "Big Family," which came apart in the face of settler protests over the lack of investment he and the HLC had made in their area. Ellicott, who had earlier resigned from the Erie Canal commission because of ill health, also was forced to resign as the manager (ibid., 346).

27. The first Erie Canal shipping company was opened on the finished portion of the canal in 1820 by two Yankees. It prospered from the start (ibid., 144).

28. In the fall of 1836, the future governor of the State and Abraham Lincoln's and Andrew Johnson's secretary of state, William Seward, was hired by the HLC to be its manager in Chautauqua County, near the Pennsylvania border. He rented a house near the Buffalo-Erie road and worked at land development for about fifteen months, until the Panic of 1837 hit the area. See Goodwin, *Team of Rivals*, 78–80.

29. Stegner and Stegner, *Marking the Sparrow's Fall*, 260.

30. Linklater, *Measuring America*, 4–5.

31. The Astor fur empire in the Pacific Northwest in the early 1800s was built on selling furs directly to China, rather than to New York or Europe.

32. Part of the reluctance to develop colonies originated with the continued troubles the English had in keeping the "wild Irish" subdued. See Howard Mumford Jones, *O Strange New World: American Culture; The Formative Years* (New York: Viking, 1964), chap. 5.

33. In 1774, for instance, Jefferson criticized the King for wanting to introduce feudal

land tenures into America, along with the exactions that accompanied them. See Robert Middlekrauft, *The Glorious Cause: The American Revolution, 1763–1789*, rev. ed. (New York: Oxford University Press, 2005), 242.

34. Even so, by 1675, politically connected individuals managed to accumulate holdings of over 130,000 acres, presumably for speculation. See Bernard Bailyn, *The Peopling of British North America: An Introduction* (New York: Alfred A. Knopf, 1986), 67–8.

35. Billington, *Westward Expansion*, 101–2.

36. Ziff, *Puritanism in America*, 83–4, 208–9, 225–6. This eventuality never came to pass, but the British Privy Council did make one policy that had some relevance in post-Revolutionary America. In 1773, the British government restricted the colonial legislatures' right to allocate land, requiring that plots be sold directly and that the sale occur after proper surveys were made, at auction, and with the prices paid made public (Bailyn, *Peopling of British North America*, 73).

37. Billington, *Westward Expansion*, 243.

38. It said, in part: "LAND SURVEYING: Of all kinds, according to the best methods known….Distinct and accurate Plans of Farms furnished….Apply to Henry D. Thoreau, near the depot, Concord Mass."

39. One of the measures of length, a rod, was defined in the 1500s as a "length of exactly 16…shoes from sixteen men, short and tall, one after the other, as they came out of church"; see Joseph T. Stuart, "The Initial Points of Michigan," *Turning the Horizon* 3 (1, 2005): 12.

40. Davis, *Frontier Illinois*, 92.

41. A few quaint usages are still left of this system in North America. Windsor, Nova Scotia, is still referred to locally as the shiretown of Hants County, and Nova Scotian county council chairpersons are called reeves. The old Anglo-Saxon word shire-reeve, meaning the chief officer of a shire, was transformed into the office of sheriff, still in common usage for a county's chief law enforcement officer.

42. This basic outline is derived from Billington, *Westward Expansion*, 78–9, and from a genealogy website for New Hampshire and Vermont settlers, http://www.skinner-webb.com.

43. Billington, *Westward Expansion*, 204–5.

44. The squatters refused to obey government eviction orders and simply melted into the woods at the approach of any soldiers. This drama in futility was to be re-enacted all across the frontier (Hurt, *Ohio Frontier*, 144, 146). As the land became organized, squatters clamored for a pre-emption law giving them the right to keep what they were using. Where there were land auctions, squatting became a rational response, as such people often did not have the capital to outbid speculators and the rich for good lands. At first, in 1807, there were attempts to threaten squatters with fines or prison (Linklater, *Measuring America*, 166), but then a reaction set in, and a law was passed in 1813 allowing a 160-acre pre-emption. Other laws on squatting were passed in 1830 and 1841 (Davis, *Frontier Illinois*, 143; Landes, *Wealth and Poverty of Nations*, 319–20).

45. This contained what are now the seven States of Ohio, Indiana, Illinois, Michigan, Wisconsin, Iowa, and Minnesota.

46. Davis, *Frontier Illinois*, 93.

47. Part of the value of the Gunter chain in early America was that it allowed for a conversion of the base 4 Anglo-Saxon method of measurement to the base 10 used in modern times.

48. Linklater (*Measuring America*, 2) opens with a present-day trip to the "Point of Beginning."

49. Much of his team was from New England, and the first township to be surveyed was called Township 5 of the First Range. Jared Mansfield, the surveyor general from 1803 to 1812, established a First Principal Meridian on the western edge of Ohio that served as a base for a more regular survey westward than the patches of granted lands and different types of survey constants in Ohio would allow (ibid., 161). There are many descriptions of the process in the literature; see, for instance, Billington, *Westward Expansion*, 205.

50. Davis, *Frontier Illinois*, 94. The Ohio Company's advice played a large part in the design of the Northwest Ordinance governing the Territory.

51. For an account of how the ordinance was drafted and the dealmaking that went behind it, see Dunbar and May, *Michigan*, 94–5. Some of the thinking behind the ordinance is spelled out in Linklater, *Measuring America*, 82–3.

52. Hurt, *Ohio Frontier*, 156–7.

53. Ibid., 168.

54. One of the key players was Rufus Putnam, a military officer in the French and Indian War and an American general in the Revolutionary War. Between these wars, he was a millwright and a self-educated surveyor, doing surveys of the Gulf Coast and the Mississippi River lands, where New England veterans had been allotted land by the British. During the Revolution, Washington made him his chief engineer in New York; he proposed unsuccessfully that Congress create a Corps of Engineers. After the war, he surveyed parts of Maine and was one of the founders of the Ohio Company. He had been told about the area by Washington a decade earlier. Putnam went to the purchased land with the first group of settlers. In 1796, he was made the first surveyor general of the United States, holding the position until 1803.

55. Hurt, *Ohio Frontier*, 156.

56. Davis, *Frontier Illinois*, 260. Squatting tended to gradually die out in the more orderly North, but in the South the inability to settle land claims and provide a regular means of guaranteeing ownership led to its increase, with the consequent disinterest by squatters in property development (Linklater, *Measuring America*, 152).

57. For a brief summary, see John S. Dailey, "Fabric of Surveying: Ohio Lands and Survey Systems," *American Surveyor*, November 30, 2004; available online at http://www.amerisurv.com/content/view/3962/150/.

58. The money was invested to assist schools in the State; the proceeds were used for this purpose at least until the late twentieth century.

59. See Ted Heineman, "Surveying the Western Reserve," *Riverside Cemetery Journal* (2009); available online at http://riversidecemeteryjournal.com/Events/Events/page195.html.

60. Miller and Wheeler, *Cleveland*, 7–8.

61. Besides personal security, there was the additional complication that the sale by Connecticut seemed to transfer governing jurisdiction to the Company, something not

resolved in favor of the federal government until 1800. Also, a number of Pennsylvania speculators had sold property in the Reserve and resultant title disputes then had to be settled (Hurt, *Ohio Frontier*, 164–6; see also Billington, *Westward Expansion*, 254).

62. Hurt, *Ohio Frontier*, 164–9. It had done almost nothing to develop the land (Pierson, *Tocqueville in America*, 10).

63. The last straw of what Jefferson called "dabbling in the federal filth" (using depreciated money to speculate in land) was the Symmes failure, on lands that today are around Cincinnati (Hurt, *Ohio Frontier*, 164).

64. For instance, four offices were established in Ohio in 1800 (Johnson, *History of the American People*, 296).

65. Hutchins died in 1789, before the Seven Ranges survey was complete. A 1796 act finalized the system for surveying and selling land in the United States along the lines laid down by Hutchins. It even specified the use of Gunter's chain in measurement and applied to all land measurement in the Northwest (Linklater, *Measuring America*, 142).

66. Details of the 1785 Land Ordinance may be seen online at http://www.ohiohistorycentral.org/entry.php?rec=174.

67. With the first year's subsistence, housing costs, and fencing, the total could reach $2,000 a year, a sum that was out of reach for most of those coming to the frontier (Linklater, *Measuring America*, 163–6). By 1832, the minimum was reduced to forty acres.

68. Davis, *Frontier Illinois*, 157–8.

69. Hurt, *Ohio Frontier*, 168–9, 173.

70. Section 16 in each township was to be used for public education, and sections 8, 11, 26, and 29 were reserved for Revolutionary War veterans, a scheme later abandoned as the warrants for veterans were used up (Dunbar and May, *Michigan*, 94; Hurt, *Ohio Frontier*, 144).

71. The problem was that government prices for adjoining land kept the profit potential down. Town swindles were a favored alternative (Billington, *Westward Expansion*, 285).

72. Linklater, *Measuring America*, 174.

73. Faragher, *Rereading Frederick Jackson Turner*, 49–52.

74. See Dunbar and May, *Michigan*, 154–5; another excellent source is the Michigan Museum of Surveying, in Lansing.

75. A concise version of the process is Stuart, "Initial Points of Michigan," 12–13.

76. Dunbar and May, *Michigan*, 154.

77. The terms were good: 5 percent down to hold the land for forty days, when another 20 percent was due and the rest on a mortgage, interest free for three years. Reversion could occur only if the purchaser had been delinquent for five years, so the mortgage was effectively for eight years (Davis, *Frontier Illinois*, 157–8).

78. The first sales were auctions in Detroit in 1818 (Dunbar and May, *Michigan*, 155).

79. The first steamship on the lakes was the *Walk-in-the-Water*, launched in 1818.

80. A number of acts were signed in the late 1800s to assist settlers in taking less-desirable pieces of land; see Stegner and Stegner, *Marking the Sparrow's Fall*, 173; and Johnson, *History of the American People*, 525.

81. Linklater, *Measuring America*, 226.

82. Land grants and railroads make a complex and interesting story; see, for instance, ibid., 183–7.

83. Ibid., 224.

Chapter 19

1. Rachel Carson, *Silent Spring* (1962; New York: Houghton Mifflin, 2002), 3.

2. Quoted in Pierre Berton, *Niagara: a History of the* Falls (Toronto: McClelland & Stewart, 1992), 419.

3. Patrick McGreevy, *Imagining Niagara: The Meaning and Making of Niagara Falls* (Amherst: University of Massachusetts Press, 1994), 2–3.

4. Berton, *Niagara*, 44–5.

5. Ibid., 46–7. Porter still was not averse to charging visitors for access and the building of a bridge to Goat Island.

6. William R. Irwin, *The New Niagara: Tourism, Technology, and the Landscape of Niagara Falls, 1776–1917* (University Park: Pennsylvania State University Press, 1996), 27–9; and McGreevy, *Imagining Niagara*, 108

7. McGreevy, *Imagining Niagara*, 108.

8. Irwin, *New Niagara*, 29. The discrepancy got wider until, by 1890, there were 5,000 people in Niagara Falls, but 250,000 in Buffalo (McGreevy, *Imagining Niagara*, 110).

9. McGreevy, *Imagining Niagara*, 34. It was done at the urging of a Canadian archbishop.

10. Ibid., 37.

11. Ibid., 84.

12. One version has it that the group and the plan were devised in the prestigious Century Club in New York City. See Ginger G. Strand, *Inventing Niagara: Beauty, Power, and Lies* (New York: Simon & Schuster, 2008), 142–3. Another places it in a stroll on Goat Island in 1869 (Berton, *Niagara*, 178–80).

13. Strand, *Inventing Niagara*, 148. A similar Canadian process was also under way. There had been considerable cooperation between advocates on both sides of the border (Berton, *Niagara*, 182–6).

14. Irwin, *New Niagara*, 67. By 1887, both the American and Canadian shores of the Falls and lower river had been reserved for parkland.

15. Ibid., 84.

16. Ibid., 76–7.

17. Ibid., 227.

18. Ibid., 99.

19. Berton, *Niagara*, 202–28.

20. The quiet elegant hum of the turbines became a popular tourist destination in itself (Irwin, *New Niagara*, 122; see also Strand, *Inventing Niagara*, 167).

21. McGreevy, *Imagining Niagara*, 107, 113.

22. The just reelected President William McKinley was assassinated there while standing in a reception line.

23. He claimed to have single-handedly settled 15,000 people in what is now Guthrie, Oklahoma, during the 1880s land rush there; *Niagara Falls Gazette*, March 3, 1942.

24. Based on a calculation of five people per horsepower generated by the canal and used by associated factories (Irwin, *New Niagara*, 144).

25. Ibid., 145.

26. McGreevy, *Imagining Niagara*, 120–1.

27. Ibid., 121.

28. David Germain, "Dream that drowned in Love Canal," *Buffalo News*, June 6, 1993.

29. Love had planned his canal to be a ship canal as well, to compete with the Welland Canal on the Canadian side, which had been completed in 1829 as a means of keeping British lake trade from being diverted from the Lake Ontario/St. Lawrence River system to the Erie Canal.

30. Strand, *Inventing Niagara*, 172–3.

31. Tim Dowling, *King Camp Gillette, 1855–1932: Inventor of the Disposable Culture* (London: Faber & Faber, 2002).

32. Berton, *Niagara*, 231–4.

33. McGreevy, *Imagining Niagara*, 105

34. Originally called the "United Company" and in later writings the "World Corporation"; see Irwin, *New Niagara*, 142.

35. McGreevy, *Imagining Niagara*, 125

36. Irwin. *The New Niagara*, 101, 161. An earlier exhibition in Frankfurt, Germany, in 1891 had been powered by alternating current sent 100 miles over wires. *Ibid.*, 109

37. McGreevy, *Imagining Niagara*, 131

38. Ibid., p.135

39. Today, the amount of water going over the Falls can be adjusted to provide a pretty scene by day, with extra water being diverted at night for power generation.

40. The Robert Moses Parkway was built along the river, effectively cutting off Olmstead's Reservation from the rest of the city except for a pedestrian overpass. Strand, *Inventing Niagara*, 263.

41. Ibid., 174–5.

42. Ibid., 225–9, 243–4. Radioactive sludge and waste, including dead animals used in plutonium experiments, was sent from all over the United States to the Model City area for storage.

43. Ibid., 176, 179. The site of a large munitions factory built during the Second World War has been turned into Chemical Waste Management's "Model City Facility," the only hazardous waste dump in the Northeast (Strand, *Inventing Niagara*, 241).

44. Michael Desmond, "'Model community' dream haunts Love Canal residents," *Buffalo Courier-Express*, August 6, 1978.

45. Lois Marie Gibbs, *Love Canal: The Story Continues* (Gabriola Island, BC: New Society Publishers, 1998), 22.

46. Strand, *Inventing Niagara*, 195.

47. A 1948 *Saturday Evening Post* article noted the smog over the town, which "is sniffed happily by the industrially-minded" (quoted in ibid., *Inventing Niagara*, 179).

48. Gibbs, *Love Canal*, 21.

49. Rick Atkinson, *Day of Battle* (New York: Henry Holt, 2007), 448.

50. Much of the biographical material on Carson is condensed from Linda Lear's "Introduction" to Carson, *Silent Spring.*

51. See, for instance, a retrospective by John Tierney, "Fateful voice of a generation drowns out real science," *New York Times,* June 5, 2007.

52. See Smithsonian Institution Archives, "Robert H. Goddard: American Rocket Pioneer" (Washington, DC, n.d.), available online at http://siarchives.si.edu/history/exhibits/stories/robert-h-goddard-american-rocket-pioneer; see also Clark University, Archives and Special Collections, "Robert Hutchings Goddard Biographical Note" (Worcester, MA, n.d.), available online at http://www.clarku.edu/research/archives/goddard/bio_note.cfm; and National Museum of the US Air Force, "Dr. Robert H. Goddard: 'The Father of Modern Rocketry'" (Wright-Patterson AFB, OH, 2009), available online at http://www.nationalmuseum.af.mil/factsheets/factsheet.asp?id=12374.

53. The *New York Times* published a retraction of a mocking editorial page article almost fifty years later, after the launch of the Apollo XI mission; see "A severe strain on credulity," *New York Times,* January 13, 1920, 12; and "A correction," *New York Times,* July 17, 1969, 43.

54. Bob Ward, *Dr. Space: The Life of Werner von Braun* (Annapolis, MD: Naval Institute Press, 2005), 226.

55. See, for instance, Foreman, *Confessions of an Eco-Warrior* , 196; Kaplan, *Empire Wilderness,* 296; and Bishop Jr., *Epitaph for a Desert Anarchist,* 205.

56. Gibbs, *Love Canal,* 91.

57. Ibid., 47. About this time, the local media became aware of the problems, but the health authorities continued to deny there were any issues with smell and seepage, even though the State had ordered corrective action be taken; see Berton, *Niagara,* 417–18.

58. Ibid., 419–20.

59. The story of the residents' organizing and pressuring government to take appropriate action is the core of Gibbs, *Love Canal*; see also Berton, *Niagara,* 420-439 for a condensed version.

60. Gibbs, *Love Canal,* 9.

61. Strand, *Inventing Niagara,* 224–5. None of the Niagara area sites was ever mentioned in connection with the Manhattan Project, but they played a key role in it.

62. The first apparently was in Telluride, Colorado, in 1891, and the second in Germany shortly thereafter; see Wershler-Henry, *Iron Whim,* 270.

63. McGreevy, *Imagining Niagara,* 159.

Chapter 20

1. David Ewen, ed., *American Popular Songs from the Revolutionary War to the Present* (New York: Random House, 1966, 239. Contrary to what one might think, "Buffalo Gals" is not a "cowboy song," but refers to the prostitutes at the terminus of the Erie Canal.

2. Quoted in the history of the Carrier company on the Web site of its current owners, United Technologies, http://www.corp.carrier.com/vgn-ext-templating/v/index.jsp?vgnextoid=4455d66bdcb08010VgnVCM100000cb890b80RCRD&cpsextcurrchannel=1.

3. Raymond Arsenault, "The End of the Long Hot Summer: The Air Conditioner in Southern Culture," *Journal of Southern History* 50 (4, 1984): 598.

4. David Shi, "Air conditioning — It's made the south what it is," *Independent-Mail* (Anderson, SC), June 24, 2007; available online at http://www.independentmail.com/news/2007/jun/24/air-conditioning-its-made-south-what-it/.

5. The layout was similar to that of Washington, DC, not surprising given the involvement of Joseph Ellicott's brother Andrew in the planning of that city; see David M. Ellis and Sherri Goldstein Cash, *New York State: An Illustrated History* (Sun Valley: CA: American History Press, 2008), 166.

6. Berstein, *Wedding of the Waters*, 286–91. In 1810, DeWitt Clinton, then mayor of New York City, scorned Buffalo in 1810 for having "5 lawyers and no church."

7. The *Walk-in-the-Water*, a combination of sailing ship and steamer able to accommodate a hundred passengers, was built in Black Rock in 1818. She embarked on her maiden voyage on August 23 that year with twenty-nine passengers, who paid $24 each to go to Detroit, a trip that took 44 hours and 10 minutes, including stops at Erie, Pennsylvania, and Cleveland and Sandusky, Ohio. She was destroyed in a storm on November 21,1821, though her engine was saved and reused. See Rockne P. Smith, *Our "Downriver" River: Nautical History and Tales of the Lower Detroit River* (Gibraltar, MI: Rockne Smith, 1997), 11–12.

8. Berstein, *Wedding of the Waters*, 349.

9. For a good survey of the growth of Buffalo industry until the early twentieth century, see Robert Holder, "The Beginnings of Buffalo Industry," in *Adventures in Western New York History*, vol. 5 (Buffalo, NY: Buffalo and Erie County Historical Society, n.d.); available online at http://bechsed.nylearns.org/pdf/The_Beginning_of_Buffalo_Industry.pdf.

10. See Levy, *American Vertigo*, 40.

11. The prime source is Margaret Ingels, *Willis Haviland Carrier, Father of Air Conditioning* (New York: Arno Press, 1972), chap. 1.

12. Ingels might not have been a great biographer, but she deserves a biography herself; see Marsha E. Ackermann, *Cool Comfort: America's Romance with Air-Conditioning* (Washington, DC: Smithsonian Institution Press, 2002), 105–6. Ingels was the first woman to receive a mechanical engineering degree from an American university, graduating in 1916 from the University of Kentucky, the same university Willis Carrier's lifelong friend Irvine Lyle attended. She spent four years with Carrier before joining the research staff of the American Society of Heating and Ventilating Engineers. She returned to Carrier in 1929, and retired in 1952; see the history of the Carrier company on the Web site of its current owners, United Technologies, at http://www.corp.carrier.com/vgn-ext-templating/v/index.jsp?vgnextoid=35070d9653b08010VgnVCM100000cb890b80RCRD&cpsextcurrchannel=1.

13. Cooper, *Air-Conditioning America*, 192n34.

14. http://freepages.genealogy.rootsweb.ancestry.com/~carrier/gramps/ppl/1/c/bca41937b2232acb6c1.html.

15. Ingels, *Willis Haviland Carrier*, 6. A different source places the marriage in 1893 in Buffalo; see M. Tifft, "A Partial Record of the Descendants of John Tefft, of Portsmouth,

Rhode Island, and a Nearly Complete Record of the Descendants of Text of John Tifft, of Nassau, New York" (Buffalo, NY: Peter Paul Book Company, 1896); available online at http://www.archive.org/stream/apartialrecorddootiffgoog/apartialrecorddootiffgoog_djvu.txt, 372.

16. Tifft, "Partial Record," 405.

17. Oscar E. Anderson Jr., *Refrigeration in America* (Princeton, NJ: Princeton University Press, 1953), 71. The Egyptians may have been the first to recognize that a fan dipped in water could produce cool air as a result of evaporation. Carrier air conditioned the Egyptian parliament in 1937 with more modern technology. Apparently, the Chinese had invented a rotary fan by A.D. 200 and a water-powered one 500 years later, but the concept was then "lost" until reinvented around 1800.

18. So-called swamp coolers were commonly used to cool rooms and patios in Arizona until the late 1980s,

19. The first export of New England ice was from Boston to Martinique in 1806; see Anderson Jr., *Refrigeration in America*, chap. 1. See also Barbara Krasner-Khait, "The Impact of Refrigeration," *History Magazine*, February-March 2000; and Bill Bryson, *At Home: A Short History of Personal Life* (New York: Doubleday, 2010), 71–3.

20. Cooper, *Air-Conditioning America*, 8. Gorrie received a patent for his process in 1851, but died before he was able to commercialize it. See James V. Warren, "John Gorrie," *Transactions of the American Clinical and Climatological Association* 93 (1982): 183–8; available online at http://www.ncbi.nlm.nih.gov/pmc/articles/PMC2279554/pdf/tacca00096-0227.pdf?tool=pmcentrez. See also Anderson Jr., *Refrigeration in America*, 71.

21. C.P. Arora, *Refrigeration and Air Conditioning* (1981; New York: McGraw-Hill, 2000), 477.

22. Cooper, *Air-Conditioning America*, 13–15.

23. Oliver (*History of American Technology*, 415) claims, however, that neither of these projects resulted in a satisfactory outcome.

24. Cooper, *Air-Conditioning America*, 1. The debate over natural ventilation versus a standardized, air-conditioned environment would continue. Yet standardization and mechanization have been central to the Yankee experience.

25. She had been taken in by the Carriers after the death of her father, Harry Moor, one of Carrier's associates in a number of his inventions, including the centrifugal compressor. See the history of the Carrier company on the Web site of its current owners, United Technologies, http://www.fatherofcool.carrier.com/corp/details/0,,CLI1_DIV51_ETI3579,00.html.

26. Much of the material on Carrier's career is taken from Ingels, *Willis Haviland Carrier*.

27. Cooper, *Air-Conditioning America*, 23.

28. In fact, although the term had not been invented yet, the first air conditioner went into service in Brooklyn on July 17, 1902; Carrier would not receive his patent until 1906. See Arora, *Refrigeration and Air Conditioning*, 477.

29. Spraying air had been tried in textile mills as early as 1838; see Oliver, *History of American Technology*, 415.

30. Carrier's first client for this apparatus was a bank in La Crosse, Wisconsin. By 1907, his apparatus had been installed in a silk mill in Yokohama, Japan.

31. The term "air-conditioning" had been popularized in 1906 by Stuart Cramer, a specialist in air problems related to textile mills. I. Hardeman, a Buffalo Forge engineer, picked up the term while installing one of CACC's systems in a North Carolina textile mill and it found its way into the new company's name (Cooper, *Air-Conditioning America*, 19). Cramer had used the term for years, but it only became popular after he mentioned it in a paper delivered to the National Cotton Manufacturing Society; see Samuel E. Kronick, "Air Conditioning: A Conflict of Environmental Control" (course material prepared for STS-001-Technology in American History, Massachusetts Institute of Technology, May 14, 2009), 2.

32. Kronick, "Air Conditioning," 2–3.

33. Ibid., 5. "Cool comfort" in movie theaters allowed for multiple showings per day in the summer, as well as providing hot customers with a cool bonus besides the entertainment. As a kid in the 1950s going to the Della Theater in Flint, Michigan, for a double feature and some serials on hot Saturday afternoons, I remember the "coolth" almost more than I do the plots of Ming the Merciless.

34. Ackermann, *Cool Comfort*, 51–7.

35. See William P. Barrett, "Willis Carrier's Ghost," *Forbes*, May 29, 2000, 152–62; and Cooper, Air-Conditioning America, 137.

36. Ackermann, *Cool Comfort*, 58–61. Martin (*Railroads Triumphant*, 89) illustrates some of the social complexities this engendered and solved. See also Anderson Jr., *Refrigeration in America*, 308.

37. A possible example is the USS *Perch*, launched in 1936 and lost in the Pacific; see Kevin Denlay, "On Eternal Patrol: The Unexpected Discovery of the USS Perch SS-176," *Advanced Diver Magazine*, n.d.; available online at http://www.advanceddivermagazine.com/articles/perch/perch.html.

38. Ackermann, *Cool Comfort*, 81.

39. Quoted in ibid., 80.

40. Technically, it was not a room or home air conditioner but a winter gas heater with an automatic humidifier (Cooper, *Air-Conditioning America*, 114). The first true window unit was left to others to create.

41. Annual production rose from 1,000 units in 1945 to 1.3 million units in 1956 (ibid., 143). For a history of the expansion of air conditioning in housing in the latter half of the twentieth century, see Jeff Biddle, "Explaining the Spread of Air-conditioning, 1955–1980," *Explorations in Economic History* 45 (4, 2008): 402–23.

42. J.D. McConnell, "Power Distribution in Textile Plants," *Electrical Engineering* 66 (7, 1947): 667–9.

43. Raymond A. Mohl, ed,. *Searching for the Sunbelt: Historical Perspectives on a Region* (Knoxville: University of Tennessee Press, 1990); Ackermann, *Cool Comfort*, 162.

44. Shi, "Air-conditioning." In 1951, the Carrier company built demonstration subdivision houses in Louisiana, Virginia, and Texas to spur air-conditioned home sales in the South.

45. Kronick, "Air Conditioning," 6.

46. Stan Cox, "AC: It's not as cool as you think," *Los Angeles Times*, July 18, 2010; available online at http://articles.latimes.com/print/2010/jul/18/opinion/la-oe-cox-ac-20100718.

47. Shi, "Air-conditioning."

48. Arsenault, "End of the Long Hot Summer," 598–9.

49. Southern commentators have often speculated on what urban growth in the region might have looked like without air conditioning — none of them is positive

50. A *New York Times* editorial noticed this as early as 1970; see Arsenault, "End of the Long Hot Summer," 599.

51. Quoted in ibid., 628.

52. "The Most Influential Innovations of the Millennium," *Wall Street Journal*, 11 January 11, 1999.

53. Russell Hitchings and Shu Jun Lee, "Air Conditioning and the Material Culture of Routine Human Encasement: The Case of Young People in Contemporary Singapore," *Journal of Material Culture* 13 (3, 2008): 254–5.

54. The world's largest producer of air-conditioning units is now LG Electronics in South Korea.

55. One example is the popularity of rum as a drink. Formerly restricted in popularity to certain geographic areas, the rise in Caribbean tourism due to air conditioning has spread its use to new areas. See "All about Rum," Beverage Testing Institute; available online at http://www.tastings.com/spirits/rum.html.

56. In the early 1950s, *House Beautiful* magazine launched a campaign advocating a more diverse and open ventilation for house design that did not require air conditioning; its influence, however, was minimal (Ackermann, *Cool Comfort*, 111–22).

57. Ackermann, *Cool Comfort*, 138–9. For a review of Miller's book, see Dan Geddes, "Henry Miller's 'On the Road'," *Satirist*, July 2002; available online at http://www.the-satirist.com/books/Air-Conditioned-Nightmare.html. The inspiration for the title of his book might have come from his visit of Frank Lloyd Wright's air-conditioned Johnson Wax Building in Racine, Wisconsin; he hated it.

58. Quoted in Arsenault, "End of the Long Hot Summer," 606–7n34. For a survey of air conditioning Washington, see Ackermann, *Cool Comfort*, chap. 4. Roberts (*History of the English-Speaking Peoples since 1900*, 186) notes that novelist Gore Vidal said much the same thing earlier.

Chapter 21

1. Salsbury, *State, the Investor, and the Railroad*, 111.

2. Gould made his statement in the late 1860s; quoted in Martin, *Railroads Triumphant*, 373. The company's capital, whenever there were earnings, was retained and used to improve the railroad or buy other lines. The Erie paid some dividends in the 1850s and then not until 1942. Johnson (*History of the American People*, 551) suggests that one was paid in 1873, after Gould was forced out.

3. Gordon, *Passage to Union*, 250.

4. Much of this section parallels and is enlightened by Crump, *Brief History of the Age of Steam*, chaps. 1–7.

5. Even the US Congress took an interest in marine railways in the 1820s (Gordon, *Passage to Union*, 14).

6. Harlow, *Steelways of New England*, 18. Crump (*Brief History of the Age of Steam*, 28) suggests this method was in use in German mines as early as the fifteenth century. A similar gravity railroad was developed near Chester, Pennsylvania, in 1810 to move granite blocks; another was made in 1810 near Lewiston, New York, for military haulage; see John Westwood and Ian Wood, *A Historical Atlas of North American Railroads* (Edison, NJ: Chartwell Books, 2007), 28.

7. Dalzell, *Enterprising Elite*, 85.

8. There is an interesting description of Josiah Quincy and his railroad in Louise Hall Tharp, *Until Victory: Horace Mann and Mary Peabody* (Boston: Little Brown, 1953), 62–3. The horse moved around a capstan that had chains attached to two cars, an empty one being pulled up and a full one being lowered. See also Paterson H. Browne, *Quincy* (Charleston, SC: Arcadia Publications, 2004), 10–11; Carter, *When Railroads Were New*, 12–13; Douglas, *All Aboard*, 18; and Harlow, *Steelways of New England*, 25–7, 31–5. Incorporation was seen as a way to expropriate land for the railroad's right-of-way and to use the power of government to enforce a monopoly on the route, which was necessary to entice investors. The Quincy, though, was not a real railroad but what the British, who had been using them for a long time, called "a tramway"; see Charles Francis Adams Jr.,. *Railroads: Their Origin and Problems* (1878; New York, London: G.P. Putnam's Sons, 1987), 37. See also Salsbury, *State, the Investor, and the Railroad*, 45. The Quincy was, for a time, a tourist attraction as well.

9. The earliest guided and wheeled transportation is in a 1430 military drawing in which a rutted path is used to move wagons of earth from trenches to the top of ramparts. The modern railroad evolved from the idea of putting wooden strips in the ruts made by coal wagons as they hauled their contents down to water and boats. The idea of a mechanical way to move the cars came later; see Crump, *Brief History of the Age of Steam*, 22–52; and John Moody, *The Railroad Builders* (New Haven, CT: Yale University Press, 1919), 2–3.

10. The use of rails in mines goes back a long way — even the ancient Greeks apparently used "rutways" to reduce friction; see Douglas, *All Aboard*, 14; and Gordon, *Great Game*, 73.

11. Crump, *Brief History of the Age of Steam*, 52.

12. For the reaction in Boston, see Harlow, *Steelways of New England*, 3–20; and Gordon, *Passage to Union*, 16.

13. Crump, *Brief History of the Age of Steam*, 180.

14. Oddly, although the steamboat created an instant sensation and was soon put to use as far west as the Mississippi River, it did not seem to create any logical connection to land travel in the public's mind for another twenty years; see Douglas, *All Aboard*, 8, 13. Henry Shreve was piloting a steamboat up and down the Ohio-Mississippi system as early as 1814, and also went as far north as present-day Galena, Illinois, to trade for lead with the local Indians (Crump, *Brief History of the Age of Steam*, 79–82. A steamboat finally made it all the way upriver to the present site of Minneapolis in 1823, two short years before the opening of the Erie Canal (Martin, *Railroads Triumphant*, 9-10). Even so, the growth of

trade in the interior of the country was such that steamboat traffic on the rivers making up the Mississippi system constituted three-fifths of the steam power used in the 1840s; in the 1850s, New Orleans surpassed New York in the volume of shipping, with half of American exports moving through its ports (Crump, *Brief History of the Age of Steam*, 82).

15. Before the Erie Canal was built, the natural outlet for the produce of farmers in western New York was northeast on Lake Ontario to Montreal. Canadian traders opened up stores on the lakeshore before the War of 1812 and bought potash and salt. When Thomas Jefferson proclaimed an embargo on foreign trade in 1808, a "Potash Rebellion" developed in the area and trade northwards jumped. By 1815, the British were building canals around the rapids on the St. Lawrence River and launching steamships to capture the western New York trade. Transportation costs are estimated to have been 30 percent lower to Montreal than to Albany for Western wheat (Koeppel, *Bond of Union*, 52, 72, 83, 85, 114, 118).

16. Horses and mules were used in British mines to pull rail cars as early as 1806. A quick synopsis of the diffusion of steam technology can be found in Martin, *Railroads Triumphant*, 12–14.

17. Weightman, *Industrial Revolutionaries*, 101–2.

18. See Crump, *Brief History of the Age of Steam*, 148–53; Douglas, *All Aboard*, 17; and Hobsbawm, *Age of Revolution*, 44–5. Its builder, George Stephenson, established the rails at 4 feet 8½ inches wide, the width used in the mines he knew and also the width, more or less, of Britain's ancient Roman roads. It is still the standard width used by most of the world's railroads today. For a Boston account of the opening of the Stockton and Darlington, see Salsbury, *State, the Investor, and the Railroad*, 44–5.

19. Gordon, *Business of America*, 92. Clearly, if a steam engine could propel a boat, there was no reason it could not also propel a carriage; see idem, *Great Game*, 74; and Hobsbawm, *Age of Revolution*, 44–5. For an account of the Boston reporting on the progress of the Manchester and Liverpool, see Salsbury, *State, the Investor, and the Railroad*, 54–6. Horatio Allen, son of an engineering professor at Union college in Schenectady, New York, was commissioned to go to Britain in early 1828 to investigate the use of steam for railroading by the Delaware and Hudson canal company. In 1829, he shipped four locomotives, the first seen in America, to the company for testing, but little came of it. Allen then went to South Carolina to become the chief engineer for the new Charleston and Hamburg Railroad; see Carter, *When Railroads Were New*, 12–23.

20. Crump, *Brief History of the Age of Steam*, 154–63; and Harlow, *Steelways of New England*, 61–2. Three locomotives were imported into Pennsylvania for mine haulage in 1829; though only one was temporarily used, it was enough to show the practicality of steam.

21. Adams Jr., *Railroads*, 3, 5.

22. Berstein (*Wedding of the Waters*, 344) has a somewhat garbled rendition of this early railroad. A broader context for the creation of early railroads in New York can be found in Billington, *Westward Expansion*, 328–30.

23. Martin, *Railroads Triumphant*, 15. The steamship line was part of Cornelius Vanderbilt's growing coastal empire (Adams Jr., *Railroads*, 15). It was opened as a steam railway in 1830, and had to compete with a new canal project up the Potomac from

Washington, DC (Crump, *Brief History of the Age of Steam*, 98–9). In August 1830, the Baltimore and Ohio held time trials between a horse and a steam engine; the horse won when the engine broke down (Adams Jr., *Railroads*, 43–5). For a chronicle of the failure of governments and entrepreneurs to open up a Potomac-Ohio connection to compete effectively with the Erie Canal and the northern railroads, see Goodrich, *Government Promotion*, 75–9.

24. Harlow, *Steelways of New England*, 76–7.

25. Ibid., 86–91. The first locomotive to run on the line was the "Patrick," named after Lowell executive Patrick Tracy Jackson. The first ride was rather exciting, if only because of the sparks from the smokestack that singed everyone's clothes (Douglas, *All Aboard*, 30–1). It took three hours to cover the forty-two miles (Carter, *When Railroads Were New*, 8–9). By 1838, the shop had made thirt-eight locomotives for various lines throughout the country; see Weible and Lowell Historical Society, *Continuing Revolution*, 149. Most of the work on this and other New England railroads was done by imported Irish workers (Struik, *Yankee Science in the Making*, 310).

26. Until the late 1830s, corporations could only be created by an act of the State Legislature, called a "charter." The charter spelled out the objectives, powers, and capital requirements of the new company. Because these were State charters, companies whose lines crossed State boundaries had to gain a charter in the next State as well and treat the extension as a different railroad. This obviously created another nuisance for companies that were taking on a regional or national scope (Gordon, *Passage to Union*, 108). For an account of the conflicting regional interests that led to Massachusetts staying out of the railroad business, unlike the State canal efforts in New York and Pennsylvania, see Salsbury, *State, the Investor, and the Railroad*, 62–79.

27. Harlow, *Steelways of New England*, 78. Adams (*Railroads*, 66) says 1835; others say 1834. It may be that Adams was referring to the official opening, while the others mean the line's physical completion.

28. Martin, *Railroads Triumphant*, 15. The rest of New England followed suit — one estimate is that 60 percent of all the railroad mileage constructed in the 1840s was in that part of the country (Douglas, *All Aboard*, 34–5).

29. Patrick Jackson, who had become manager of the Waltham mill on the death of Lowell, became the superintendant of the Worcester line (Dalzell, *Enterprising Elite*, 88).

30. Harlow, *Steelways of New England*, 123–30. Salsbury (*State, the Investor, and the Railroad*, 146–55) estimates the Western received $5 million in loans from Massachusetts and the City of Albany and $1 million in stock purchases by Massachusetts, out of a total invested of $8 million. Three directors out of nine were appointed by the Commonwealth, which held significant amounts of common stock (Mazlish, *Railroad and the Space Program*, 173). The loans by Massachusetts and the City of Albany were repaid (Goodrich, *Government Promotion*, 126–8).

31. Salsbury, *State, the Investor, and the Railroad*, 172. The line cost $8 million to build; even twenty years later, few mills and ironworking facilities were capitalized at more than $2 million (Chandler Jr., *Visible Hand*, 89).

32. Fares were not cheap: a $5.50 ticket to go from Boston to Albany would be the equivalent of about $550 in today's money (Gordon, *Great Game*, 74; for a sample of fares,

see Gordon, *Passage to Union*, 83). In the early 1800s, it took a steamboat as long as thirty-two hours to go from New York City to Albany; thirty years later, that time should have been significantly reduced, but the railroad time was still faster (Douglas, *All Aboard*, 9).

33. Quoted in Harlow, *Steelways of New England*, 131–2.

34. Ibid., 134. George Whistler was also in charge of constructing this line. He left for Russia in 1842 to construct the line from St. Petersburg to Moscow and died there some years later.

35. Chandler Jr., *Visible Hand*, 91. Yankee rail executives dominated the railroad business until at least 1890, and "Boston capital" was ubiquitous in the West (Martin, *Railroads Triumphant*, 212).

36. Though there was an attempt in the 1850s by the Western to take over the New York Central, in 1900 the latter, Corning's and Vanderbilt's creation, took over what came to be known as the Boston and Albany (Harlow, *Steelways of New England*, 330–1).

37. From a report by the New York City Board of Aldermen, 1842 (quoted in ibid., 132–3).

38. Canal boats could be rafted down the river to New York in groups of fifty or more, bypassing the need to transship freight at Albany, as the Western required (Crump, *Brief History of the Age of Steam*, 95).

39. Salsbury, *State, the Investor, and the Railroad*, 284–5. There was an attempt by the leadership of the Western Railroad in 1854–55 to interest Boston investors in acquiring control of the Grand Central Railroad in New York to gain control of the trade with the West, but there was a lack of interest. See also Berstein, *Wedding of the Waters*, 368. The smaller railroads in New England were never coordinated or amalgamated until after the Civil War, and therefore had only a local purpose (Martin, *Railroads Triumphant*, 44).

40. Carter, *When Railroads Were New*, 75–6.

41. Adams Jr., *Railroads*, 39. The chief engineer of what was then the longest railroad in the world was Horatio Allen, who had been active in the promotion of steam transportation between the Delaware and Hudson Rivers. His father was a graduate of Brown University in Rhode Island and taught at Union College in Schenectady, New York. The basic story of the creation of the line is in Carter, *When Railroads Were New*, 23–5.

42. The engine was made at a foundry in West Point, New York (Adams Jr., *Railroads*, 39). The first such trip on the line took place on Christmas Day 1830, the train hauling 141 passengers six miles (Paquet, *Wanderlust*, 101).

43. Carter, *When Railroads Were New*, 77.

44. Ibid., 76–7. Henry and Maria's son, George, graduated from Yale in 1854 and had a career in Chicago real estate; see Yale University, *Obituary Record of Graduates of Yale University Deceased during the Academical Year ending in June, 1905* (New Haven, CT); available online at http://mssa.library.yale.edu/obituary_record/1859_1924/1904-05.pdf. By the 1820s, the Ramapo Works were producing both textiles and iron fittings, primarily for the local and New York City markets (Jeremy, *Transatlantic Industrial Revolution*, 102).

45. After the opening of the Erie Canal, the State Legislature authorized a survey to be made for a road through the southern part of the State, but the terrain was so rough that the project was abandoned. Political opposition from north of New York's "Southern

Tier" also contributed to its failure; see Moody, *Railroad Builders*, 66–7; and Fox and Remini, *Decline of Aristocracy*, 334–5.

46. Douglas (*All Aboard*, 24) refers to a "mild epidemic" of enthusiasm.

47. At first, the project seemed too large for one company, but Lord managed to convince everyone that it should be done that way (Carter, *When Railroads Were New*, 78–9).

48. Charles Francis Adams Jr., *Chapter of Erie* (Boston: Fields, Osgood, 1869), 1. The promoters had to raise $1 million in subscriptions before the charter could come into effect (ibid., 76–7, 80; see also Gordon, *Business of America*, 92).

49. Its charter said it had to have $10 million in subscriptions, with 5 percent paid in before it could operate (Carter, *When Railroads Were New*, 79).

50. Douglas, *All Aboard*, 38–40.

51. Adams, Jr., *Chapter of Erie*, 1. The survey was apparently carried out by DeWitt Clinton Jr. (Moody, *Railroad Builders*, 67).

52. Other entrepreneurs built a connection from Ramapo to Jersey City, saving Erie passengers a lot of time. The line could not literally connect with the Erie, but Ramapo was a small town. In 1851, the rules were changed and the Erie leased the connection (Carter, *When Railroads Were New*, 102–3).

53. Gordon, *Business of America*, 92–4

54. The freight from the West that would make Buffalo, but not Dunkirk, was flour. Although wheat was carried by lake steamer, it was more profitable to transport it to the eastern end of Lake Erie and there to have it milled, in huge quantities, into flour and shipped onward. Flour was a more valuable product than wheat, and the railroads did better by it (Martin, *Railroads Triumphant*, 168).

55. See, for instance, Bethany Anderson, "A Political Fire," *Quarto* 29 (spring-summer 2008): 5–7; and Carter, *When Railroads Were New*, 80–1.

56. Moody, *Railroad Builders*, 68.

57. The story is told in Carter, *When Railroads Were New*, 82–4, 94–5.

58. Chandler (*Railroads*, 45–6, 56) explains how these syndicates made their money; see also Carter, *When Railroads Were New*, 93. Alexander Divan, a lawyer from Elmira, New York, who had been involved in a line south from Elmira to Pennsylvania in the 1830s, proposed the first construction syndicate. He was involved in railroads, especially the Erie, almost until he died in the 1890s. The most infamous construction syndicate was the Crédit Mobilier of America, which was created by the management of the Union Pacific as it proceeded with construction across the Great Plains from 1867 to 1869. A scandal broke when it was found that the syndicate had been distributing some of its shares to congressmen to secure less onerous legislation governing the railroad. The vice president was implicated as well, and was dropped from Ulysses S. Grant's re-election team in 1872.

59. Upon completion, the State cancelled $6 million in Erie bonds that it held, thus helping relieve the line's heavy indebtedness (Goodrich, *Government Promotion*, 59).

60. The story is told in detail in Carter, *When Railroads Were New*, 77–8, 95–100. See also Burrows and Wallace, *Gotham*, 655–7.

61. Martin, *Railroads Triumphant*, 207; see also Moody, *Railroad Builders*, 70.

62. Gordon, *Business of America*, 94. There may have been other dividends before the late 1850s, but, after that, none until 1942. Things only got more difficult during the Civil War, as expansion and financial depredations raised the debt load to $16 million (Adams, Jr., *Chapter of Erie*, 2).

63. Carter, *When Railroads Were New*, 78

64. Gordon, *Business of America*, 94.

65. Douglas, *All Aboard*, 38–40. Although the Erie had no local industry of any size along its route, as did the New York Central and the Pennsylvania Railroad, its two chief competitors, it did gain a certain advantage from the discovery of oil in Pennsylvania, just south of its line.

66. Harlow, *Steelways of New England*, chap. 20. Another study of railroad executives active in the 1840–90 period includes material on sixty-one men, thirty of whom were from New England and five others from upstate New York, which demonstrates the Yankee influence on railroad development; see Thomas C. Cochran, *Railroad Leaders, 1845–1890: The Business Mind in Action* (New York: Russell & Russell, 1965), appendix.

67. During the English Civil War, Corning's ancestors had supported Oliver Cromwell and left for New England around the time of the Restoration. A short, useful family history is given in Kennedy, *O Albany!*, 328–31; see also Paul Grundahl, *Major Erastus Corning: Albany Icon, Albany Enigma* (Albany: SUNY Press, 2007), chap. 2; and Irene Neu, *Erastus Corning: Merchant and Financier, 1794–1872* (Ithaca, NY: Cornell University Press, 1960).

68. Corning's ironworks provided the plate that was used in the construction of the USS *Monitor*, which fought the CSS *Virginia* (formerly the *Merrimac*) during the Civil War, in the first of the modern iron-plated naval battles.

69. Featherstonehaugh led an interesting and varied life. He was apparently a wealthy British cattle breeder living in Duanesburg, west of Albany, who posted a proposal to incorporate a railway in an Albany paper in 1825. He became a director of the railroad when the charter was issued in 1826. He then suffered a number of personal losses, including a spouse and children, and left the area before 1831. He moved to Philadelphia, where he became interested in science and geology. He was appointed a government geologist and went off to the (then) Southwest — Missouri and Arkansas. In 1840, he returned to England, where he was appointed one of the British commissioners tasked with settling the boundary dispute between Canada and America. Then he was appointed British consul for the Seine department in France, and was central in helping King Louis-Philippe escape to Britain in the Revolution of 1848. See Brian Solomon with Mike Schafer, *New York Central Railroad* (St. Paul, MN: Voyager Press, 2007), 14-16; and George P. Merrill, *The First One Hundred Years of American Geology* (1925, repr. New York: Hafner Publishing, 1964, 1969), 136–8.

70. Grundahl, *Major Erastus Corning*, 21; Martin, *Railroads Triumphant*, 14. See also Fox and Remini, *Decline of Aristocracy*, 306.

71. For descriptions of the locomotive, see Crump, *Brief History of the Age of Steam*, 184; and Oliver, *History of American Technology*, 184. The line and the locomotive were designed by one of the most creative engineers of the times, John Jervis, who came from a Long Island Yankee family that moved to Rome, New York, shortly after 1800. He worked on the Erie Canal, then became chief engineer for the Delaware and Hudson

Canal; it was he who sent Horatio Allen to Britain to obtain designs for a locomotive. He then moved to Albany and the Hudson and Mohawk line. Jervis was later involved with many railroad projects in the West and with the development of the New York City water supply. See, for instance, Hawke, *Nuts and Bolts of the Past*, 224–5.

72. The Albany station was at "The Point," where the Western Turnpike, now US 20, met Madison Street. The toll booth for this turnpike was two blocks to the west of the station's location; it was eventually relocated to the waterfront (Kennedy, *O Albany!*, 117, who also claims that the railroad did not open until September 24). It reduced travel time from a full day on the turnpike to only three hours. It had no freight cars until 1840. Within sixty days of its opening, an Albany paper contained notices of intent to charter requiring an aggregate capital of $22 million. By 1836, 141 railroad charters had been issued, though only thirty lines were ever built; see Carter, *When Railroads Were New*, 150–6; and Klein, *Empire State*, 312.

73. An earlier railway from Schenectady to Saratoga went into operation in 1832.

74. Troy made an attempt to secure Western freight transshipment from Albany by building the Schenectady-Troy line in 1842 and running it as a municipal enterprise, but Albany, in turn, pledged $250,000 for upgrades to the Mohawk and Hudson and successfully prevented this diversion. The State of New York prevented further action of this sort in its 1846 Constitution (Goodrich, *Government Promotion*, 58–9).

75. Carter, *When Railroads Were New*, 168. Twenty-five years later, a passenger could go from New York City to Chicago in around thirty hours (Stiles, *First Tycoon*, 510).

76. Winter also affected the railroads. Since the Mohawk and Hudson was a bypass of the falls on the Mohawk River, there were no passengers when the canal was not operating.

77. These changes were not as threatening as they might seem to us today. Before the Civil War, the ability of engines to pull high tonnages was improving, but not great. Without steel rails, something not common until the 1870s, heavy weights could damage the railbeds. Finally, Western produce haulage tended to be concentrated in the fall of the year, when the canal had not yet frozen over. The railroads could haul manufactured products all year round and the sales cycle for these tended to be in the winter and spring, when the canal was inoperable; see Salsbury, *State, the Investor, and the Railroad*, 28, 277. A transplanted Vermont Yankee, Henry Wells, entered the shipping business first on the Erie Canal and then on the New York Central. He eventually partnered with William Fargo and John Butterfield to create American Express and then Wells Fargo (Koeppel, *Bond of Union*, 393).

78. At the dinner that followed the first trip on the Mohawk and Hudson, a number of toasts were given, one of which Corning must have heard, if he did not utter it: "The Buffalo Railroad — may we soon breakfast in Utica, dine in Rochester, and sup with our friends on Lake Erie!" (Adams Jr., *Railroads*, 52).

79. An engaging version of the story of the growth and consolidation of the New York Central is told in Carter, *When Railroads Were New*, 157–84.

80. The New York Central gave most of its business to the steamship line that ran between Albany and New York. As there was no rail connection to the rail lines on the east bank of the Hudson until 1866, it made sense, if the cars had to be unloaded, to give

the business to the ships, whose cost and prices were lower than those of the rail lines. During the winter, the traffic was transshipped to the rail line (Stiles, *First Tycoon*, 382–3).

81. Martin, *Railroads Triumphant*, 247. Corning also provided much of the hardware, including ironwork, to these railroads. He normally took no salary for his positions, but his business with the railroads was most advantageous to him.

82. The Lake Shore Railroad and the Michigan Central were completed at about the same time, giving the New York Central access to Chicago (ibid., 17–18).

83. The canal, which still operates today, was built in 1855; within five years, a quarter of the pig iron made in America came from iron mined along the shores of Lake Superior. It also caused iron production to move from eastern to western Pennsylvania, closer to the Lake ports where the iron ore was offloaded onto rail cars (Chandler, *Railroads*, 32).

84. Corning retired from the presidency in April 1864 (Stiles, *First Tycoon*, 406). "Commodore" Cornelius Vanderbilt acquired control of the line between New York City and Albany and then took control of the New York Central (idem, chaps. 14–17; see also Burrows and Wallace, *Gotham*, 112). The first board meeting of the combined railway was in January 1870.

85. Schlereth, *Victorian America*, 22.

86. Crump, *Brief History of the Age of Steam*, 136.

87. Dee Brown, *Hear that Lonesome Whistle Blow: Railroads in the West* (New York: Touchstone, 1994), 29–32.

88. The Southern Pacific reached east from Los Angeles in 1876, the Santa Fe reached the coast in 1885, as did the Canadian Pacific at Vancouver. The Northern Pacific reached Tacoma, Washington, in 1887, and the Great Northern reached Seattle in 1893. The Canadian Pacific was the only railway to cross the continent from tidewater to tidewater: all of the American lines met somewhere in the Midwest. See Crump, *Brief History of the Age of Steam*, 202; and Gordon, *Passage to Union*, 189.

89. Crump, *Brief History of the Age of Steam*, 216.

90. Martin, *Railroads Triumphant*, 18, 326–7.

91. See William Cronon, *Nature's Metropolis: Chicago and the Great West* (New York: W.W. Norton, 1991), 80–1.

92. Martin, *Railroads Triumphant*, 246. The system was developed in tandem with the geographically faster-growing telegraph system, which had many of the same requirements; see Lubrano, *Telegraph*, 86.

93. Chandler Jr., *Visible Hand*, 79, 94.

94. Ibid., 96–7; and Salsbury, *State, the Investor, and the Railroad*, 186–9.

95. Alfred D. Chandler, Jr., *The Railroads, the Nation's First Big Business; Sources and Readings* (New York: Harcourt, Brace & World, 1965), 98–9; for a good example of what organizational challenges were presented to a railroad of the mid-1800s, see pp.102–25. See also Mazlish, *Railroad and the Space Program*, 24–5.

96. Chandler Jr., *Visible Hand*, 101–4.

97. Martin, *Railroads Triumphant*, 273.

98. Chandler Jr., *Visible Hand*, 120. This need led to the early development of business machines.

99. Ibid., 188.

100. Gordon, *Passage to Union*, 273.
101. Cronon, *Nature's Metropolis*, 85–6.
102. Chandler Jr., *Railroads*, 97.
103. Stiles, *First Tycoon*, 403.
104. Chandler, Jr., *Railroads*, 50–1. With the exception of lines crossing the mountains in the western part of the region, little finance for the construction of New England railroads came from bonds, which were used only if there was a miscalculation in the cost of a line. Sometimes the capital subscribed in shares ended up being more than the cost of construction — a happy result (50).
105. Ibid., 44. A refusal to pay the subscriptions might lead to a lawsuit as the company tried to collect its pledges. Abraham Lincoln, as a corporate lawyer in the 1850s, had experience with these. See James W. Ely Jr., "Abraham Lincoln as a Railroad Attorney" (paper delivered at the Indiana Historical Society, 2005 Railroad Symposium: Lincoln and the Railroads), available online at http://www.indianahistory.org/our-services/books-publications/railroad-symposia-essays-1/Abe%20Lincoln%20as%20a%20Railroad%20Attorney.pdf.
106. As Goodrich (*Government Promotion*, 6–7) notes, "[t]hese programs were indeed large enough to provide in their time a leading example of 'the modern tendency to extend the activity of the state into industry'." Goodrich is quoting Guy Stevens Callender, who, in 1902, went on to ask why this should have taken place "here in America, where of all places in the world we should least expect to find it." The question is still a challenging one to all who regard nineteenth-century America as the historic stronghold of *laissez-faire* capitalism. Carter (*When Railroads Were New*, 7–11) points to three reasons for federal and State governmental action in infrastructure: the risk of extending works through unpopulated or lightly populated areas, the lack of local capital and existence of the frontier, and the newness of large-scale corporate organization and finance.
107. Gordon, *Passage to Union*, 4–5.
108. For an account of the scrip issued by Massachusetts for the Western Railraod, see Salsbury, *State, the Investor, and the Railroad*, 143.
109. Because bonds were seen as having more reliable income, as they were "first in line" to be repaid in the case of company failure, they were more welcome by distant investors. Most railroads outside the Northeast were financed by the use of bonds (Chandler, Jr., *Railroads*, 44). Sometimes, railroad company directors might become the syndicate, thereby profiting not only from the construction of the line, but also from its subsequent operation. Erastus Corning's hardware and iron foundry businesses benefited from the construction of the railroad from Albany to Buffalo, as, probably, did the Pierson iron businesses from the construction of the Erie.
110. Ibid., 44–5. The modern investment bank made its appearance as Wall Street became the national stock exchange. The Morgans, a Yankee family that settled in New York and London, became preeminent in this field.
111. In 1835, three railroads were listed on the exchange, rising to ten in 1840 and to thirty-eight by 1850. By the outbreak of the Civil War, they accounted for a third of the activity on the exchange (Gordon, *Great Game*, 76).
112. There was considerable concern over this issue by the directors of the Boston

and Worcester and Western Railroads, but the lure of high (and safe) profits from textile manufacturing tended to force early private financing efforts into the Wall Street market. By 1835, banks held 20 percent of Boston and Worcester stock and another 45 percent was traded on Wall Street (Salsbury, *State, the Investor, and the Railroad*, 95–7).

113. Martin, *Railroads Triumphant*, 274.

114. Weissend, *Life and Times of Joseph Ellicott*, 16–18.

115. The Western sold State-owned land in Maine (Salsbury, *State, the Investor, and the Railroad*, 144).

116. Chandler, Jr., *Railroads*, 30. Despite their high profile in railroading history, land grants apparently assisted in the building of only 8 percent of the system (43).

117. Goodrich, *Government Promotion*, 268–9; and author's estimates. The federal contributions increased after the Civil War with the building of the transcontinental lines, but the government proportion of investment relative to that supplied privately declined in relative value. Canal investment from 1815 on tended to rely on public sources more extensively than did the railroads (Goodrich, *Government Promotion*, 268–71).

118. Martin, *Railroads Triumphant*, 195.

119. George W. Wilson, "US Intercity Passenger Transportation Policy 1930–1991, An Interpretive Essay," in Canada, Royal Commission on National Passenger Transportation Directions, *Final Report*, vol. 3 (Ottawa, 1992), 184. See also Douglas, *All Aboard*, 69.

120. Stott, "Artisans and Capitalist Development," 103–4; and Beatty, *Colossus*, 63. The ton/miles hauled by railroads grew from 10 billion in 1865 to 79 billion in 1890 (Douglas, *All Aboard*, 146)

121. Chandler, Jr., *Railroads*, 8.

122. Mazlish, *Railroad and the Space Program*, 120.

123. Quoted in Gordon, *Great Game*, 75.

124. Chandler, Jr., *Railroads*, 160–1.

125. It was only after considerable consolidation of lines into a few companies and the influence of financier J.P. Morgan in the 1890s that a pool arrangement that was not cheated upon was devised. Morgan could use his financial power to keep the temptation to cheat away (ibid., 162).

126. Ibid., 9–10, 159.

127. The first real civil engineering program began as a result of reforms made after the War of 1812, though the military academy was created in 1802.

128. Chandler, Jr., *Visible Hand*, 132; see also Clifford J. Schexnayder and Richard Mayo, *Construction Management Fundamentals* (Chicago: McGraw-Hill Professional, 2003), 5.

129. Schexnayder and Mayo, *Construction Management Fundamentals*, 9.

130. Chandler, Jr., *Visible Hand*, 77.

131. Douglas, *All Aboard*, 143–4; and Gordon, *Passage to Union*, 7, 244–5.

132. Martin, *Railroads Triumphant*, 49.

133. The width in the South was part of an attempt to cut off the "nationalizing" pattern created by the railroads. In 1861, when the Civil War started, there was thus no connection between the Northern and Southern lines from Washington, DC, west along the Ohio River (Gordon, *Passage to Union*, 5). As well, the Southern lines were plagued

by varying track widths; one writer estimated that, in 1860, there were eight changes of track width between Charleston, South Carolina, and Philadelphia (Hawke, *Nuts and Bolts of the Past*, 222). US railroad gauges were not standardized until 1886 (Martin, *Railroads Triumphant*, 46; see also Gordon, *Passage to Union*, 250). In the early years of railroads, cities and towns often tried to create or keep gauge differences to force passengers to stop and spend money there. The "Erie War" of the 1850s erupted when the New York Central tried to standardize the gauge through Erie, Pennsylvania, and the townspeople kept tearing up the new track. The famous *Effie Alton* case involving Lincoln was another attempt to halt railroads from providing quick, through service (Gordon, Passage to Union, 114–5, 274).

134. Douglas, *All Aboard*, 144; see also Mazlish, *Railroad and the Space Program*, 179–80.

135. Quoted in Schlereth, *Victorian America*, 30–1. The promoter of this idea was Sir Stanford Fleming, an executive of the Canadian Pacific railroad. His summer home in Halifax, now a public park, is next to our former home. There were five zones across the continent, but a number of States and cities rejected the railroads' decision for years, causing all kinds of problems. In 1918, a federal law made railroad time federal time (Gordon, *Passage to Union*, 250).

136. Benjamin F. Taylor, 1874 (quoted in Martin, *Railroads Triumphant*, 33).

137. The *Car and Locomotive Cyclopedia of American Practices*, as it is now called, was last updated in 1997; see Gordon, *Passage to Union*, 254–5; and Builders of Wooden Railway Cars...and Some Other Stuff, "The Car-Builder's Dictionary" (2007); available online at http://www.midcontinent.org/rollingstock/builders/bibliog-bldrs.htm.

138. Carter, *When Railroads Were New*, 92.

139. Chandler, Jr., *Railroads*, 32.

140. Gordon, *Passage to Union*, 347.

141. Ibid., 300. The basic network in America was finished by 1900 (Chandler, Jr., *Railroads*, 11).

142. Federal regulation began in 1887 (Chandler, Jr., *Railroads*, 185–6). The earliest railroad commissions were begun in New England around 1840 to judge disputes between landowners and the railroads' power of eminent domain. Regulation expanded from there (Mazlish, *Railroad and the Space Program*, 174). In 1825, Ohio created a Canal Commission to regulate it canals and to collect a "canal tax" that was designed to repay the State's investment in the projected canal system (Goodrich, *Government Promotion*, 134).

Chapter 22

1. Pulliam and Van Patten, *History of Education in America*, 79.

2. Lawrence A. Cremin, ed., *The Republic and the School: Horace Mann on the Education of Free Men* (New York: Teachers College of Columbia University, 1957), 87. The quote comes from Mann's Twelfth Annual Report, 1848.

3. Quoted in Leslie Allen Buhite, "The Chautauqua Lake Camp Meeting and the Chautauqua Institution" (PhD diss., Florida State University, 2007), 194; available online at http://etd.lib.fsu.edu/theses/submitted/etd-07172007-132431/unrestricted/BuhiteLDissertation.pdf.

4. Lawrence A. Cremin, *Traditions of American Education* (New York: Basic Books, 1977), parts I–II. Much of the material on early schooling is derived from these lectures.

5. Schooling was at the base of cultural continuity. The Puritans saw education as a way of leading people to acquiesce to social norms (Ziff, *Puritanism in America*, 68)

6. In 1642, Massachusetts tried to put the obligation for education on parents, but when that did not work, the law was changed to make towns responsible. The first tax for education was levied by the Town of Dedham in 1648, but tuition fees were also levied (Pulliam and Van Patten, *History of Education in America*, 94). Because there might be more than one village in a town, it was sometimes difficult for the town's inhabitants to agree on the location of the school, which led to people not paying the school tax; see Robert L. Church, *Education in the United States: An Interpretive History* (New York: Free Press, 1976), 9.

7. Pulliam and Van Patten, *History of Education in America*, 92.

8. William Hayes, *Horace Mann's Vision of the Public Schools: Is It Still Relevant?* (Lanham, MD: Rowman and Littlefield Education, 2006), 2.

9. Ibid., 3, 8. Others mention that boys went in the winter and girls in the summer.

10. Pulliam and Van Patten, *History of Education in America*, 135.

11. In 1648, the apprentices were assured that they, too, would get book learning (Ziff, *Puritanism in America*, 164).

12. Other parts of society had a similar interest in education. In 1821, Seth Y. Wells, a former instructor at the Hudson Academy, was made superintendent of the Shaker schools. He opened them to outside inspection, and the Shakers received their appropriate share of State education funding (Andrews, *People Called Shakers*, 189).

13. In 1789, when the public school system in Boston was created, the rules for girls restricted them to "penny schools" and to the months from April to October (Hecker, *Short History of Women's Rights*, 169).

14. Pulliam and Van Patten, *History of Education in America*, 136, 147. This was five years after the female seminary was opened in Troy, New York. By the Civil War, there were three hundred high schools in America, one-third of them in Massachusetts.

15. Ziff, *Puritanism in America*, 116.

16. There were plans to build Yale as early as 1641, but protests from Massachusetts to protect Harvard postponed it. Dartmouth's founder was a Connecticut missionary among the Indians (Lee, *Yankees of Connecticut*, 155); Princeton, too, though in New Jersey, was also a Yankee creation of thirty families from New Haven, Connecticut, that settled in Newark and formed the College of New Jersey, later renamed (154).

17. Oliver (*History of American Technology*, 58) says the first press was owned by Stephen Day and the publication was called *The Freeman's Oath*. Ziff (*Puritanism in America*, 164), however, says the first press was established 1638 and that most of its output was religious books, though it also published poetry and history.

18. Joel Spring, *The American School, 1642–1985* (New York: Longman, 1986), 4–6.

19. Oliver, *History of American Technology*, 59.

20. William S. Reece, "The First Hundred Years of Printing in British North America: Printers and Collectors" (Worcester, MA: American Antiquarian Society, 1990); available online at http://www.reeseco.com/papers/first100.htm. Daniel Webster started his

career as a schoolmaster, then went on to publish and personally sell copies of his spelling and grammar texts.

21. The aim of one Connecticut grammar school in 1684 was "the education of hopeful youth in the Latin tongue, and other learned languages so far as to prepare such youths for the college and public service of the country in church, and commonwealth" (Spring, *American School*, 7).

22. Two interesting and useful biographies are Mary Peabody Mann, *Life of Horace Mann*, new ed. (Boston: Lee and Shepard, 1888); and Tharp, *Until Victory*. Many of the personal details in this section come from these sources.

23. Cremin, *Traditions of American Education*, 59–60.

24. Mann had worked with Massachusetts education pioneer, James Carter, among others, to get the Board of Education established (Pulliam and Van Patten, *History of Education in America*, 141).

25. See, for instance, Church, *Education in the United States*, 61–104.

26. Cremin, *American Education*, 138.

27. Charles Hoyt and R. Clyde Ford, *John D. Pierce, Founder of the Michigan School System: A Study of Education in the Old Northwest* (Ypsilanti, MI: Scharf Tag and Box Co., 1905), chap. 8; available online at https://archive.org/stream/johndpiercefound00hoyt#page/n7/mode/2up. See also Cremin, *American Education*, 141, 161–2.

28. Mann was approached by the American Sunday School Union to designate a series of their materials for use as school texts, but he turned them down as too sectarian, leading to considerable bitterness. See Bob Pepperman Taylor, *Horace Mann's Troubling Legacy: The Education of Democratic Citizens* (Lawrence: University Press of Kansas, 2010), 48. As of 1855, of 50,890 libraries in the country, 30,000 were Sunday school libraries and 18,000 were district school libraries. These averaged at best about 200 volumes each. The remaining 4,800 libraries were larger entities, usually at universities (Cremin, *American Education*, 306).

29. Mann's journal entry for the day he accepted the position includes; "Henceforth, so long as I hold this office, I devote myself to the supremest welfare of mankind on earth....I have faith in the improvability of the race" (Spring, *American School*, 84). The texts of these reports have been collected in Cremin, *Republic and the School*.

30. This is the inscription on his statue near the Massachusetts Statehouse; see Taylor, *Horace Mann's Troubling Legacy*, 9.

31. Mann, *Life of Horace Mann*, 108.

32. Pulliam and Van Patten, *History of Education in America*, 141. Mann got Dwight to offer a $10,000 "challenge grant" to fund a normal school if the State would match the amount (Tharp, *Until Victory*, 145, 174).

33. Cremin (*American Education*, 142) notes that Mann gave schooling "its essential meaning, both in educational terms and in broader political terms." Mann was a Whig at the time, and was inclined to promote industrial progress and more centralized control of institutions over the Jeffersonian Democratic preference for agriculture and local autonomy. Today, these basic stances in American politics seem to have switched partisan homes (see Taylor, *Horace Mann's Troubling Legacy*, 68). See also Church, *Education in the United States*, 55–6; and Cremin, *American Education*, 138.

34. Mann did a lot, but left a lot for others to do. In 1850, the proportion of white children in school ranged from 1 percent in California to 5 percent in the Southern States to 32 percent in Vermont (Hayes, *Horace Mann's Vision*, 29).

35. Ibid., 16.

36. He could have led a number of State-supported institutions in the West instead, but apparently chose this small, private institution because he saw higher education not as a part of the democratic citizenship struggle, but as a way of forming a wise elite (Taylor, *Horace Mann's Troubling Legacy*, 78–9).

37. Tharp, *Until Victory*, 308.

38. The school could have been started as part of a land-promotion scheme, as Mann (*Life of Horace Mann*, 517) seems to suggest. Another problem was that the female students wanted the same rights as the men in terms of their social activity, which continually upset the villagers (524).

39. Hayes, *Horace Mann's Vision*, 17.

40. Antioch's location is either in, or immediately next to, the Virginia Military District, a part of Ohio given over to that State for allocation to its Revolutionary War veterans.

41. Pulliam and Van Patten, *History of Education in America*, 143. See also Ruth M. Baylor, *Elizabeth Palmer Peabody: Kindergarten Pioneer* (Philadelphia: University of Pennsylvania Press, 1965).

42. Schlereth, *Victorian America*, 253.

43. *American Journal of Education for the Year 1826*, vol. 1 (Boston: Wait, Greene, 1826), 526; available online at http://books.google.com/books?id=sAUUAAAAIAAJ&printsec=frontcover&dq=American+Journal+of+education&source=bl&ots=HbfYzEd5p6&sig=7IU7pRz2Z_O4bKeDpCi6IIXCL9E&hl=en&ei=e1KOTdT6PPSz0QH-k6C_Cw&sa=X&oi=book_result&ct=result&resnum=9&ved=0CE0Q6AEwCA#v=onepage&q&f=false.

44. Holbrook had been a student of Professor Silliman at Yale, as had been Samuel F.B. Morse. See also "Josiah Holbrook," in *Columbia Electronic Encyclopedia*, 6th ed. (New York: Columbia University Press, 2012; available online at http://www.infoplease.com/ce6/people/A0823960.html#ixzz1HjzwfwXg.

45. The Lyceum in ancient Athens was a garden with walkways adjacent to the Temple of Apollo Lyceus, where Aristotle taught in the fourth century BC. The word was used indiscriminately to mean a social practice, a specific local association, a building constructed for the use of a Lyceum, and the program or performance; see Angela G. Ray, *The Lyceum in Public Culture in the Nineteenth Century United States* (East Lansing: Michigan State University Press, 2005), 3. In French, a *lycée* is a high school. Holbrook was aware of mechanics' institutes formed in Boston in 1795 (Oliver, *History of American Technology*, 154), in Scotland about the 1820s for similar purposes, and the creation of a Lyceum in Gardner, Maine, in 1823 (Cremin, *American Education*, 312–15). See also James Truslow Adams, *Frontiers of American Culture: A Study of Adult Education* (New York: Scribners, 1944), 12.

46. Cremin, *American Education*, 312.

47. John E. Tapia, *Circuit Chautauqua: From Rural Education to Popular Entertain-

ment in Early Twentieth Century America (Jefferson, NC: McFarland and Company, 1997), 12–13.

48. It was extremely active, with 784 lectures, 105 debates, and 14 concerts in its early years (Cremin, *American Education*, 316).

49. Ray, *Lyceum in Public Culture*, 20–1.

50. Ibid., 2.

51. Ibid., 5.

52. Adams, *Frontiers of American Culture*, 36.

53. There are suggestions that others were started earlier, in 1751 or 1769, but Raikes's effort became the clear founding line.

54. In Connecticut, Sunday schools were created in 1816, just before the State changed its constitution to disestablish the Congregational church, effectively taking formal religious instruction out of its schools. The Congregational church as a whole began to advocate for Sunday schools in 1825. See Cross, *Burned-Over District*, 128.

55. Mann had opposed the inclusion of their materials in the common school texts as being too sectarian (Taylor, *Horace Mann's Troubling Legacy*, 48).

56. Buhite, "Chautauqua Lake Camp Meeting," 159.

57. Jeffrey Simpson, *Chautauqua: An American Utopia* (Chautauqua NY: Chautauqua Institution, 1999), 17. William Least Heat-Moon has an interesting story of his experiences in the area; see *River Horse*, 77–8.

58. The place was originally incorporated as the Fair Point Sunday School Assembly, then renamed the Chautauqua Lake Sunday School Assembly, then shortened to the Chautauqua Assembly, and, after the turn of the twentieth century, changed to the Chautauqua Institution (Buhite, "Chautauqua Lake Camp Meeting," 149).

59. Chautauqua Institution, "Welcome to Chautauqua: A Walking Tour Guide of the Chautauqua Institution" (n.d.), 1.

60. Schlereth, *Victorian America*, 253–4.

61. One of Vincent's correspondents suggested that the camp meeting format might be used, not as an evangelical device, but to expand the views of those already devout in their beliefs. It was an easy step to take to see one such group as those who volunteered to teach Sunday school; Buhite, "Chautauqua Lake Camp Meeting," 144; see also Andrew C. Rieser, *The Chautauqua Moment: Protestants, Progressives, and the Culture of Modern Liberalism* (New York: Columbia University Press, 2003), 100. There is some controversy over whether Miller had been one of the "incorporators" of the Chautauqua Lake Camp Meeting Association that bought and developed the site from 1871 on (Simpson, *Chautauqua*, 32). Miller, whose wife had attended a Fair Point meeting one summer, was a trustee of the Association. There is no question that the Association had been poorly managed and was in danger of going bankrupt (Buhite, "Chautauqua Lake Camp Meeting," chap. 3; and Rieser, *Chautauqua Moment*, 90–1).

62. Joseph E. Gould, *The Chautauqua Movement: An Episode in the Continuing American Revolution* (New York: State University of New York, 1961), 4.

63. Camp meetings, by this point, had evolved from "temporary towns" for rural people into "temporary rural vacation spots for urbanites." To a degree, their success depended on access by railroad. Also, the word "vacation" was not in common use before

the Civil War; see Cindy S. Aron, *Working at Play: A History of Vacations in the United States* (New York: Oxford University Press, 2001), 32–3.

64. Ibid., 6–9. The Puritan approach was that work glorified God, even though being saved or not was out of one's control. Even after this connection was lost, the notion of idle relaxation was regarded badly; Chautauqua solved this for Sunday school teachers.

65. This whole approach is best described in ibid., especially, 9, 34–42, 101–26; see also Paquet, *Wanderlust*, 199.

66. Aron, *Working at Play*, 111.

67. Townsend, a Methodist minister, had become disaffected with Methodist theology by the mid-1880s and had moved toward a more liberal interpretation. He created an Independent Congregational Church in Jamestown, New York, at the eastern end of Lake Chautauqua, and in 1886 he started another chautauqua at Lakewood, just west of Jamestown, that was based on a more liberal theology than that promoted by Vincent and the Methodists just up the lake. His chautauqua enjoyed a number of years of success, due in part to people's hunger for the combination of morality and culture, but it did not affect the influence and fortunes of the original. See, for instance, "Rev. James G. Townsend" (Clarion: Clarion University of Pennsylvania, 2008), available online at http://www.clarion.edu/618/; and Len Faulk, "Historical Significance of Rev. James G. Townsend's Lakeside School of New Theology, *New Theology Herald*, and Liberal Religious Writings" (presentation to the Chautauqua County Historical Society, June 23, 2010), available at http://www.mcclurgmuseum.org/collection/library/lecture_list/lecture_list.html.

68. The Association had purchased the land in 1866 with help from a local oil-rich family, and development money was raised from share subscriptions. A Jamestown author who had gone to the huge Round Lake, New York, camp meeting tried to interest its organizers to come to Chautauqua for their 1872 camp meeting, but was turned down. Declining attendance in 1873 and bad weather in 1874 further discouraged them (Buhite, "Chautauqua Lake Camp Meeting," chap. 3).

69. Ellis and Cash, *New York State*, 114. Simpson (*Chautauqua*, 33–4) estimates that four thousand were there, from twenty-five States, England, Scotland, Ireland, and India, to hear eight sermons, twenty-two lectures on Sunday school theory and practice and seven on Biblical history and geography. A final exam was given to two hundred people at the end of the program. Paquet (*Wanderlust*, 200), says five hundred came.

70. The Association deeded the grounds and buildings to the Chautauqua Institution in 1876.

71. Simpson, *Chautauqua*, 37

72. Ibid., 38.

73. See Huey B. Long, "Adult Education in the Oneida Community: A Pattern for the Chautauqua Assembly," *Journal of the Midwest History of Education Society* 22 (1995): 203–15.

74. Both Miller and Vincent were unhappy about the evangelical and emotional side of Methodism and tried to make the Assembly as unlike a traditional camp meeting as possible (Simpson, *Chautauqua*, 33).

75. Rieser, *Chautauqua Moment*, 37.

76. Ellis and Cash, *New York State*, 114.

77. Rieser, *Chautauqua Moment*, 45.
78. Simpson, *Chautauqua*, 17; and Tapia, *Circuit Chautauqua*, 20–2.
79. Brown, *American West*, 183.
80. Rieser, *Chautauqua Moment*, 10.
81. Wish, *Society and Thought in America*, 114.
82. Rieser, *Chautauqua Moment*, chap. 5.
83. Ibid., 191–2. Ida Tarbell, the famous journalist, was effectively the managing editor of the CLSC magazine during much of the 1880s.
84. Ibid., 180–4.
85. Ibid., 3–4. As well, Rieser notes (103) that Chautauqua tried to disassociate itself from the Lyceums and Mechanics' Institutes, but was not successful.
86. Simpson, *Chautauqua*, 51–5. Rainey later would have a falling-out with Chautauqua when he tried to move the educational elements of the university program as well as the CLSC to Chicago. The board refused, and he suffered a double defeat when faculty at the new University of Chicago resisted his attempt to set up similar programs as a part of its structure. Even so, many Chicago professors, as well as others from elsewhere, including John Dewey and Frederick Jackson Turner, took part in the Chautauqua summer programs; see, for instance, Gould, *Chautauqua Movement*, 27–60.
87. No advocate of idleness, Vincent nevertheless had to remind the visitors to "be careful not to overtax yourself. Do not go to every thing" (Paquet, *Wanderlust*, 200).
88. Quoted in Buhite, "Chautauqua Lake Camp Meeting," 161–2.
89. Simpson, *Chautauqua*, 47.
90. Clement Studebaker, the car manufacturer, followed Miller as president upon his death in 1899. After Studebaker's death in 1906, John Heyl Vincent's son, George, became president. Other prominent supporters included Henry Ford, Thomas Edison (who married Miller's daughter), Andrew Carnegie, Harvey Firestone, poet William Cullen Bryant, Charles Welch of grape juice fame, and the Massey brothers, Canadian industrialists (Vincent Massey would become governor general of Canada; Chester Massey married John Heyl Vincent's half-sister). See Kathleen Crocker and Jane Currie, *The Chautauqua Institution, 1874–1974* (Charleston, SC: Arcadia Publishing, 2001), 11, 30, 110, 123.
91. Oddly enough, one of Chautauqua's major supporters, who rescued the Institution from receivership in 1936, was John D. Rockefeller. His most famous critic, muckraker Ida Tarbell, whose father was a Pennsylvania oil producer, began her career as a reporter for the Institution's newspaper, the *Chautauquan*. See Simpson, *Chautauqua*, 34; Chernow, *Titan*, 437; and Crocker and Currie, *Chautauqua Institution*, 40. Joseph Gould (*Chautauqua Movement*, vii) sees Chautauqua as having "given discipline and direction to angry and inchoate movements of social protest of the times….a response to an unspoken demand for 'something better'."
92. A Catholic chautauqua was formed in 1892, followed by a Jewish one in 1897, and an African American one in 1906, which met only once (Rieser, *Chautauqua Moment*, 124, 159).
93. Ibid., 48–57.
94. Aron, *Working at Play*, 125. Active "playing" seemed to become accepted as nonslothful.

95. From 1860 to 1900, the proportion of the American population that was urban doubled from 20 percent to 40 percent (Simpson, *Chautauqua*, 26).

96. Ibid., 18.

97. This is the title of a collection of remembrances about the circuit chautauquas by a performer; see Gay MacLaren, *Morally We Roll Along* (Boston: Little, Brown, 1938).

98. Rieser, *Chautauqua Moment*, 38.

99. Gladstone Historical Society, *Gladstone Chautauqua, 1894–1927: A Centennial Remembrance* (Gladstone, OR, 1994). Other examples can be found in Tapia, *Circuit Chautauqua*, 22–5.

100. Lakeside, located between Toledo and Cleveland on the Marblehead Peninsula, is a half-square-mile vacation community founded by the Methodists in 1873. It still has a "season" and about six hundred residents.

101. Generally speaking, if an American town has the word "Grove" in its name, it was probably the site of many camp meetings in the nineteenth century.

102. Bay View went through many of the same stages of physical and program development as did Chautauqua, though with a decade's distance. It was a camp meeting ground and a Sunday school teachers' venue and had a more general chautauqua program format; see Mary Jane Doerr, *Bay View: An American Idea* (Allegan Forest, MI: Priscilla Press, 2010).

103. Much of the circuit chautauqua story presented here is derived from Tapia, *Circuit Chautauqua*; and Robert A. McCown, "Records of the Redpath Chautauqua," *Books at Iowa* 19 (1973): 8–23.

104. Sheilagh S. Jameson, with Nola B. Ericson, *Chautauqua in Canada* (Calgary: Glenbow-Alberta Institute, 1979), 12.

105. McCown, "Records of the Redpath Chautauqua."

106. Adams, *Frontiers of American Culture*, 133. The most popular speaker on the chautauqua circuit was William Jennings Bryan.

107. Hobson, *Remembering America*, 202.

108. McCown, "Records of the Redpath Chautauqua,"

109. Ibid.

110. Simon Schama, *A History of Britain: The British Wars: 1663-1776* (Toronto: McClelland & Stewart, 2002, 458.

111. William M. Keith, *Democracy as Discussion: Civic Education and the American Forum Movement* (Lanham, MD: Lexington Books, 2007), 213.

Bibliography

Abbey, Edward. *The Brave Cowboy: An Old Tale in a New Time*. New York: Dodd, Mead, 1956.

———. *Desert Solitaire: A Season in the Wilderness*. New York: McGraw-Hill, 1968.

———. "In Defense of the Redneck." in Edward Abbey, *Abbey's Road*. New York: Dutton, 1979.

———. *The Monkey Wrench Gang*. New York: Lippincott, Williams & Wilkins, 1975.

Ackermann, Marsha E. *Cool Comfort: America's Romance with Air-Conditioning*. Washington, DC: Smithsonian Institution Press, 2002.

Adams Jr., Charles Francis. *Chapter of Erie*. Boston: Fields, Osgood, 1869.

———. *Railroads: Their Origin and Problems*. 1878; New York: G.P. Putnam's Sons, 1987.

Adams, Henry. *The Education of Henry Adams*. New York: Random House, 1999.

Adams, James Truslow. *Frontiers of American Culture: A Study of Adult Education*. New York: Scribners, 1944.

Aitken, Hugh G.J. *Taylorism at Watertown Arsenal: Scientific Management in Action, 1908–1915*. Cambridge, MA: Harvard University Press, 1960.

Aldridge, Alfred Owen. *Jonathan Edwards*. New York: Washington Square Press, 1964.

Aleshire, Peter. *Arizona Highways*, April 2005.

Algeo, Matthew. *Harry Truman's Excellent Adventure: The True Story of a Great American Road Trip*. Chicago: Chicago Review Press, 2009.

Allen, Michael Patrick, and Nicholas L. Parsons, "The Institutionalization of Fame: Achievement, Recognition and Cultural Consecration in Baseball." *American Sociological Review* 71. 5, 2006.: 808–25.

American Chemical Society. "National Historic Chemical Landmarks: The Life of Russell Marker." Washington, DC, 2004. Available online at http://acswebcontent.acs.org/landmarks/marker/life.html.

American Society of Mechanical Engineers. "National Mechanical Engineering Heritage Site: Milam Building." San Antonio, TX, 1991. Available online at http://files.asme.org/ASMEORG/Communities/History/Landmarks/5595.pdf.

Amory, Cleveland. *The Proper Bostonians*. Orleans, MA: Parnassus Imprints, 1984.

Anderson, M.B., and D.K. Bartlett. "Address and Sermon on the Death of William Abelard Reynolds with other memorial papers." Rochester, 1872, Available online at http://books.google.ca/books?id=o3YPAAAAYAAJ&dq=Abelard+reynolds&printsec=frontcover&source=bl&ots=FcsGOyScBg&sig=XBSo3U8UYqBIEsptmvt_h5XdhI&hl=en&ei=SDTnSpnJKcbDlAenyb3_Bw&sa=X&oi=book_result&ct=result&resnum=6&ved=oCBoQ6AEwBQ#v=onepage&q=&f=false.

Anderson Jr., Oscar E. *Refrigeration in America*. Princeton, NJ: Princeton University Press, 1953.

Anderson, Sherwood. *Winesburg, Ohio: A Group of Tales of Small-town Ohio Life*. New York: B.W. Huebsch, 1919.

Anderson, Virginia DeJohn. "The Origins of New England Culture." *William and Mary Quarterly* 48. 2, 1991.: 231–7.

Andrews, Edward Deming. *The People Called Shakers*. 1953; New York: Dover, 1963.

Aron, Cindy S. *Working at Play: A History of Vacations in the United States*. New York: Oxford University Press, 2001.

Arrington, Leonard J. *Great Basin Kingdom: An Economic History of the Latter-day Saints, 1830–1900*. 1958; repr. Cambridge, MA: Harvard University Press, 1993.

Arora, C.P. *Refrigeration and Air Conditioning*. 1981; New York: McGraw-Hill, 2000.

Arsenault, Raymond. "The End of the Long Hot Summer: The Air Conditioner in Southern Culture." *Journal of Southern History* 50. 4, 1984.: 597–628.

Asbell, Bernard. *The Pill: A Biography of the Drug that Changed the World*. New York: Random House, 1995.

Ashton, Thomas Southcliffe, and Pat Hudson. *The Industrial Revolution, 1760–1830*. Oxford: Oxford University Press, 1997.

Astor, Gerald, et al. *The Baseball Hall of Fame 50th Anniversary Book*. Englewood Cliffs, NJ: Prentice-Hall, 1988.

Atkinson, Rick. *Day of Battle*. New York: Henry Holt, 2007.

Bacon, Margaret F. *Valiant Friend: A Life of Lucretia Mott*. Philadelphia: Quaker Books, 1999.

Bailyn, Bernard. *The Peopling of British North America: An Introduction*. New York: Alfred A. Knopf, 1986.

———. *Voyagers to the West: A Passage in the Peopling of America on the Eve of the Revolution*. New York: Vintage Books, 1986.

Bain, David Haward. *The Old Iron Road: An Epic of Rails, Roads, and the Urge to Go West*. New York: Penguin, 2005.

Baker, Carlos. *Emerson among the Eccentrics: A Group Portrait*. New York: Viking, 1996.

Baker, Jean H. *Sisters: The Lives of America's Suffragists*. New York: Hill and Wang, 2005.

Barrett, William P. "Willis Carrier's Ghost." *Forbes*, May 29, 2000, 152–62.

Baylor, Ruth M. *Elizabeth Palmer Peabody: Kindergarten Pioneer.* Philadelphia: University of Pennsylvania Press, 1965.

Beard, Mary. *Woman as a Force in History: A Study in Transitions and Realities.* New York: Macmillan, 1946.

Beatty, Jack. *Colossus: How the Corporation Changed America.* New York: Broadway Books, 2001.

Beauchamp, K.G., and Institution of Electrical Engineers. *History of Telegraphy.* London: Institution of Electrical Engineers, 2001.

Bellah, Robert N., et al. *Habits of the Heart: Individualism and Commitment in American Life.* New York: Harper and Row, 1985.

Berners-Lee, Tim. *Weaving the Web: Original Design and Ultimate Destiny of the World-Wide Web.* New York: HarperCollins, 1999, 2000.

Berstein, Peter L. *Wedding of the Waters: The Erie Canal and the Making of a Great Nation.* New York: W.W. Norton, 2005.

Berton, Pierre. *Niagara: A History of the Falls.* Toronto: McClelland & Stewart, 1992.

Biddle, Jeff. "Explaining the Spread of Air-conditioning, 1955–1980." *Explorations in Economic History* 45 (4, 2008): 402–23.

Billington, Ray Allen. *The Far Western Frontier, 1830–1860.* New York: Harper, 1956.

———. *Westward Expansion: A History of the American Frontier,* 4th ed. New York: Macmillan, 1974.

Bishop Jr., James. *Epitaph for a Desert Anarchist: The Life and Legacy of Edward Abbey.* New York: Touchstone, Simon and Schuster, 1995.

Bloom, Harold. *The American Religion: The Emergence of the Post-Christian Nation.* New York: Simon & Schuster, 1992.

Boudway, Ira. "A Problem Like LeBron." Bloomberg *Business Week,* June 21–27, 2010, 7–9.

Boyle, T. Coraghessan. "A voice griping in the wilderness." *New York Times,* February 10, 2002.

Braudel, Fernand. *The Wheels of Commerce, Civilization and Capitalism 15th–18th Century,* vol. 2. New York: Harper and Row, 1979, 1982.

Brayer, Elizabeth. *George Eastman: A Biography.* 1996; Rochester, NY: University of Rochester Press, 2006.

Brinkley, Douglas. *The Majic Bus: An American Odyssey.* San Diego: Harcourt Brace and Company, 1993.

Brogan, Hugh. *Alexis de Tocqueville: A Life.* New Haven, CT: Yale University Press, 2006.

Brooks, David. *On Paradise Drive: How We Love Now (And Always Have) in the Future Tense.* New York: Simon & Schuster, 2004.

———. "The responsibility deficit." *New York Times,* September 23, 2010.

Brown, Dee. *The American West.* New York: Charles Scribner's Sons, 1994.

———. *Hear that Lonesome Whistle Blow: Railroads in the West.* New York: Touchstone, 1994.

Brown, Lynn Elaine, and Steven Sass. "The Transition from a Mill-Based to a Knowledge-Based Economy: New England, 1940–2000." In *Engines of Enterprise: An Economic History of New England*, ed. Peter Temin. Cambridge, MA: Harvard University Press, 2000.

Brown, Richard D. *Massachusetts: A Bicentennial History*. New York: Norton, 1978.

Browne, Paterson H. *Quincy*. Charleston, SC: Arcadia Publications, 2004.

Brownlee, W. Elliott, and Mary M. Brownlee, *Women in the American Economy*. New Haven, CT: Yale University Press, 1976.

Bryson, Bill. *At Home: A Short History of Personal Life*. New York: Doubleday, 2010.

———. *A Walk in the Woods*. New York: Broadway Books, 1998.

Buhite, Leslie Allen. "The Chautauqua Lake Camp Meeting and the Chautauqua Institution." PhD diss., Florida State University, 2007. Available online at http://etd.lib.fsu.edu/theses/submitted/etd-07172007-132431/unrestricted/BuhiteLDissertation.pdf.

Builders of Wooden Railway Cars..and Some Other Stuff. "The Car-Builder's Dictionary." 2007. Available online at http://www.midcontinent.org/rollingstock/builders/bibliog-bldrs.htm.

Burnett, Constance Buel. *Five for Freedom*. New York: Greenwood Press, 1968.

Burns, Amy Stechler, and Ken Burns. *The Shakers: Hands to Work, Hearts to God*. New York: Aperture Foundation, 1987.

Burrows, Edwin G., and Mike Wallace. *Gotham: A History of New York City to 1898*. New York: Oxford University Press, 1999.

Busch, Norma. *In the Eyes of the Law: Women, Marriage, and Property in Nineteenth-Century New York*. Ithaca, NY: Cornell University Press, 1982.

Bushman, Richard L. *From Puritan to Yankee: Character and the Social Order in Connecticut, 1690–1765*. New York: Norton, 1967.

Butler, Jon. *Awash in a Sea of Faith: Christianizing the American People*. Cambridge, MA: Harvard University Press, 1990.

Camp Meeting.Dot. Org. "What Is a Camp Meeting?" n.d. Available online at http://www.campmeeting.org/whatis.htm.

Carroll, Peter N. *Puritanism and the Wilderness: The Intellectual Significance of the New England Frontier, 1629–1700*. New York: Columbia University Press, 1969.

Carson, Rachel. *Silent Spring*. 1962; New York: Houghton Mifflin, 2002.

Carter, Charles Frederick. *When Railroads Were New*. New York: Simmons-Boardman, 1926.

Cashman, Sean Dennis. *America in the Gilded Age: From the Death of Lincoln to the Rise of Theodore Roosevelt*. New York: New York University Press, 1984.

Chafets, Zev. *Cooperstown Confidential*. New York: Bloomsbury, 2009.

Chandler, Jr., Alfred D. *The Railroads, the Nation's First Big Business; Sources and Readings*. New York: Harcourt, Brace & World, 1965.

———. *The Visible Hand: The Managerial Revolution in American Business*. Cambridge, MA: Belknap Press of Harvard University Press, 1977.

Chang, M.C. "Recollections of 40 Years at the Worcester Foundation for Experimental Biology." *Physiologist* 28 (5): 400–1.

Chapman, J. Wilbur. *The Life and Work of Dwight Lyman Moody.* 1900. Available online at http://www.biblebelievers.com/moody/06.html.

Chautauqua Institution. "Welcome to Chautauqua: A Walking Tour Guide of the Chautauqua Institution." n.d.

Chernow, Ron. *Titan: The Life of John D. Rockefeller, Sr.* New York: Random House, 1998.

Chesler, Ellen. *Woman of Valor: Margaret Sanger and the Birth Control Movement in America.* New York: Anchor Books, 1993.

Church, Robert L. *Education in the United States: An Interpretive History.* New York: Free Press, 1976.

Clark, Christopher. *The Roots of Rural Capitalism: Western Massachusetts, 1780–1860.* Ithaca, NY: Cornell University Press 1990.

Clark University. Archives and Special Collections. "Robert Hutchings Goddard Biographical Note." Worcester, MA, n.d. Available online at http://www.clarku.edu/research/archives/goddard/bio_note.cfm.

Cochran, Thomas C. *Railroad Leaders, 1845–1890: The Business Mind in Action.* New York: Russell & Russell, 1965.

Cohen, Isaac. *American Management and British Labor: A Comparative Study of the Cotton Spinning Industry.* New York: Greenwood Press, 1990.

Coleman, Margaret. "Homemaker as Worker in the United States." *Challenge* 41 (6) 1998: 75–87.

"Combined Oral Contraceptive Pill." *Wikipedia: The Free Encyclopedia.* Available online at http://en.wikipedia.org/wiki/Combined_oral_contraceptive_pill, accessed January 7, 2014.

Common Sense for Drug Policy, "Drug Offenders in the Correctional System"; available online at http://drugwarfacts.org/cms/node/63.

Conforti, Joseph A. *Imagining New England: Explanations of Regional Identity from the Pilgrims to the Mid-Twentieth Century.* Chapel Hill: University of North Carolina Press, 2001.

———. *Saints and Strangers: New England in British North America.* Baltimore: Johns Hopkins University Press, 2006.

Cooper, Gail. *Air-Conditioning America: Engineers and the Controlled Environment, 1900–1960.* Baltimore: Johns Hopkins University Press, 1998.

Cox, Stan. "AC: It's not as cool as you think." *Los Angeles Times*, July 18, 2010. Available online at http://articles.latimes.com/print/2010/jul/18/opinion/la-oe-cox-ac-20100718.

Cremin, Lawrence A. *American Education: The National Experience, 1783–1876.* New York: Harper and Row, 1980.

———, ed. *The Republic and the School: Horace Mann on the Education of Free Men.* New York: Teachers College of Columbia University, 1957.

———. *Traditions of American Education.* New York: Basic Books, 1977.

Cringley, Robert X. *Accidental Empires: How the Boys of Silicon Valley Make Their Millions, Battle Foreign Competition, and Still Can't Get a Date.* New York: Harper, 1996.

Crocker, Kathleen, and Jane Currie. *The Chautauqua Institution, 1874–1974.* Charleston, SC: Arcadia Publishing, 2001.

Cromwell, Otelia. *Lucretia Mott.* Cambridge, MA: Harvard University Press, 1958.

Cronon, William. *Nature's Metropolis: Chicago and the Great West.* New York: W.W. Norton, 1991.

———. *Uncommon Ground: Rethinking the Human Place in Nature.* New York: W.W. Norton, 1996.

Cross, Whitney R. *The Burned-over District; The Social and Intellectual History of Enthusiastic Religion in Western New York, 1800–1850.* New York: Harper and Row, 1950.

Crump, Thomas. *A Brief History of the Age of Steam: The Power that Drove the Industrial Revolution.* New York: Carroll & Graf, 2007.

C-SPAN and Ann Bentzel. *C-SPAN's Traveling Tocqueville's America.* Baltimore: Johns Hopkins University Press, 1988.

Cullen, Jim. *The Art of Democracy: A Concise History of Popular Culture in the United States.* New York: Monthly Review Press, 1996.

Dailey, John S. "Fabric of Surveying: Ohio Lands and Survey Systems." *American Surveyor*, November 30, 2004. Available online at www.amerisurv.com/content/view/3962/150/.

Dalzell, Robert F. *Enterprising Elite: The Boston Associates and the World They Made.* Cambridge, MA: Harvard University Press, 1987.

Damrosch, Leo. *Tocqueville's Discovery of America.* New York: Farrar, Straus & Giroux, 2010

Davies, Pete. *American Road: The Story of an Epic Transcontinental Journey at the Dawn of the Motor Age.* New York: Henry Holt and Company, 2002.

Davis, James E. *Frontier Illinois.* Bloomington: Indiana University Press, 1998.

Davis, John P. *Corporations.* 1905; New York: Capricorn Books, 1961

Denlay, Kevin. "On Eternal Patrol: The Unexpected Discovery of the USS Perch SS-176." *Advanced Diver Magazine*, n.d. Available online at http://www.advanceddivermagazine.com/articles/perch/perch.html.

Desmond, Michael. "'Model community' dream haunts Love Canal residents." *Buffalo Courier-Express*, August 6, 1978.

de Soto, Hernando. *The Mystery of Capital: Why Capitalism Succeeds in the West, but Fails Everywhere Else.* New York: Basic Books, 2000.

DeVol, Ross, and Anita Charuwon. *California's Position in Technology and Science: A Comparative Benchmarking and Assessment.* Santa Monica, CA: Milken Institute, 2008.

DeVoto, Bernard. *The Year of Decision, 1846.* 1942; New York: St. Martins Press, 2000.

Diamond, Jared M. *Guns, Germs, and Steel: The Fates of Human Societies.* New York: W.W. Norton, 1997.

Diogenes Laërtius. *Lives of Eminent Philosophers,* Book 2, trans. R.D. Hicks. Cambridge MA: Harvard University Press, 1925.

Doerr, Mary Jane. *Bay View: An American Idea.* Allegan Forest, MI: Priscilla Press, 2010.

Dorfner, John J. *Kerouac: Visions of Lowell.* Raleigh NC: Cooper Street Publications, 1993.

Douglas, George H. *All Aboard: The Railroad in American Life.* New York: Smithmark, 1996.

Dowling, Tim. *King Camp Gillette, 1855–1932: Inventor of the Disposable Culture.* London: Faber & Faber, 2002.

Downey, Kirsten. *The Woman Behind the New Deal: The Life and legacy of Frances Perkins.* New York: Anchor Books, 2010.

Drucker, Peter F. *The Concept of the Corporation.* New York: John Day, 1946.

Dublin, Thomas, and United States National Park Service. *Lowell: The Story of an Industrial City — A Guide to Lowell National Historical Park and Lowell Heritage State Park, Lowell, Massachusetts.* Washington, DC: Department of the Interior, 1992.

Dunbar, Willis F., and George S. May. *Michigan: A History of the Wolverine State,* 2nd ed. Grand Rapids, MI: William B. Eerdmans, 1965.

Duncan, Dayton. *Miles from Nowhere: In Search of the American Frontier,* 2nd. ed. New York: Penguin, 1994.

Earls, Alan R. *Route 128 and the Birth of the Age of High Tech.* Charleston, SC: Arcadia, 2002.

Elazar, Daniel J. *Cities of the Prairie: The Metropolitan Frontier and American Politics.* New York: Basic Books, 1970.

Ellis, David M., and Sherri Goldstein Cash. *New York State: An Illustrated History.* Sun Valley: CA: American History Press, 2008.

Ely Jr., James W. "Abraham Lincoln as a Railroad Attorney." Paper delivered at the Indiana Historical Society, 2005 Railroad Symposium: Lincoln and the Railroads. Available online at http://www.indianahistory.org/our-services/books-publications/railroad-symposia-essays-1/Abe%20Lincoln%20as%20a%20Railroad%20Attorney.pdf.

Evans, Sara M. *Born for Liberty; A History of Women in America.* New York: Free Press, 1997.

Ewen, David, ed. *American Popular Songs from the Revolutionary War to the Present.* New York: Random House, 1966.

Fainsod, Merle. *Smolensk under Soviet Rule.* Boston: Unwin Hyman, 1989.

Faragher, John Mack. *Rereading Frederick Jackson Turner: "The Significance of the Frontier in American History" and Other Essays.* New York: Henry Holt, 1994.

Federal Bureau of Investigation. "Edward Abbey." Washington, DC, n.d. Available online at http://www.fbi.gov/fbi-search#output=xml_no_dtd&client=google-csbe&cx=004748461833896749646%3Ae41lgwqry7w&cof=FORID%3A10%3BNB%3A1&ie=UTF-8&siteurl=www.fbi.gov%2F&q=Edward+Abbey.

Fields, Armond. *Katharine Dexter McCormick: Pioneer for Women's Rights.* Westport, CT: Praeger, 2003.

Finch, Christopher. *Highways to Heaven: The Autobiography of America.* New York: HarperCollins, 1992.

Finke, Roger, and Rodney Stark. *The Churching of America, 1776–1990.* New Brunswick, NJ: Rutgers University Press, 1992.

Finney, Charles G. *Charles G. Finney: An Autobiography.* Westwood, NJ: Fleming H. Revell, 1876, 1908.

Firestation. "A Brief History of Leopold Desk Company." Burlington, IA, n.d. Available online at http://burlingtonfirestation.com/leopold.htm.

Fischer, David Hackett. *Albion's Seed: Four British Folkways in America.* New York: Oxford, 1989.

Fishwick, Marshall W. *Great Awakenings: Popular Religion and Popular Culture.* New York: Haworth Press, 1995.

Fitzpatrick, Ellen. *History's Memory: Writing America's Past, 1880–1980.* Cambridge, MA: Harvard University Press, 2004.

Fordham University. "Modern History Sourcebook: The Declaration of Sentiments, Seneca Falls Conference, 1848." Available online at http://www.fordham.edu/halsall/mod/Senecafalls.html.

Foreman, Dave. *Confessions of an Eco-Warrior.* New York: Crown Trade Paperbacks, 1991.

Foreman, John. *Frommer's Comprehensive Travel Guide to New York State.* New York: Prentice-Hall Travel, 1994.

Foster, Lawrence. *Religion and Sexuality.* New York: Oxford University Press, 1981.

Fowler Jr., William. "Marine Insurance in Boston: The Early Years of the Boston Marine Insurance Company, 1799–1807." In *Entrepreneurs: The Boston Business Community, 1700–1850,* ed. C.E. Wright and K.P. Viens. Boston: Massachusetts Historical Society, 1997.

Fox, Dixon Ryan, and Robert V. Remini. *The Decline of Aristocracy in the Politics of New York: 1801–1840.* New York: Harper & Row, 1965.

"Framing (Construction): Platform Framing." *Wikipedia: The Free Encyclopedia.* Available online at http://en.wikipedia.org/wiki/Platform_framing#Platform_framing, accessed January 7, 2014.

Franklin, Benjamin. *The Autobiography of Benjamin Franklin.* 1817; New York: Simon and Schuster/Touchstone, 1962.

Freiberger, Paul, and Michael Swaine. *Fire in the Valley: The Making of the Personal Computer,* 2nd ed. New York: McGraw-Hill, 2000.

Friends of Watertown Free Publications and Watertown Historical Society. *Watertown.* Charleston, SC: Arcadia Publications, 2002.

Frost, Elizabeth, and Kathryn Cullen-DuPont. *Women's Suffrage in America: An Eyewitness History.* New York: Facts on File, 1992.

Geddes, Dan. "Henry Miller's 'On the Road'." *Satirist,* July 2002. Available online at http://www.thesatirist.com/books/Air-Conditioned-Nightmare.html.

Germain, David. "Dream that drowned in Love Canal." *Buffalo News,* June 6, 1993.

Gibbs, Lois Marie. *Love Canal: The Story Continues.* Gabriola Island, BC: New Society Publishers, 1998.

Gilder, George *Microcosm: The Quantum Revolution in Economics and Technology.* New York: Simon & Schuster, 1989.

Gilje, Paul A. "The Rise of Capitalism in the Early Republic." In *Wages of Independence: Capitalism in the Early American Republic,* ed. Paul A. Gilje. Lanham, MD: Rowman & Littlefield, 1997.

Gladstone Historical Society. *Gladstone Chautauqua, 1894–1927: A Centennial Remembrance.* Gladstone, OR, 1994.

Gompers, Paul, and Joshua Lerner, eds. *The Venture Capital Cycle,* 2nd ed. Cambridge, MA: MIT Press, 2004.

Goodrich, Carter. *Government Promotion of American Canals and Railroads, 1800–1890.* New York: Columbia University Press, 1960.

Goodwin, Doris Kearns. *Team of Rivals: The Political Genius of Abraham Lincoln.* New York: Simon & Schuster, 2005.

Gordon, John Steele. *The Business of America.* New York: Walker Publishing, 2001.

———. *The Great Game: The Emergence of Wall Street as a World Power, 1653–2000.* New York: Scribner, 1999.

Gordon, Robert B., and Patrick M. Malone. *The Texture of Industry: An Archaeological View of the Industrialization of North America.* New York: Oxford University Press, 1994.

Gordon, Sarah. *Passage to Union: How the Railroads Transformed American Life, 1829–1929.* Chicago: Ivan R. Dee, 1996.

Gould, Joseph E. *The Chautauqua Movement: An Episode in the Continuing American Revolution.* New York: State University of New York, 1961.

Gray, Susan E. *The Yankee West: Community Life on the Michigan Frontier.* Chapel Hill: University of North Carolina Press, 1996.

Grills, Russell A. *Upland Idyll: Images of Cazenovia, New York, 1860–1900.* Albany: New York State, Office of Parks & Historic Preservation, 1993.

Grundahl, Paul. *Major Erastus Corning: Albany Icon, Albany Enigma.* Albany: SUNY Press, 2007.

Gunther, John. *Inside U.S.A.* New York: Harper & Brothers, 1947.

Hafner, Katie, and Matthew Lyon. *Where Wizards Stay Up Late: The Origins of the Internet.* New York: Simon & Schuster Paperbacks, 2006.

Hall, Peter Dobkin. "What the Merchants Did with Their Money." In *Entrepreneurs: The Boston Business Community, 1700–1850,* ed. C.E. Wright and K.P. Viens. Boston: Massachusetts Historical Society, 1997.

Hallowell, Anna Davis. *James and Lucretia Mott: Life and Letters.* Boston: Houghton Mifflin, 1884.

Hamill, Susan Pace. "From Special Privilege to General Utility: A Continuation of Willard Hurst's Study of Corporation." *American University Law Review* 49 (1999): 81–180.

Hansen, Debra Gold. *Strained Sisterhood: Gender and Class in the Boston Female Anti-Slavery Society.* Amherst: University of Massachusetts Press, 1993.

Hardin, George, ed. *A History of Herkimer County.* Syracuse, NY: D. Mason, 1893. Available online at http://www.herkimer.nygenweb.net/remington/remington-fam5.html.

Hare, Lloyd C.M. *The Greatest American Woman: Lucretia Mott.* New York: American Historical Society, 1937.

Harlick, David. *No Place Distant.* Washington, DC: Island Press, 2002.

Harlow, Alvin F. *Steelways of New England.* New York: Creative Age Press, 1946.

Hartog, H. *Public Property and Private Power: The Corporation of the City of New York in American Law, 1730–1870.* Chapel Hill: University of North Carolina Press, 1983.

Hartwell, David G., and Kathryn Cramer. *The Hard SF Renaissance.* New York: Tor, 2002.

Hatch, Nathan O. *The Democratization of American Christianity.* New Haven, CT: Yale University Press, 1989.

Hatch, Nathan O., and Harry S. Stout. "Introduction," in *Jonathan Edwards and the American Experience*, ed. Nathan O. Hatch and Harry S. Stout. New York: Oxford University Press, 1988.

Hawke, David Freeman. *Nuts and Bolts of the Past: A History of American Technology, 1776–1860.* New York: Harper & Row, 1988.

Hayes, William. *Horace Mann's Vision of the Public Schools: Is It Still Relevant?* Lanham, MD: Rowman and Littlefield Education, 2006.

Hecker, Eugene A. *A Short History of Women's Rights.* Westport, CT: Greenwood Press, 1914.

Heineman, Ted. "Surveying the Western Rserve." *Riverside Cemetery Journal*, 2009. Available online at http://riversidecemeteryjournal.com/Events/Events/page195.html.

Hendrickson, Robert. *The Grand Emporiums: The Illustrated History of America's Great Department Stores.* New York: Stein and Day, 1980.

Hewitt, Nancy A. *Women's Activism and Social Change: Rochester, New York, 1822–1872.* Ithaca, NY: Cornell University Press, 1984.

Hindle, Brooke. *America's Wooden Age: Aspects of Its Early Technology.* Tarrytown, NY: Sleepy Hollow Restorations, 1975.

Hine, Robert V. *Community on the American Frontier: Separate but Not Alone.* Norman: University of Oklahoma Press, 1980.

Hitchings, Russell, and Shu Jun Lee. "Air Conditioning and the Material Culture of Routine Human Encasement: The Case of Young People in Contemporary Singapore." *Journal of Material Culture* 13 (3, 2008): 251–65.

Hobsbawm, Eric. *The Age of Capital, 1848–1875*. London: Cardinal, 1991.

———. *The Age of Revolution, 1789–1848*. New York: Barnes & Noble, 1962.

Hobson, Archie. *Remembering America: A Sampler of the WPA American Guide Series*. New York: Collier Macmillan, 1987.

Hodgson, Godfrey. *A Great and Godly Adventure*. New York: Public Affairs, 2006.

Hofstadter, Richard. "William Leggett, Spokesman of Jacksonian Democracy." *Political Science Quarterly* 58 (4, 1943): 581–94.

Holbrook, Stewart H. *The Old Post Road: The Story of the Boston Post Road*. New York: McGraw-Hill, 1962.

Holder, Robert. "The Beginnings of Buffalo Industry." In *Adventures in Western New York History*, vol. 5. Buffalo, NY: Buffalo and Erie County Historical Society, n.d. Available online at http://bechsed.nylearns.org/pdf/The_Beginning_of_Buffalo_Industry.pdf.

Horgam Paul. *Great River: The Rio Grande in North American History*. New York: Wesleyan University Press, 1984.

How Products Are Made. "Christopher Latham Sholes Biography. 1819–1890." Available online at http://www.madehow.com/inventorbios/13/Christopher-Latham-Sholes.html.

Hoyt, Charles, and R. Clyde Ford. *John D. Pierce, Founder of the Michigan School System: A Study of Education in the Old Northwest*. Ypsilanti, MI: Scharf Tag and Box Co., 1905. Available online at https://archive.org/stream/johndpiercefound00hoyt#page/n7/mode/2up.

Hugill, Peter J. *Upstate Arcadia: Landscape, Aesthetics, and the Triumph of Social Differentiation in America*. Lanham, MD: Rowman & Littlefield, 1995.

Hurt, R. Douglas. *The Ohio Frontier: Crucible of the Old Northwest, 1720–1830*. Bloomington: Indiana University Press, 1996.

"Hutchinson Family Singers." *Wikipedia: The Free Encyclopedia*. Available online at http://en.wikipedia.org/wiki/Hutchinson_Family_Singers, accessed January 9, 2014.

Hymowitz, Carol, and Michaele Weissman. *A History of Women in America*. New York: Bantam, 1978.

Ierley, Merritt. *Traveling the National Road: Across the Centuries on America's First Highway*. Woodstock, NY: Overlook Press, 1990.

Ingels, Margaret. *Willis Haviland Carrier, Father of Air Conditioning*. New York: Arno Press, 1972.

Ingle, Dwight J. "Gregory Goodwin Pincus, 1903–1967: A Biographical Memoir." In *Biographical Memoirs*, vol. 64. Washington DC: National Academy of Sciences, 1971. Available online at http://books.nap.edu/html/biomems/gpincus.pdf.

Innes, Stephen. *Creating the Commonwealth: The Economic Culture of Puritan New England*. New York: Norton, 1995.

Irwin, William R. *The New Niagara: Tourism, Technology, and the Landscape of Niagara Falls, 1776–1917*. University Park: Pennsylvania State University Press, 1996.

Isenberg, Nancy. *Sex and Citizenship in Antebellum America.* Chapel Hill: University of North Carolina Press, 1998.

James, E.W. "Making and Unmaking a National System of Marked Roads." *American Highways* 12 (4, 1933): 16–18.

"James Rand, Jr." *Wikipedia: The Free Encyclopedia.* Available online at http://en.wikipedia.org/wiki/James_Rand,_Jr., accessed January 8, 2014.

Jameson, Sheilagh S., with Nola B. Ericson. *Chautauqua in Canada.* Calgary: Glenbow-Alberta Institute, 1979.

Jensen, Jamie. *Road Trip USA.* Berkeley, CA: Avalon Travel Publishing, annual.

Jeremy, David J. *Transatlantic Industrial Revolution: The Diffusion of Textile Technologies between Britain and America, 1790–1830.* Oxford: Blackwell, 1981.

Johnson, Curtis D. *Islands of Holiness: Rural Religion in Upstate New York, 1790–1860.* Ithaca, NY: Cornell University Press, 1989.

Johnson, Paul. *A History of the American People.* London: Phoenix, 1997.

———. *A Shopkeeper's Millennium: Society and Revivals in Rochester New York, 1815–1837.* New York: Hill and Wang, 1978.

Jones, Howard Mumford. *O Strange New World: American Culture; The Formative Years.* New York: Viking, 1964.

Kahney, Leander. *The Cult of Mac.* San Francisco: No Starch Press, 2004.

Kamensky, Jane. *The Exchange Artist: A Tale of High-Flying Speculation and America's First Banking Collapse.* New York: Viking, 2008.

Kanigel, Robert. *The One Best Way: Frederick Winslow Taylor and the Enigma of Efficiency.* New York: Viking Penguin, 1997.

Kaplan, Robert D. *An Empire Wilderness: Travels into America's Future.* New York: Random House, 1998.

Katzenbach, Jon R., and Douglas K. Smith. "The Wisdom of Teams," *Small Business Reports* 18 (7, 1993): 68–71.

Keith, William M. *Democracy as Discussion: Civic Education and the American Forum Movement.* Lanham, MD: Lexington Books, 2007.

Kennan, George F. *Sketches from a Life.* New York: Pantheon Books, 1989.

Kennedy, William. *O Albany!.* New York: Penguin, 1983.

Kephart, William M. *Extraordinary Groups: The Sociology of Unconventional Life-Styles.* New York: St. Martin's Press, 1976.

Kerouac, Jack. *On The Road.* 1957; New York: Penguin, 1991.

Kessler, W.C. "A Statistical Study of the New York General Incorporation Act of 1811." *Journal of Political Economy* 48 (6, 1940): 877–82.

Kidder, Tracy. *Soul of a New Machine.* New York: Avon Books, 1981.

Kingsolver, Barbara. *High Tide in Tucson: Essays from Now or Never.* New York: HarperCollins, 1995.

Kinsella, W.P. *Shoeless Joe.* 1982; New York: Mariner Books, 1999.

Klees, Emerson. *The Crucible of Ferment: New York's "Psychic Highway".* New York: Friends of the Finger Lakes Publishing, 2001.

———. *People of the Finger Lakes Region, the Heart of New York State.* New York: Friends of the Finger Lakes Publishing, 1995.

Klein, Milton M. *The Empire State: A History of New York.* Ithaca, NY: Cornell University Press, 2001.

Klepp, Susan E. *Revolutionary Conceptions: Women, Fertility, and Family Limitation in America, 1760–1820.* Chapel Hill: University of North Carolina Press, 2009.

Koeppel, Gerard. *Bond of Union: Building the Erie Canal and the American Empire.* Cambridge, MA: Da Capo Press, 2009.

Krasner-Khait, Barbara. "The Impact of Refrigeration." *History Magazine*, February-March 2000.

Krugman, Paul. "The Future of New England." In *Engines of Enterprise: An Economic History of New England*, ed. Peter Temin. Cambridge, MA: Harvard University Press, 2000.

Labaree, Benjamin W. "The Making of an Empire: Boston and Essex County, 1790–1850." In *Entrepreneurs: The Boston Business Community, 1700–1850*, ed. C.E. Wright and K.P. Viens. Boston: Massachusetts Historical Society, 1997.

Labov, William. *Principles of Linguistic Change*, 3 v. New York: Wiley-Blackwell, 1994; reprinted 2010.

Lackey, Kris. *Road Frames: The American Highway Narrative.* Lincoln: University of Nebraska Press, 1997.

Lader, Lawrence, and Milton Meltzer. *Margaret Sanger: Pioneer of Birth Control.* New York: Thomas Y. Crowell, 1969.

Lamoreaux, Naomi. "The Partnership Form of Organization: Its Popularity in Early Nineteenth Century Boston." In *Entrepreneurs: The Boston Business Community, 1700–1850*, ed. C.E. Wright and K.P. Viens. Boston: Massachusetts Historical Society, 1997.

Lampe, David. *The Massachusetts Miracle: High Technology and Economic Revitalization.* Cambridge, MA: MIT Press, 1988.

Landes, David S. *Revolution in Time: Clocks and the Making of the Modern World.* New York: Barnes & Noble, 1993.

———. *The Wealth and Poverty of Nations: Why Some Are So Rich and Some So Poor.* New York: W.W. Norton, 1998.

Lang, Amy Schrager. "'A Flood of Errors': Chauncy and Edwards in the Great Awakening." In *Jonathan Edwards and the American Experience*, ed. Nathan O. Hatch and Harry S. Stout. New York: Oxford University Press, 1988.

Laurence, Kevin. "The *Exciting* History of Carbon Paper!" 1995. Available online at http://www.kevinlaurence.net/essays/cc.php.

Leach, William. *Land of Desire: Merchants, Power, and the Rise of a New American Culture.* New York: Vintage, 1993.

———. *True Love and Perfect Union: The Feminist Reform of Sex and Society.* Middletown, CT: Wesleyan University Press, 1980.

Lears, Jackson. *Fables of Abundance: A Cultural History of Advertising in America*. New York: Basic Books, 1994.

Leary, Lewis. "Introduction." In Benjamin Franklin, *The Autobiography of Benjamin Franklin*. 1817; New York: Simon and Schuster/Touchstone, 1962.

Least Heat-Moon, William. *Blue Highways: A Journey into America*. Boston: Little, Brown, 1999.

———. *River Horse: A Voyage Across America*. New York: Houghton Mifflin, 1999.

Lee, W. Storrs. *The Yankees of Connecticut*. New York: Holt, 1957.

Leighly, John. "Town Names of Colonial New England in the West." *Annals of the Association of American Geographers* 68 (2, 1978): 233–48.

Leland, John. *Why Kerouac Matters: The Lessons of On the Road. They're Not What You Think*. New York: Viking, 2007.

Leopold, Aldo, et al. *Aldo Leopold's Southwest*. Albuquerque: University of New Mexico Press, 1995.

———. *A Sand County Almanac*. New York: Ballantine Books, 1966.

Levin, David. "Edwards, Franklin, and Cotton Mather: A Meditation on Character and Reputation." In *Jonathan Edwards and the American Experience*, ed. Nathan O. Hatch and Harry S. Stout. New York: Oxford University Press, 1988.

Levy, Bernard-Henri. *American Vertigo: Traveling in the Footsteps of Tocqueville* [New York: Random House, 2006.

Lewis, Sinclair. *Babbitt*. 1922; repr. Mineola, NY: Dover Publications, 2003.

Lewis, Tom. *Divided Highways: Building the Interstate Highways Transforming American Life*. New York: Penguin Group, 1997.

Lienhard, John H. *Inventing Modern: Growing Up with x-Rays, Skyscrapers, and Tailfins*. New York: Oxford University Press, 2003.

———. "Willis Carrier." *Engines of Our Ingenuity* 688. Available online at http://www.uh.edu/engines/epi688.htm.

Linklater, Andro. *Measuring America: How the United States Was Shaped by the Greatest Land Sale in History*. New York: PLUME Penguin Group, 2003.

Lipset, Seymour Martin. *American Exceptionalism: A Double-Edged Sword*. New York: W.W. Norton, 1996.

Long, Huey B. "Adult Education in the Oneida Community: A Pattern for the Chautauqua Assembly." *Journal of the Midwest History of Education Society* 22 (1995): 203–15.

London, Bill. *Country Roads of Idaho*. Castine, ME: Country Roads Press, 1995.

Lorbiecki, Marybeth. *Aldo Leopold: A Fierce Green Fire*. Guilford, CT: Falcon, 2005.

Lubow, Lisa B. "From Carpenter to Capitalist: The Business of Building in Postrevolutionary Boston." In *Entrepreneurs: The Boston Business Community, 1700–1850*, ed. C.E. Wright and K.P. Viens. Boston: Massachusetts Historical Society, 1997.

Lubrano, Annteresa. *The Telegraph: How Technology Innovation Caused Social Change*. New York: Garland Publishing, 1997.

Lundmark, Torbjorn. *Quirky QWERTY: A Biography of the Typewriter & Its Many Characters*. New York: Penguin, 2002.

MacDougall, Hugh C. "James Fenimore Cooper: Pioneer of the Environmental Movement." James Fenimore Cooper Society, 1999. Available online at www.oneonta.edu/external/cooper/articles/informal/hugh-environment.html.

MacLaren, Gay. *Morally We Roll Along*. Boston: Little, Brown, 1938.

Mann, Mary Peabody. *Life of Horace Mann*, new ed. Boston: Lee and Shepard, 1888.

Markoff, John. *What the Dormouse Said: How the Sixties Counterculture Shaped the Personal Computer Industry*. New York: Viking, 2005.

Marks, Lara V. *Sexual Chemistry: A History of the Contraceptive Pill*. New Haven, CT: Yale University Press, 2001.

Marryat, Frederick. *A Diary in America: With Remarks on Its Institutions*. Philadelphia: Carey & Hart, 1839.

Marshall, Walter P. *Ezra Cornell, His Contributions to Western Union and to Cornell University*. New York: Newcomen Society in North America, 1951.

Martin, Albro. *Railroads Triumphant: The Growth, Rejection, and Rebirth of a Vital American Force*. New York: Oxford University Press, 1992.

Mathews, Catharine van Cortlandt. *Andrew Ellicott: His Life and Letters*. 1908; repr. New York: Ralph Roberts, 2001.

Matson, Cathy. "Capitalizing Hope: Economic Thought in the Early National Economy." In *Wages of Independence: Capitalism in the Early American Republic*, ed. Paul A. Gilje. Lanham, MD: Rowman & Littlefield, 1997.

May, Elaine Tyler. *America and the Pill: A History of Promise, Peril, and Liberation*. New York: Basic Books, 2010.

May, Henry F. "Jonathan Edwards and America." In *Jonathan Edwards and the American Experience*, ed. Nathan O. Hatch and Harry S. Stout. New York: Oxford University Press, 1988.

Maynard, W. Barksdale. *Walden Pond: A History*. New York: Oxford University Press, 2004.

Mazlish, Bruce. *The Railroad and the Space Program; An Exploration in Historical Analogy*. Cambridge, MA: MIT Press, 1965.

McConnell, J.D. "Power Distribution in Textile Plants." *Electrical Engineering* 66 (7, 1947): 667–9.

McCown, Robert A. "Records of the Redpath Chautauqua." *Books at Iowa* 19 (1973): 8–23.

McCracken, John A. "Editorial: Reflections on the 50th Anniversary of the Birth Control Pill." *Biology of Reproduction* 83 (4, 2010): 684–6.

McGreevy, Patrick. *Imagining Niagara: The Meaning and Making of Niagara Falls*. Amherst: University of Massachusetts Press, 1994.

McHugh, Paul. "The Pure Jones of It." In *The Road Within: True Stories of Transformation and the Soul*, ed. Sean O'Reilly et al. San Francisco: Travelers' Tales, 2002.

McLaughlin, Loretta. *The Pill, John Rock, and the Catholic Church: The Biography of a Revolution.* Boston: Little, Brown, 1982.

McMillen, Sally G. *Seneca Falls and the Origins of the Women's Rights Movement.* New York: Oxford University Press, 2008.

McNichol, Dan. *The Roads that Built America: The Incredible Story of the U.S. Interstate System.* New York: Sterling, 2006.

McPherson, James M. *Battle Cry of Freedom: The Civil War Era.* New York: Oxford University Press, 1988.

Melder, E. Keith. *Beginnings of Sisterhood: The American Woman's Rights Movement, 1800–1850.* New York: Schocken Books, 1977.

Menand, Louis. *The Metaphysical Club: A Story of Ideas in America.* New York: Farrar, Straus and Giroux, 2001.

Merrill, George P. *The First One Hundred Years of American Geology.* 1925, repr. New York: Hafner Publishing, 1964, 1969.

Metcalf, Allan. *OK: The Improbable History of America's Greatest Word.* New York: Oxford University Press, 2011.

Micklethwait, John, and Adrian Wooldridge. *The Company: A Short History of a Revolutionary Idea.* New York: Modern Library Chronicles Book, 2003.

Middlekrauft, Robert. *The Glorious Cause: The American Revolution, 1763–1789,* rev. ed. New York: Oxford University Press, 2005.

Miller, Carol Poh, and Robert A. Wheeler. *Cleveland: A Concise History, 1796–1996,* 2nd ed. Bloomington: Indiana University Press, 1997.

Miller, Char. *Pioneers of Conservation: Gifford Pinchot and the Making of Modern Environmentalism.* Washington, DC: Island Press, 2001.

Miller, Stephen. *The Peculiar Life of Sundays.* Cambridge, MA: Harvard University Press, 2008.

Mohl, Raymond A., ed. *Searching for the Sunbelt: Historical Perspectives on a Region.* Knoxville: University of Tennessee Press, 1990.

Monroe County Library System. "The Reynolds Arcade." Rochester, NY, n.d. Available online at http://www.rochester.lib.ny.us/rochimag/architecture/LostRochester/Reynolds/Reynolds.htm.

Moody, John. *The Railroad Builders.* New Haven, CT: Yale University Press, 1919.

Morgan, Edmund. *The Puritan Family: Religion and Domestic Relations in Seventeenth Century New England.* New York: Harper Torchbook, 1944, rev. ed. 1966.

Morgan, Ted. *A Shovel of Stars: The Making of the American West, 1800 to the Present.* New York: Simon & Schuster, 1995.

Morris, Edmund. *Theodore Rex.* New York: Modern Library, 2002.

Morris, Peter. *But Didn't We Have Fun? An Informal History of Baseball's Pioneer Era, 1843–1870.* Chicago: Ivan R. Dee, 2008.

Moyer, Paul B. *Wild Yankees: The Struggle for Independence along Pennsylvania's Revolutionary Frontier.* Ithaca, NY: Cornell University Press, 2007.

Mulloney, Stephen. *Traces of Thoreau: A Cape Cod Journey*. Boston: Northeastern University Press, 1998.

Munroe, Alfred. "Concord and the Telegraph." Paper presented to the Concord Antiquarian Society, Concord, MA, January 6, 1902. Available online at http://www.archive.org/stream/preliminariesofcootolm/preliminariesofcootolm_djvu.txt.

Murphy, Cait. *Crazy '08: How a Cast of Cranks, Rogues, Boneheads, and Magnates Created the Greatest Year in Baseball History*. New York: Smithsonian/Collins, 2007.

Myerson, Joel, ed. *The Brook Farm Book: A Collection of First-Hand Accounts of the Community*. New York: Garland, 1987.

myTypewriter.com. "Jack Kerouac 1922–1969." 2006. Available online at http://www.mytypewriter.com/authors/featured/kerouac.html.

Nace, Ted. *Gangs of America: The Rise of Corporate Power and the Disabling of Democracy*. San Francisco: Berrett-Koehler, 2005.

Nash, Roderick. *Wilderness and the American Mind*. New Haven, CT: Yale University Press, 1967.

National Museum of the US Air Force. "Dr. Robert H. Goddard: 'The Father of Modern Rocketry'." Wright-Patterson AFB, OH, 2009. Available online at http://www.nationalmuseum.af.mil/factsheets/factsheet.asp?id=12374.

Nelson, Mac. *Twenty West: The Great Road across America*. Albany: State University of New York Press, 2008.

Neu, Irene. *Erastus Corning: Merchant and Financier, 1794–1872*. Ithaca, NY: Cornell University Press, 1960.

Newell, Margaret Ellen. "The Birth of New England in the Atlantic Economy: From Its Beginning to 1770." In *Engines of Enterprise: An Economic History of New England*, ed. Peter Temin. Cambridge, MA: Harvard University Press, 2000.

———. "A Revolution in Economic Thought: Currency and Economic Development in Eighteenth-Century Massachusetts." In *Entrepreneurs: The Boston Business Community, 1700–1850*, ed. C.E. Wright and K.P. Viens. Boston: Massachusetts Historical Society, 1997.

Noll, Mark. *The Scandal of the Evangelical Mind*. Grand Rapids, MI: William B. Eerdmans, 1994.

Nordhoff, Charles. *The Communistic Societies of the United States*. 1875; New York: Schocken, 1965.

Oliver, John W. *A History of American Technology*. New York: Ronald Press, 1956.

Olmstead, Frederick Law. *Civilizing American Cities: Writings on City Landscapes*, ed. S.B. Sutton. New York: Da Capo Press, 1997.

———. *The Cotton Kingdom: A Traveler's Observations on Cotton and Slavery in the American Slave States*, ed. Arthur M. Schlesinger. New York: Knopf, 1953.

Oslin, George P. *The Story of Telecommunications*. Macon, GA.: Mercer University Press, 1992.

Padfield, Peter. *Maritime Power: The Struggle for Freedom.* Woodstock, NY: Overlook Press, 2003.

Panati, Charles. *Extraordinary Origins of Everyday Things.* New York: Perennial Library, 1987.

Paquet, Laura Byrne. *Wanderlust: A Social History of Travel.* Fredericton, NB: Goose Lane Editions, 2007.

Parker, Margaret Terrell. *Lowell: A Study of Industrial Development.* Port Washington, NY: Kennikat Press, 1940.

Patton, Phil. *Open Road: A Celebration of the American Highway.* New York: Simon and Schuster, 1986.

Peddle, Michael T. "Planned Industrial and Commercial Developments in the United States: A Review of the History, Literature, and Empirical Evidence Regarding Industrial Parks and Research Parks." *Economic Development Quarterly* 7 (1, 1993): 107–24.

Petersen, Charles. "Google and Money." *New York Review of Books* 57 (19, 2010).

Pierson, George Wilson. *Tocqueville in America.* Baltimore: Johns Hopkins University Press, 1996.

Pincus, Gregory G. *The Control of Fertility.* New York: Academic Press, 1965.

Pirsig, Robert M. *Zen and the Art of Motorcycle Maintenance.* New York: Bantam, 1974.

Plooij, D. *The Pilgrim Fathers from a Dutch Point of View.* New York: New York University Press, 1932.

Power, Richard Lyle. *Planting Cornbelt Culture: The Impress of the Upland Southerner and the Yankee in the Old Northwest.* Indianapolis: Indiana Historical Society, 1953.

Pratt, John Webb. *Religion, Politics, and Diversity: The Church-State Theme in New York History.* Ithaca, NY: Cornell University Press, 1967.

Prude, Jonathan. "Capitalism, Industrialization, and the Factory in Post-revolutionary America." In *Wages of Independence: Capitalism in the Early American Republic,* ed. Paul A. Gilje. Lanham, MD: Rowman & Littlefield, 1997.

Pulliam, John D., and James J. Van Patten. *A History of Education in America.* Upper Saddle River, NJ: Merrill Prentice Hall, 2003.

Pynchon, Thomas. *Mason and Dixon.* New York: Henry Holt, 1997.

Rabkin, Peggy A. "The Silent Feminist Revolution: Women and the Law in New York State from Blackstone to the Beginnings of the American Women's Rights Movement." PhD diss., State University of New York at Buffalo, 1975.

Rashin, Steve. *Road Swing: One Man's Journey into the Soul of American Sports.* New York: Doubleday, 1998.

Ray, Angela G. *The Lyceum in Public Culture in the Nineteenth Century United States.* East Lansing: Michigan State University Press, 2005.

Reece, William S. "The First Hundred Years of Printing in British North America: Printers and Collectors." Worcester, MA: American Antiquarian Society, 1990. Available online at http://www.reeseco.com/papers/first100.htm.

Reed, James. *The Birth Control Movement and American Society: From Private Vice to Public Virtue*. Princeton, NJ: Princeton University Press, 1978.

Reisner, Marc. *Cadillac Desert: The American West and Its Disappearing Water*. 1846; rev. ed. New York: Penguin, 1993.

Richardson, Mark. *Zen and Now: On the Trail of Robert Pirsig and the Art of Motorcycle Maintenance*. New York: Alfred A. Knopf, 2008.

Riddle, John M. *Eve's Herbs: A History of Contraception and Abortion in the West*. Cambridge, MA: Harvard University Press, 1997.

Rieser, Andrew C. *The Chautauqua Moment: Protestants, Progressives, and the Culture of Modern Liberalism*. New York: Columbia University Press, 2003.

Rivard, Paul E. *Samuel Slater: Father of American Manufactures*. Pawtucket, RI: Slater Mill Historic Site, 1974; repr. 1988.

Roberts, Andrew. *A History of the English-Speaking Peoples since 1900*. London: Weidenfeld & Nicolson, 2006.

Roland, Charles P. *Albert Sidney Johnston: Soldier of Three Republics*. Lexington: University Press of Kentucky, 2001.

Rosegrant, Susan, and David Lampe. *Route 128: Lessons from Boston's High-Tech Community*. New York: Basic Books, 1992.

Rosenberg, Nathan, and L.E. Birdzell, *How the West Grew Rich: The Economic Transformation of the Industrial World*. New York: Basic Books, 1986.

Rosenbloom, Joshua L. "The Challenges of Economic Maturity: New England, 1880–1940." In *Engines of Enterprise: An Economic History of New England*, ed. Peter Temin. Cambridge, MA: Harvard University Press, 2000.

Rothenberg, Winifred Barr. "The Invention of American Capitalism: The Economy of New England in the Federal Period." In *Engines of Enterprise: An Economic History of New England*, ed. Peter Temin. Cambridge, MA: Harvard University Press, 2000.

Rowland, Wade. *Spirit of the Web: The Age of Information from Telegraph to Internet*. Toronto: Somerville House, 1997.

Ruthven, Malise. "The Birth of Islam: A Different View." *New York Review of Books*, April 7, 2011.

Ryan, Mary. *Cradle of the Middle Class: The Family in Oneida County, New York, 1790–1865*. New York: Cambridge University Press, 1981.

Rybczynski, Witold. *City Life*. New York: Harper Perennial, 1996.

———. *A Clearing in the Distance: Frederick Law Olmstead and America in the Nineteenth Century*. New York: Scribner, 1999.

Salsbury, Stephen. *The State, the Investor, and the Railroad: The Boston and Albany, 1825–1867*. Cambridge, MA: Harvard University Press, 1967.

Sampson, Anthony. *Company Man: The Rise and Fall of Corporate Life*. New York: Random House, 1995.

Schama, Simon. *Landscape and Memory*. Toronto: Vintage Books, 1996.

Schama, Simon. *A History of Britain: The British Wars: 1663-1776* Toronto: McClelland & Stewart, 2002

Scharchburg, Richard P. *Carriages without Horses: J. Frank Duryea and the Birth of the American Automobile Industry.* Warrendale, PA: Society of Automotive Engineers, 1993.

Scherer, John L., ed. *A Shaker Legacy: The Shaker Collection at the New York State Museum*, rev. ed. Albany: New York State Education Department, 2000.

Schexnayder, Clifford J., and Richard Mayo. *Construction Management Fundamentals.* Chicago: McGraw-Hill Professional, 2003.

Schlereth, Thomas J. *Victorian America: Transformations in Everyday Life, 1876–1915.* New York: HarperCollins, 1991.

Schlesinger Jr., Arthur M. *The Age of Jackson*, 2nd ed. Old Saybrook, CT: Konecky & Konecky, 1971.

Schoenberger, Chana R. "Tarnished," *Forbes*, March 15, 2004, 80–1.

Scoville, James C. "The Taylorization of Vladimir Ilich Lenin." *Industrial Relations* 40 (4. 2002): 620–6.

Sellers, Charles. *The Market Revolution: Jacksonian America, 1815–1946.* New York: Oxford University Press, 1995.

Setterberg, Fred. *The Roads Taken: Travels through America's Literary Landscapes.* Athens: University of Georgia Press, 1993.

Shaw, Ronald R. *Erie Water West: A History of the Erie Canal, 1792–1854.* Lexington: University of Kentucky Press, 1966.

Shi, David. "Air conditioning — It's made the south what it is." *Independent-Mail* (Anderson, SC), June 24, 2007. Available online at http://www.independentmail.com/news/2007/jun/24/air-conditioning-its-made-south-what-it/.

Shorto, Russell. *The Island at the Center of the World: The Epic Story of Dutch Manhattan and the Forgotten Colony that Shaped America.* New York: Vintage Books, 2005.

Silverman, Kenneth. *Lightning Man: The Accursed Life of Samuel F.B. Morse.* New York: Alfred Knopf, 2003.

Simpson, Jeffrey. *Chautauqua: An American Utopia.* Chautauqua NY: Chautauqua Institution, 1999.

Slaski, Lisa K. "The Remington Family and Works of Ilion, NY." 2001. Available online at http://www.herkimer.nygenweb.net/remington/remingtonfam5.html.

Sloan, Alfred P., John McDonald, and Catharine Stevens. *My Years with General Motors.* New York: Doubleday/Currency, 1990.

Slotkin, Richard. *Gunfighter Nation: The Myth of the Frontier in Twentieth-Century America.* Norman: University of Oklahoma Press, 1998.

———. *Regeneration through Violence: The Mythology of the American Frontier, 1600–1860.* New York: Harper Perennial, 1996.

Smith, Merritt Roe. *Harpers Ferry Armory and the New Technology: The Challenge of Change.* Ithaca, NY: Cornell University Press, 1977.

———. "Military Entrepreneurship." In *Yankee Enterprise: The Rise of the American System of Manufactures*, ed. Otto Mayr and Robert C. Post. Washington, DC: Smithsonian Institution Press, 1981, rev. 1995.

Smith, Robert H. *Clinton's Ditch: The Erie Canal, 1825*. Del Mar, CA: C Books, 2006.

Smith, Rockne P. *Our "Downriver" River: Nautical History and Tales of the Lower Detroit River*. Gibraltar, MI: Rockne Smith, 1997.

Smithsonian Institution. National Museum of American History. "Lighting a Revolution — Lamp Inventors 1880–1940: G.E.M. Lamps." Washington, DC, 2013. Available online at http://americanhistory.si.edu/lighting/bios/whitney.htm.

———. National Portrait Gallery. "The Seneca Falls Convention, July 19–20, 1848." Washington, DC. Available online at http://www.npg.si.edu/col/seneca/senfalls1.htm.

Smithsonian Institution Archives. "Robert H. Goddard: American Rocket Pioneer." Washington, DC, n.d. Available online at http://siarchives.si.edu/history/exhibits/stories/robert-h-goddard-american-rocket-pioneer.

Snyder, Eldon E. "Sociology of Nostalgia: Sport Halls of Fame and Museums in America." *Sociology of Sport Journal* 8 (3, 1991): 228–38.

Solnit, Rebecca. *Wanderlust: A History of Walking*. New York: Penguin Books, 2001

Solomon, Brian, with Mike Schafer. *New York Central Railroad*. St. Paul, MN: Voyager Press, 2007.

Spangenburg, Ray, and Diane Moser. *The Story of America's Roads*. New York: Facts on File, 1992.

Speroff, Leon. *A Good Man: Gregory Goodwin Pincus*. Portland OR: Arnica Publishing, 2009.

Spirn, Anne Whiston. "Constructing Nature: The Legacy of Frederick Law Olmsted." In *Uncommon Ground: Rethinking the Human Place in Nature*, ed. William Cronon. New York: W.W. Norton, 1997.

Spring, Joel. *The American School, 1642–1985*. New York: Longman, 1986.

Standage, Tom. *The Victorian Internet: The Remarkable Story of the Telegraph and the Nineteenth Century's On-line Pioneers*. London: Phoenix, 1999.

Stegner, Wallace, and Page Stegner, *Marking the Sparrow's Fall: The Making of the American West*. New York: Henry Holt, 1998.

Steinbeck, John. *Travels With Charley: in Search of America*. New York: Viking: 1962.

Steinberg, Theodore. *Nature Incorporated: Industrialization of the Waters of New England*. New York: Cambridge University Press, 1991.

———. *Slide Mountain, or the Folly of Owning Nature*. Berkeley: University of California Press, 1995.

Stiles, T.J. *The First Tycoon: The Epic Life of Cornelius Vanderbilt*. New York: Alfred A. Knopf, 2009.

Stone, Alan. *How America Got On-Line: Politics, Markets, and the Revolution in Telecommunications*. Armonk, NY: M.E. Sharpe, 1997.

Stott, Richard. "Artisans and Capitalist Development." In *Wages of Independence: Capi-*

talism in the Early American Republic, ed. Paul A. Gilje Lanham, MD: Rowman & Littlefield, 1997.

Strand, Ginger G. *Inventing Niagara: Beauty, Power, and Lies*. New York: Simon & Schuster, 2008.

Struik, Dirk J. *Yankee Science in the Making*. New York: Collier Books, 1962.

Stuart, Joseph T. "The Initial Points of Michigan." *Turning the Horizon* 3 (1, 2005).

Sutter, Paul. "'A Retreat from Profit': Colonization, the Appalachian Trail, and the Social Roots of Benton MacKaye's Wilderness Advocacy." *Environmental History* 4 (4, 1999): 553–7.

Swade, Doron, and Charles Babbage. *The Difference Engine: Charles Babbage and the Quest to Build the First Computer*. New York: Viking, 2001.

Sweet, Leonard I. *The Evangelical Tradition in America*. Macon, GA: Mercer University Press, 1984.

Tapia, John E. *Circuit Chautauqua: From Rural Education to Popular Entertainment in Early Twentieth Century America*. Jefferson, NC: McFarland and Company, 1997.

Taves, Ann. *Fits, Trances, and Visions: Experiencing Religion and Exploring Experience from Wesley to James*. New Brunswick, NJ: Princeton University Press, 1999.

Taylor, Alan. *William Cooper's Town: Power and Persuasion on the Frontier of the Early American Republic*. New York: Vintage Books, 1995.

Taylor, Bob Pepperman. *Horace Mann's Troubling Legacy: The Education of Democratic Citizens*. Lawrence: University Press of Kansas, 2010.

Taylor, Frederick Winslow. *The Principles of Scientific Management*. New York: W.W. Norton, 1967.

Taylor, Nick. *American-made: The Enduring Legacy of the WPA: When FDR Put the Nation to Work*. New York: Bantam, 2008.

Temin, Peter, ed. *Engines of Enterprise: An Economic History of New England*. Cambridge, MA: Harvard University Press, 2000.

———. "The Industrialization of New England, 1830–1880." In *Engines of Enterprise: An Economic History of New England*, ed. Peter Temin. Cambridge, MA: Harvard University Press, 2000.

Tharp, Louise Hall. *Until Victory: Horace Mann and Mary Peabody*. Boston: Little Brown, 1953.

Thompson, Robert L. *Wiring a Continent: The History of the Telegraph Industry in the United States, 1832–1866*. New York: Arno Press, 1972.

Thorn, John. "Doc Adams." SABR Baseball Biography Project. Phoenix: Society for American Baseball Research, n.d. Available online at http://bioproj.sabr.org/bioproj.cfm?a=v&v=l&bid=%20639&pid=16943.

Tierney, John. "Fateful voice of a generation drowns out real science." *New York Times*, June 5, 2007.

Tifft, M. "A Partial Record of the Descendants of John Tefft, of Portsmouth, Rhode Island, and a Nearly Complete Record of the Descendants of Text of John Tifft, of Nassau, New York." Buffalo, NY: Peter Paul Book Company, 1896. Available online at http://www.archive.org/stream/apartialrecorddootiffgoog/apartialrecorddootiffgoog_djvu.txt.

Toal, Barbara Ann. *Batavia*. Charleston, SC: Arcadia Publishing, 2000.

Tone, Andrea. *Devices and Desires: A History of Contraceptives in America*. New York: Hill and Wang, 2001.

Tozer, Lowell. "A Century of Progress, 1833–1933: Technology's Triumph over Man." *American Quarterly* 4 (1, 1952): 78–81.

Traxel, David. *1898: The Birth of the American Century*. New York: Vintage Books, 1999.

Tsipis, Yanni Kosta, and David Kruh. *Building Route 128*. Charleston, SC: Arcadia, 2003.

Turner, Frederick. *Rediscovering America: John Muir in His Time and Ours*. San Francisco: Sierra Club Books, 1985.

———. *Spirit of Place: The Making of an American Literary Landscape*. San Francisco: Sierra Club Books, 1989.

Turner, Steve. *Angelheaded Hipster: A Life of Jack Kerouac*. New York: Viking, 1996.

Twain, Mark. *A Connecticut Yankee in King Arthur's Court*. 1889; New York: Washington Square Press, 1964.

Twombly, Robert, ed. *Frederick Law Olmstead: Essential Texts*. New York: W.W. Norton, 2010.

United Nations Educational, Scientific and Cultural Organization. "Science Policy and Capacity-Building: Science Parks around the World". Paris: UNESCO, n.d. Available online at http://www.unesco.org/new/en/natural-sciences/science-technology/university-industry-partnerships/science-parks-around-the-world/.

United States. Department of Transportation. Federal Highway Administration. *America's Highways: 1776–1976*. Washington, DC: US Government Printing Office, 1977.

———. Library of Congress. "American Memory: The Evolution of the Conservation Movement, 1850–1920." Washington, DC.

"The United States Numbered System." *American Highways* 11 (3, 1932): 16.

Uselding, Paul. "Measuring Techniques and Manufacturing Practice." In *Yankee Enterprise: The Rise of the American System of Manufactures*, ed. Otto Mayr and Robert C. Post. Washington DC: Smithsonian Institution Press, 1981, rev. 1995.

Van Dulken, Stephen. *Inventing the 20th Century: 100 Inventions that Shaped the World from the Airplane to the Zipper*. New York: Barnes & Noble, 2007.

Vidal, Gore. *Burr*. New York: Bantam Books, 1974.

———. *Empire*. New York: Random House, 1987.

Von Drehle, David. *Triangle: The Fire that Changed America*. New York: Grove Press, 2003.

Ward, Bob. *Dr. Space: The Life of Werner von Braun*. Annapolis, MD: Naval Institute Press, 2005.

Warren, James V. "John Gorrie." *Transactions of the American Clinical and Climato-*

logical Association 93 (1982): 183–8. Available online at http://www.ncbi.nlm.nih.gov/pmc/articles/PMC2279554/pdf/tacca00096-0227.pdf?tool=pmcentrez.

Waterfield, Robin. *Xenophon's Retreat*. London: Faber and Faber, 2006.

Watkin, Elizabeth Siegel. *On the Pill: A Social History of Oral Contraceptives, 1950–1970*. Baltimore: Johns Hopkins University Press, 1988.

Watts, Linda S. *The Encyclopedia of American Folklore*. New York: Facts on File, 2007.

Webb, Jim. *Born Fighting: How the Scots-Irish Shaped America*. New York: Broadway Books, 2004.

Weber, Donald. "The Recovery of Jonathan Edwards," in *Jonathan Edwards and the American Experience*, ed. Nathan O. Hatch and Harry S. Stout. New York: Oxford University Press, 1988.

Weber, Max. *The Protestant Ethic and the Spirit of Capitalism*. London: Allen and Unwin, 1930; repr. Routledge Classics, 2001.

Weible, Robert. *The World of the Industrial Revolution: Comparative and International Aspects of Industrialization*. Lanham, MD: Rowman & Littlefield, 1987.

Weible, Robert, and Lowell Historical Society. *The Continuing Revolution: A History of Lowell, Massachusetts*. Lowell, MA: Lowell Historical Society, 1991.

Weightman, Gavin. *The Industrial Revolutionaries: The Making of the Modern World, 1776–1914*. New York: Grove Press, 2007.

Weingroff, Richard F. "From Names to Numbers: The Origins of the U.S. Numbered Highway System." Washington, DC: Department of Transportation, Federal Highways Administration, n.d.

Weissend, Patrick. *The Life and Times of Joseph Ellicott*. Batavia, NY: Holland Land Purchase Historical Society, 2002–3.

Wershler-Henry, Darren. *The Iron Whim: A Fragmented History of Typewriting*. Toronto: McClelland & Stewart, 2005.

Westwood, John, and Ian Wood. *A Historical Atlas of North American Railroads*. Edison, NJ: Chartwell Books, 2007.

Whyte, Jack. *The Skystone*. Toronto: Viking Canada, 1992.

Wills, Garry. *Head and Heart: A History of Christianity in America*. New York: Penguin Books, 2007.

———. *Henry Adams and the Making of America*. New York: Houghton Mifflin Harcourt, 2005.

Wilson, George W. "US Intercity Passenger Transportation Policy 1930–1991, An Interpretive Essay." In Canada, Royal Commission on National Passenger Transportation Directions, *Final Report*, vol. 3. Ottawa, 1992.

Winston, Brian. *Media Technology and Society: A History*. London: Routledge, 1998.

Wish, Harvey. *Society and Thought in Modern America: A Social and Intellectual History of the American People from 1865*, 2nd ed. New York: David McKay, 1962.

Wollstonecraft, Mary. *A Vindication of the Rights of Woman*, 2nd. ed., ed. Carol H. Poston. New York: W.W. Norton, 1988.

Wood, Gordon S. "The Enemy Is Us: Democratic Capitalism in the Early Republic."

In *Wages of Independence: Capitalism in the Early American Republic*, ed. Paul A. Gilje. Lanham, MD: Rowman & Littlefield, 1997.

Woodward, Colin. *The Lobster Coast: Rebels, Rustications, and the Struggle for a Forgotten Frontier.* New York: Viking, 2004.

Yale University. *Obituary Record of Graduates of Yale University Deceased during the Academical Year ending in June, 1905.* New Haven, CT. Available online at http://mssa.library.yale.edu/obituary_record/1859_1924/1904-05.pdf.

Young, Ralph F., ed. *Dissent in America: Voices That Shaped a Nation.* New York: Pearson Longman, 2006.

Zboray, Ronald J., and Mary Saracino Zboray. "The Boston Book Trades, 1789–1850: A Statistical and Geographical Analysis." In *Entrepreneurs: The Boston Business Community, 1700–1850*, ed. C.E. Wright and K.P. Viens. Boston: Massachusetts Historical Society, 1997.

Zelinsky, Wilbur. *The Cultural Geography of the United States.* Englewood Cliffs, NJ: Prentice-Hall, 1973.

Ziff, Larzer. *Puritanism in America: New Culture in a New World.* New York: Viking, 1973.

Index

A

A Sand County Almanac (by Aldo Leopold), 116
Abbey, Edward ("Cactus Ed"), 103, 116–120
Abortion statistics, 127–128
Adams, Abigail, 229
Advanced Research Projects Agency (ARPA), 275–276
Air conditioning, 12, 38, 40, 308–320, 360
Albany, NY, 155 et seq.
Allston, George Washington, 266
American Industrial Revolution birthplace, 39
American road book literature, generally, xix–xx
American System, defined, 153
Americanization of English Puritans. See *Transformation of Puritan into Yankee*
Ames Manufacturing Company, 152
Anabasis, Zenaphon's, xviii
Anarchist Edward Abbey, 116–120
Andreessen, Marc, 277
Anti-Slavery "Fairs," 233
Appleton, Nathan, 76, 80
Arches National Monument, 118
Architecture of factories, 152
Arkwright, Richard, 67, 73–74
Armour, Philip, 200
Artaxerxes, Persian Emperor, xviii
Artificial ice or refrigeration, 312
Artists and painters
    Allston, George Washington, 266
    Cole, Thomas, 106
    Morse, Samuel F.B., 265, 361
    Russell, Charles Marion, 105
Associated Press syndicate formation, 269
Author's road books influence. See *Introduction, Yankee Values and Roads*
Automobiles
    Battery driven, 144–145
    Duryeas' car, producing, 142–143, 145
    Electric motors, 144
    Growing importance of motorized vehicles, 36
    Interchangeable parts movement, 142–145
    Internal combustion engine, 143–144
    National Parks, excluding in, 119
    Otto-cycle gasoline engine, 143–145
    Racing, 143–144
    Refining oil for fuel, 145
    Steam-driven engine, 144
    Types of engines, generally, 143–144, 265

B

Ball games
    Baseball. See Baseball
    Basketball, 2, 179
    Football, 2, 80
    Sports performances. See Sports Performances
    Teams of Boston and New England, 2
Bankruptcy. See Insolvency and Bankruptcy
Banks and banking
    Democratizing capitalism
        Banking, 156, 161, 165–168
        Banking incorporation, 163, 166, 168
        Free banking, 156, 161, 165–168

Industrial revolution, 69
National bank and financial network, 70, 166–168, 272
Telegraph use by bankers, 272
Baseball
   Acculturation of Americans to organizations, 178–179
   Carew, Rod, 179
   Characteristics of teams, 178
   Cooperstown. *See* Cooperstown, NY
   Doubleday Field, 172
   Everyman as the hero, 174–177
   Fenway Park, 108, xix
   Field of Deams movie, xxiv
   Founder, 40
   Glory or victory?, 179–180
   Hall of Fame, 171–173, 179–180, 228
   Kinsella's novel *Shoeless Joe*, xxiv
   Moneyball movie, 179
   Name of team, 2
   Oakland Athletics Baseball Team, 179
   Organizations and the individual, 173–174, 177–180
   Origins, 171–173
   Shoeless Joe, xxiv
Basketball
   Jordan, Michael, 179
   Name of team, 2
Batavia, NY, 279–280, 282
Bausch and Lomb, 265
Baxter, George, 266
Beat Generation, xxi
Berkshire Hills, 39, 41, 324
Bicycle, 32, 36, 110, 142, 156, 350
Birth control and family planning movement
   Generally, 122–140
   Abortion restriction or prohibition, 123–124, 128
   Abortion statistics, 127–128
   Aftermath of the "Pill," 138–140
   Bureau of Social Hygiene, Rockefeller-funded, 135
   Chang, M.C., 137
   Clinics, 127, 134–135
   Comstock, Henry, 126–128
   The Control of Fertility (by Pincus), 130
   Development of oral contraceptive "the Pill," generally, 122, 131–132, 135–138
   Enovid, 137–138
   FDA approval of Envoid for menstrual disorders, 137
   FDA approval of the "Pill," 132–133, 132–137
   Hecter, Oscar, 130
   Historical perspective, 122–129
   Hormone research, 129–131, 135–137
   Human trials for the "Pill" development, 137–138
   Impact of the "Pill" on society, 138–140
   Influence of family planning, 128
   Laurentian Hormone Conference, 130
   Mail distribution of information or devices prohibition, 126–127
   Marker, Russell, 136
   McCormick, Katharine Dexter, 130–134, 137–138, 140
   Mid-nineteenth-century factors influencing attitudes about family and population growth, 124–125
   Pincus, Gregory Goodwin (Goody), 122, 129–133, 135, 137
   Plants that control pregnancies and female irregularities, 124, 136
   Post-Civil war activism influence, 125–129

Regulating menstrual irregularities, 136–139
Rock, John, 137
Rockefeller-funded Bureau of Social Hygiene, 135
Sanger, Margaret Higgins, 127, 130–140
Searle pharmaceutical, 137–138
Statistics, fertility rates, 123, 128, 139
Strength of the "Pill" today, 138
Syntex, 136
Synthesizing estrogen and progesterone, 136
Thalidomide crisis, 133
Venereal diseases, 127
In vitro fertilization, 137
Worcester Foundation for Experimental Biology (WFEB), 122, 130–133, 137
Black Hawk, 113, 197
Black Rock, NY, 295, 309
Blackstone River Valley
Generally, 73–75
National Heritage Corridor, establishment, 39
Blue Highways (by William Least Heat-Moon), xxiii–xxiv
Blue highways, scenic, 38, xix, xxiii–xxiv
Blue Ridge parkway, 37
Boone, Daniel, 105
Born Fighting (by Senator James Webb), 198
Boston Celtics, 2
Boston Patriots, 2
Boston Post Road, 38–39
Boston Red Sox, 2
Braudel, Fernand, 157
Brinkley, Douglas, xxi, xxiv–xxv
Brook Farm, 102, 225, 247, 257–259, 262
Bryson, Bill, xix

Buffalo Bill, 42, 176, 296
Buffalo Forge Company, 308–310, 313, 315
Buffalo, NY
Anabaptists settlements, 247
Batavia, 280
Burned-over District, 205–206, 215
Chautauqua, 349
De Tocqueville and the Yankees, 24–26
Fast trip across the continent on US 20, 37–40
Irish immigration, 28
Lecture engagements of Horace Mann, 347
Niagara Falls, 295–297
Planning city parks, 107
Railroading, 322–323, 326–327, 329
Revivals by Charles Finney, 221
Steamships advertising passage, 292
Telegraph line to Cleveland, 270
Telegraph line to Springfield, 268
Town laid out by Ellicott (Holland Land Co.), 282–283, 308–309
Turning typewriters into computers, 188
Wanting Erie Canal built, 199
Women's rights movement, 228
Bullets. *See* Firearms (using, manufacturing, and selling)
Bunker Hill Memorial, building, 322
Burned-over District, 205, 215–219, 234, 243
Butler, Nicholas Murray, 168
Butterfield, John, 268, 270

C

Calvinism, 208, 213, 216, 220, 247–248
Canada
Author's background, xxiv

Canals usage, 323
Church services, 205
Circuit chautauquas, 356
Destroyed Iroquois villages inhabitants going to, 280
Disciples of Christ church, 108
Dutch or British East India Companies, 158–159
Dwight Moody speaking in, 223
Exporting to Canada via Lake Ontario and St. Lawrence River, 203
Fleming, Sir Sandford, 271
Grand Island, 295
John Muir going to, 108
Large landholdings acquired, 284
Lucretia Mott speaking in, 228
Mosaiac pattern of culture and attitudes, xxvii
New York Central Railroad extension, 329
Niagara Falls and Horseshoe Falls, 262, 295–296
Prince Edward Island., 284
Prohibition, xxi
Railroad land, selling, 293
Railroad, Canadian Pacific, 271
Railway connections to Canada, 329–330
Revolutionary War refugees, 295
Romanticism and wilderness living, 106
Tax treaty, xxiii
Yankee temperament, 9
Canyons
Activism, 103, 116–120
Glen Canyon Dam, 119
Wind River Canyon, 42
Capitalism
Albany, NY, 155 et seq.
Corporate productivity, 47–65
Democratizing. *See* Democratizing Capitalism
Growth of, 156
Labor movement, 47–60
Scandals and fights, 156
Sweatshops, 156
Capitol and the White House, air conditioning, 316
Carew, Rod, 179
Carrier, Willis, 40, 52, 308–320
Carrier's air conditioning companies, 315–316
Cassady, Neal, xxii, xxv
Catalogue shopping growth, 32
Cazenove, Theophile, 199, 281
Cazenove, Thomas, 203
Cazenovia, NY, 198–200
Cell phone, 277
Central Park in New York City, 37, 107, 297
Cerf, Vint, 276
Charters provided by Charles I, expiration of, 7
Chautauqua Institution, 349–353
Chautauqua Literary and Scientific Society (CLSC), 350–353
Chautauqua movement
Generally, 340–358
Adult education mixed with Social Gospel program, 352–353, 358
Chautauqua Literary and Scientific Society (CLSC), 350–353
Correspondence courses, 352–353
Democratizing education, 341–342, 349–358
Horace Mann, 342–345
Local chautauquas, 353–356
Lyceum movement, 345–348
Social Gospel program, 352
Spinoffs of movement, 353–358

Summer school curriculum at Institution, 352–353
Sunday schools, 348–349
Theory of, 353, 358
Town Hall meetings, 356–358
University status secured for Institution, 352
Vawter, Iowan Keith, 354–356
Willamette Valley Chautauqua Association (OR), 354
Chautauqua, NY
  Chautauqua Institution community, 350–353
  Location description, 349–350
Chicago World's Fair 1893, 107–108, 300, 353
Chicago World's Fair 1933, 140
Child laborers, 348
City on a Hill settlement concept, 5–8
Civil War era, xxvii–xxviii
Civilian Conservation Corps (CCC), 63, 113, 115
Cleveland, Grover (President), 186, 296
Cleveland, Moses, 290–291
Cleveland, OH, 40–41, 145, 270, 291, 300, 309, 329, 347
Climate and humanity, 311–317
Clothes
  Garment industry fires, 60–65
  Ready-to-wear and standardized sizes clothes, 51, 82
Coal mine operations, 323
Cody, William "Buffalo Bill," 42, 176, 296
Cold War matters
  Greek times, xviii–xix
  Kerouac's attention, xxii
  Yankee values, xxvi–xxix
Cole, Thomas, 106

Colleges and Universities
  Alabama University, 337
  Antioch College, 232, 344
  Arizona University, 118
  Boston University, 129
  Brown University, 344
  California, University of, 110
  Cazenovia College, 200
  Chicago University, 90, 107
  City of New York University, 337
  Clark University, 90, 129–130
  Columbia University, 61
  Community colleges, 91
  Connecticut, University of, 129
  Dartmouth College, NH, 342
  Engineering departments in, number of, 90–91
  Geneva Medical College, 232
  Harvard University, 8, 87, 92, 129, 337, 342
  Illinois, University of, 277
  Johns Hopkins University, 90, 301
  Land-grant colleges, 89–90, 232, 292
  Massachusetts Medical School., University of, 123
  Michigan Agricultural College (now Michigan State University), 89
  Minnesota,, University of, 277
  Mount Holyoke College, 60
  New Mexico, University of, 117
  Norwich University, VT, 337
  Oberlin College, 221–222, 232, 310, 354
  Pennsylvania College for Women in Pittsburgh, 301
  Pennsylvania University, 92
  Reed College, 192
  Rensselaer Polytechnic in Troy, New York, 337
  Research universities, 89–90, 98
  Rochester, University of, 135
  Rutgers University, 129

Stanford University, 87, 90, 107, 117, 200
Storrs Agricultural College, 129
Tufts University, 129
Union College in Schenectady, New York, 337
Vassar College, 232
Virginia Military Institute, 337
West Point, 337
Wisconsin University, 108, 115
Women's colleges, 356
Yale University, 8, 89, 337, 342
Colonial economics, 6
Colonies, origin of, 3–5
Colonization by corporation, 158–159
Colt, Samuel, 151
Colt's pistols, production of, 151
Computers
Allen, Paul, 192–193
Altair 8800, 192
Alto, 190
Apple, 192–194
Bolt, Beranek and Newman (BBN) in Cambridge, MA, 275–276
Brand, Stewart, 194
Cell phone, 277
Compaq, 97
Comparing Massachusetts's Route 128 to California's Silicon Valley, 191–192
Computer kits, 192
Criminal activity using, 195
DEC, 190
Development of the Internet, 189, 275–278
Email technology, 276
Englebart, Doug, 189
Englebart, Fred, 194
Ethernet, 190
Gates, Bill, 192–193
Heathkit, 192
Hewlett-Packard (HP), 191
Historical perspective
Computer kits and private ownership computers for personal use, 192–195
Mainframe computers, 188–189
Mini/Micro-computers and the PC revolution, 92, 96–97, 189–195
Present status and next actions, 194–195
Turning typewriters into computers, 188–189
Homebrew Computer Club in Palo Alto, 192, 194
IBM, 188, 190, 193
Internet. *See* Internet Connectivity
Jobs, Steve, 192
Lotus's spreadsheet/accounting software, 97
Mainframe computers, 188–189
Massachusetts Microelectronics Center (M2C), 190–191
Microsoft, 97, 192–193
Military, 94, 189, 275
Mini/Micro-computers and the PC revolution, 92, 96–97, 189–195
MITS, startup company in Albuquerque, New Mexico, 192
Mouse, 190
Palo Alto Research Center (PARC), 190, 194
Pentagon, 189, 275
Personal computers, 92, 95–97, 189–195
Private ownership computers for personal use, 192–195
RadioShack, 192
Rand, James H. Jr., 188
Remington Rand, 188
Routers, 275, 278

Silicon Valley in California, 190–192, 194
Sperry Rand, 188
Stanford university-related industrial/research park, 191
Terman, Fred, 191
Turning typewriters into computers, 188–189
Unisys, 189
Varian Associates, 191
Vision for, 194–195
Wang, 97
Wozniak, Steve, 192
Xerox, 190
Connecticut River valley Yankees, 146–148
Connecticut Trail, 38
Consulting firm, first, 90
Consumer goods
   Generally, 147, 152
   Air conditioners, 316–318
   Clothes, standardized sizes, 51
Contracted technology research, 90
Cook, Captain James, 45
Cooke, Morris, 57
Cooling equipment, 311–317
Cooper Union in NYC, 357
Cooper, James Fennimore, 9, 26, 104–106, 110, 170, 174–175
Cooper, William, 40, 105, 170, 200–203
Cooperstown, NY
   Generally, 169–180
   Baseball Hall of Fame, 171–173, 179–180, 228
   And the Coopers, 169–171
   Everyman as the hero, 174–177
   Founder, 40
   Glory or victory?, 179–180
   Microcosm of all frontier development, 198
   Organizations and the individual, 173–174
   Origins of baseball, 171–173
   Road from the Mohawk River to, 202
   Wanting Erie Canal built, 199
   Yankee organization, 177–179
Cornell, Ezra, 271
Crockett, Davy, 10, 13

D

De Beaumont, Gustave, 23–27, 155, 206, 250
De Soto, Hernando, 201
De Tocqueville, Alexis, 23–28, 64, 125, 155, 177, 206, 216, 231, 250
Debt, imprisonment for, 161
Defense Advanced Research Projects Agency (DARPA), 276
Defense Highway Act of 1956, 38
Definitions
   American System, 153
   Blue highways, xxiii–xxiv
   Burned-over District, 205
   Capitalism, 157
   Deseret, 253
   Freeway, 37
   Internet, 272
   Lyceum, 346
   Multicultural terminology, xxvii
   Parkway, 37
   Pioneering, 23
   Puritan, 246
   Rebel, xxviii
   Road book, xix–xx
   Settlement, 5
   Technology, 88
   Yankee, 2–3, 5, 7, 10, 359, xviii, xxvii, xxviii
   Yankee Imperialism, 3
   Yankee Road, xviii, xxix, xxvii

Democratizing capitalism
  Generally, 154–168
  Albany, NY, 156–157
  Banking, 156, 161, 165–168
  Banking incorporation, 163, 166, 168
  Braudel, Fernand, 157
  Chartering some land development companies, 159
  Colonial economics, 6
  Colonization by corporation, 158–159
  Corporate productivity, 47–65
  Counterfeiting, 166
  Credit systems, 165–168
  Defining capitalism, 157
  Failed corporations and bankruptcy, 162
  Financing America, 164–166
  Free banking, 156, 161, 165–168
  Free or automatic incorporation for businesses, 159–161
  Government agency method, 157–158
  Joint-stock corporation, 157–158
  Labor movement, 47–60
  Limited liability of corporations concept, 161–162
  Logic of incorporation, 157–158
  Model corporation evolution in textile industries, 80
  Money and credit systems, 165–168
  Monopoly concerns, 160, 163–164
  Nature of the corporation relative to rest of society, 163
  New York Stock Exchange, 162
  Organizational permutations and combinations, 157–158
  Partnership, 157–158
  Productivity, scientific management movement, 47–65
  Sole proprietorship, 157–158
  State and nation in incorporation, 163–164
  Stock trading, 162
  Trusts, 163–164
  Unions, 61–62
  Value of the corporation, 161–162
  Weber, Max, 49, 65, 157
Democratizing education
  Generally, 178, 341–342, 358
  Chautauqua movement, 349–358
  Horace Mann, impact of, 342–345
  Internet, 357
  Internet social media usage, 357
  Lyceum movement, 345–348, 358
  Sunday schools, 348–349
  Town Hall meetings, 356–358
Democratizing information, 278
Densmore, James, 183, 185
Depression-era
  Air conditioning, democratization of, 316
  Civilian Conservation Corps (CCC), 113, 115
  Migration, xxii, xxix
Deseret colony of Mormons, 251–257, 262
Desert Solitaire (by Edward Abbey), 117–118, 120
Desert, canyon, and dam activism, 103, 116–120
Dominion of New England, creation of, 7
Doubleday Field, 172
Dow, Lorenzo (preacher), 219–220
Drucker, Peter, 58–59
Duncan, Dayton, xix
Dunkirk, NY, 321–339
Dunlavy, John, 250
Duryeas' car and Interchangeable parts movement, 142–143, 145
Dylan, Bob, xxi

E

Earle, Horatio "Good Roads," 36
Earth First! activist group, 118–119
Eastman, George, 265
Edison, Thomas, 90, 144–145, 185, 190, 295
Education. See Schools and Education
Edwards, Jonathan, 208–210, 213, 225
Einstein, Albert, 313
Electrical communication technologies
   Internet. See Internet Connectivity
   Radio development and use, 273–274
   Telegraph. See Telegraph
   Telephone. See Telephones and Telephone System
   Television, development of, 274
Electrical engineering profession, 271
Electricity
   Automobiles, electric motors, 144
   Communication technologies, 273–274
   Internet network, 272
   Metropolis, 299
   Niagara Falls power, 297, 300, 306
   Operate factory machinery, 297
   Professor André-Marie Ampère lectures, 266
   Professor Benjamin Silliman's lectures, 266
   Telegraph, validating electric telegraph, 267
   Telephone, 273
   Television, 274
   Whote City, 300
Ellicott, Andrew, 203
Ellicott, Joseph, 203, 280, 282–283, 309
Elmira, New York, 188
Email of author, xv
Emerson, Ralph Waldo
   Lucretia Mott's speaking engagement, 238
   Lyceum circuit, 346–347
   Railroads, 269
   Wilderness and nature, 102–104, 106, 109–110
Emigration to New England, 3–5
Engines, types and production of, 143–145, 265
Environment
   Anarchist Edward Abbey, 116–120
   Desert, canyon, and dam activism, 103, 116–120
   Earth First! activist group, 118–119
   Love Canal, 305–307
   Marsh, George Perkins, 111, 115
   Terrestrial ecosystem, 304
   Wilderness. See Wilderness and Nature
Epilogue, 359–362
Ergonomics industry, 51–52, 58–59
Erie Canal
   Albany, NY, 156, 283
   Banking, impact on, 160
   Buffalo, NY, impact on, 309
   Development of, 25, 228, 265, 283, 309
   Education in rural areas, impact on, 343
   Effect on Yankee settlement patterns, generally, 23, 197–198
   Federal land sales, impact on, 292
   Ilion, New York, 182–183
   Lincklaen and Cooper (NY), 200, 203
   Niagara Falls, impact on, 295
   Railroad building, impact on, 323–329, 334–335
   Revivals, impact on, 215–216, 361
   Rochester, NY, 265
   Seneca Falls NY and women's rights convention, 228

Syracuse and Utica (NY), 205–206, 283
Western Reserve, 290–291
Evolution of Puritan society. See *Transformation of Puritan into Yankee*
Exodus, Book of, xviii
Expansion westward
    Generally, 196–198
    Cazenove, Theophile, 199, 281
    Cazenove, Thomas, 203
    Cazenovia, NY, 198–200
    Farther west, 203
    Lincklaen and Cooper, 200–203

F

Factory Investigating Commission (FIC), 62–63
Fainsod, Merle, 178
Family planning. See Birth Control and Family Planning Movement
Famine of 1846-50, 2
Father of all Yankees, 8
Faxton, Theodore, 268, 270
Federal Highway Act of 1921, 36, 38
Federal Writers' Project of the WPA, xxv
Field of Dreams movie, xxiv
Finney, Charles Grandison (preacher), 220–222
Fire, Triangle Waist Co.in NYC, 60–65
Firearms (using, manufacturing, and selling), 47–48, 50, 148–152, 182–184
First Day Society, 348–349
Fleming, Sir Sandford, 271
Football, 2, 80
Ford Hall in Boston, 357
Foreign countries
    Air conditioning's effect on, 318–319
    Scientific management concept, appreciation of, 57–58

Foreman, Dave, 118
Forestry management, 111–116
Franklin Institute in Philadelphia, 267
Franklin, Benjamin, 8, 22, 39, 89, 208, 312, 345, 351
Free Soil Party, 229, 240, 344–345
Freeway, 37
French-Canadians in New England, impact of, 82–83
Freon, 316–317
Frontiersman archetype, 6, 361
Fur trade, 6

G

Gale, Leonard, 266
Game management, 114–116
Gandhi's nationalist "homespun" clothing campaign, 68
Gantt, Harvey, 58
Garment industry fires, 60–65
Garrison, William Lloyd, 238, 243
Gaslights, 156, 297
Gay, Edwin F., 57
Genetics, 129
Geographic types of Americans, xxvii–xxviii
GI Bill's impact on university admissions, 191
Gila Wilderness plan, 114
Gilbreth, Frank, 58–59, 186
Glen Canyon Dam, 119
Glidden, Carlos, 182–185
Globalization
    Generally, 12, 319, 361
    Air conditioning's effect on the world, 317–320
    Birth control Pill, impact on global society, 123
    Financial crises, global, 165
    Internet connectivity, 274, 277–278

Modern global economy, general structure of, 157
Personal computer industry, 193
Social activist with global reputation, Katharine Dexter McCormick, 131
Yankee organization, communications, and products and servic, 180
Goddard, Robert, 39, 122, 302–305
Gold discovery in California, 82, 106, 108, 172, 233, 254, 330–331
Good Roads Association, 36
Goodyear Tires, shipping, 35
Gorrie, Dr. John, 312
Grauman's Metropolitan Theater in Los Angeles, air conditioning, 316
Great Genesee Road, 202, 283
Great Smoky Mountains parkway, 37
Great Western Pike, 39–40, 202, 295
Greek road book, xviii–xix
Greenpeace organization, 113
Griffin and Little, 90
Gunther, John, xxv
Gutenberg, 184

H

Halls of Fame
   Baseball, 171–173, 179–180, 228
   Historical perspective, 173
   Visiting on road trip, xix
   Women's, 228
Hamilton, Alexander
   National Bank system, 70, 166–168, 272
   Urban industrial vision, 70, 74, 81, 163
Harper's Ferry and Armory, 47, 147, 149–150
Hawaii, 45

Health issues
   Impact of the "Pill" on society, 138–140
   Love Canal incident in NY, 305–307
   Mail service of health information or devices, 126–127
   Requiring doctors to provide information on health risks, 139
   Venereal disease, 127
Heat and humidity control, 311–317
Hecter, Oscar, 130
Hero in American society, concept, 173–180
Hetch Hetchy Reservoir, 110, 119
Highway 61, 37
Highway numbering system, 36–37
Highway Trust Fund, 38
Hoagland-Pincus Conference Center of the University of Massachusetts Medical School, 122
Hoagland. Hudson, 122
Holbrook, Josiah, 346
Holbrook's Lyceum movement, 346
Holland Land Company
   Batavia, 282–283
   Buffalo, 309
   Cazenovia, 199–200, 203
   Erie Canal Commission, 335
   Great Western Pike toll companies, 40
   Incorporating, 282
   Museum in Batavia, 280
   Purpose, 159
Holland Land Company Museum, 280
Homer's *Odyssey*, xviii
Homestead Act, 90
Hormone research, 129–131, 135–137
Horseshoe Falls, Canada, 295
Hotel rooms, air conditioning, 316
Human behavior and motions, measurement of, 59

I

IBM, 188, 190, 193
Ilion, New York, 182, 184, 189
Immigrants, arrival of, 28, 360
Imprisonment for debt, 161
Industrial research, 90
Industrial revolution, American
    Generally, 66–83
    Blackstone River Valley, 73–75
    Cottage industries development, 71–72
    Decline, 82–83
    First large-scale planned industrial community, 81
    Ironworks, establishment of, 71
    Lowell, Francis Cabot, 75–78
    Lowell, MA mills, 78–82
    New England and industry, 71–73
    New York Manufacturing Company, 73
    Public support and involvement in economic development, 71–72
    Shipping trade, 71
    Slater, Samuel, 73–75
    Steam power, effect of, 82
    Waltham, MA mills, 75–78
Industrial revolution, British, 67–68, 70, 82
Industry management, 47–60
Inside U.S.A. (by John Gunther), xxv
Insolvency and bankruptcy
    Failed corporations and bankruptcy, 162
    Imprisonment for debt, 161
    Telegraph, growing insolvency of telegraph lines, 269–271
Interchangeable parts movement
    Generally, 50, 141–153
    American System concept, 152–153
    Architecture of factories, 152
    Arms factory and armories, 50, 148–152
    Arrows, producing, 145
    Assembly line, 50, 152
    Automobiles, 142–145
    Building the Yankee world ("The American System"), 152–153
    Choosing a technology, 143–145
    Connecticut River valley Yankees, 146–148
    Consumer goods, generally, 147, 152
    Duryeas' car, 142–143
    Factories as interchangeable, 152
    Labor as interchangeable, 152–153
    Machine tools, importance of, 151–152
    Military armaments and ordinances, 148–152
    Nail-making machine, 152
    People as interchangeable, 51, 152
    Reason for interchangeable parts quest, 145–146
    Samuel Colt and his pistols, 50, 151
    Textile products, producing, 145–147
    Wood building construction, balloon-type, 50, 152
Internal combustion engine, 143–144
International Workers of the World, 134
Internet connectivity
    See also Computers
    Generally, 271–278
    Addresses and domains, assigning, 276–277
    Advanced Research Projects Agency (ARPA)., 275–276
    American Standard Code for Information Interchange (ANSCII), 276–277
    Andreessen, Marc, 277
    Browser, 277
    Cerf, Vint, 276

CERN research facility in Switzerland, 277
Computers' impact on, 194–195
Corporate Intranet, 278
Cost of, 278
Defense Advanced Research Projects Agency (DARPA), 276
Definition of "Internet," 272
Development of the Internet, 189, 275–278
E-democracy movement, 357
Email and MAILBOX technology, 276
Father of the Internet, 276
Fifth form as combining forms 1 thru 4 plus more, 274
Global monopoly, 278
ICANN, 276
Impact of, 277–278
Internet Explorer, 277
Internet Service Providers, 277
Intranet, corporate, 278
Metcalfe, Bob, 278
Metcalfe's Law, 278
Mosaic, Internet Explorer, and Netscape, 277
Netscape, 277
Pentagon, 189, 275
Radio as the third form of "Internet," 273–274
Social media used as Town Hall, 357
TCP/IP (Transmission Control Protocol/Internet Protocol), 276
Telegraph as the first form of "Internet," 265, 272, 274
Telephone as the second form of "Internet," 272, 274
Television as the fourth form of "Internet," 272, 274
Tim Berners-Lee at the CERN, 277
Transmission Control Protocol/Internet Protocol (TCP/IP), 276
World Wide Web, 276–277
Interstate highways system, 36–38, 119
Introduction, Yankee values and roads
  Generally, xvii et seq.
  American road books, xix–xx
  Author's road books influence
    Generally, xx et seq.
    At age 19 (nineteen), Jack Kerouac, xx–xxii
    At age 21 (twenty-nine), Robert Pirsig, xxii–xxiii
    At age 39 (thirty-nine), William Kinsella, xxiv
    At age 39 (thirty-nine), William Least Heat-Moon, xxiii–xxiv
    At age 49 (forty-nine), William Kinsella, xxiv
    At age 59 (fifty-nine), Douglas Brinkley, xxiv–xxv
    At age 69 (sixty-nine), Works Progress Administration (WPA) guides, xxv–xxvi
  Cold War influence, xxvi–xxix
  Greek road book influence, xviii–xix
  Works Progress Administration (WPA) guides, xxv–xxvi
Iroquois, 18, 23, 146, 197, 280, 283

J

Jackson, Patrick Tracy, 81
Jefferson, Thomas, 8, 21, 69–70, 75, 89, 166, 170, 229, 241, 291
Jensen, Jamie, xxv–xxvi
Jones, John, 185

K

Kenmore Square, 30, 38
Kendall, Amos, 268

Kerosene, 145, 147, 156, 271
Kerouac, Jack, xix–xxii
Kesey, Ken, 194
Kinsella, William, xix, xxiv
Knox, General Henry, 47
Kodak camera, 265
Kyoto Protocol, 113

L

Labor movement, 47–60
Lackey, Kris, xix
Lake Chautauqua, NY, 349
Land development companies
   Advantages of private land companies, 203
   Chartering, reason for, 159
   Holland Land Company. *See* Holland Land Company
Land rush to the Far West, creation of, 292
Land sale, world's greatest
   Generally, 279–293
   Batavia, NY, 280–283
   Chart, Western New York HLC Lands, 283
   Federal land sales, 291–293
   Land as an asset, 284–286
   Model for the West, the Seven Ranges as a, 286–291
   Northwest Ordinance, 289
   Public Land Survey System (PLSS), 287–290
   Railroads, federal land grants to, 292–293
   Western Reserve area, 290–291
Land speculation
   Batavia, NY, 280–281
   Cazenovia, 198–199
   Colonization by corporation, 159
   Cooperstown and William Cooper, 170, 198
   Dow, Lorenzo, 220
   Movement westward, 285–286
   Puritans, 6
   Reynolds Arcade, 265
   Romanticism, 105
   Western Reserve area, 289, 291
   Yankee colonization patterns, 22–23
Larry Bird's Shoes, 2
Laser printer invention, 190
Laurentian Hormone Conference, 130
Least Heat-Moon, William, 44–45, xix, xxiii–xxiv
Leatherstocking novels (James Fennimore Cooper), 105
Lee, Ann, 247–251
Leggett, William, 160–161
Leopold Memorial Reserve, 116
Leopold, Aldo, 85, 103, 108, 113–116, 119
Limited access roads, 37
Lincklaen, John, 199–203
Lincoln Highway, 31, 34, xxix
Little, Arthur D., 90
Livestock "wars," 292
Lord, Eleazar, 326–328
Love Canal disaster in NY, 305–307
Love, Henry, 305–307
Lowell, Francis Cabot, 75–78
Lowell, Massachusetts
   Generally, 78–82
   Kerouac influence, xx–xxi
   Textile workers' strike 1912, evacuation of children, 134
Lyceum tours and circuits
   Boston Lyceum Bureau, 347
   Chautauquas, 356–358
   Defined, 346
   Holbrook's Lyceum movement, 346

Horace Mann raising money on the Lyceum circuit, 345
Movement, generally, 345–348, 358
Redpath Lyceum Bureau, 347, 354, 356

M

Macadamized road surfaces, 322
Machine tools
    Famous machine tool factories, 152
    Importance of, 151–152
    Merrimack Manufacturing Company, 79–80
Magnetic Telegraph Company, 268
Mail service, 32, 38–39, 126–127, 278
Main Streets, generally, 38–39
Majic Bus, The (by Douglas Brinkley), xxiv–xxv
Making of the universal Yankee nation, xxviii
Malthus, Thomas, 124
Man and Nature (by George Perkins Marsh), 111
Management of business and all social activities, 47–60
Mann, Horace, 332, 340, 342–347, 358
Manufacturing businesses
    American system model, 81
    Ames Manufacturing Company, 152
    Architecture of factories, 152
    Bicycles, 32
    Clockmaking, 147
    Consumer goods, generally, 147
    Contrasting Connecticut River Valley Yankee from other machinists, 147–148
    De-skilling the shop floor, 51
    Development of, 146–148
    Equipment size and placement, 51–52
    Ergonomics of plant, 51–52, 59
    First true factory in US, 73
    Historical perspective, 71–73
    Impact of machine tools, 151–152
    Industrial research, 90
    Industrial revolution. *See* Industrial Revolution
    Machine-tool manufacturers, 152
    Managing operations and organization, 50–60
    Mass production.. *See* Interchangeable Parts Movement
    Merrimack Manufacturing Company, 79
    Migration South, 317
    Military ordinance and guns, 148–151
    Nails, making, 152
    New York Manufacturing Company, 73
    Productivity, scientific measurement and management, 47–65
    Society for Useful Manufactures, incorporation of, 163
    Temperature in factories, 312, 314–317, 320
    Terry, Eli, 147
    Textiles. *See* Textile Industries
    Wood building construction, balloon-type, 152
Marconi, Guglielmo, 274
Marietta, Ohio, 289
Marker, Russell, 136
Marsh, George Perkins, 111, 115
Marshall, Robert, 115
Mass production movement. *See* Interchangeable Parts Movement
Mass-production organizations, impact of, 186
Massachusetts Bay Company, 3–4, 6, 159, 246, 284–285, 342
Massachusetts Institute of Technology (MIT), 32, 86–87, 90–92, 131, 276

McAdam, John Loudon, 322
McCormick, Katharine Dexter, 130–134, 137–138, 140
McCormick, Stanley, 132–133
Meatpacker Philip Armour, 200
Melting pot notion of culture and attitudes, xxvii
Merrimack River mills, 78–80, 89, 103, 118, 361
Metcalfe, Bob, 278
Metcalfe's Law, 278
Middlesex Canal, 78–79
Midgely, Dr. Thomas, 316
Migration of industry and people South, 317
Military
  Generally, 361–362
  Colt pistols, 151
  Computer growth, impact on, 94
  Expenditures on southern bases, effect of, 317
  Internet connectivity development and the Pentagon, 189, 275
  Manufacturing armaments and ordinance, 148–151
  Motorized convoy, transporting, 35
  Sex and birth control matters, 127
  Telegraph, use of, 268
Mimeograph machines, 185
MIT (Massachusetts Institute of Technology), 32, 86–87, 90–92, 131, 276
Money and credit systems
  Generally, 165–168
  Greenbacks, 168
  Horace Mann raising money on the Lyceum circuit, 345
  Industrial revolution, 69
Moneyball (movie), 179
Moody Bible Institute in Chicago, 224
Moody, Dwight (preacher), 222–224
Moriarity, Dean, xxii, xxv

Mormons, 251–257, 262
Morrill Act, 89–90, 232, 292
Morrill, Justin (Senator), 89
Morris, Robert, 166, 203, 281
Morse code, 273, 277
Morse, Samuel F.B., 265–271, 361
Mosaic notion of culture and attitudes, xxvii
Mother Road (US 66), 31, xxix
Movies
  Field of Dreams, xxiv
  It's a Wonderful Life, 168
  The Lonely and the Brave, 117
  Moneyball, 179
Muir, John, 102–103, 108–112, 116, 119
Multicultural terminology, xxvii
Muskets. *See* Firearms (using, manufacturing, and selling)

N

Nail-making machine, 152
Naming roads, xxix
National Anti-Slavery Society, 237–238
National bank and financial network, 70, 166–168, 272
National Road, 163, xxix
National road system, 36–37
Natty Bumppo in Cooper's novels, 105–106, 113, 118
New Deal, 60, 63–64, 157, 172
New Lebanon, NY, 39, 248–249
New Machine (group of engineers), 58
New York and Mississippi Valley Printing Telegraph Company (NYMV), 270–271
New York capitalism. *See* Capitalism
New York City
  Central Park, 37, 107, 297
  Compared to Upstate New York, 156
New York Stock Exchange, 162

New York Yankees ball team, 2
Niagara Falls, NY
   Generally, 294–307
   Dumping grounds, as, 300–301, 306
   Egregious development, 300–301
   Love Canal disaster, 305–307
   Tourism and industry, 296–301
Niagara Reservation, 107, 296–297
Niskayuna, NY, 155, 246–248, 250
Noyes, John Humphrey, 259–262
Noyes, John Philip, 217
Noyes, Robert, 96
Numerical routes system, 36–37
NYMV (New York and Mississippi Valley Printing Telegraph Company), 270–271

O

O'Reilly, Henry, 268, 270
Oakland Athletics Baseball Team, 179
Obscenity, mail distribution of information or devices prohibition, 126
Octane rating for gasoline, 136
Odyssey, Homer's, xviii
Ohio Company, 159
Oil refining, 145
Olmstead, Frederick Law, 37, 106–109, 111, 259, 296–297, 301
On the Road (by Jack Kerouac), xx–xxii
Oneida colony of the Perfectionists, 217, 243, 247, 259–262
Organizations and the individual, paradigm, 173–180
Otto, Nikolas, 143–145
Ownership of land in colonial era, 6

P

Pacific Railroad Act, 90
Page numbering machine, 185
Painters. *See* Artists and Painters
Palo Alto Research Center (PARC), 190
Paradise, Sal, xix, xxii, xxv
Parkman, Francis, 106
Parks, 37, 83, 106–109, 111, 259, 296–297, 301
Parkways system, 37
Penn, William, 159
Pennsylvania Turnpike, 37
Perfectionists utopia, 217, 243, 247, 259–262
Perfectionists, the Oneida community of, 217, 243, 247, 259–262
Perkins, Frances (Fannie), 60–65
Perkins, Jacob, 152
Pierce, John D., 343
Piermont, NY, 326
Pierson, Jeremiah, 326
Pilgrims, 3–4
Pinchot, Gifford, 107–108, 110–116, 119, 307
Pincus, Gregory Goodwin (Goody), 122, 129–133, 135, 137
Pioneer experience, 196–198
Pirsig, Robert, xix, xxii–xxiii
Pistols. *See* Firearms (using, manufacturing, and selling)
Plymouth colony
   First integrated woolen mill, 78
   Merger with Massachusetts Bay, 4–5, 159, 285
   Origins, 3–4
   Separatist tradition, 246–247
   Virginia company origin, 159
   Yankee population growth, 18
Plymouth Company, 3–4, 289
Pope, Colonel Albert Augustus, 32
Post Road, 38
Potash or "pot-ash," 71–72, 199, 201–202, 291, 361
Pratt and Whitney in Hartford (machine-tool manufactorers), 152

Printing industry in New England, 342
Productivity, scientific measurement and management, 47–65
Pulteney Associates, 199, 202, 281
Puritans, historical perspective
   Compared to Mormons, 255
   Converting the West, 212–218
   Defining "Puritan," 246
   Education emphasis, 125, 341, 361
   Equality of all people, 229
   Internal migration, 18, 34
   Interstate system, 38
   Organizing life, 358
   Population size, 123
   Undesirability for Puritans, 284
   Wilderness, relationship to, 104, 106
   Yankee culture incorporated Puritan tradition, 346
   Yankee origins, 3–8

Q

Quakers, 4, 211, 233, 243, 247
Quincy Market/Faneuil Hall area, 2
QWERTY format of keyboards, 184, 195

R

Radar warning system, 92
Radio development and use, 273–274
Raikes, Robert, 348
Railroads and railroading
   Generally, 321–339
   Air conditioning railroad cars, 316
   Boston and Lowell railroad, 81
   California railroad king, Leland Stanford, 200
   Canadian Pacific Railroad, 271
   Completion of transcontinental railroad in 1869, 106
   Disadvantages of other modes of transportation, 322–323
   Dunkirk, NY, 322
   Effects of railroad
      Corporate organization, 332–333
      Distribution, 335–337
      Finance, 333–335
      Maturity and decline of railroad, 338–339
      Political union, 330–332
      Technology, 337–338
      Wilderness and freedom, 106
   Erie Railroad, building the, 326–328
   Free incorporation's effect on creating, 160–161
   Interstate highways' effect on, 38
   Land grants of acreage, financing by, 292–293
   Lord, Eleazar, 326–328
   Military motorized convoy, moving, 35
   Mohawk and Hudson Railroad, 328–329
   New York Central Railroad, building the, 328–330
   Pacific Railroad Act, 90
   Pierson, Henry and Helen, 326–327, 330
   Schedules and delays, 314, 332
   Selling railroad land, 293
   South Carolina Railroad, 326, 331
   Steam railroad, 323–324
   Time zones, creation of, 271
   Use of rails for transporting goods, 322–323
   Ventures into railroading, 322
   Western Railroad, 324–325
Ramapo, NY, 326–327
Rand, James H. Jr., 188
Reagan, Ronald, xxiii
Records

Sports performances, 179
Redpath Lyceum Bureau, 347, 354, 356
Redpath, James, 347–348
Refining oil, 145
Refrigerant and Freon, 316–317
Religion and religious differences
   Act of Toleration 1689, 8
   Anabaptists (Mennonites, Amish, and Hutterites), 247
   Burned-over District, 205, 215–219, 234, 243
   Camp meetings and evangelist preachers, 219–224
   Communism, religious, 250
   Contraceptives, 139
   Converting the West, 212–215
   Democratic christianity, 224–226
   Disestablishment and competition, 210–212
   Dow, Lorenzo (preacher), 219–220
   Finney, Charles Grandison (preacher), 220–222
   German Pietists, 247
   Head and heart: Jonathan Edwards, 206–210
   Historical perspective, 3–5, 203–226, 243
   Moody, Dwight (preacher), 222–224
   Quakers, 4, 211, 233, 243, 247
   Schools, church denomination, 348–349
   Separatist tradition. *See* Separatist Tradition
   Shakers. *See* Shakers
   Women's rights movement, 233
Remington Company, 152, 182–184
Remington Rand, 188
Remington, Eliphalet, 50, 182–184, 186
Remington, Philo, 183
Research and development (R&D), 90
Research universities, 90, 98
Revolution in 1688 bringing William and Mary to power, 7–8
Reynolds Arcade, 265, 269–272
Reynolds, Abelard, 265
Rhode Island
   Religious dissenters founding of, 5
Rifles. *See* Firearms (using, manufacturing, and selling)
Road building
   Federal money, 35–37
   Interstate system, 37–38
   War surplus construction equipment used for, 35
Road surfaces, 322
Road themes, xix–xx
Road to Zeus, xx
Road Trip U.S.A. (by Jamie Jensen), xxv–xxvi
Rochester, Nathaniel, 265
Rochester, NY
   Telegraphy and the five Internets, 265–278
   Women's rights movement, Convention of 1853, 235–236, 239, 242
Rockefeller, John D., 94, 145, 265, 300, 349, 352, 362
Rockefeller, Nelson, 155
Rockets and rocketry development, 39, 122, 302–305
Rockwell Museum, 39
Romanticism movement, 105–106
Route 128, 92, 96, 98, 189, 191–192, 278
Route 66, 37
Routes' numbering system, 36–37
Rural Free Delivery (RFD), 32
Russell, Charles Marion, 105

S

Safety in the workplace, 60–65
Sanger, Margaret Higgins, 127, 130–140

Schenectady, NY, 25, 90, 155–156, 246, 310, 326, 329

Schools and education
- Adult education, 352–353, 358
- Attending or teaching at Harvard or Yale, 8
- Chautauqua movement. See Chautauqua Movement
- Church denomination schools, 348–349
- Colleges. See Colleges and Universities
- Common schools movement, 343–344
- Comprehensive State educational system, 343
- Control of, 341
- Correspondence courses, 352–353
- Democratization of education. See democratizing Education
- Development of a universal and organized school system, 178
- Education methods, generally, 341–342
- Financing of, 341
- Free public education, 344, 348, 357–358
- General structure of American education, 345
- Horace Mann, impact of, 342–345
- Internet, 357
- Love Canal incident, 305–307
- Lyceum movement. See Lyceum Tours and Circuits
- National Office of Public Instruction, 344
- Printed works as key to education, 342
- Pursuit of both individual and organizational goals, 178
- Summer school curriculum, 352–353
- Sunday schools, 348–349
- Teacher qualifications and profession, 341, 343–344, 349
- Town Hall meetings, 356–358
- Types of schools in New England, 342
- Universities. See Colleges and Universities

Science parks, 87

Scientific management movement, 47–60

Scope of book, 359–362, xxviii–xxix

Scope of influence of Yankee culture, xxviii

Selden, George, 144, 265

Selection of travel route, importance of, xxvi

Semi Automatic Ground Environment (SAGE), 92

Seneca Falls, NY and the women's rights movement, 228–244

Separatist tradition
- Generally, 246–247, 262–263
- Brook Farm, 102, 225, 247, 257–259, 262
- Deseret colony of Mormons, 251–257, 262
- Perfectionists, the Oneida community of, 217, 243, 247, 259–262
- Shakers. See Shakers

Settlement
- Americanization of English Puritans. See *Transformation of Puritan into Yankee*
- Charters provided by Charles I, expiration of, 7
- *City on a Hill* concept, 5–8
- Erie Canal, 23, 197–198
- Patterns of Yankee settlement, 17–23, 28, 361
- Puritan idea of, 5–6

Sewing machines, 152, 183–184

Shakers
    Generally, 243, 246–251, 262–263
    Activities, 147
    Berkshire Hills, 39
    Chastity belief, 217
    Compared to Mormons, 252, 260
    Coupling faith with social action, 217
    Disestablishment, 212
    Equality of the sexes, 217, 233
    Niskayuna, NY, 155–156, 246
    Perfectionism, 217
Shipping trade, 71, 361
Shoeless Joe (by William Kinsella), xxiv
Sholes, Christopher, 183–185
Shop floor, 57
Shop Management (by Frederick Taylor), 53
Shrewsbury, MA, 122, 130
Sibley, Hiram, 270–271, 278
Sierra Club, 109–110, 119
Skyscrapers, air conditioning, 316
Slater, Samuel, 73–75, 152, 348
Smith, F.O.J. "Fog," 267–268
Smolensk Under Soviet Rule (by Merle Fainsod), 178
Snow, George Washington, 152
Social reform, 60–65
Social security, 63
Society of American Foresters, 112
Soule, Samuel, 185
Southern United States
    Banking, 167
    Interstate highways' effect on creating the South, 38
    Migration of industry and people South, 317–318
Soviet Union
    Demise of, xxviii
    Factory managers dealing with peasant recruits, 178

Smolensk Under Soviet Rule (by Merle Fainsod), 178
    Technology, xxvi–xxvii
Space flight matters, 275, 302–305, 362
Sperry Rand, 188
Sports performances
    Baseball Hall of Fame, 171–173, 179–180, 228
    Glory vs victory, 179–180
    Movies
        Field of Dreams, xxiv
        Moneyball, 179
Springfield Armory, 144, 152, 183
Springfield, MA
    Arms factory, 147–150, 152, 183
    Duryeas car, 141–142
    Interchangeable parts, 50
    Mail service, 39
    Marriage of military and commercial interests, 94
    Ordnance dumps established, 47
    Steam-driven car developed, 144
Springfield, railroad connecting Albany to, 268, 324
Springsteen, Bruce, xxi–xxii
St. Louis World's Fair 1904, 313
Stagecoaches, 26, 40, 268, 277
Standard Oil Company, 145, 163
Stanford, Leland, 200
Statistics
    Fertility rates, 123, 128, 139
    Lowell, MA, 80–81
    Motorized vehicles, 36
    Sports performances, 179
    Textile manufacturing, 80–83
Steam-driven engine, 144
Steam-heating systems, 313
Steel factories, 53
Steinbeck, John, xix, xxvi
Stepping Westward (by Sally Tisdale), xix

Stock trading, 162–163
Stoddard, Reverend Solomon, 208
Streetlights, 297
Strike
   Lowell, MA textile workers' strike 1912, evacuation of children, 134
   Triangle waist Co. (NYC) fire, 61
   Watertown strike, 53–54
Submarines, air conditioning, 316
Sugar production, 68, 71, 164, 202, 229
Sullivan, General John, 280
Sunday school movement, 348–349

T

Tanglewood drama and music sites, 39
Taxation in the colonies, 7
Taylor, Frederick W., 47–60, 65
Technology
   Air conditioning, 12, 38, 40, 308–320, 360
   Computers. *See* Computers
   Electrical communications. *See* Electrical Communication Technologies
   Engines, types and production of, 143–145, 265
   Firearms (using, manufacturing, and selling), 50, 148–151, 182–184
   Interchangeable parts. *See* Interchangeable Parts Movement
   Internet. *See* Internet Connectivity
   Mass-production, impact of, 186
   Mechanical writing and its impact, 182–188
   Mimeograph machines, 185
   Page numbering machine, 185
   Railroads, effects of, 337–338
   Soviet economy, xxvi–xxvii
   Space race, 275
   Standardized pieces, impact of, 186
   Steam-heating systems, 313
   Toshiba scandal, xxvi–xxvii
   Typing machines, 182–189
Telegraph
   Generally, 265–274, 278
   Associated Press syndicate formation, 269
   Canada, 271
   Cornell, Ezra, 271
   Development of, 265–272
   Dominance of the telephone, Western Union's response to, 273
   Electrical engineering profession, seed for, 269–271
   Fees and charges, 269–271
   First "Internet," 272
   Growing insolvency of telegraph lines, 269–271
   Internet as version of telegraph, 265, 272, 274
   James Densmore's attempt to sell his shares in Western Union, 185
   Kendall, Amos, 268
   Leasing telegraph line in another state, 163
   Limited participants, 278
   Magnetic Telegraph Company, 268
   Marconi, Guglielmo, 274
   Morse code, 273, 277
   New York and Mississippi Valley Printing Telegraph Company (NYMV), 270–271
   New York Stock Exchange, impact on, 162
   News stories, sharing, 269, 347
   O'Reilly, Henry, 268
   Radio, impact of, 273–274
   Railroad consumer goods delivery, 336
   Railroad finance issues, 335

Railroad schedules and delays, 314, 332
Samuel F.B. Morse fame, 265–271, 361
Sibley, Hiram, 270–271, 278
Smith, F.O.J. "Fog," 267–268
Stockholders, impact on, 162, 272
Telephone dominance, Western Union's response to, 273
Transformative effect on other industries, 271–272
Typing technology, impact of, 185
Validation of, 267
Western Union, 90, 185, 265, 271–273
Women workers, 272
Women's convention, publication of, 239
Telephones and telephone system
   Ability of multi-site organizations to be managed as single units, 186
   Albany, NY, 156
   Automatic exchange, 273
   Bell Telephone system in-house facilities, 90
   Cell phone, 277
   Conference call, 274
   Development of, 272–274, 277
   Dominance over telegraph, 273
   Growth of local roads, 32
   Historical perspective, 272–278, 356
   Internet era, status at beginning of, 276–277
   Ma Bell, 273
   MCI, 276
   Radio, impact of, 273–274
   As the second form of "Internet," 272, 274
   Western Union as model for monopoly, 271
   Wireless, 277
Television
   Development of, 274

Town hall meetings, 357
Temperature control, 311–317
Testimony of Christ's Second Appearing (1810), 250
Textile industries
   Air conditioning of factories, 317
   Albany, NY area, 156
   Blackstone River Valley, 73–75
   Child laborers, 348
   Cottage industries development, 71–72
   Decline of New England mills, 82–83
   First all-electric textile mill, 317
   First power loom, 76
   Garment industry fires, 60–65
   Growth of non-textile industries, 50
   Historical perspective, 67, 71–83
   Integrated factories with large numbers of workers and professional managers, transition to, 78
   Integrating spinning and weaving, 76–77
   Interchangeable parts movement, producing textiles, 145–147
   Interstate incorporation, 80
   Labor relations, 82
   Lowell, MA, 78–82, 134
   Lowell, MA workers' strike 1912, evacuation of children, 134
   Merrimack River mills, 78–80, 89, 103, 118, 361
   Model corporation evolution, 80
   Principles of mill management, 75
   Ready-to-wear and standardized sizes clothes, 51, 82
   Slater, Samuel, 73–75
   Sunday schools for child laborers, 348
   Waltham, MA mill, 75–78
The Brave Cowboy (by Edward Abbey), 117

The Control of Fertility (by Gregory Pincus), 130
The Majic Bus (by Douglas Brinkley), xxiv–xxv
The Monkey Wrench Gang (by Edward Abbey), 117, 119
The Principles of Scientific Management (by Frederick Taylor), 54, 56
Tiffit family, 310–311
Time consciousness movement, 47–60
Tisdale, Sally, xix
Tomlinson, Ray, 276
Toshiba scandal, xxvi–xxvii
Tourism and the wilderness, 117, 120
Town Hall meetings, 356–358
Transcendentalists
   Brook Farm, 247, 257
   Emerson, Ralph Waldo, 106
   Fuller, Margaret, 232
   Muir, John, 110
   Thoreau, Henry David, 102
Transformation of Puritan into Yankee
   Generally, 5–8
   Act of Toleration 1689, impact of, 8
   Assertion of English control over colonies, impact of, 7–8
   Conflict over how to settle the wilderness, 6
   Wealth of settlers, impact of, 6–7
Transistors, development of, 274
Transportation methods
   Historical perspective, 322–323
   Railroads. *See* Railroads and Railroading
Triangle Waist Co. fire in NYC, 60–65
Troy, NY, 155–156, 221
Trucking service and transport, 35
Turner, Frederick Jackson, 8, 116
Twain, Mark, 187
Typewriter
   Development and sales, 182–185

IBM, 188, 190
Impact of mechanical writing, 185–188, 195
Mark Twain's use of, 187
Mini/Microcomputers and the PC revolution, 189–195
QWERTY format of keyboards, 184, 195
Soles and Glidden typewriter, 183–185
Turning rifles into typewriters, 152, 182–184
Turning typewriters into computers, 188–189
Women, used by, 186–187
Writers' use of, 187–188
Xerox, 190
Typing classes, 187

U

UN Scientific Conference on the Conservation and Utilization of Resources, 113
Unemployment insurance, 63
Unions, 61–62
United Society of Believers. *See* Shakers
University of Chicago, 352
University status, Chautauqua Institution, 352
Urban parks, 37, 107–108
US 101, 44
US 20
   Generally, 30–45, 360
   Albany, 156
   Bend, Oregon, 44
   Bighorn Basin, 42
   Boise, Idaho, 43
   Boston, MA, 38
   Casper, Wyoming, 42
   Cody, Wyoming, 42

Continental Divide, 43
Craters of the Moon National Monument, 43
End of, 44–45
Fast trip across the continent on, 38–45
Great Western Pike, 39–40, 202, 295
Harney Lakes, Oregon, 43–44
Idaho, 43
Interstate system, 37–38
Making of the universal Yankee nation, xxviii
Missouri River, 42
Montana, 43
Nebraska, 31, 42, 331
Nicknamed the Yankee Road, xxix
Oregon, 42–44
Oregon Trail's Goodale Cutoff, 43
Overview of route across the country, 38–45
Seneca Falls, NY, 228
Shrewsbury, MA, 122, 130
Sioux City, Iowa as halfway point, 42
Snake River Basin, 43
Start of in Boston, MA, 38
Three Sisters peak, Oregon, 44
Western end of, 44–45
Willamette Valley, Oregon, 44
Wind River Canyon, 42
Yellowstone Park, 42–43
US 66, 31, xxix
US 95, 37, 43

V

Vail, Alfred, 266–267
Vail, Theodore, 273
Vaux, Calvert, 37, 107, 297
Vawter, Iowan Keith, 354–356
Virginia Company, 159
Voting, 8

W

Wages
 For productivity, 53
 Waltham mill factory, 78
Wagner, Robert, 63
Walden Pond, 99–103, 105, 110, 118
Walk-in-the-Water (1st steamship on Great Lakes), 309
Washington and New Orleans Telegraph Company, 268
Water rights, 292
Watertown Arsenal, 47–48, 53, 59
Watertown,, MA, 46–48, 53, 59, 81
Watervliet,, NY, 47, 247
Wealth of settlers, 6–7, 69
Webb, Senator James, 198
Weber, Max, 49, 65, 157
Website
 Companion for this book, xv
West Point, 337
Western Union, 90, 185, 265, 271–273
WFEB. See Worcester Foundation for Experimental Biology (WFEB)
White City, 107, 300, 353
White House
 Air conditioning of, 316
 First female US cabinet member, Frances Perkins, 63
 Using social media for Town Hall discussions, 357
Wilderness Act of 1964, 115
Wilderness and nature
 Generally, 99–120
 Abbey, Edward ("Cactus Ed"), 103, 116–120
 Conflict over how to settle, 6
 Cooper, James Fennimore, 105–106
 Desert, canyon, and dam activism, 103, 116–120
 Earth First! activist group, 118–119

Emerson, Ralph Waldo, 102–104, 106, 109–110
End of the frontier, 106
First formal wilderness area, 114
Forestry management, 111–116
Foundation of Wilderness Society, 115
Game management, 114–116
Historian Francis Parkman, 106
Impact in colonial era, 5–8
Muir, John, 102–103, 108–112, 116, 119
Romanticism movement, 105–106
Russell, Charles Marion, 105
Sierra Club, 109–110, 119
Thoreau, Henry David, 102–104
Walden Pond and wilderness, 99–102
Wilderness Society, 115
Williams, Roger, 4, 207, 214
Wolff, Alfred, 312
Women, employment of
Garment industry fires, 60–65
Telegraph operators, 272
Typewriter factories, 186–187
Urbanism, effect of, 233
Wages in estate separate from spouse, 236
Women's Hall of Fame, 228
Women's rights movement
Adams, Abigail, 229
Anthony, Susan B., 234, 236, 242–244
Anti-Slavery "Fairs," 233
Birth control, 129, 134
Chautauqua movement, 352
Child, Lydia Maria, 233
Commencement of, 228
Convention of 1848 in Seneca Falls, NY, 228–229, 234, 239, 241, 243
Convention of 1850 in Worcester, MA, 241
Convention of 1851 in Akron, Ohio, 234, 239
Convention of 1853 in Rochester, NY, 235–236, 239, 242
Coverture concept, 212, 230, 234–235, 361
Davis, Paulina Wright, 241
Declaration of 1848 Convention in Seneca Falls, NY, 241
Earnings Act (NY 1860), 236
History of the Woman Rights Movement (1881), 243
Involvement in voluntary organizations, 233
Legality and equality, 134, 229–236, 241, 243
Married Women's Property Rights Act (NY 1848), 229, 234–236, 241
Mott, Lucretia, 228–229, 236–240, 244
Overview, 129, 134, 243–244
Perkins, Frances (Fannie), 60–65
Religion and social causes, 233
Social change factors, 231–234, 243
Stanton, Elizabeth Cady, 228, 236, 238–244
Stanton, Henry, 228–229, 240
Stone, Lucy, 242
Suffrage Amendment of 1920, 236
Suffrage to male ex-slaves dispute, 242–243
Truth, Sojourner, 234
Urbanization's effect on, 233
Voting right, 129, 236, 243
Women's Trade Union League (WTUL), 61
Women's Trade Union League (WTUL), 61
Wood building construction, balloon-type, 152
Worcester Foundation for Experimen-

tal Biology (WFEB), 122, 130–133, 137
Worcester, MA
  Blackstone River Valley National Heritage Corridor, 39
  Rocketry development, 39, 122, 303
Worker's compensation, 63
Working conditions, Triangle Waist Co fire, 60–65
Workplace organization, efficiency and management, 50–60, 82
Works Progress Administration (WPA) guides, xxv–xxvi
World Anti-Slavery Convention, London 1840, 228, 238, 240
World War II, impact of scientific management concept, 58–59

X

Xenophon (Athenian officer), xviii–xix, xx
Xerox, 190

Y

Yankee
  Defined, 2–3, 5, 7, 10, 359, xviii, xxvii, xxviii
  George M. Cohan's "The Yanks are coming," 360
  Imperialism, 3
  Making of the universal Yankee nation (US 20), xxviii
  Origins, 3–5
  Peddlers, 10, 13, 72, 150, 209, 336, 366
  Transformation of Puritan into Yankee, 5–8
Yankee Road, 359–362, xviii, xxix, xxvii
Yosemite National Park, 107–110
Young Men's Christian Association (YMCA), 223
Youngs, Benjamin Seth, 250

Z

Zen and the Art of Motorcycle Maintenance (by Robert Pirsig), xxii–xxiii

James D. McNiven

James D. McNiven is Professor Emeritus at Dalhousie University and Senior Policy Research Advisor with Canmac Economics Ltd. He was the Fulbright Research Professor at Michigan State University's Canadian Studies Center in 2010-11.

Until his retirement from Dalhousie, he held the R. A. Jodrey Chair in Commerce in the School of Business Administration and was a Professor of Public Administration. From 1988 to 1994, he was the Dean of the Faculty of Management at Dalhousie. Prior to that, he was the Deputy Minister of Development for the Province of Nova Scotia (1981-88) and the President of the Atlantic Provinces Economic Council (1977-81).

He has been the CEO of a small technology company and has been a member of a number of corporate and government boards, including the Blue Cross of Atlantic Canada and the federal government's International Trade Advisory Committee. He was a member of the federal government's Royal Commission on National Passenger Transportation.

Dr. McNiven has a PhD from the University of Michigan. He has written widely on public policy and economic development issues and is the co-author of three books. His most recent research work has been about the relationship of demographic changes to Canadian regional economic development. He still teaches at Dalhousie on a part-time basis and is a columnist with factsandopinions.com. a news site out of Vancouver. He can be reached at j.mcniven@gmail.com

CPSIA information can be obtained at www.ICGtesting.com
Printed in the USA
BVOW03s2030220616

453031BV00004B/84/P

9 781627 871419